Nanobiotechnology

Edited by
C. M. Niemeyer and
C. A. Mirkin

Further Titles of Interest

V. Balzani, A. Credi, M. Venturi

Molecular Devices and Machines

2003, ISBN 3-527-30506-8

M. Schliwa (ed.)

Molecular Motors

2002, ISBN 3-527-30594-7

Ch. Zander, J. Enderlein, R. A. Keller (eds.)

Single Molecule Detection in Solution

2002, ISBN 3-527-40310-8

E. Bäuerlein (ed.)

Biomineralization –
From Biology to Biotechnology and Medical Application

2000, ISBN 3-527-29987-4

Nanobiotechnology

Concepts, Applications and Perspectives

Edited by
Christof M. Niemeyer and Chad A. Mirkin

WILEY-VCH Verlag GmbH & Co. KGaA

Edited by

Prof. Dr. Christof M. Niemeyer
Universität Dortmund, Fachbereich Chemie
Biologisch-Chemische Mikrostrukturtechnik
Otto-Hahn-Str. 6
44227 Dortmund
Germany
cmn@chemie.uni-dortmund.de

Prof. Dr. Chad A. Mirkin
Department of Chemistry &
Institute for Nanotechnology
Northwestern University
2145 Sheridan Road
Evanston, IL 60208-3113
USA
camirkin@chem.northwestern.edu

1st Edition 2004
  1st Reprint 2004
  2nd Reprint 2004
  3rd Reprint 2005
  4 th Reprint 2005
  5 th Reprint 2007

Cover illustration
Malign (top) and normal cells (bottom) on pillar
interfaces which sense cellular forces. In the middle
illustration, the molecular distribution of integrin
(green) and actin (red) is shown. All micrographs
were kindly provided by W. Roos, J. Ulmer, and
J.P. Spatz (University of Heidelberg, Germany).

Library of Congress Card No.: applied for
British Library Cataloguing-in-Publication Data
A catalogue record for this book is available
from the British Library.

Bibliographic information published by
Die Deutsche Bibliothek
Die Deutsche Bibliothek lists this publication
in the Deutsche Nationalbibliografie;
detailed bibliographic data is available in the
Internet at http://dnb.ddb.de.

© 2004 WILEY-VCH Verlag GmbH & Co.
KGaA, Weinheim

Printed in Singapore
Printed on acid-free paper

Typesetting   Hagedorn Kommunikation,
Viernheim
Printing and Bookbinding   Markono Print Media
Pte Ltd, Singapore

ISBN-13: 978-3-527-30658-9
ISBN-10: 3-527-30658-7 ✓

Contents

Part II   Protein-based Nanostructures

Preface

Nanobiotechnology is a young and rapidly evolving field of research at the crossroads of biotechnology and nanoscience, two interdisciplinary areas each of which combines advances in science and engineering. Although often considered one of the key technologies of the 21st century, nanobiotechnology is still in a fairly embryonic state. Topical areas of research are still being defined, and the entire scope of technological applications cannot be imagined. At present, nanobiotechnology is a field that concerns the utilization of biological systems optimized through evolution, such as cells, cellular components, nucleic acids, and proteins, to fabricate functional nanostructured and mesoscopic architectures comprised of organic and inorganic materials. Nanobiotechnology also concerns the refinement and application of instruments, originally designed to generate and manipulate nanostructured materials, to basic and applied studies of fundamental biological processes.

This book is intended to provide the first systematic and comprehensive framework of specific research topics in nanobiotechnology. To this end, the current state-of-the-art has been accumulated in 27 chapters, all of them written by experts in their fields. Each of the chapters consists of three sections, (i) an overview which gives a brief but comprehensive survey on the topic, (ii) a methods section which orients the reader to the most important techniques relevant for the specific topic discussed, and (iii) an outlook discussing academic and commercial applications as well as experimental challenges to be solved.

Nanobiotechnology: Concepts, Applications and Perspectives combines contributions from analytical, bioorganic, and bioinorganic chemistry, physics, molecular and cell biology, and materials science in an attempt to give the reader a feel for the full scope of current and potential future developments. The articles in this volume clearly emphasize the high degree of interdisciplinary research that forms the backbone of this joint-venture of biotechnology and nanoscience.

The book is divided into four main sections. The first concerns interphase systems pertaining to biocompatible inorganic devices for medical implants, microfluidic systems for handling biological components in analytical lab-on-a-chip applications, and microelectronic silicon substrates for the investigation and manipulation of neuronal cells. Moreover, two chapters describe methodologies regarding the microcontact printing of proteins and the use of nanostructured substrates to study basic principles of cell adhesion.

The second section is devoted to protein-based nanostructures. Individual chapters concern the use of specific proteins, such as S-layers to be used as building blocks and templates for generating functional nanostructures, bacteriorhodopsin for photochromic applications, protein nanopores as nanoscopic cavities for analytical and synthetic tasks, and biomolecular motors for the translocation of cargo in synthetic environments. The use of a variety of functional proteins as transducers and amplifiers of biomolecular recognition events is described in the chapters on nanobioelectronic devices and polymer nanocontainers. Contributions concerning the microbial production of inorganic nanoparticles and magnetosomes as well as the discussion of genetic approaches to generate proteins for the specific organization of particles provide insight into the body of classical biotechnology, implemented in nanobiotechnology.

In the third section, DNA-based nanostructures are described, beginning with semisynthetic conjugates of DNA and proteins, which link the advantages of nucleic acids to the unlimited functionality of proteins. Three contributions concern the use of the topographic and electrostatic properties of DNA and proteins for the templated growth of inorganic materials. Hybrid conjugates of gold nanoparticles and DNA oligomers are described with a focus on their applications in the high sensitivity analyses of nucleic acids. Finally, the use of pure DNA molecules for applications in nanomechanics and computing is discussed.

The fourth section deals with the area of nanoanalytics, which currently includes the majority of commercial products in nanobiotechnology. In particular, four chapters describe the use of metal or semiconductor nanoparticles, supplemented with nucleic acid- and protein-based recognition groups, for biolabeling, histochemical applications and for signal enhancement in optical detection methods. Nanoparticles are also employed as carriers for genetic material in the non-viral transfection of cells. To exemplify the use of modern nano-instrumentation for the study of biological systems, two chapters describe the use of the scanning probe microscope, the key instrument in nanotechnologies, for investigating biomolecular structure, conformation and reactivity.

The purpose of *Nanobiotechnology: Concepts, Applications and Perspectives* is to provide both a broad survey of the field and also instruction and inspiration to all levels of scientists, from novices to those intimately engaged in this new and exciting field of research. Although the collection of articles addresses numerous scientific and technical challenges ahead, the future of nanobiotechnology is bright and appears to be limited, at present, only by imagination.

Dortmund, November 2003   Christof M. Niemeyer
Evanston, November 2003   Chad A. Mirkin

Contributors

Absar Ahmad
Biochemical Sciences Divison
National Chemical Laboratory
411 008 Pune
India

Udo Bakowsky
Department of Biopharmaceutics and
Pharmaceutical Technology
Saarland University
Im Stadtwald
66123 Saarbrücken
Germany

Holger Bartos
STEAG microParts GmbH
Hauert 7
44227 Dortmund
Germany

Hagan Bayley
Department of Chemistry
University of Oxford
Mansfield Road
Oxford, OX1 3TA
UK

Dennis A. Bazylinski
Department of Physics
California Polytechnic State University
San Luis Obispo, CA 93407
USA

Samantha M. Benito
Department of Chemistry
University of Basel
Klingelbergstrasse 80
4056 Basel
Switzerland

Orit Braha
Department of Chemistry
University of Oxford
Mansfield Road
Oxford, OX1 3TA
UK

Erez Braun
Department of Physics
Solid State Institute
Technion-Israel Institute
of Technology
32000 Haifa
Israel

Stanley Brown
Department of Molecular
Cell Biology
University of Copenhagen
Øster Farimagsgade 2A
1353 Copenhagen K
Denmark

Stephen Cheley
Texas A&M University
Health Science Center
Medical Biochemistry and Genetics
440 Reynolds Medical Building
College Station, TX 77843-1114
USA

Signe Danielsen
Norwegian University
of Science and Technology
Department of Physics
Høgskoleringen 5
7491 Trondheim
Norway

Emmanuel Delamarche
IBM Research
Zürich Research Laboratory
Säumerstrasse 4
8803 Rüschlikon
Switzerland

Stefan Diez
Max Planck Institute of Molecular
Cell Biology and Genetics
Pfotenhauerstrasse 108
01307 Dresden
Germany

Timothy J. Drake
Center for Research at the
Bio-nano Interface
Department of Chemistry,
McKnight Brain Institute,
University of Florida,
Gainesville, FL 32611
USA

Eva-Maria Egelseer
Center for Ultrastructure Research
and Ludwig Boltzmann Institute
for Molecular Nanotechnology
University of Natural Resources
and Applied Life Sciences
Gregor-Mendel-Straße 33
1180 Wien
Austria

Mahnaz El-Kouedi
Department of Chemistry,
The Pennsylvania State University
University Park, PA 16802
USA

Richard B. Frankel
Department of Physics
California Polytechnic State University
San Luis Obispo, CA 93407
USA

Xiaohu Gao
Department of Biomedical Engineering
Emory University School of Medicine
1639 Pierce Drive
Atlanta, GA 30322
USA

Friedrich Götz
Gelsenkirchen University of Applied
Sciences
Neidenburger Str. 43
45877 Gelsenkirchen
Germany

Alexandra Graff
Department of Chemistry
University of Basel
Klingelbergstrasse 80
4056 Basel
Switzerland

Li-Qun Gu
Texas A&M University
Health Science Center
Medical Biochemistry and Genetics
440 Reynolds Medical Building
College Station, TX 77843-1114
USA

Gerhard W. Hacker
Research Institute for Frontier
Questions of Medicine and
Biotechnology
St. Johanns-Hospital
Landeskliniken Salzburg
Muellner Hauptstr. 48
5020 Salzburg
Austria

James F. Hainfeld
Brookhaven National Laboratory
Department of Biology
Upton, NY 11973
USA

Norbert Hampp
Fachbereich Chemie
Philipps-Universität Marburg
Hans-Meerwein-Straße, Geb. H
35032 Marburg
Germany

Helen G. Hansma
Physics Department
University of California
Santa Barbara, CA 93106
USA

John H. Helenius
Max Planck Institute of Molecular
Cell Biology and Genetics
Pfotenhauerstrasse 108
01307 Dresden
Germany

Jonathon Howard
Max Planck Institute of Molecular
Cell Biology and Genetics
Pfotenhauerstrasse 108
01307 Dresden
Germany

Eugenii Katz
Department of Organic Chemistry
Hebrew University
Givat Ram
91904 Jerusalem
Israel

Christine D. Keating
Department of Chemistry,
The Pennsylvania State University
University Park, PA 16802
USA

M. Islam Khan
Biochemical Sciences Divison
National Chemical Laboratory
411 008 Pune
India

Rajiv Kumar
Catalysis Divison
National Chemical Laboratory
411 008 Pune
India

M. N. V. Ravi Kumar
Department of Pharmaceutics
NIPER
SAS Nagar, Sector 67
160 062 Mohali
India

Claus-Michael Lehr
Department of Biopharmaceutics
and Pharmaceutical Technology
Saarland University
Im Stadtwald
66123 Saarbrücken
Germany

Stephen Mann
School of Chemistry
University of Bristol
Bristol BS8 1TS
UK

Eric Mayes
NanoMagnetics Ltd.
108 Longmead Road
Bristol BS16 7FG
UK

Wolfgang Meier
Department of Chemistry
University of Basel
Klingelbergstrasse 80
4056 Basel
Switzerland

Michael Mertig
Technische Universität Dresden
Institut für Werkstoffwissenschaft
01062 Dresden
Germany

Chad A. Mirkin
Department of Chemistry &
Institute for Nanotechnology
Northwestern University
2145 Sheridan Road
Evanston, IL 60208-3113
USA

Shuming Nie
Department of Biomedical Engineering
Emory University School of Medicine
1639 Pierce Drive
Atlanta, GA 30322
USA

Christof M. Niemeyer
Universität Dortmund
Fachbereich Chemie
Biologisch-Chemische
Mikrostrukturtechnik
Otto-Hahn-Str. 6
44227 Dortmund
Germany

Dieter Oesterhelt
Max-Planck Institute for Biochemistry
Am Klopferspitz 18A
82152 Planegg-Martinsried
Germany

Andreas Offenhäusser
Institute for Thin Films and Interfaces,
Bio- and Chemosensors
Research Centre Jülich
52425 Jülich
Germany

Emin Oroudjev
Department of Physics
University of California
Santa Barbara, CA 93106
USA

Ralf-Peter Peters
STEAG microParts GmbH
Hauert 7
44227 Dortmund
Germany

Wolfgang Pompe
Technische Universität Dresden
Institut für Werkstoffwissenschaft
01062 Dresden
Germany

Richard D. Powell
Nanoprobes, Incorporated
95 Horseblock Road
Yaphank, NY 11980-9710
USA

Dietmar Pum
Center for Ultrastructure Research
and Ludwig Boltzmann Institute
for Molecular Nanotechnology
University of Natural Resources
and Applied Life Sciences
Gregor-Mendel-Straße 33
1180 Wien
Austria

Murali Sastry
Materials Chemistry Divison
National Chemical Laboratory
411 008 Pune
India

Thomas Sawitowski
Institut für Anorganische Chemie
Universität GH Essen
Universitätsstr. 5-7
45117 Essen
Germany

Bernhard Schuster
Center for Ultrastructure Research
and Ludwig Boltzmann Institute
for Molecular Nanotechnology
University of Natural Resources
and Applied Life Sciences
Gregor-Mendel-Straße 33
1180 Wien
Austria

Nadrian C. Seeman
Department of Chemistry
New York University
New York, NY 10003
USA

Markus Seitz
Department of Applied Physics
Ludwig-Maximilians-Universität
Amalienstrasse 54
80799 München
Germany

Uri Sivan
Department of Physics
Solid State Institute
Technion-Israel Institute of Technology
32000 Haifa
Israel

Uwe B. Sleytr
Center for Ultrastructure Research
and Ludwig Boltzmann Institute
for Molecular Nanotechnology
University of Natural Resources
and Applied Life Sciences
Gregor-Mendel-Straße 33
1180 Wien
Austria

Joachim P. Spatz
Institut für Physikalische Chemie
Universität Heidelberg
INF 253
69120 Heidelberg
Germany

Weihong Tan
Center for Research at the
Bio-nano Interface
Department of Chemistry
McKnight Brain Institute
University of Florida
Gainesville, FL 32611
USA

C. Shad Thaxton
Northwestern University
2145 Sheridan Road
Evanston, IL 60208
USA

Corinne Verbert
Department of Chemistry
University of Basel
Klingelbergstrasse 80
4056 Basel
Switzerland

Angela K. Vogt
Max-Planck Institute
for Polymer Research
Ackermannweg 10
55228 Mainz
Germany

Itamar Willner
Department of Organic Chemistry
Hebrew University
Givat Ram
91904 Jerusalem
Israel

Xiaojun Julia Zhao
Center for Research at the
Bio-nano Interface
Department of Chemistry,
McKnight Brain Institute,
University of Florida,
Gainesville, FL 32611
USA

1
Biocompatible Inorganic Devices

Thomas Sawitowski

1.1
Introduction

New technologies have always been a major driving force in medical device technology [1], and it is largely due to the high economical and social value of modern medical devices that new materials and processes are incorporated at a very early stage into new products. Taking this into consideration, the emergence of nanotechnology over the past few years has had an immediate influence on medical device technology [2]. This stems from the fact that, by changing the size of commonly known materials, new properties arise that can be used in many areas of today's technologies [3–8]. For this reason, nanotechnology can be termed an "enabling" technology. It is a highly interdisciplinary field of material science which involves chemists, physicist, biologists, engineers, and physicians to name just a few. At the border of those disciplines lie new problems and innovative solutions; for example, new implant coatings which were originally developed to improve the wear resistance of tools are now used as a biocompatible coating on stents [9]. Likewise, many other examples have been developed and currently available on a commercial basis. Some of these combine even more of the "smart" properties of nanosized material, including the enabling of drug delivery and improvement of biocompatibility [10].

1.2
Implant Coatings

Implants can be classified as either permanent or temporary devices. Examples of permanent implants include seeds, hip joints, stents, nails, and dental implants, while catheters or needles are perfect examples of temporarily implanted devices. Each year, the number of implants implanted is directly related to the occurrence of the diseases treated this way. The major diseases of the western countries relate to the cardiovascular system, or to cancer. In Germany during 1998, circulatory diseases (including myocardial infarction) caused more than 500 000 deaths, or 58 % of all deaths [11]. Likewise, more than 210 000 people died from cancer, representing 25 % of all deaths in Germany that year.

Nanobiotechnology. Edited by Christof Niemeyer, Chad Mirkin
Copyright © 2004 WILEY-VCH Verlag GmbH & Co. KgaA, Weinheim
ISBN 3-527-30658-7

Taking these facts into consideration, there is clearly a great demand for improved treatment of these diseases, and this involves the use of modern medical device technology. This includes the concept of local rather than systemic treatment. Local treatment is the basis of every implant used. For example, when treating diseased vessels with stents the implant is inserted very precisely into the stenotic area; the same holds true for seeds used to treat prostate cancer. Hence, implants represent the ideal carrier for drugs to be delivered locally to the site of implantation. Different carrier systems are currently being evaluated, including polymers, dendrimers, sol–gel-coatings, or other porous inorganic materials. In this chapter, the focus will be on nanoscale inorganic materials for use in local drug delivery rather than polymeric materials, among which nanoporous alumina is one of the most interesting that is currently being used in cardiology and oncology. In a more general approach, specific units such as implantable capsules and pumps can be used as carrier technologies [12], while MEMS devices [13] can be applied to deliver drugs locally. All of these devices must fulfil a vast number of criteria before being used in humans [14]. The basic safety and efficacy requirements can be subdivided into biocompatibility, which can be further subdivided into tissue or hemic compatibility. For these reasons it is difficult to present a complete and comprehensive overview of all inorganic medical devices, and so for technical, medical, and economic reasons the focus here will be on stents and seeds.

1.2.1
Stents

As mentioned previously, the current most important causes of death in modern western countries are vascular diseases and myocardial infarction. Various risk factors such as low-density lipoproteins, high blood pressure, nicotine abuse, and diabetes lead to changes in the vessel and, in turn, a narrowing of the free lumen. In time, this causes angina pectoris or perhaps sudden events such as myocardial infarction. There are three invasive methods to treat these stenoses. In cardiology, the treatment of coronary artery disease has long been limited to bypass surgery to circumvent the stenotic region. With this technique, veins are taken for example from the patient's legs and used to bypass the narrowed region, thus re-enabling blood flow. Such a treatment represents a significant burden for the patient, and one of the most intriguing aspects of modern medical technologies is the opportunity to reduce both patient burden and treatment costs by using minimally invasive methods. A minimally invasive approach would be to use a catheter, at the end of which is mounted a small balloon. The catheter is placed into the narrowed lesion and inflated by applying external pressure. In this way, the material which is causing the narrowing is pressed into the vessel wall, and this leads to an increase in the lumen free space. In 1977, Andreas Grüntzig [15] performed the first percutaneous transluminal balloon angioplasty (PTCA), when he opened up a narrowed vessel using a small, expandable balloon. To date, PTCA treatment remains one of the most successful applications of minimally invasive medical technologies. However, in addition to the many advantages of this technique, inevitably there are also some drawbacks, the most important being an elastic recoil of the vessel caused either by plaque material remaining in the blood stream or by hard plaque material which cannot be removed with the balloon [16].

In these cases an additional mechanical support is needed to improve treatment outcome, and it is for this reason that small metallic meshes – called "stents" – are implanted into the lesion. The stents may be made from different materials, and are mounted on the balloon which, when inflated, also inflates the stent; this occurs when a stainless steel stent is used. Alternatively, the stent may self-expand when it is released from the delivery system; this occurs when a stent is made from Nitinol®, an alloy made from nickel and titanium. Many different designs of stent are currently available on the market [17]. Some are made of tubes which are laser-cut to build a tubular mesh, while others are made from planar meshes that are laser-welded along the long axis to build the stent. Three typical examples of laser-cut stents are illustrated in Figure 1.1. In all of these stents the design consists of a rather rigid metal structure in order to ensure mechanical stability.

The rigid structure guarantees a good mechanical stability of the stent on the balloon when finally it is implanted into the vessel. These rigid areas are interconnected by flexible parts that ensure stability of the stent as it is moved through the vessel towards the narrowed lesion. In addition, in most cases the lesion to be treated is not a straight part of the vessel but is more often rather curved and roughened on the inside. Here, the flexible parts ensure that the stent matches the contours of the inner lumen as well as possible.

The initial outcome of stent implantation showed a great improvement compared with PTCA. However, adverse factors such as mechanical stress and/or damage to the arterial wall, heavy metal ion dissolution (e. g., nickel, chromium, or molybdenum) from the implant, and disturbing turbulences in the bloodstream means that a positive outcome is not guaranteed for all patients. Thus, 30–40 % of the vessels develop a re-narrowing (restenosis), mainly as a result of proliferating smooth muscle cells (SMCs) resembling scar tissue. In clinical trials, bypass surgery was compared with PTCA and PTCA plus stent, but the problems of restenosis ultimately reduced the positive initial outcome associated with stent implantation [18]. Restenosis can be seen as a problem of poor biocompatibility involving of course not only biochemical but also mechanical and technical aspects of the implantation.

Bearing in mind the huge number of patients suffering from vascular diseases, there is clearly a great demand for an improved biocompatibility of stents. To achieve this, many different types of coatings have been applied to stents, including silicon carbide (SiC), dia-

Figure 1.1 Different stent types: Left: Genius MEGAFlex coronary stent; Middle: Small Vessel stent (BiodivYsio SV); Right: Terumo Tsunami.

mond-like carbon (DLC), turbostratic carbon (TC), gold, iridium oxide, aluminum oxide, and many different polymeric coatings which may, or may not, be biodegradable [10]. The initial goal has been to improve the limited biocompatibility of stents made from stainless steel or Nitinol by the use of passive coatings; this prevents heavy metal ion dissolution from the stents. In one study, the risk of restenosis was compared in patients with and without a nickel allergy [19]. The results indicated that patients who were allergic towards nickel or molybdenum were more likely to suffer in-stent restenosis than those without such hypersensitivity. These allergic reactions may trigger the fibroproliferative or inflammatory responses characteristic of in-stent restenosis. Almost all coatings made from inorganic materials target the dissolution of nickel, chromium, and other metals from stents. One of the very first coatings to be applied was a thin layer of gold, though the idea of passivating the surface by chemical means generally fails. In a clinical trial involving more than 700 patients, gold-coated stents caused a significant increase in neointimal hyperplasia compared with stainless steel stents. Thus, gold-coated stents were associated with a considerable increase in the risk of restenosis over the first year after stenting [20]. The precise mechanism by which gold coating causes intimal hyperplasia is still hypothetical, though it is possible that mechanical damage of the coating leads to particle formation and hence inflammation. On the other hand, it is possible that as a stent is dilated, due to the significant strain and stress cracks might occur in the coating. In this way, local elements of steel and gold are formed which might even increase the dissolution of certain metal ions.

Silicon carbide has also been applied to stents to cover the surface and reduce intimal hyperplasia. Initial data obtained from animals suggested that the coating shows a significantly lower platelet and leukocyte adhesion at the surface of the SiC-coated tantalum stent compared with the surface of stainless-steel stents [21].

The first clinical trials conducted with silicon carbide-coated 316L stainless steel stents showed a better outcome after 6-month follow-up than uncoated stainless steel stents with respect to binary restenosis rate. In a selected group of patients, the implantation of a coated coronary stent showed a high incidence of immediate success and an absence of in-hospital cardiac events. A significant reduction of major adverse cardiac events and a reduced reintervention rate compared to conventional stents was also observed by others [22].

Carbon with a mixed hybridization state between 2 and 3 has also been applied as a barrier layer. This type of coating has been named turbostratic carbon (TC), but it has not shown a major benefit in a clinical trial comparing bare and TC-coated stents. Another suggested approach is to use titanium-nitride-oxide-coated and iridium oxide-coated stents, and initial results with these have been seen as promising [23].

Until now, it has not been clear what influences the biological response to certain surfaces. As material, stent design, biology, chemistry, and physics represent a complex system, it is difficult to elucidate clear-cut interactions. Although passive coatings on stents have the potential to improve biocompatibility, the next major step towards reducing intimal hyperplasia is to bind drugs locally onto the stents themselves in order to overcome the ultimate problems of restenosis.

Besides certain polymeric coatings, only one inorganic coating has yet proved capable of taking-up and releasing drugs from implant surfaces, and this is nanoporous alumina.

This coating, which consists of an amorphous alumina ceramic with pores in the order of 5 to 500 nm diameter, can be used to store and release drugs locally [9]. The material is formed using the well-established process of anodizing aluminum in different electrolytes [24]. The oxidation is normally carried out in diluted acidic electrolytes (typically of oxalic or phosphoric acid) by applying potentials in the order of a few tens up to a few hundreds of volts, and a direct current. With this process, the aluminum surface is converted into an amorphous aluminum oxide and hydroxide surface, which can be best described as a Boehmite composition of aluminum oxide [25].

The first step of the oxidation process is the formation of a dense layer of oxide on the metal surface, the thickness of which is dependent upon the applied potential. The high electrical field, together with some initial surface perturbation (coming from the natural surface roughness or from grain boundaries for example), causes the first pores to be formed. In this high-electrical field regime, the oxide crystal lattice is deformed at slight perturbations and the electrolytes dissolve the oxide more rapidly, causing pores to be formed. While the electrical field determines the oxide formation and dissolution, the pore geometry may be controlled by the electrical field and thus by the potential applied in the process of anodizing aluminum [26]. Ultimately, a structure similar to that shown in Figure 1.2 can be obtained, with the pores ordered parallel to each other and perpendicular to the substrate surface [27].

At this stage, a thin oxide layer – the so-called barrier layer – remains at the bottom of the pores. As a rule of thumb, it can be said that for each Volt of anodic potential the pore diameter increases by 1.5 nm. So, by applying 10 Volts, pores in the order of 10–15 nm are formed [28]. The pores are packed hexagonally, with an amorphous Boehmite forming the pore wall in between. Pore densities can reach values up to 10^{11} pores cm^{-2}, while the porosity always remains the same (~30%) because small pores are packed more densely compared with larger ones [29]. The pore length is more or less controlled by the electrical

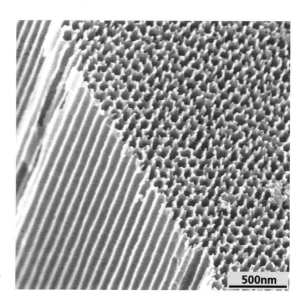

Figure 1.2 Cross-sectional scanning electron microscopy (SEM) image of a nanoporous alumina membrane. The pore diameter is ~120 nm.

500nm

charge, which is proportional to the time of anodic oxidation. Increasing this time leads to an increase in oxide layer thickness until an equilibrium is reached between oxide formation and porous layer dissolution in the electrolyte; layer thicknesses up to ~100 µm are common. For implants such as stents not made from aluminum, the material must first be coated with aluminum in a physical vapor deposition (PVD) process. During this process, the stents are mounted on a stent holder which is rotated in the aluminum plasma being coated at the same time. The thickness of the coating can be varied by a few hundred nanometers and some microns. In a second step, this layer is electrochemically converted into a nanoporous ceramic by using the above-described methods [30]. Depending on the conditions, different porosities and pore sizes can be achieved (Figure 1.3).

The release kinetics for a specific drug can be varied to a certain degree by changing the layer thickness and varying the pore sizes. Nevertheless, the solubility of the drug is also a very important aspect for the release time and the release kinetics. Until now, no hydrophilic drug has been applied to stents because there is clearly no delivery platform suitable for the retained release of these drugs.

In order to achieve local drug delivery, there is a need for the new technologies to bind a certain amount of the substance onto a stent and assure sustained release over a few days and up to a few weeks. For this reason, stents are typically coated with a 200–500-nm thin metallic layer of aluminum metal, which is converted afterwards into the nanoporous alumina. Stents carrying 40 µg of a hydrophobic drug showed a release of this drug in phosphate-buffered saline at 37 °C over 6 days, with good reproducibility and a narrow standard deviation.

The amount of drug which is fixed onto the stent is limited by the nature of the pore system. Assuming a 1 µm-thick coating with a porosity of 40 % on a stent having 2-cm^2 surface area, it can be calculated that ~80 µg of drug can be fixed. Coated stents of this type, with and without 40 µg drug have been tested for their influence on intimal hyperplasia in the carotid artery of rabbits. The animals were sacrificed after 4 weeks and the morphology of the vessel at the region of implantation was investigated.

0.5 µm

Figure 1.3 Surface SEM image of a nanoporous alumina-coated stent. The pore diameter is ~20–50 nm.

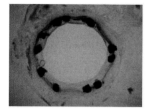 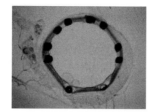 

Figure 1.4 Histological analysis of different stents implanted in the carotid artery of rabbits. Left: bare stent control; middle: nanoporous alumina-coated stent; right: nanoporous coated and drug-carrying stent.

In-vivo serum concentrations of the drug applied to nanoporous coated stents were also determined. The serum level was ~50 ng mL^{-1} (far below any toxic concentration) at 10 minutes after implantation, and later fell exponentially to zero within 2 days. It can be assumed that the drug is released to the adjacent tissue over a significantly longer time period as diffusion processes are changed significantly in direct tissue contact.

Three histological slides of a noncoated stent, a coated stent, and a coated stent with drug loading are shown in Figure 1.4.

In the illustration (Figure 1.4), the stents can still be seen as small square-like, black areas in the tissue slices. These histological sections show clearly that the coating itself markedly reduced neointimal growth, and this effect is enhanced in combination with an anti-proliferative drug.

Quantitative analysis of the histomorphological findings showed a markedly reduced formation of neointima compared with the control for both the coated and the drug-releasing stents. The coating itself reduced tissue growth by 43%, and the reduction in neointimal thickness was even more pronounced above the stent struts. Again, the bare stent outcome was markedly improved by either the drug coating or the coating itself [31]. The same concept was evaluated in a smaller clinical trial involving approximately 50 patients, and showed the safety of nanoporous alumina-coated stents to date.

With this development of biocompatible stent coatings, including the potential of sustained drug release, a new chapter in bioengineering has been opened. The motivation arising from preventing or at least reducing restenosis is high not only with respect to patient care but also to the economic burden that restenosis places on healthcare systems. On the basis of this new stent technology, it is estimated that the US stent market will grow from its current value of approximately US$2.5 billion to US$3.8 billion by 2005 [32]. The technique of using drug-eluting stents should also lead to a minimization of cost-intensive bypass surgery, thus producing further overall cost reductions.

1.2.2
Seeds

The concept of local drug delivery can be extended to the irradiation of malignant tissue, the most prominent examples being devices used to irradiate tissue locally; these are the so-called "seeds". A cut-away diagram showing the internal structure of a typical seed is

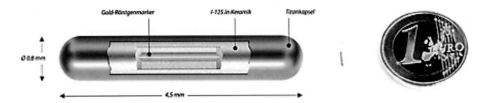

Figure 1.5 Left: Cut-away diagram of an iodine-125 seed. Inside the titanium capsule is a ceramic carrier which carries the ^{125}I. In addition, there is an inner core of gold wire, which ensures X-ray opacity. Right: Comparative size of current seeds used to treat prostate cancer.

shown in Figure 1.5 (left) [33]. The length of such an implant is ~5 mm, and its diameter 0.8 mm.

The implant consists of a capsule that surrounds the inner carrier of the radionuclide shown (see Figure 1.5; left) [33]. Seeds are used to provide radiation therapy to prevent or reduce the growth of tumors which do not form – or have not yet formed – metastases; an example is early prostate cancer. Typical nuclides used in cancer therapy are ^{125}I and ^{103}Pd. These sources are also used in cardiology to irradiate the restenotic area [34]. Instead of using drugs (as described previously), the radiation prevents the smooth muscle cells (SMCs) from further growth. In this treatment, the seeds are further encapsulated in a special delivery catheter which is brought to the narrowed region for a certain time before or after implanting a stent. For any given seed in cardiology or oncology, the carrier consists of either a ceramic or metallic wire or tube which on occasion is chemically modified to bind the desired isotope. For example, in order to bind ^{125}I (which is a gamma-emitter with a half-life of 60 days), either silver is deposited onto the carrier, or the carrier is completely made from thin silver wire. By using a chemical precipitation reaction, the ^{125}I is bound onto the surface of the carrier. The chemistry occurring is determined by the very small amounts of the isotope needed to ensure correct activity of the seed. For example, to fix a typical therapeutic activity of 40 MBq (1 Bq = 1 Becquerel, which is equal to 1 decay s^{-1}) ^{125}I, only 60 ng of the pure ^{125}I isotope is needed to react with the silver carrier. Even the extremely low solubility of silver iodide cannot be easily attained when working with such small amounts, and terms such as insoluble' become meaningless in that range.

When using ceramic carriers, heavy metals such as gold must be applied, either as wires or very small dots to ensure the X-ray visibility of the seeds. So ultimately, the device consists of an X-ray-dense marker, a specific carrier for the radionuclide, and a laser-welded titanium or stainless steel capsule [35].

The capsule is an absolute requirement as no leakage of radioactive materials must occur. This leakage would lead to a systemic contamination of the patient, and cause damage to healthy tissue. Iodine isotopes in particular are known to concentrate in the thyroid gland and lead to severe problems. For this reason, seeds must undergo very different tests to ensure their heat stability and mechanical stability and safety during implantation, especially with regard to leak-tightness [35]. In addition to these safety regulations, the biocompatibility must also be ensured. At first glance, it appears contradictory that a radioactive implant must be biocompatible, but if one takes into consideration that the tumor tissue will be destroyed and ultimately replaced by healthy tissue, then biocompatibility of the implant after complete decay of the radioactive material in necessary. In summary, implant safety is the first important issue, followed closely by biocompatibility and finally processing of the radioactive seed. The point to be raised is, how can nanotechnology improve these issues?

The major key point of handling radioactive compounds is to increase the efficiency of the process. Radioactive nuclides are expensive to produce, handle and dispose of, as the need for protection and safety is huge. Materials in the nanometer scale consist of a very large fraction of the surfaces where the reaction between the radionuclide and the carrier take place. The following account focuses on ^{125}I seeds in order to enlighten this point. ^{125}I is available from a variety of different sources, usually as a byproduct of fission processes inside radioactive reactors [36]. After separation and production of a silver iodate (III) solution, reduction of the silver leads to the formation of iodine anions that are able to react with silver-cations. As the silver is fixed on a solid carrier, the silver surface significantly determines the rate of reaction and the final yield. A nanoscale silver surface is, for those reasons, of great benefit. Nanoporous alumina offers a natural nanometer-size cavity in which silver can be deposited. This process is very well established; starting from silver nitrate solution, silver wires can be formed starting at the base of the pores by applying an alternating current [37]. The oxide is a weak conductor during the cathodic half-cycle, but blocks the current in the anodic half-cycle. In addition, by using an alternating current, polarization of the surface can be reduced to a minimum. A cross-section through an anodized and silver-filled wire is shown in Figure 1.6.

The topographical image on the left in Figure 1.6 shows the remaining aluminum wire at the bottom, with the oxide layer on the top. The silver wires cannot yet be distinguished from the surrounding oxide. An energy dispersive X-ray analysis (EDX) clearly shows the presence of silver wires in the lower third of the alumina layer. The silver nanowires react easily with the ^{125}I.

The yield of the labeling process with ^{125}I using the silver nanowires may be up to 98 %, the maximum capability of uptake is far above that of any therapeutic dose and, due to the special porous structure, the radioactive compound is bound deep inside an inert, ceramic-type layer.

This last-mentioned feature, together with the fact that every single pore takes up only a very small part of the total activity, provides additional safety. While all commercial seeds need to be encapsulated into a laser-welded capsule to prevent any undesired release of radioactive material, dividing the activity into very small fractions and burying them in a strong matrix eliminates the need for an additional capsule. The critical amount derived from ISO standards [35] to be released from a single seed is ~200 Bq per day; when using the above-described system, the measured value is ~2 Bq per day. In case of any potential damage, either during application or afterwards, a disrupted Ti-capsule will surely release all the activity in a short time, whereas a broken nanodevice will probably lead to several thousand opened pores that release less than 1 ppm of the iodine into the body, while the remainder is retained safely inside the structure. Another important aspect of the nanostructured implant surface is the fact that the activity is very homogeneously distributed over the implant surface. On the surface, the pores (10^{10} per cm^2) act as a kind of nano

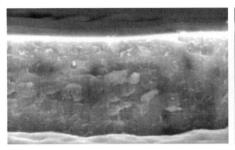 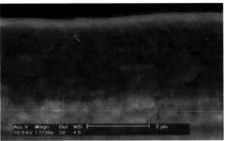

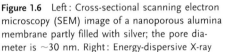

Figure 1.6 Left: Cross-sectional scanning electron microscopy (SEM) image of a nanoporous alumina membrane partly filled with silver; the pore diameter is ~30 nm. Right: Energy-dispersive X-ray (EDX) measurement of the same area. Purple the aluminum (either oxide or metal), and orange the silver-rich areas.

test-tube. Hence, even when only 1 % of all the pores are filled with radioactive material, 10^9 pores remain unfilled. For this reason, dosimetry measurements show a perfect image of the seed geometry itself, making it easy for physicians to calculate very accurately the individual doses for each patient, thereby minimizing side effects in the surrounding tissue. Cell tests using MC3T3 cells have been performed on silver-free specimens in order to prove the biocompatibility of the porous oxide itself. The growth rate of the cells have been determined in comparison to polystyrene standard and to a negative control including sodium azide. The cell-growth curves recorded for cells grown on nanoporous alumina and polystyrene as control did not differ with respect to their slope for both materials, thus indicating similar growth rates on the different substrates. However, when sodium azide was added as a negative control the growth rate was close to zero, clearly indicating that cells growth is stopped in nonbiocompatible environments.

So, the question of how nanotechnology can improve local brachytherapy by the use of seeds must be answered in different ways. First, the safety of the device should be further improved, as must be the production process and technical characteristics of the single seeds. All of this can be achieved by using the nanoporous surface acting as a carrier for the activity, but without causing a loss in biocompatibility when compared with current seed technologies.

1.3
Conclusion

Progress in medical device technology is clearly linked to progress in materials science technology, and new materials which have been developed for very different applications have influenced the design and also the mechanical, chemical, and biological properties of implants. In all cases, the implant interacts in a highly sophisticated manner through its surface, thus becoming an interface with the surrounding tissue. Hence, biocompatibility is a real issue for any given device. Nanotechnology can, in certain very well-defined areas, improve the biocompatibility of implants either passively by the use of thin films, or actively by releasing therapeutic agents from implant surfaces. As thin-film technology is a well-established technology, it can be claimed that active devices are the main area of nanotechnology. Ensuring that drugs can be fixed and subsequently released from soft- or hard-tissue implants such as stents, seeds, or hip-joints can lead to major improvements in the current concepts of passive thin-film technology. In this way, undesirable side effects can be avoided and cost-intensive drugs can be used more efficiently, thereby reducing both patient burden and the economics of the healthcare system. One major area of application for such modern devices is that of stents in cardiology. Rather than perform bypass surgery, a minimally invasive intervention can ultimately lead to the same outcome for the patient, but with less surgical trauma and a reduced time in hospital. Among the many concepts currently proceeding through scientific development phases, it seems inevitable that nanotechnology will ultimately become medical routine.

References

[1] Herman, W. A., Marlowe, D. E., Rudolph, H. Future trends in medical device technology: results of an expert survey. Rockville (MD), Food and Drug Administration. **1998**.

[2] Anderson, M. K., "Dreaming about Nanomedicine", *Wired Magazine*, November **2000**.

[3] Merkle, R. C., "Nanotechnology and Medicine", *Advances in Anti-Aging Medicine*, December **1994**.

[4] Morris M., "Nanorobots may be the Ultimate Medical Devices", Medical Device Outlook, January **2001**.

[5] V. Altstädt, W. Bleck, W. Drachma, A. Dammanns, A. Grunwald, H. Harig, H. Hofmann, W. A. Kaysser, C. J. Langenbach, J. Müssig, K.-M. Nigge, R. Renz, T. Sawitowski, G. Schmid, V. Thole, Neue Materialien für Innovative Produkte. – Entwicklungstrends und gesellschaftliche Relevanz, Europäische Akademie zur Erforschung von Folgen wissenschaftlich-technischer Entwicklungen Bad Neuenahr-Ahrweiler GmbH, Springer-Verlag, Heidelberg, **1999**.

[6] R. W. Siegel, Nanophase materials, *Encycl. Appl. Phys.* **1994**, *173*.

[7] J. H. Fendler (ed.), *Nanoparticles and Nanostructured Films*, Wiley-VCH, Weinheim, **1998**.

[8] G. A. Ozin, Nanochemistry, *Adv. Mater.* **1992**, *4/10*, 612.

[9] H. Wieneke, T. Sawitowski, S. Wendt, A. Fischer, O. Dirch, I. A. Karoussos, R. Erbel, Stent coatings – what are the real differences?, *Herz* **2002**, *27*, 518–526.

[10] H. Wieneke, T. Sawitowski, S. Wendt, O. Dirch, Y. L. Gu, U. Dahmen, A. Fischer, R. Erbel, Kermaische Nanobeschichtung von koronaren Stents zur Reduktion der neointimlane Proliferation, *Z. Kardiol* **2002**, *91 (Suppl. 1)*, 66.

[11] Press release from the Federal Institute for Statistics, **1999**, Germany.

[12] M. Richter, R. Linnemann, P. Woias, Robust design of gas and liquid micropumps, *Sensors Actuators* **1998**, *A68*, 480–486.

[13] D. Maillefer, H. van Lintel, G. Rey-Mermet, R. Hirschi, A high-performance silicon micropump for an implantable drug delivery system, Proc. 12th IEEE MEMS 1999, Technical Digest, Orlando, FL, USA, January **1999**, pp. 541–546.

[14] European Guideline 93/42/EWG, **1993**.

[15] Grüntzig, A. R., Senning, A., Siegenthaler, W. E., Non-operative dilatation of coronary artery stenosis – percutaneous transluminal coronary angioplasty, *N. Engl. J. Med.* **1979**, *301*, 61–67.

[16] Fingerle, J., Johnson, R., Clowes, A. W., Majesky, M. W., Reidy, M. A., Role of platelets in smooth muscle cell proliferation and migration after injury. Endothelial and smooth muscle growth in chronically denuded vessels, *Lab. Invest.* **1986**, *54*, 293–303.

[17] Rogers, C., Edelman, E., Endovascular stent design dictates experimental restenosis and thrombosis, *Circulation* **1995**, *91*, 2995–3001.

[18] Topol, E. J., Leya, F., Pinkerton, C. A., Whilow, P. L., Hofling, B., Simonton, C. A., Masden, R. R., Serruys, P. W., Leon, M. B., Williams, D. O., King, S. B., Mark, D. B., Isner, J. M., Holmes, D. R., Ellis, S. J., Lee, K. L., Keeler, G. P., Berdon, L. G., Hinohara, T., Califf, R., A comparison of directional atherectomy with, coronary angioplasty in patients with coronary disease: the Caveat Study Group, *N. Engl. J. Med.* **1993**, *329*, 221–225.

[19] R. Koster, D. Vieluf, M. Kiehn, M. Sommerauer, J. Kahler, S. Baldus, T. Meinertz, C. W. Hamm, *Lancet* **2000**, *356*, 1895–1897.

[20] A. Kastrati, A. Schömig, J. Dirschinger, J. Mehilli, N. von Welser, J. Pache, H. Schühlen, T. Schilling, C. Schmitt, F.-J. Neumann, *Circulation* **2000**, *101*, 2478.

[21] Monnink, S. H., van Boven, A. J., Peels, H. O., Tigchelaar, I., de Kam, P. J., Crijns, H. J., van Oeveren, W. J., *Invest. Med.* **1999**, *47*, 304–310.

[22] N. T. Duda, R. T. Tumelero, A. P. Tognon, *Am. J. Cardiol.* **2002**, *90* (Suppl. 6A), 75H.

[23] C. Caussin, M. Hamon, D. Carrié, P. Commeau, G. Grollier, J. Puel, B. Lancelin, *Am. J. Cardiol.* **2002**, *90* (Suppl. 6A), 77H.

[24] J. P. O'Sullivan, G. C. Wood, *Proc. R. Soc. London* **1970**, *317*, 511.

[25] R. Kniep, P. Lamparter, S. Steeb, *Adv. Mater.* **1989**, *1/7*, 975.

[26] D. D. Macdonald, *J. Electrochem. Soc.* **1993**, *140/3*, L27.

[27] J. W. Diggle, T. C. Downie, C. W. Goulding, *Chem. Rev.* **1969**, *69*, 365.

[28] M. M. Lohrengel, *Mater. Sci. Eng.* **1993**, *6*, 241.

[29] K. Wefers, C. Misra, Oxides and Hydroxides of Aluminium, Alcoa Laboratories, **1987**.

[30] G. E. Thompson, G. C. Wood, *Treatise of Materials Science and Technology*, **1983**, *23*, 205.

[31] H. Wieneke, O. Dirsch, T. Sawitowski, Y.L. Gu, H. Brauer, U. Dahmen, A. Fischer, S. Wnendt, R. Erbel, *Cath. and Card. Interv.* **2003**, *60*, 399–407.

[32] A. Machaoui, P. Grewe, A. Fischer (eds), *Koronarstenting*, Steinkopf, Darmstadt, **2001**, pp. 290–297.

[33] Company brochure, EZAG, Berlin.

[34] R. Robertson, Encapsulated low-energy brachytherapy sources, US Patent No. 6099458.

[35] ISO 9978, Strahlenschutz – Geschlossene radioaktive Quellen – Dichtheitsprüfung.

[36] K. L. Lathrop, P. V. Harper, Argonne Cancer Research Hospital Magazine, *12*, **1962**.

[37] C. A. Foss, G. L. Hornyak, J. A. Stockert, C. R. Martin, *J. Phys. Chem.* **1994**, *98*, 2963.

2
Microfluidics Meets Nano:
Lab-on-a-Chip Devices and their Potential for Nanobiotechnology

Holger Bartos, Friedrich Goetz, and *Ralf-Peter Peters*

2.1
Introduction

Microfluidic devices and integrated chemical measurement systems were among the first ideas when the investigation of nonmicroelectronic applications of microfabrication technology was started more than two decades ago. In 1979, an integrated gas chromatograph was fabricated on a 2-inch (5-cm) silicon wafer [1]. Concepts and first applications of miniaturized total analysis systems emerged in 1990 [2]. During the past decade, array technologies and microfluidics have become commercially available in biochips for genomics and proteomics. It is expected that many more applications will appear on the market in the near future, as these devices are presently under development in many companies world-wide.

It should be noted that the structures used in microfluidic and in Lab-on-a-Chip devices are not nanostructures, but are in the micrometer to even millimeter range. However, bionanotechnology requires a microfluidic platform technology as an interface to the macroworld: for self-assembled monolayers; for the handling of nanoparticles, cells or nanobarcodes; and to monitor and control cellular machinery.

On the other hand, nanobiotechnology will enable novel microfluidic platforms due to the integration of nanostructures, nanocoatings or nanoactuators, by the integration of nanoporous membranes, and by integrating detection and measurement techniques such as nanoelectrodes, nanooptics, and patch–clamp arrays.

2.2
Overview

2.2.1
Definition and History

A microfluidic chip is defined as an assembly of microstructures on a common substrate, used for the manipulation of fluids (gases and/or liquids).

A Lab-on-a-Chip device is a combination and integration of fluidic elements, sensor components and detection elements to perform the complete sequence of a chemical

Nanobiotechnology. Edited by Christof Niemeyer, Chad Mirkin
Copyright © 2004 WILEY-VCH Verlag GmbH & Co. KgaA, Weinheim
ISBN 3-527-30658-7

reaction or analysis, including sample preparation, reactions, separation, and detection. This chapter focuses on Lab-on-a-Chip devices for Life Science applications, and does not cover microreactors for chemical synthesis [3].

Both, microfluidic as well as Lab-on-a-Chip devices, were part of the vision when micro-fabrication technology – which had emerged from the fabrication tools for microelectronic devices – was first applied to problems in mechanics, optics, and fluidics. Among the first examples were a gas chromatograph developed at Stanford University [1], and pioneering work on inkjet printheads at IBM in the late 1970s [4]. The inkjet printhead has become one of the commercially most successful fluidic applications of this new technology, which was called "MEMS" (Micro Electro Mechanical Systems) in the U.S. and "Microsystem Technology" in Europe.

Many discrete microfluidic devices, such as microvalves [5], micropumps [6, 7], flow sensors [8], and chemical and biological sensors [9] were developed, but the benefits of miniaturization are best taken advantage of when these devices are integrated into a fluidic system. Intensive work on Lab-on-a-Chip systems was started in the early 1990s [10–12], and today integrated microfluidic devices are established in laboratory equipment for biomedical research and starting to penetrate the diagnostic market for point-of-care and laboratory automation applications.

2.2.2
Advantages of Microfluidic Devices

Microfluidics offer advantages both from a technical as well as from an economical view-point. When the dimension of fluidic structures are scaled down to the micrometer region, the surface to volume ratio of the fluids involved increases dramatically, and surface effects start to dominate volume effects. For the fluid flow in microstructures this leads to well-defined flow characteristics, as the flow is strictly laminar and turbulence can only appear in very limited regions around sharp edges.

Due to the absence of reasonable turbulence, mixing of different fluids can only be achieved by diffusion, or by specially designed fluidic mixing elements. Moreover, due to the scaling factors of diffusion and heat conduction, the equilibrium conditions can be reached much faster.

The small sample volumes involved are of enormous advantage especially for highly parallel applications, like array devices used in genomics, proteomics, and drug discovery. The reduction in the amount of substance required for each reaction leads to significant cost reductions for these types of applications. Another advantage associated with small sample volumes is that minimally invasive methods are sufficient for taking samples, for example of blood or interstitial fluids.

These small volumes can be precisely controlled by taking advantage of microfluidics. In some cases, this is achieved just by a proper definition of the geometric dimensions of the corresponding channels, wells, and reactors. Another method to define precisely small fluid volumes is droplet generation; this is a separate application field of microfluidics, with important products such as inkjet printheads or drug delivery systems. Array spotters are another product of this kind, used in the immobilization process of nucleic acids, antibodies, etc., and will be described in section 2.3.5.

The large surface implies a high reaction efficiency, as the surface areas which may be coated with catalysts or enzymes are large compared to the reaction volume. Furthermore, due to the large surface to volume ratio, capillary forces dominate volume forces such as gravity, and may advantageously be used for fluid transport in single-use devices. Finally, integration and the mass-fabrication capabilities of microfabrication technology make the application of microfluidics economically attractive.

2.2.3
Concepts for Microfluidic Devices

For microfluidic chips, two main organization principles are used in integrating the fluidic elements on the chip.

One principle is parallelization; this is used when the same type of reaction has to be performed in parallel many times. Examples are array type of chips found in DNA analysis, proteomics and high-throughput screening. Parallelization can lead to dramatic cost advantages in three ways: First, the manufacturing cost for a device with many integrated reaction wells is much lower than that for many devices for just one reaction; second, all reactions are performed in parallel, saving labor cost and time; and third, parallel reactions are an ideal input for laboratory automation and information processing of the assay results.

One very basic application of the parallelization principle are nanotiterplates, an extension of the well-established micro plate technology into the nanoliter region. A review of nanotiterplates is given in Ref. [13].

As an example of parallelization, an array of 250 µm × 250 µm wide, 500 µm deep, bottomless wells is shown in Figure 2.1. One chip will carry up to 100 000 of these reaction wells. This Living Chip™ technology was developed at MIT and commercialized by BioTrove, Inc.

The second basic organization principle is sequential integration. Here, several fluidic structures, each designed to perform one step in a processing sequence, are integrated on

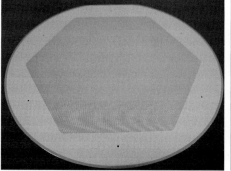

Figure 2.1 The 100K Living Chip™ plate (left) and a detail view of the 50-nl wells filled with liquid (right). Massively parallel reactions may be initiated by stacking of chips; applications include drug discovery, genomics, and proteomics. (Courtesy BioTrove, Inc.)

one substrate and interconnected by a channel network to provide the transport of the fluids between the processing steps. The fluids will pass the processing steps in a sequential manner. The fluidic structures involved are channels, mixers, reaction chambers, detection chambers, sample and waste reservoirs, microvalves, micropumps, microsensors, heating zones, and many others; for a detailed description see section 2.3.2. Some of the fluidic structures may also have electrical, mechanical, or optical functions and the corresponding elements and interfaces; these may also be integrated into the microfluidic chip, or added in a discrete way. In many cases, complete fluidic components, for example micropumps, are added as discrete components to the microfluidic device. Recently, attempts have been started to standardize such elements with respect to size and input/ output terminals, to create standard building blocks for modular fluidic devices [14].

One important example of sequential organization is that of micro Total Analysis Systems (µTAS). These are fluidic systems which are integrated on one substrate and are intended to perform the total sequence of a chemical analysis, having been developed in several laboratories worldwide [15, 16]. Recent results are found in the proceedings of the annual conference on this topic [17]. A first application was an integrated system to monitor the glucose content in a fermentation process [18]. Another, very important application for microfluidic devices are PCR reactions [19, 20]. An example of a commercially available system, which performs sequentially the preparation, amplification, and detection of DNA, is described in Ref. [21].

The sequential organization scheme is also represented by capillary electrophoresis chips [22] (see also Figure 2.5). With dielectrophoresis, cells and particles in a weak electrolyte solution may be moved and collected using the forces induced by travelling, rotating, or alternating electrical fields; a review is given in Ref. [23].

Some microfluidic devices combine both organizational principles. Array-type fluidic chips will, in most cases, require a channel network for fluid transport, and more complex reaction sequences require more than one reaction site. On the other hand, in most cases it is favorable to include parallel processing in sequential arrangements. A combination of 96 wells, together with a fluid distribution network, on a single chip is shown in Figure 2.2.

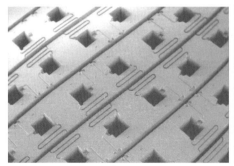 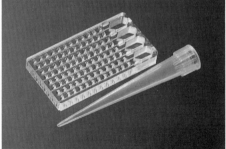

Figure 2.2 Microtiterplate "Lilliput" for bacteria identification and antibiotics susceptibility tests. Samples are distributed and dosed via a microfluidic network into 96 reaction cavities by capillary forces. (Courtesy STEAG microParts and Merlin Diagnostika.)

2.2.4
Fluid Transport

Obviously, one important aspect of microfluidic devices is the fluidic transport. One or several fluids must be transported to reaction sites, and often a sequence of transport actions at defined times is required. To achieve the transport, two types of mechanisms are used.

In *actively driven transport*, active fluidic elements such as pumps and valves are used to achieve the transport. These may be external elements, but in some cases they are part of the fluidic device, either by adding them as discrete elements, or by integrating them into the fluidic device. These active devices require an outside energy supply to operate, and this can be either electrical, pneumatic, or mechanical. This may require an electrical network to be part of the fluidic system. In one example (the Mixed Circuit Board (MCB) concept, [24]), printed circuit boards have been chosen as the basis for the fluidic device, carrying both the fluid microchannels as well as the electrical network. The fluidic elements, like discrete electrical components, are then assembled on this MCB, which requires both electrical and fluidic interconnections. In the case of integrated active fluidic elements (e. g., a piezo-driven membrane pump), it may be of advantage to allow the fluidic structures such as the pumping chamber, membrane and input and output valves, be part of the fluidic element. The drive elements like the piezo could then be placed on a separate drive plate which is attached to the fluidic chip, for example by clamping, during the operation. In this case, the fluidic chip could be a single-use, disposable device, while the more expensive drive plate would be re-usable.

Another means of actively providing fluidic transport is the use of *mechanical forces*. In the case of centrifugal forces, the fluidic structures are usually on a CD-like substrate, which is placed on a spinning device which resembles a laboratory centrifuge. The fluidic transport can be triggered by correct selection of rotational speed, position on the substrate, and channel width. Commercially available platforms include the "LabCD" [25] and the "Gyrolab™ microlaboratory" (Figure 2.3).

A major advantage of microfluidics is that fluidic transport can also be achieved in a *passive* manner. In this case, capillary forces are used to transport the fluid to the reaction sites. As mentioned above, capillary forces can be large compared to volume forces in microfluidics. However, to make use of this effect it is essential that the surface of the fluidic structures is hydrophilic with respect to the fluids to be transported; this may require a surface modification of the material (see section 2.3.4).

By correct design of the fluidic structures, the flow front in the device can be controlled, and this allows the transport times and volumes of the fluids transported to be set to desired values. For a continuous flow through the device for a long period, larger "waste reservoirs" are required at the end of the channel network. Locally hydrophobic areas in the channels may be used to stop the flow at defined positions.

Capillary fluid transport is not reversible, and once the complete fluidic network is filled, the flow stops. Hence, this transport mechanism is well-suited for priming of the device, or for single-use, disposable devices. On the other hand, neither an active (and often expensive) element nor an energy source are needed. This reduces the manufacturing costs and enables the use of microfluidic disposable devices, e. g. in point of care diagnostics and patient self-testing.

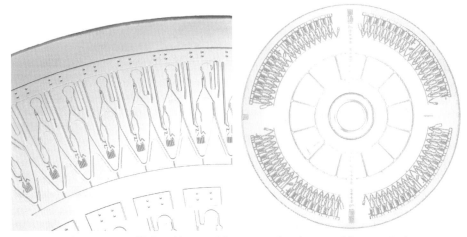

Figure 2.3   Left : Injection-molded CD-like microlaboratory. Right : Close up of the microfluidic structures. This system enables functions such as volume definition, chromatography, and enzymatic reactions to be conducted. (Courtesy Gyros AB.)

2.2.5
Stacking and Sealing

Except for some very basic array configurations, all microfluidic devices require a top cover to create enclosed structures (e. g., channels, reservoirs). This can be achieved by sealing the top side of the substrate carrying the fluidic structures with a foil, a cover plate, or by stacking several microstructured fluidic plates.

Sealing with a thin, and often optically transparent foil is a cost-effective procedure and allows easy access to the fluid, for example when optical methods such as fluorescence are used as detection methods. Furthermore, special materials may be selected for the foil, such as foils with high diffusion coefficients for gases, thereby allowing oxygen supply to cells in the chips.

Cover plates may carry fluidic structures themselves, complementing the fluidic network on the base substrate. One simple example are through-holes in the cover plate which are used for input and output ports of the fluidic device.

Stacking of several microstructured plates is of advantage for more complex fluidic devices because it is extending integration into the vertical direction. With stacking, multilayer fluidic interconnections can be created, and many fluidic devices are much easier to build when vertical integration is used. One example is that of micropumps, where the pumping chamber and valve seats may be on one plate, the membrane and the valve lids on a second, and the driver and input and output on other plates. A very early example of the stacking principle can be found in the above-mentioned realization of a µTAS system [18, 26].

If no sealing is used for simple array devices, then hydrophobic surface properties between the spots may be used to concentrate fluids at the (hydrophilic) spot areas in the form of droplets, thus avoiding cross-talk between different reaction sites.

2.3
Methods

2.3.1
Materials for the Manufacture of Microfluidic Components

Three types of materials are common for microfluidic and Lab-on-a-Chip devices: silicon, glass, and polymer materials.

2.3.1.1 **Silicon**

Silicon is the dominant material in microelectronics, and knowledge of micromachining of this material has been accumulated for several decades. Because of this, silicon has also been the dominant material used in nonelectronic applications of MEMS and, in the past, also in microfluidics. In fact, the known micromachining methods for silicon are well-suited for the generation of high-precision fluidic structures. For example, channels with square or v-shaped cross-sections can be easily generated.

Among the advantages of silicon are the simple generation of an inert surface (SiO_2) by oxidation, high-temperature stability, high chemical resistance to organic solvents and acids, well-established bonding processes, an extensive knowledge about coatings, and its well-defined and excellent mechanical properties as a single crystal material.

Silicon may be the material of choice if electric functions such as heaters and sensors are required as part of the microfluidic component. These can be easily integrated into the silicon substrate using standard microelectronics fabrication technology.

The disadvantages of silicon are the nonideal surface for many biochemical applications, and the high price for material and processing. Silicon is a relatively expensive, single-crystal material, and the process equipment, process materials from microchip technology are very expensive. As fluid chips tend to be much larger than electronic chips, this may lead to high manufacturing costs per chip. Another cost disadvantage is that the batch processing sequence used for silicon is more complicated than the one-step replication methods used for polymers. In the silicon process, the alignment of subsequent layers must be carried out for each wafer in production. In polymer replication technology, the alignment is necessary only during the production of the master, eliminating this error source once the master has been correctly manufactured.

Furthermore, silicon cannot be used for applications involving electrical fields (e.g., capillary electrophoresis) due to its low electrical resistance.

2.3.1.2 **Glass**

Glass is another important material for the production of microfluidic components and systems; borosilicate types of glass are often used. Some of these glasses, such as Borofloat or Pyrex 7740, have thermal expansion coefficients which are matched to that of silicon, and are used together with silicon in stacked arrangements, for example as transparent cover plates. These glasses can be bonded to silicon by anodic bonding, without the need of a bonding material.

The advantages of glass are its high chemical resistance, excellent thermal and mechanical stability, and optical transparency. In many cases, glass is well-suited as a surface for

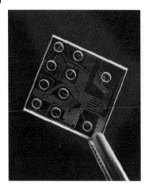

Figure 2.4 Capillary electrophoresis chip for nucleic acid separation. (Courtesy Caliper Technologies, Inc.)

biological and chemical reactions. There is an extensive knowledge about inorganic and organic coatings with glass as a base material. Auxiliary electric functions (e. g., heaters) may be added using the well-established procedures of thin film technology. Glass is also well-suited for electroosmotic flow applications and capillary electrophoresis [27]. An example is shown in Figure 2.4.

Although glass as a base material is less expensive than silicon, batch fabrication, optical polishing steps, and the micropatterning steps will lead to comparatively high production costs. Micromachining procedures for glass are much less developed than for silicon, and in most cases isotropic etching is used. High-aspect ratio and multilevel structures are difficult to manufacture, and this restricts the use of glass to simple applications such as array chips, single depth channel networks (e. g., capillary channels), or intermediate and cover plates in stacked arrangements.

The photostructurable glass FOTURAN (Schott) allows the fabrication of high-aspect ratio fluidic structures, but the disadvantages are high substrate and processing costs, and compared to other materials a high surface roughness of the structures.

2.3.1.3 Polymers

Polymers are the third type of material used in the manufacture of microfluidic devices [28]. The main benefit of polymer materials is based on simple and cost-effective replication methods such as injection molding or hot embossing, because this allows the manufacture of all microstructures of the device in one manufacturing cycle. The capabilities of these manufacturing processes in the micro and nano regime are illustrated by the manufacturing of CDs and DVDs, where a 120 mm-sized device, including metallization and printing, can be manufactured for much less than 1$ – dramatically less than for a silicon or glass device of the same size. However, these replication methods require the manufacture of a master structure, which is used as a tool in the replication step. As the manufacturing cost of the mastering is considerable, these methods only make sense for high-volume applications, where at least a few 100 000 parts are manufactured, and the mastering cost can be shared by many replicated parts.

Another advantage of polymers is the broad range of materials suited for these manufacturing methods, including PMMA, PS, PC, cyclic olefins, PEEK, POM, elastomers, and others. This allows a choice to be made of the material properties suitable for the specific application. Typical properties of the material that may be of fundamental importance include optical transparency, autofluorescence, thermal expansion coefficients, and stiffness. A summary of the properties for a range of materials is provided in Ref. [28].

Convenient sealing methods, such as lamination, ultrasonic welding, laser welding, gluing or thermal bonding, are available for polymer devices.

One disadvantage of polymer materials is a reduced thermal stability, as these devices can only be operated at temperatures below the glass transition temperature. This also limits processes to coat or functionalize the polymer surface. Another disadvantage is the reduced stability against organic solvents, acids, and bases.

A variety of methods for chemically modifying the plastic surface, and functionalizing the surface have been published, and extensive work is under way in that field. Most polymer surfaces are not hydrophilic with respect to the fluids used in nanobiotechnology, and will require a suitable surface modification (e. g., plasma polymerization) if capillary forces are to be used for fluid transport.

2.3.2
Fluidic Structures

The most basic fluidic structures to build microfluidic devices are microchannels. These channels provide the fluidic interconnection network between the fluidic elements of the device, but may have additional functions, like the channels in capillary electrophoresis and other separation techniques. Various shapes for the channel cross-section are used, including rectangular, v-shaped, and round. The shape of the cross-section may be determined by the fabrication method; a review is provided in Ref. [29]. In many cases, the upper half of the channel contour is flat due to the sealing of the channels by a flat cover. Interesting exceptions to this are silicon nitride channels with a round cross-section buried underneath the surface of a silicon substrate [30], or round PDMS channels.

Important parameters of the microchannels include surface roughness and the aspect ratio of the structure, which is defined as the ratio of depth to width in the case of a channel. High-aspect ratio channels have a high surface to volume ratio and consume less floor space on the microfluidic chip. Channel widths commonly vary between the millimeter to the micrometer range; aspect ratios up to 10 are used. One microfluidic device may carry channels of different widths and aspect ratios, for different purposes. For example, auxiliary channels are used in capillary devices, with a much smaller diameter than the fluid channels, to allow the air to exit from the device when it is filled with fluid by capillary forces.

Other important structures are reaction/detection chambers, and sample and waste reservoirs. These are larger, well-type structures, with dimensions often in the millimeter range, designed to hold the correct amount of fluid.

In the case of reaction chambers, it is often advantageous to generate a high surface to volume ratio. This can be achieved by using auxiliary structures (Figure 2.5), by folding up a channel in a meander-like form, or by using a porous, nanostructured surface [31].

One special, but important, case of microreaction chambers is that of microcompartments used in array-type microfluidic devices for parallel processing, such as DNA chips and nanotiter plates. These are designed to hold fluid volumes in the order of

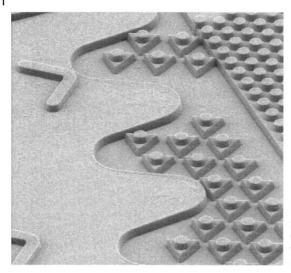

Figure 2.5 Scanning electron micrograph of auxiliary structures in a microfluidic chamber. These are designed to generate a large surface to volume ratio. (Courtesy STEAG microParts.)

10 nL to several hundreds of nL. In the most simple case, these compartments will be not a spatial microstructure at all, but a spot on a flat surface carrying immobilized reagents, with good wetting properties, and separated from neighboring spots by hydrophobic regions. In the case of nanotiterplates, the bottom of the compartments may be a thin membrane, so that optical detection techniques can be applied through the membrane from the bottom side. For applications in combinatorial chemistry, where the possibility to wash and filter reagents is essential, these membranes may be patterned to contain pores in the nanometer or micrometer range.

Active and passive valves are needed to block the fluid flow in a controlled manner. Valves may be used as discrete devices, or integrated into the fluidic chip. Technically, cantilever and diaphragm-type valves are used; reviews are provided in Refs. [32, 33]. Fluidic diodes, which do not have any moving parts [34], are also used; these do not block the flow in one direction completely, but provide a large difference in impedance. Moreover, they are easy to integrate into the system.

Propagation of fluids in the chip is achieved by the use of micropumps which, as in the case of valves, may be either external or integrated into the microfluidic device. Technically, most micropumps are membrane-actuated pumps, using pneumatic, thermopneumatic, piezoelectric, electrostatic, bimetallic, or shape-memory effects for actuation. Some electric field-actuated pumps (electrohydrodynamic and electroosmotic) and micro gear pumps are also available; an overview is provided in Refs. [32, 33].

As flow in microfluidic devices is strictly laminar, mixing must be initiated using a specially designed element, a micromixer. Most micromixers are static mixers, which are exclusively based on the diffusion of the liquids to be mixed. Diffusion requires time, and this must be provided in the microsystem by using long, parallel flow regions and having large interfaces between the liquids to be mixed. This is often achieved by multiple splitting of the fluid strand, and recombining. Methods to go beyond laminar mixing include the use of microbeads [35] and chaotic mixing using relief structures at the channel bottom [36]. A review of mixers is provided in Ref. [33].

Other fluidic structures in microfluidic devices include sensors for physical parameters such as pressure, temperature, and flow, as well as chemical sensors and biosensors. Such elements are found in Lab-on-a-Chip devices, while in single-use disposable fluidic devices these more expensive systems will not be part of the fluidic device.

2.3.3
Fabrication Methods

The fluidic structures are fabricated using standard methods of microfabrication. These are well-documented in standard textbooks of microsystem technology, for example by Menz and Mohr [37] or Madou [38], and are beyond the scope of this chapter. Other summaries of fabrication technology, more specific for the application to fluid devices, may be found in Refs. [32, 39].

These microfabrication methods are also used to manufacture the master tool for the microreplication of polymers. The master is usually a metal (or in some cases a silicon) tool. The master structure is the inverse of the fluidic structure to be generated in the replication process. Channels, for example, will be a line on the master.

Practical, marketable fluidic devices are generally multilevel structures. This means that a device will not carry structures of one common structural depth only, but will have channels, wells, and reservoirs with a variety of structural depths. This cannot usually be achieved with one single fabrication step, nor by using just one fabrication technology, and in practice a combination of different fabrication technologies, each suitable for the generation of structures of the desired size, shape, and precision, will have to be applied. For example, in the fabrication sequence of a replication master, the channel structures for small channel diameters could be fabricated by lithography and electroplating, the more coarse channels by milling, and through holes by spark erosion.

2.3.4
Surface Modifications

Modifications of the surface of the device are essential for the designed functionality of microfluidic devices in (nano)biotechnology. Often, these modifications are to be achieved locally, and therefore different areas of the device will require different modifications. These modifications must be achieved on all surfaces, including the sidewalls of high-aspect ratio microstructures, for example in deep channels.

The objectives for modifications of the fluidic device surface include the modification of wetting characteristics (hydrophobic/hydrophilic), increased biocompatibility, reducing or eliminating solute interactions with the device surfaces, modifying electroosmotic flow, immobilizing the reagents, enzymes, antibodies, proteins, DNA, etc. to carry out chemical reactions or detection mechanisms, or to provide a proper surface for immobilization, increasing the surface area for catalytic reactions, and tethering sieving matrices or stationary phases for separation devices.

The surface modifications may be achieved by a variety of techniques, including CVD and PVD methods, spin coating and solution casting, plasma processes (e. g., plasma etching and plasma polymerization), grafting, chemical self-assembly, the Langmuir–Blodgett technique, printing, and others. In some cases, these surface modifications will involve nanotechnology. The thickness of the modification layer is in the nanometer range; thicker layers might modify the device geometry, and its function, and so for the objective of the functionalization, often only a few monolayers are sufficient.

For example, when multilayer films containing ordered layers of protein species are assembled by means of alternate electrostatic absorption with positively charged PEI, PAH, chitosan or with negatively charged PSS, DNA and heparin, the enzymatic activity of the films does not increase with layer number for more than 10–15 layers [40].

Requirements which the surface modifications must meet include good adhesion, chemical stability against the media used in the device, and a time stability which is better than the lifetime of the device.

One very important surface modification is that of modifying the wetting characteristics of the surface. As the interfacial tensions cannot be monitored directly, measurement of the contact angle between the surface and a droplet of liquid is widely used to characterize the wetting characteristics of the surface.

Materials such as glass, Si and SiO_2 have many OH-groups on their surfaces, and this causes hydrophilic behavior. Especially in the case of silicon, the wettability depends strongly on the pre-treatment and history of the surface. Hydrophobic surfaces may be produced using octadecyltrichlorosilane (OCTS), and hydrophilic behavior may be stabilized using hexamethydisilazane (HMDS).

Polymer surfaces are hydrophobic in most cases. Hydrophilic surfaces may be easily generated using O_2 plasma treatment, but such surfaces are stable only for a few days. More stable surface modifications are obtained by plasma polymerization of layers involving OH-groups at the surface.

The wetting characteristics of the surface may also be modified by a nanostructured surface. This principle of nanobiotechnology is found in nature, for example, in the cuticular structure of leaf surfaces [41] and in fractal surfaces [42]. Such water-repelling surfaces have self-cleaning properties (the Lotus effect), as particles on nanostructured hydrophobic surfaces are more readily wetted and washed away (Figure 2.6).

Large surface areas are required for both catalytic reactions and separation assays, and this may be achieved by coating microfluidic chips with a porous material. In the case of silicon, porous silicon with pore sizes in the nanometer to micrometer range may be generated.

Another important surface functionalization is the binding of specific molecules to designated areas of the chip. Such applications include DNA-, proteomics-, cell-, and tissue-chips. Generally, by using various surface chemistries, linkers for such molecules must be provided in designated areas, while the remaining surface should be nonbinding.

Methods to immobilize the specific molecules include adsorption, crosslinking, covalent binding, microencapsulation, and entrapment. A thin, sputtered gold film can be used to immobilize a dense molecular film of thiols [43], providing a high density of alkyl groups as binding sites for surface reactions.

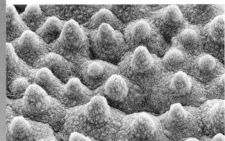

Figure 2.6 Left: *Nelumbo nucifera*, the Lotus flower. Right: a double-structured surface optimized for self-cleaning. Contact areas are minimized through the combination of micro- (cells) and nanostructures (wax crystals). (Courtesy University of Bonn.)

One example of polymer substrates is the building of a functional chemical scaffold on PMMA using an ethylene diamine foundation [44]. In this way, various materials such as oligonucleotides, enzymes, or stationary phases may be attached to the device surface.

2.3.5
Spotting

For array-type microfluidic devices, large numbers of molecules must be collected and placed either in defined microvessels, as with libraries in solution in nanotiter plates, or at defined spots on the surface of a carrier for probe molecules being immobilized on the substrate. This requires the microdispensing of a large variety of (different) fluids in drop volumes down to the picoliter range, with spot sizes and drop distances down to some 10 μm. Special devices to accomplish this task – the "spotters" – have been developed, and a review of spotting methods is provided in Ref. [45].

Dedicated capture spots with optimized wetting characteristics for the dispensed liquid, and non-wetting bars between these spots, may support the array production using spotters.

The main types of spotting methods currently in use include pin-based spotting, ink-jet spotting, photolithographic synthesis, electronic addressing, and stamping.

In pin-based spotting, an array of metal (tungsten) pins picks up a small volume each by dipping into a well plate, and then transfers it when touched down onto the substrate.

Ink-jet spotting uses proven technology from piezoelectric printheads of ink-jet print technology. Large arrays of heads are used for spotting with good control of drop sizes down to the pL range, at high speed.

Photolithographic synthesis is a method developed by Affymetrix [46], where capture probes are synthesized directly on the chip. Photolithography masks the direct, light-sensitive removal of protective groups from hydroxyls in the exposed regions. This allows specific protected nucleotides to attach to these hydroxyls, after which the process is repeated for the next nucleotide.

In the electronic addressing method developed by Nanogen [47], use is made of the fact that the biologic target material is usually either positively or negatively charged. By setting voltage potentials at the test sites of the array, the target can be attracted and docked at these sites. However, this method requires full semiconductor processing in the manufacture of the array.

Another spotting method currently under development is that of micro contact printing (see Chapter 3). Elastomeric stamps with posts in the μm size region are used to deliver either the reagent of choice, or a deprotecting agent, to the spots.

2.3.6
Detection Mechanisms

For most microfluidic applications, detection devices are not integrated into the fluidic chip, but form part of a separate (in many cases highly automated) handling and detection system. In this way, the system can be re-used for the evaluation of a large number of chips.

One problem associated with detection in microfluidic devices is the small sample volume. For example, due to the small dimensions of the system, the optical pathlength for absorption measurements is also likely to be very small.

Commonly used detection methods include absorption (ultra-violet, optical, infra-red), fluorescence, luminescence, electrochemical, thermal or electrical conductivity, and others. Several miniaturized or even microstructured detection systems are available, one example being that of micromolded microspectrophotometers [48].

2.4
Outlook

During the past few years, microsystem technologies – and especially microfluidics for Life Sciences applications – have been identified as the enabling technology of the 21st century. A variety of biomicrosystems has been developed, and research and commercialization efforts on bioMEMS, biochips and Lab-on-a-Chip devices are booming. Today, less than two decades after MEMS technology first emerged, nanotechnology – again, often focused on biology – has begun to attract the interest of the research community. Currently, it is considered that nanobiotechnology will have at least the same impact as bioMEMS technology has had in the past.

On one hand, due to a top-down approach and a continuous shift of technical limits (e. g., resolution), an extension from microstructures to nanostructures has been anticipated. Nanofluidics, nanooptics, nanomechanics, and nanoelectronics will be disciplines that derive naturally from their larger counterparts, not predecessors. When favorable for the envisioned assay, nanochannels, nanocavities, nanoposts and other structural features will be used instead of (or in combination with) microstructures [49], and nanoelectrodes or nanooptical structures will enable further progress in detection technologies and sensitivity.

On the other hand, due to a bottom-up approach, nanosystems will need microfluidic devices as a physical interface to instruments or humans. A variety of examples for the bottom-up approach, often in combination with novel, nanoanalytical characterization methods, are described in detail in other chapters of this book. Most functionality will be created when bottom-up and top-down strategies are combined, whereupon microfluidics and nanobiotechnology will emerge towards integrated systems. Nanostructures such as self-assembled systems or biomimetic surfaces, nanocoatings, nanopores, nanoactuators, nanoparticles, nanocomposites, nanobarcodes or nanoelectrodes will enable novel microfluidic devices for Life Science applications such as drug discovery, diagnostics, and therapy.

Nanoparticles and nanocoatings have been already established for microfluidic devices. Commercially available lateral flow immunoassays involve biofunctionalized particles in the nano range, and magnetic nanoparticles are used for the purification of biomolecules such as cells or nucleic acids. Nanobarcodes – sub-µm-sized metal particles functionalized with biomolecules, comprise freestanding, cylindrically shaped metal nanoparticles that are self-encoded with sub-µm stripes. Intrinsic differences in reflectivity between adjacent metal stripes (e. g., gold and silver) of the nanobarcodes allow individual particles to be identified by conventional optical microscopy. Nanobarcode particles are thus the nanoscale equivalent of conventional bar codes, and are used to decode the sample bound to the functionalized particle surface; for details, see Chapter 26.

Nanocoatings are derived from "conventional" surface chemistry, and also have been found in nature. One of the most impressive examples of biomimicry has been the Lotus effect. This phenomenon of superhydrophobic, self-cleaning surfaces which is seen not only in the Lotus flower but also in many other leaves (e. g., cabbage, reeds, Indian cress, tulips) as well as in animals (e. g., wings of butterflies and dragonflies), has been explored in detail by W. Barthlott and others [41]. The self-cleaning property is connected with a microstructured surface as well as with a coating of water-repellent waxy crystals. Besides inorganic contamination, organic contaminations such as spores, bacteria or algae play an important role in plants. An elegant way to cope with this is to use the Lotus effect, which prevents pathogens from binding to the leaf surface. As the Lotus effect is based solely on physico-chemical properties and is not bound to a living system, artificial self-cleaning surfaces have been successfully manufactured, and such devices for Life Science applications are currently being tested. Nanocoatings with other functionalities are also under development; for example, for guided migration, spreading, growth and differentiation of cells in culture, for the enhanced integrity of biological samples, or for a controlled release of embedded drugs.

Both nanocoatings and nanostructures are currently being evaluated for tissue engineering [50]. Another approach to mimic nature is that of molecular imprinting technology (MIT), which can be described as making artificial 'locks' for 'molecular keys'. Although molecular imprinting was used as early as the 1930s by Polyakov to selectively capture various additives in a silica matrix, progress has been comparably slow. Recently, a team of chemists at the University of Illinois developed a way of creating artificial antibodies by using a process in which a single molecular template is imprinted into a single macromolecule – a highly branched polymer called a dendrimer. Upon removal of the template, a synthetic molecular shell is created, which can bind specifically shaped molecules and can, like a natural antibody, reject others [51].

In principle, the molecular key may be any type of molecule, ranging from small molecules (e. g., drugs, amino acids, steroid hormones) to larger molecules (e. g., nucleic acids, proteins). Large molecular assemblies such as cells and viruses may also be perceived, though the difficulty of making the imprinted materials increases with the size of the selected key molecule. A combination of MIT and future Lab-on-a-Chip devices promises many advantages for Life Science applications, although in this case the period between proof-of-principle and commercialization is likely to be long.

The use of nanopores in Life Science applications leads to another interesting field of research. Current investigations on nanopore membranes include patch–clamp arrays, biocapsules for biosensor protection, and drug delivery systems, for example nanopore membranes as functional parts of subcutaneous implants or microparticles with nanopores, such as porous silicon particles. Nanopores are also currently under investigation for use in haplotyping, SNP detection, and DNA sequencing [52, 53]. A detailed overview is provided in Chapter 7.

An additional impact on microfluidic devices is expected from nanomechanics. One such embodiment is that of silicon cantilevers in a Lab-on-a-Chip; these are a few hundred nanometers thick, and have biomolecules (e. g., antibodies) attached to one side. The binding of protein molecules to the capture antibodies causes the cantilevers to bend, and this can be monitored either electronically or optically [20]. Another class of actuators which, as in muscles, harnesses molecular deformations to generate meso- and macroscopic forces and displacement, are the conductive or electroactive polymers (EAPs) [54, 55]. These materials, which undergo large conformational changes in response to electrical or chemical stimuli, might be well suited for actuators, regulators, valves or sensors of future bioMEMS, respectively bioNEMS generations.

References

[1] S. C. Terry, J. H. Hermann, J. B. Angell, *IEEE Trans. Elec. Dev.* **1979**, *26*, 1880.

[2] A. Manz, N. Graber, *Sensors and Actuators* **1990**, *B1*, 244.

[3] Detailed information about microreaction technology may be found in the proceedings on the annual International Conference on Microreaction Technology, W. Ehrfeld (ed.), *Microreaction Technology: Industrial Prospects*, Springer, Berlin **2000**.

[4] E. Bassous, H. H. Taub, L. Kuhn, *Appl. Phys. Lett.* **1977**, *31*, 135.

[5] A. Emmer, M. Jansson, J. Roeraade, U. Lindberg, *J. Microcolumn Separation* **1992**, *4*, 13.

[6] R. Zengerle, A. Richter, H. Sandmaier, Proc IEEE Micro Electro Mechanical Systems, Travemünde, Germany, **1992**, 19.

[7] F. C. M. van der Pol, H. T. G. van Lintel, M. Elwenspoek, J. H. J. Fluitman, *Sensors and Actuators* **1990**, *A21*, 198.

[8] S. T. Cho, K. D. Wise, Proc Transducers 1991, Sans Francisco, USA, **1991**, 400.

[9] M. de la Guardia, *Microchim. Acta* **1995**, *120*, 243.

[10] S. Shoji, S. Nakagawa, M. Esashi, *Sensors and Actuators* **1990**, *A21*, 189.

[11] M. Richter, A. Prak, J. Naundorf, M. Eberl, H. Leewis, W. Woias, A. Steckenborn, Proc Transducers 1997, Chicago, USA, **1997**, 303.

[12] B. H. van der Schoot, S. Jeanneret, A. van den Berg, N. F. de Rooij, *Sensors and Actuators* **1993**, *B15*, 211.

[13] G. Mayer, K. Wohlfahrt, A. Schober, J. M. Köhler, Nanotiterplates for screening and synthesis, in: J. M. Köhler, T. Mejevaia, H. P. Saluz (eds), *Microsystem Technology: A Powerful Tool for Biomolecular Studies*, Birkhäuser-Verlag, Basel, **1999**, 75.

[14] V. Grosser, H. Reichl, H. Kergel, M. Schünemann, A Fabrication Framework for Modular Microsystems, *MST News* **2000**, *1*, 4.

[15] H. M. Widmer, *Analytical Methods and Instrumentation*, Special Issue μTAS '96, 1996, 3.

[16] A. van den Berg, P. Bergveld, *Analytical Methods and Instrumentation*, Special Issue μTAS '96, 1996, 9.

[17] Y. Baba et al. (eds), *Micro Total Analysis Systems 2002*, Kluwer Academic Publishers, **2002**.

[18] H. Lüdi, M. B. Garn, S. D. Hämmerli, A. Manz, H. M. Widmer, *J. Biotechnol.* **1992**, *25*, 75.

[19] M. U. Kopp, A. J. de Mello, A. Manz, *Science* **1998**, *280*, 1046.

[20] J. Liu, M. Enzensberger, S. Quake, *Electrophoresis* **2002**, *23*, 1531.

[21] M. T. Taylor, P. Belgrader, R. Joshi, G. A. Kintz, M. A. Northrup, *Micro Total Analysis Systems* **2001**, 670.

[22] A. Manz, D. J. Harrison, E. Verpoorte, J. C. Fettiger, A. Paulus, H. Lüdi, H. M. Widmer, *J. Chromatogr.* **1992**, *593*, 1926.

[23] M. Koch, A. Evans, A. Brunnschweiler, *Microfluidic Technology and Applications*, Research Studies Press, Philadelphia, **2000**.

[24] T. S. J. Lammerink, V. L. Spiering, M. Elwenspoek, J. H. J. Fluitman, A. van den Berg, Proc MEMS 1996, San Diego, USA, **1996**, 389.

[25] D. C. Duffy, H. L. Gillis, J. Lin, N. F. Sheppard, Jr., G. J. Kellog, *Anal. Chem.* **1999**, *71*, 20, 4669.

[26] M. Busch, J. Schmidt, A. Rothen, C. Leist, B. Sonnleitner, E. Verpoorte, Proc 2nd Int Symp μTAS, Basel, **1996**.

[27] G. J. M. Bruin, *Electrophoresis* **2000**, *21*, 3931.

[28] H. Becker, L. E. Locascio, *Talanta* **2002**, *56*, 267.

[29] J. G. E. Gardeniers, R. W. Tjerkstra, A. van den Berg, Fabrication and Application of Silicon-based Microchannels, in: W. Ehrfeld (ed.), *Microreaction Technology: Industrial Prospects*, Springer, Berlin, **2000**, 36.

[30] R. W. Tjerkstra, M. J. deBoer, J. B. Berenschot, J. G. E. Gardeniers, M. C. Elwenspoek, A. van den Berg, Proc IEEE Workshop on MEMS, Nagoya, Japan, **1997**, 147.

[31] J. Drott, K. Lindstrom, L. Rosengren, T. L. Aurell, *J. Micromech. Microeng.* **1997**, *7*, 14.

[32] M. Koch, A. Evans, A. Brunnschweiler, *Microfluidic Technology and Applications*, Research Studies Press, **2000**, Chapter 6.3.

[33] S. Howitz, Components and systems for microliquid handling, in : J. M. Köhler, T. Mejevaia, H. P. Saluz (eds), *Microsystem Technology : A Powerful Tool for Biomolecular Studies*, Birkhäuser Verlag, Basel, **1999**, 31.

[34] T. Gerlach, H. Wurmus, K. Helmut, Working principle and performance of the dynamic micropump, *Sensors and Actuators* **1995**, *50*, 135–140.

[35] G. H. Seong, R. M. Crooks, *J. Am. Chem. Soc.* **2002**, *124*, 13360.

[36] A. D. Stroock, S. K. W. Dertinger, A. Ajdari, I. Mezic, H. A. Stone, G. M. Whitesides, *Science* **2002**, *295*, 647.

[37] W. Menz, J. Mohr, O. Paul, *Microsystem Technology*, Wiley-VCH, Weinheim, **2001**.

[38] M. J. Madou, *Fundamentals of Microfabrication : The Science of Miniaturization*, CRC Press, Boca Raton, **2002**.

[39] M. Gad-el-Hak (ed.), *The MEMS Handbook*, CRC Press, Boca Raton, Chapter 16, **2002**.

[40] M. Onda, K. Ariga, T. Kunitake, *J. Ferment. Bioeng.* **1999**, *87*, 87.

[41] C. Neinhuis, W. Barthlott, *Ann. Bot.* **1979**, *79*, 667.

[42] S. Shibuichi, T. Onda, N. Satoh, K. Tsuij, *J. Phys. Chem.* **1996**, *100*, 19512.

[43] M. Mrksich, G. M. Whitesides, *Trends Biotechnol.* **1995**, *13*, 228.

[44] S. Soper, in : J. Göttert (ed.), *2001 CAMD Summer School Micro-and Nanotechnologies*, Chapter 10, Baton Rouge, CAMD/LSU Publishers, **2001**.

[45] M. Pirrung, *Angew Chem Int Ed*, **2002**, *41*, 1276.

[46] M. J. Heller, DNA Microarray Technology : Devices, Systems and Applications, *Annu. Rev. Biomed. Eng.* **2002**, *4*, 159.

[47] R. Sosnowski, M. J. Heller, E. Tu, A. H. Forster, R. Radtkey, Active microelectronic array system for DNA hybridisation, genotyping and pharmacogenomic applications, *Psychiatr. Genet.* **2002**, *12*, 181.

[48] D. Brennan, *Infrared Physics Technol.* **2002**, *43*, 69–76.

[49] S. R. Quake, A. Scherer, From Micro to Nano Fabrication with Soft Materials, *Science* **2000**, *290*, 1536.

[50] S. Bhatia, *Microfabrication in Tissue Engineering and Bioartificial Organs*, Kluwer, Boston, **1999**.

[51] S. C. Zimmerman, M. S. Wendland, N. A. Rakow, I. Zharov, K. S. Suslick, Synthetic Hosts by Monomolecular Imprinting Inside Dendrimers, *Nature* **2002**, *418*, 399–403.

[52] A. Marziali, M. A. Akeson, New DNA Sequencing Methods, *Annu. Rev. Biomed. Eng.* **2001**, *3*, 195–223.

[53] D. W. Deamer, D. Branton, Characterisation of nucleic acids by nanopore analysis, *Chem. Res.* **2002**, *35*, 817–825.

[54] Y. Bar-Cohen (ed.), *Electroactive Polymer Actuators as Artificial Muscles*, SPIE Press, Bellingham, **2001**.

[55] J. D. Madden, P. G. Madden, I. W. Hunter, Conductive Polymer Actuators as Engineering Materials, SPIE 9th Annual Symposium on electroactive materials and structures, San Diego, CA, USA, Vol. 4695, pp. 424–434, March 18–21, **2002**.

3
Microcontact Printing of Proteins

Emmanuel Delamarche

3.1
Introduction

Biomolecules on surfaces have applications that range from medical diagnostics, analytical chemistry, and culturing and studying cells on surfaces, to synthesizing or engineering DNA, carbohydrates, polypeptides, or proteins. Defining patterns of biomolecules – and of proteins in particular – on surfaces is no simple task considering how complex and fragile these molecules can be. Photolithography – the art of structuring surfaces at lateral scales of less than 1 micrometer – was used recently to create DNA microarrays. Photolithography affords the capability of synthesizing strands of DNA using lithographic masks and photochemistry, but it is unlikely that a similar approach would permit the fabrication of arrays of proteins, which cannot be synthesized block-by-block at present. In photolithography, UV light, organic solvents, photoresists, and resist developers can compromise the structure and function of even a simple protein. For these reasons, novel approaches to patterning proteins include defining regions on surfaces that attract, bind, or repel proteins from solution, or in a more direct manner, by delivering small volumes of a solution of protein to a surface using drop-on-demand systems or microfluidic devices [1–3].

This chapter describes the patterning of proteins on surfaces by means of microcontact printing (μCP), where proteins are applied like ink to the surface of a stamp and transferred to a substrate by printing. Microcontact printing was originally developed by Whitesides and coworkers at Harvard for printing alkanethiols on gold with spatial control [4]. Many variants of μCP were later developed and collectively termed "soft lithography" [5].

The central element in μCP is the stamp. This is a silicone-based elastomer that is microstructured by curing liquid prepolymers of poly(dimethylsiloxane) (PDMS) on a lithographically fabricated master (or mold). Once cured, the stamp is peeled off the mold by hand; the stamp then bears an inverted pattern of that of the mold. One mold can be used to replicate many stamps. The relative softness of the stamp, compared to that of a lithographic mask, allows it to follow the contours of surfaces onto which it is applied. It is the work of adhesion between the stamp and the substrate that drives the spreading of the initial zones of contact at the expense of an elastic adaptation of the stamp [6]. In μCP – and soft lithography in general – the contact between the elastomer and a substrate occurs at the molecular scale and is termed "conformal"; it ensures the homogeneous transfer of ink from the stamp to the printed areas of the substrate [4].

PDMS materials have the following features. They are transparent to optical light and even UV down to ~240 nm, resistant to many chemicals and pH environments, good electrical insulators, thermally stable, nontoxic, and can have tailored mechanical properties using various degrees of crosslinking and amounts of resin fillers [7]. A PDMS stamp can be a simple piece cut from a PDMS slab or accurately molded and mounted on a stiff backplane [8, 9]. It can be composed of PDMS layers having different mechanical characteristics, or shaped like a paint roller [10].

Handling stamps with tweezers and printing by hand is sufficient for most needs of experimentalists. Mounting a stamp on a printing tool, however, provides the ability to vary and control the pressure applied during printing, and to align the stamp with the substrate. PDMS stamps have advancing and receding contact angles with water of ~115° and ~95° and thus are hydrophobic and promote the spontaneous deposition of proteins from solution [11]. This deposition is nonspecific and self-limiting to a monolayer of proteins if the stamp is rinsed after the inking step. An important difference between microcontact printing proteins on surfaces and alkanethiols on noble metals is the limited amount of proteins on the stamp. Alkanethiols can diffuse inside a PDMS stamp in sufficient amounts for multiple prints, whereas reinking stamps with proteins is necessary

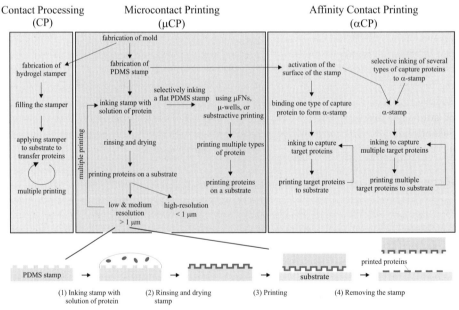

Figure 3.1 Three related methods can pattern proteins from a stamp to a surface. In contact processing (left), a hydrogel stamp mediates the diffusion of proteins from its bulk to a surface. Microcontact printing (center) utilizes an elastomeric stamp inked with proteins to print the proteins on a substrate without having a liquid. The stamp in affinity contact printing (right) is derivatized with capture proteins, which allows it to be selectively inked with target proteins released to a substrate during printing.

after each print unless hydrogel "stampers" are used [12]. These stamps can carry a large reserve of protein solution used for the contact processing (CP) of substrates. PDMS stamps derivatized with biomolecules provide the basis for selective inking strategies and lead to the affinity-contact printing (αCP) technique [13]. Figure 3.1 delineates the operations that CP, μCP, and αCP involve. These techniques are described in detail in the following sections.

3.2
Strategies for Printing Proteins on Surfaces

3.2.1
Contact Processing with Hydrogel Stamps

Contact processing (left-hand panel in Figure 3.1) mimics the deposition of proteins from an aqueous environment to a surface by utilizing a hydrogel swollen with a solution of protein [12, 14]. The proteins can diffuse through this hydrophilic matrix and adsorb onto the substrate without uncontrolled spreading. The stamp consists of two parts. The first is a reservoir above the hydrogel containing proteins dissolved in a biological buffer. The second is the hydrogel that makes contact with the substrate and mediates the transport of proteins to the substrate. A stamp that has a hydrogel made of poly(6-acryloyl-β-O-methyl-galactopyranoside), for example, embedded in a fine capillary can pattern proteins with a resolution of ~20 μm [14]. Hydrogels having a refined composition and a greater degree of crosslinking exhibited better mechanical resistance, and were patterned by replication of a mold [15]. The latter approach should allow a protein to be patterned on a surface with a resolution better than 20 μm. CP based on hydrogel stamps has interesting features. First, biomolecules remain in a biological buffer until the stamp is removed and the substrate dried. Denaturation of proteins in this case should be minimal, and may be similar to that of proteins adsorbed from solution onto polystyrene microtiter plates. Second, it is straightforward to reuse such stamps for multiple CP experiments [14].

3.2.2
Microcontact Printing

Microcontact printing of proteins uses PDMS stamps replicated from a mold (middle panel in Figure 3.1). Inking the stamp with proteins is simple, and analogous to depositing a layer of capture antibody (Ab) on polystyrene for conducting a solid-phase immunoassay. The duration of inking and the concentration of protein in the ink solution determine the coverage of protein obtained on the stamp [16]. Inking a stamp can be local and/or involve multiple types of proteins when the stamp is locally exposed using a microfluidic network (μFN) or microcontainers to one or more solutions of protein [17]. The transfer of proteins can be remarkably homogeneous and effective, depending on the wetting properties of the substrate [18]. The large area of interaction of proteins with substrates and their high molecular weight account for the high-resolution potential of μCP of proteins. At the limit, single protein molecules can be printed as arrays on a surface [16], whereas the diffusion of alkanethiols on noble metals or the reactivity of silanes with themselves limit the practical resolution achieved for microcontact printing self-assembled monolayers on surfaces. Microcontact printing proteins on surfaces appears to be limited by the resolution and mechanical stability of the patterns on the stamp. Stamps made of Sylgard 184 and using masters prepared using rapid prototyping or photolithography can have micrometer-sized patterns on fields even larger than 10 cm^2 [19]. Microcontact printing proteins with arbitrary patterns and submicrometer resolution benefits from the use of a PDMS elastomer stiffer than Sylgard 184 and masters patterned using electron-beam lithography [8].

3.2.3
Affinity-Contact Printing

Tailoring the surface chemistry of stamps to ink a particular type of biomolecule is crucial for αCP (right-hand panel in Figure 3.1). The chemical stability of silicone elastomers is both an advantage for preparing chemically resistant stamps and an obstacle to modifying the surface of PDMS stamps. Exposing PDMS to an oxygen-based plasma forms a glassy silica-like surface layer [20]. The oxidized layer is a few nanometers thick and contains silanol groups (–Si–OH), which are useful for anchoring organosilanes [21]. Oxidized PDMS can thus be derivatized similarly to glass or SiO$_2$ in a few chemical steps using silane monolayers and with crosslinkers for proteins [22]. Affinity-contact printing is the technique of covalently immobilizing ligand biomolecules onto a PDMS stamp, and using them to ink a stamp selectively with receptor molecules. A stamp for αCP is roughly analogous to a chromatography column due to its ability to extract proteins selectively from a mixture, although releasing them involves printing them onto a surface [13]. Biomolecules that are naturally present in crude solutions and have a function on a surface are ideal candidates for applications of αCP. Cell adhesion molecules is one example that has already been demonstrated, but αCP could well be extended to a large variety of biomolecules for which ligands exist. Stamps in αCP are reusable and may include sites of different affinity to capture and print multiple types of protein in parallel [23].

3.3
Microcontact Printing Polypeptides and Proteins

Many different types of proteins can be inked from an aqueous solution onto a hydrophobic silicon rubber such as PDMS [24]. Hydrophobic polymers in general promote the deposition of proteins from solution through a variety of interactions, and slight or pronounced conformational changes of the protein structure can accompany this adsorption process. The kinetics of formation of a layer of protein on hydrophobic surfaces is often compared to a Langmuir-type isotherm: the rate of deposition of the protein molecules scales with their concentration in the bulk of the solution and reaches a plateau when all sites on the substrate become occupied [25]. Hydrophobic substrates, as a general rule, have stronger interactions with hydrophobic proteins, and their adsorption process is less influenced by the pH and ionic strength of the solution and by the isoelectric point of the protein than when polar or charged substrates are employed [25]. The size of the protein does not seem to play an important role on their inking behavior. A wide range of proteins in terms of structure and functions has been microcontact printed, which includes cytochrome c (12.5 kDa) [11], streptavidin and bovine serum albumin (BSA; ~60 kDa) [26–28], protein A and immunoglobulins G (150 kDa) [11], glucose oxidase (160 kDa) [29], laminin (~210 kDa) [30], and fibronectin (440 kDa) [31]. It is sometimes necessary to employ stamps with a hydrophilic surface to ink hydrophilic polypeptides such as polylysine (with MW ranging from 38 to 135 kDa) [32] or lipid bilayers [33]. In other cases, small biomolecules such as amino-derivatized biotins were inked and printed onto surfaces reactive to amino groups [34, 35]. In general, the derivatization of biomolecules with thiol groups allows the printing of biomolecules on gold substrates [36], where patterning by printing can be complemented by the adsorption of other types of molecules from solution. The chemisorption of small biomolecules on surfaces might be necessary for efficient transfer from the stamp and to prevent rinsing the printing molecules during subsequent steps.

3.3.1
Printing One Type of Biomolecule

Immunoglobulins G (IgGs) are interesting candidate molecules for µCP: these Abs are useful on surfaces for heterogeneous immunoassays. Their numerous disulfide linkages make them robust, they adsorb from biological buffers to PDMS in a nonreversible manner [24], and they can be conjugated to fluorescent centers, metal particles, enzymes, or ligands such as biotin. Fluorescence microscopy is a versatile method to follow the results of microcontact printing IgGs onto a glass surface (Figure 3.2). There, TRITC-labeled anti-chicken Abs were inked everywhere on the stamp but transferred to glass only in the regions of contact [11]. The patterns on the glass are accurate and correspond to zones of the stamp where the inked proteins are missing. The contrast of the 1 µm-wide features in the

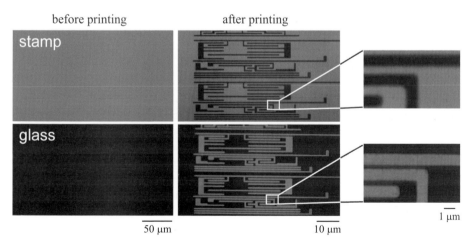

Figure 3.2 Microcontact printing proteins on glass. Fluorescence microscopy images revealing TRITC-labeled chicken Abs on a stamp after inking and accurately transferred in the regions of contact to a glass substrate. Reproduced with permission from Ref. [11]. (Copyright 1998 American Chemical Society.)

pattern is high and accurate and, as no fluorescence above background is measured in the nonprinted regions, it is clear that no transfer of Ab occurred in the recessed areas of the stamp. This might not always be the case, because small features have limited mechanical stability [6]. Demolding the stamp from the mold, capillary effects during inking and drying the stamp, and the printing itself may compromise the mechanical stability of patterns [37]. Implementing support structures in the design of the pattern, controlling the forces exerted during printing and affixing a stiff backplane to the stamp improve the stability of patterns. Stamps can be very large and have features measuring from micrometers to centimeters, making it possible to print proteins of one kind on large substrates to pattern cells indirectly. Examples include microcontact printing fibronectin [31], polylysine [30, 38, 39], laminin [40], and adhesion peptides [41].

3.3.2
Substrates

Substrates for biomolecules cover a wide range of materials, from simple glass slides to complex functional microelectronic devices or sensors. Having conformal contact between the stamp and substrate during printing is the first requirement for microcontact printing biomolecules. For this reason, the substrate should not be too rough [6], or have too prominent structures [39]. Polystyrene, poly(styrene terephthalate), glass, amphiphilic comb polymers, Si wafers, and substrates covered with a thin evaporated metal and/or a self-assembled monolayer can be microcontact printed with proteins and stamps made of Sylgard 184 [11, 29, 34, 42]. The printing time does not seem to play a role, and takes the few seconds necessary to propagate the initial contact to the entire substrate. The details of how and why proteins transfer from a stamp to a surface were intriguing until the recent discovery that the difference in wettability by water between the stamp and the surface determines whether transfer occurs [18]. Proteins tend to transfer when the substrate is more wettable, or has a higher work of adhesion for water, than the stamp. In this respect, the chemical composition of the surface does not seem to play a particular role other than defining the wettability of the surface (Figure 3.3). The surface of the stamp can be derivatized with fluorinated silanes to raise the wettability threshold of the substrate below which transfer remains effective, for example. A remarkable incidence of printing proteins occurs with poly(ethylene glycol) (PEG)-derivatized surfaces [18]. Surfaces covered with a sufficient density of PEGs resist the deposition of proteins from solution because in order to interact with the PEG layer, proteins have to remove water solvating EG repeat units and reduce the number of possible conformations of the PEG chains [43]. Both of these requirements are energetically unfavorable to the deposition of proteins from solution onto PEG-treated surfaces [44]. The mechanism accounting for the transfer of protein in μCP might thus involve the dry state of the PEG layer during printing [43], the local pressure exerted by the stamp at the line front propagation of the conformal contact [45], or some contamination of the PEG layer by low-molecular-weight silicone residues from the stamp. The deposition of proteins from solution or by printing exhibits antagonistic behaviors: proteins are more difficult to print on a hydrophobic surface than on a hydrophilic one whereas the opposite situation generally occurs in solution with, as an extreme case, PEG surfaces, which are protein-repellant [46].

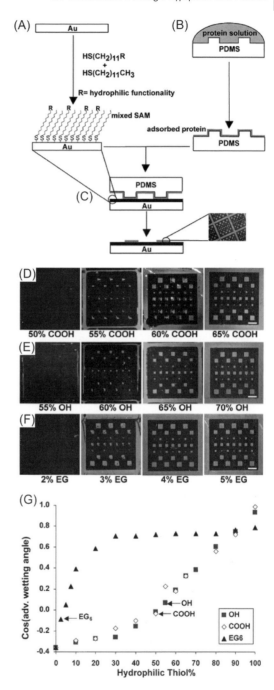

Figure 3.3 Influence of the wettability of the substrate by water on the degree of transfer of proteins from a PDMS stamp to a Au surface. (A) The substrate is derivatized with SAMs comprising variable mole fractions of two constituents having different endgroups. (B) Proteins are adsorbed from solution onto the stamp. (C) The resulting printed patterns are analyzed using fluorescence microscopy. The fluorescence micrograph corresponds to fluorescently labeled proteins printed onto a 100% COOH-terminated SAM. The transfer of proteins followed on SAMs having hydrophilic components functionalized with (D) COOH, (E) OH, or (F) EG correlates with the wettability of the mixed SAM (G). Figure kindly provided by J. L. Tan, J. Tien and C. S. Chen, and reprinted with permission from Ref. [18]. (Copyright 2002 American Chemical Society.)

Printing proteins is not limited to the patterning of planar substrates but is possible on curved surfaces, structured surfaces, and over large areas [5]. A stamp can be molded directly curved or planar and then curved and rolled over a surface [10]. Large stamps (≥ 10 cm) can be molded with a pattern having an accuracy of better than 1 μm [47]. The mechanical properties of stamps can be varied from 1 MPa (Young's modulus) to over 30 MPa by adjusting the formulation of the polymer with respect to its average molecular weight between junctions, the junction functionality, and the density and size of filler particles added to the polymer [8]. The hardness of a stamp, its work of adhesion with the substrate, the pressure applied during printing, and the topography and work adhesion of the substrate all determine whether conformal contact will occur. The stability of features on the stamp might be compromised, however, when the stamp is made too soft and pressed too hard during printing [6, 9, 48].

3.3.3
Resolution and Contrast of the Patterns

High resolution in lithography refers to patterning features of arbitrary shape at a length scale where it becomes crucial to optimize all parameters of the technique (e.g., conditions for exposing and developing the resist, transfer of the resist pattern onto the substrate). Electron-beam lithography has a high-resolution regime for making features < 100 nm, photolithography for features < 250 nm, and μCP for features < 500 nm. In conventional lithography, shrinking the dimensions of patterns is driven by the necessity to improve the performance of integrated circuits at invariant or lower cost. The resolution of lithographic techniques limits the smallest sizes of components made today. It will be the physics of tomorrow's devices that will ultimately be the limiting obstacle to further integration. Patterning biomolecules has a different paradigm for the resolution limit than conventional lithography because single functional elements, an enzyme for example, are available but do not have to be constructed. Microcontact printing meets several requirements that are necessary to place single proteins at predefined positions on a surface: (i) it is possible to fabricate Si molds with features as small as 40 nm using electron-beam lithography [16]; (ii) PDMS-replicated structures can be 80 nm and even smaller [9,47]; (iii) proteins remain in the areas of contact, unlike alkanethiols and monolayer-forming molecules which generally diffuse away from the initial printed zones on the substrate when an excess ink is present on the stamp; and (iv) the solution of protein used to ink the stamp can be diluted to limit the number of proteins inked per feature on the stamp [16].

Figure 3.4 shows high-resolution patterns of Abs on Si and glass and how a high-resolution stamp can look. Each feature in the atomic-force microscopy (AFM) image in Figure 3.4A comprises ∼1000 Abs of the same type that were printed on a Si wafer using a PDMS stamp made of Sylgard 184 [17]. The structures have a width of 500 nm and an edge resolution better than 50 nm. The contrast of the patterns seems perfect because no Ab is present outside of the printed areas. An excellent contrast, together with specific binding events between printed ligands and receptors from solution, are desirable for high-sensitivity biological assays. The photography in Figure 4B shows a 8×4 cm^2 stamp composed of a 30 μm-thick layer of PDMS attached to a flexible glass backplane

A ~1000 antibodies per μm^2

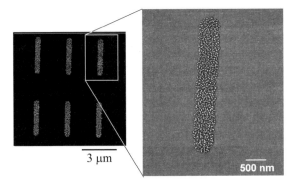

B high-resolution PDMS stamp

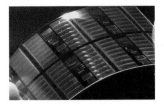

C toward printing single proteins

Figure 3.4 Microcontact printing proteins at high resolution. (A) AFM images showing that only ~1000 chicken Abs were printed onto a Si wafer in each element of this pattern. (B) High-resolution μCP is best done using stamps harder than Sylgard 184 and having a flexible glass backplane. (C) AFM images showing rabbit Abs printed on glass as a mesh comprising 100 nm-wide lines (left) or on regularly spaced areas that have none, one or a few Ab molecules (right).

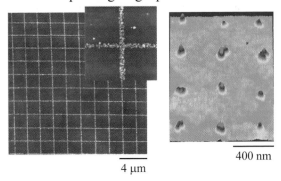

100 μm thick [8]. The PDMS layer of this stamp has numerous fields with 250 nm-wide lines, is about five times harder than Sylgard 184, but is also more brittle. The patterns are consequently more stable against collapse, and the glass backplane contributes significantly to the long-range accuracy of the pattern while making the stamp simple to handle, mount and align on a printer tool [47]. The surface tension of the polymer is an important limiting factor for the resolution of μCP. Features as small as 5 nm can be written by electron-beam lithography in PMMA and developed. PMMA is brittle, however, and hence not soft enough to form a reliable contact over surfaces. Unmolding even relatively hard PDMS stamps (having a Young's modulus > 12 MPa) from a master structured with

40 nm-wide lines results in their broadening by 20 nm owing to surface tension effects [16]. Such a broadening is relatively less important for stamps with features ≥ 100 nm, and can be compensated in the electron-beam lithography layout. The stability of small features on stamps limits the freedom of design for high-resolution patterns. Dots, lines, and meshes do not have all the same mechanical stability against pressure; some recessed areas may collapse during printing. Incorporating support structures with micrometer dimensions around the high-resolution fields can remedy these problems [6]. Another strategy is to transfer the resist pattern into the Si master with a reactive ion etching, where the etch rates depend on the geometry of the features. When large structures are made deeper in the master than smaller structures, a larger part of the load during printing is exerted on the larger structures without inducing collapse of the smallest features [16].

The AFM images in Figure 3.4C correspond to Ab molecules microcontact printed from a PDMS stamp (material B) [8] onto glass using a mesh of 100-nm-wide lines (left image) and 100-nm hemispherical posts [16]. Both the posts and the lines were 60 nm high. The detail of a mesh reveals that two to four Ab molecules define the width of the lines. In the case of posts, one to three Ab molecules occupy each visible printed site, and the statistical analysis of larger printed zones revealed that sites could have none, one or a few printed Ab molecules [16]. There, the resolution limit for microcontact printing a single molecule is reached while still leaving space for improvement to form homogeneous arrays having only one biomolecule per site. A high concentration of protein in the ink, a long inking time, a further reduction of the dimensions of the posts, and a substrate with a high work of adhesion could help printing large arrays of single protein.

3.4
Activity of Printed Biomolecules

Many studies emphasize that while the adsorption of a protein on a surface is simple to perform, it is nevertheless a complicated phenomenon in which the biological activity of the immobilized biomolecule might be lost or significantly altered [25]. Microcontact printing biomolecules harbors this risk twice: when proteins are inked onto the stamp, and when they are printed. In principle, the deposition of proteins from solution onto PDMS should be analogous to that of proteins on hydrophobic surfaces [24]. The second concern is more difficult to weigh. Transferring a protein by printing implies that the adhesion of the protein with the substrate overcomes that of the protein with the stamp. At the limit, this might create a mechanical stress on the protein and could lead to irreversible conformational changes. It might be interesting to evaluate the yield of transfer as a function of the peeling rate to better characterize the transfer mechanism [49]. Comparing the activity of different types of biomolecules printed or adsorbed onto polystyrene suggests that enzymes are more susceptible to denaturation during printing than during adsorption from solution [11]. A layer of printed proteinase K displayed half the activity of a layer deposited from solution. Abs are more robust against loss of function; the ability of printed polyclonal Abs to capture antigens from solution was similar to that where the captured Abs were adsorbed. Printed monoclonal Abs seemed to have a $\sim 10\%$ loss of capture efficiency compared to ones adsorbed from solution.

The surface activity of microcontact printed biomolecules belonging to three important biological classes is illustrated in Figure 3.5. Cell adhesion molecules are ideal candidates for printing biomolecules because these molecules are useful on surfaces as they can direct the adhesion and growth of cells to specific regions of a substrate. Moreover, many are "simple" polypeptides. Polylysine [30, 32, 38, 39], laminin [50], polylysine fused with laminin [40], fibronectin [31], specific adhesion peptides [41, 42], and neuron-glial cell adhesion molecules [13] (NgCAM), for example, were microcontact printed to promote or guide the attachment of cells to surfaces. In other examples, Ab–cell interactions were used to pattern cells on patterns of printed Abs [51, 52]. In some instances, the stamp was made hydrophilic and the substrate activated with a crosslinker to increase, respectively, the inking and transfer efficiency. The left-hand fluorescence image in Figure 3.5A corresponds to the immunostaining of the adhesion peptide PA22-2 that was microcontact printed onto a glass surface activated with amine-reactive crosslinkers [41]. The phase-contrast image (right) shows that the printed pattern of peptide was suitable to grow viable hippocampal neurons in the printed regions specifically. These results illustrate well the conservation of the function of printed adhesion molecules. The capability of printing a pattern in registry with structures predefined on a substrate opens the way to placing cells wherever desired on a complex surface to study their function and to form networks of immobilized cells [39, 50, 53]. The next example (Figure 3.5B) is a printed polyclonal Ab, which serves as antigen to bind polyclonal Abs from solution [11]. AFM reveals the printed regions of a Si surface, each of which comprises ~1000 molecules of chicken Ab molecules. Blocking the free Si surface with BSA is the next step necessary to prevent nonspecific deposition of proteins during the recognition step. After the blocking step, the Si surface is either covered with BSA or printed Abs; the prints are no longer visible. Recognition of the printed chicken Abs by anti-chicken Abs faithfully reflects the printed pattern. This experiment is an example of a highly miniaturized surface immunoassay in which the printed antigens were recognized by their specific Abs. Enzymes are probably more fragile than Abs and suffer more from a random orientation on a surface with respect to their activity than antigens for polyclonal Abs. The activity of printed enzymes can be evaluated using flat stamps, polystyrene substrates and colorimetric measurements [11]. It can be useful, however, to keep enzymatic products near their sites of production and to assess the activity of the enzymes with high spatial resolution. This is possible by using precipitating fluorescent products that accumulate in the regions of the substrate having enzymatic activity (Figure 3.5C) [17]. Real-time analysis of the development of fluorescence on the printed sites is even possible with this method. At least a part of the alkaline phosphatases printed on the glass surface in Figure 3.5C are active. This suggests that enzyme-linked immunosorbent assays can be performed using captured Abs that are printed. Interestingly, the same type of reporter enzyme can unambiguously reveal an ensemble of binding events, which are discernible through their localization. The challenge in this case remains to derivatize a surface with several types of protein.

A printing cell adhesion proteins

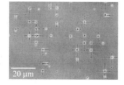

B printing antigens/antibodies

• antigen (chicken Ab)   • BSA   ■ α-chicken antibody

C printing enzymes

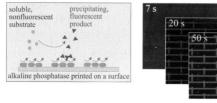

soluble, nonfluorescent substrate   precipitating, fluorescent product

alkaline phosphatase printed on a surface

7 s   20 s   50 s

200 μm

Figure 3.5 Microcontact printed proteins preserve a sufficiently high degree of activity for (A) promoting cell adhesion, (B) performing immunoassays, and (C) performing surface enzymatic catalysis.
(A) The adhesion peptide PA22-2, which was printed onto a thiol-reactive surface, was immunostained (left-hand fluorescence micrograph) in two steps using an anti-PA22-2 Ab and a secondary Ab labeled with fluorescein, and is useful for attaching hippocampal neurons with spatial control (phase-contrast micrograph on the right). (B) These AFM images illustrate three steps of an immunoassay in which chicken Abs were printed on a Si wafer (left), BSA was adsorbed from solution during the blocking step (middle), and the printed Abs were recognized by anti-chicken Abs. (C) This fluorescence image shows the deposition from solution of a fluorescent product made insoluble by printed alkaline phosphatase. The images in (A) were kindly provided by Offenhäusser et al. and reprinted from Ref. [41]. (Copyright 2000 with permission from Elsevier Science.)

3.5
Printing Multiple Types of Proteins

3.5.1
Additive and Subtractive Printing

An obvious application for microcontact printing proteins is the preparation of protein microarrays that can be used to screen different analytes in parallel while conserving reagents and still obtaining high-quality signals [54–56]. The simplest method to place two types of protein on a surface is to print one and adsorb the other from solution, as has been done with two different types of Abs [11]. This strategy requires that the printed layer of protein be complete enough to limit the adsorption of Abs from solution into the printed regions. The fabrication of arrays comprising n types of protein is possible using additive patterning steps: once a substrate is microcontact-printed, more proteins can be printed next to or over the previously printed ones (Figure 3.6A) [17, 57]. The transfer of proteins from a hydrophobic stamp to a more wettable substrate accounts for this finding

A printing 2 proteins consecutively

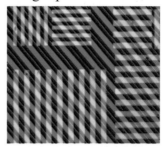

B printing 3 proteins simultaneously

Figure 3.6 Fluorescence microscopy images illustrating three strategies for printing several types of protein on a surface. (A) Two fluorescently labeled proteins were printed subsequently on a glass surface. Proteins on the stamp transferred during the second print both to glass and to the lines of proteins already patterned. (B) This pattern includes two fluorescently labeled protein and BSA printed simultaneously from a flat stamp. First, BSA was inked homogeneously by adsorption from solution onto the stamp and patterned by subtractive printing. Proteins labeled with fluorescein isothiocyanate (FITC) were then adsorbed in the regions complementary to the BSA pattern. This was repeated to remove BSA and the second protein along lines that were filled with TRITC-labeled protein adsorbed from solution. (C) A stamp was inked with different lines of proteins using independent channels of a µFN and printed onto a polystyrene surface in one step. Reproduced from Ref. [17].

C inking and printing up to 16 proteins

50 µm

because surfaces covered with printed proteins are more hydrophilic than PDMS stamps. This ability to stack proteins on top of each other is peculiar to µCP and might be useful for constructing protein-based architectures. It is also possible to place a variety of proteins on a substrate without the need for precise alignment during printing. Stamps with parallel lines can, for example, be inked and printed with a rotation between each print [17, 57]. Subtractive approaches are also possible (Figure 3.6B). One strategy is to ink a flat stamp homogeneously and to remove a subset of the proteins by printing them onto a structured surface. The sites on the stamp made free for adsorption are then covered by proteins from solution and the operation can be repeated [17]. An original way to pattern a surface with multiple types of protein is to fabricate a three-dimensional stamp and ink the different layers of the stamp with different types of proteins. Applying increasing pressure to the stamp brings each layer of the stamp successively in contact with the substrate [58]. The different patterns of protein are inherently aligned, and the difficulty of fabricating and inking the stamp might be compensated for by the relatively simple printing operation.

3.5.2
Parallel Inking and Printing of Multiple Proteins

Serial methods are simple, but probably not suitable, to pattern substrates with a large number of different proteins. A parallel inking approach of a stamp using a μFN can solve the problem of inking a stamp with different types of proteins (Figure 3.6C) [17, 59]. With such a strategy, a μFN having an ensemble of independent flowing zones is placed on a flat PDMS stamp [60, 61]. Sealing the microchannels results from the conformal contact between the μFN and the stamp. When solutions of proteins are flushed through the microchannels, proteins deposit in the areas of the stamp exposed to the conduits. Filling a μFN can be done serially, or with an array of dispensing heads or tips. The deposition of protein on the PDMS might be as fast as a few seconds when it is not limited by the mass transport of proteins from solution or the depletion of proteins from the channels. There, inking the stamp with or without a μFN could involve pin spotting, drop-on-demand, or microinjection techniques. Prefilling individual wells of a structured surface and applying it to a PDMS surface is also suited to locally derivatize a stamp with different types of proteins [23].

3.5.3
Affinity-Contact Printing

The inking of a stamp with a large number of different types of protein before each print can quickly become cumbersome when it is desirable to print a large series of substrate with the same pattern of protein. The ability to attach biomolecules to PDMS covalently yields the opportunity to define zones on a stamp that can actively bind a target molecule from an ink. Crosslinking capture molecules (e.g., Abs or antigens) to a stamp allows inking the stamp by exposing it to a solution containing targets for the surface-bound capture molecules. This inking strategy, termed αCP, is analogous to capturing a protein on a column for affinity chromatography, although release of the captured species can occur during printing (Figure 3.7A, see p. 46) [13]. The idealized view of αCP is to have an affinity stamp (α-stamp) with multiple sites for capturing target molecules from solution in parallel. This could be used for many cycles of capture and printing. The fluorescence image in Figure 3.7B corresponds to the detection of fluorescently labeled Abs that were inked onto an α-stamp and printed on a glass substrate. This stamp had two types of binding sites (antigens) that were crosslinked to PDMS with spatial control by means of subtractive printing and ordinary printing [23]. The α-stamp used to print the Abs in Figure 3.7C was prepared by attaching capture antigens to a stamp activated with a crosslinker for proteins using a μFN. The yellow line corresponds to the printing of TRITC and FITC-labeled Abs that were simultaneously captured on a line of protein A immobilized on the stamp.

Affinity contact printing, in particular when it employs stamps having multiple affinity sites, is both powerful and challenging. It is powerful because the ink can be a complex solution of biomolecules, the stamp is reusable, and patterns can have high resolution, as in μCP. The difficulty of αCP lies in the preparation of the α-stamp because the PDMS surface must be derivatized with crosslinkers and the quality of the patterns may degrade when preparation involves a large number of steps. The capture and release of radioactive or fluorescent proteins using αCP demonstrated the specificity of the capture event and the reusability of the α-stamp for at least 10 cycles [13]. Neuron-glial cell adhesion molecules (NgCAM), which are 200 kDa transmembrane proteins present at a concentration of ~1 μg mL^{-1} in membrane homogenates of chicken brain, were captured by monoclonal Abs of an α-stamp and patterned on a polystyrene surface. The patterned surface appeared to be suitable for the attachment and growth of dorsal root glial neurons (Figure 3.7D and E), which was not the case where polystyrene was exposed directly to a nonpurified source of NgCAM [13]. Among all the printing methods reviewed here, αCP might be the most powerful method owing to the selective inking step and the reusability of α-stamps. It also has, in principle, the potential to print biomolecules with a defined orientation on a surface.

3.6
Methods

3.6.1
Molds and Stamps

Molds for preparing stamps are most often Si wafers patterned with a combination of photolithography and reactive ion etching. Reactive ion etching Si, instead of using directly the pattern of photoresist as the mold, prolongs the lifetime of molds. Si molds can be washed and cleaned; passivation of the Si surface is necessary before pouring PDMS with a release layer. Passivation can be done in situ after reactive ion etching or using a simple dessicator that can be evacuated. Typical release agents are fluorinated silanes. High-resolution molds are Si wafers patterned using electron-beam lithography. The density and resolution of the high-resolution features determine the price of these molds, which easily reaches $1000 per written cm^2 for features having a critical dimension of 100 nm or less. Rapid prototyping is suitable for the preparation of stamps with a resolution of ~25 μm [62] and only requires access to a high-resolution printer (5000 dpi). Replicating molds to make stamps and other operations such as inking stamps and printing are best done in a clean room or using a laminar flow bench to minimize the contamination of surfaces by dust particles.

Sylgard 184 (Dow Corning) is used to prepare stamps in many cases, and comprises two prepolymers which, once mixed at a ratio of 1:10 (catalyst and hydridosiloxanes:vinyl-functionalized siloxanes), are poured on the master and cured at 60 °C overnight. The formulation of harder, mechanically more stable PDMS is sometimes necessary when stamps have tall, isolated features [8]. Stamps can be cured on a backplane made of a thin steel

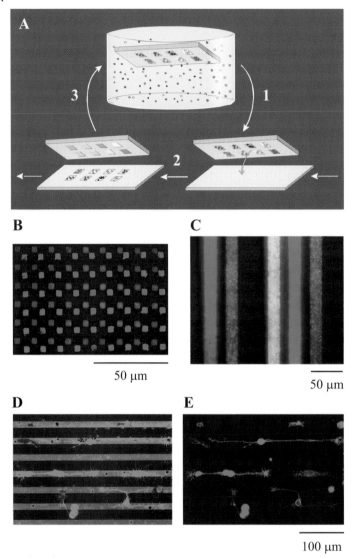

Figure 3.7 Affinity contact printing. (A) Stamps for αCP are prederivatized with capture sites and alternatively inked selectively and used for releasing the captured molecules on a substrate during the printing step. The fluorescence microscopy images correspond to (B) fluorescently labeled Abs co-captured and co-printed on glass using an α-stamp that had two types of capture sites; (C) fluorescently labeled proteins captured on lines of an α-stamp and printed on glass; (D) the immunofluorescent detection of NgCAM that was patterned on polystyrene using an α-stamp decorated with lines of anti-NgCAM mAbs; and (E) the staining by immunofluorescence of neurons which attached to and developed on the printed pattern of NgCAM.

plate or glass sheet [47], or on another PDMS layer [63]. Backplanes are typically flexible, but nevertheless give dimensional and long-range stability to stamps, which can then be mounted on a mask aligner, a printer, or handled by hand more conveniently. Design rules to make stamps are described jointly with the analytical description of the formation of conformal contact between stamps and substrates [6, 48]. This information is valuable to estimate the mechanical stability of stamps against pressure and to predict whether conformal contact will occur on rough surfaces or surfaces having topography.

Hydrogel stamps are composed of a polymer that is crosslinked to the desired value (2–4 %) and embedded in a microcapillary [12] or patterned by photocuring the hydrogel precursor sandwiched between a slide and a mold [15]. Both the inking and printing with hydrogel stamps rely on the diffusion of proteins through the hydrogel medium.

3.6.2
Surface Chemistry of Stamps

PDMS is hydrophobic and promotes the adsorption of proteins from solution in a manner analogous to polystyrene. Modifying the surface chemistry of stamps is necessary in two cases. Polypeptides and homogeneously polar biomolecules require stamps to have a hydrophilic surface for inking [38]. Affinity stamps must be derivatized with a ligand specific for biomolecule targets [13]. Exposing a PDMS stamp to an O_2-based plasma creates a silica-like layer on PDMS in a self-limiting manner. Stamps should be freshly inked (within ~5 min) after the plasma treatment to prevent the recovery of their hydrophobic character [20, 22, 46]. This hydrophobic recovery originates from the migration of low-MW silicone residues from the bulk to the surface. The plasma-induced scission of some polymer chains might also create mobile residues. It is impractical to extract these residues from the prepolymer components or after polymerization. Instead, plasma-treated stamps can be kept under water for long periods of time (more than days). Anchoring crosslinkers for proteins on plasma-treated stamps in one or more steps permits attaching covalently ligands onto the stamps [23]. Unreacted crosslinkers can be quenched with chemicals or reacted with noninterfering proteins such as BSA.

3.6.3
Inking Methods

The time to ink a stamp with a full monolayer of protein can be relatively long: up to 30 min at room temperature to obtain a monolayer of Ab using a concentration in phosphate-buffered saline (PBS) of 1 mg mL^{-1}, and 45 min with a solution of 5 µg mL^{-1} [11, 16]. The stamp is rinsed and dried after the inking step and then placed in contact with a substrate. The inking of hydrogel stamps with 1 mg mL^{-1} solutions of Abs in PBS takes similar times, and might even be faster if the gel is initially dry [12]. Shortening the inking time of PDMS stamps and localizing the inking is possible using µFNs [59]. In this case, the adsorption of proteins to the stamp might not be mass transport-limited, and is local in the regions of the stamp exposed to the channels. Localized inking is equally possible using microcontainers. These are small reservoirs microfabricated in Si, for example, and filled with the same, or different, solutions of proteins by hand or using pipetting robots [23]. Subtractive inking corresponds to inking entirely a flat PDMS stamp with proteins and transferring a subset by printing on a structured target. This strategy can remove proteins from areas of the stamp that become free for the inking of other proteins from solution [17, 23, 57]. This method can be repeated to form patterns with different types of protein next to each other but it requires an alignment step. Inking a stamp for αCP is analogous to linking a protein to a chromatography column via NH_2 residues. Crosslinking protocols are usually well detailed by chemical suppliers or reviewed elsewhere [64].

3.6.4
Treatments of Substrates

Surfaces more wettable by water than PDMS stamps are, in principle, suited for printing proteins [18]. Otherwise, they can be derivatized appropriately using plasma deposition techniques, oxidizing methods, or by grafting ultrathin organic layers. The wettability criteria may not apply when hydrophilic stamps are used to print certain polar biomolecules. In this case, derivatization of the substrate might also be necessary. Polylysine has been printed on glass directly [38] and on glass derivatized with glutaraldehyde [32], biotin on amine-reactive substrates [34, 35, 65, 66], and the adhesion peptide PA22-2 on a thiol-reactive surface [41]. In principle, crosslinkers can be attached to many types of substrates to bind proteins from stamps with high efficiency.

3.6.5
Printing

Handling a stamp with tweezers is the simplest approach to microcontact print proteins on surfaces both at low and high resolution. Typically, the stamp is brought close to the surface at an angle and set down gradually to ensure that conformal contact propagates from the initial contact areas without trapping air. Occasionally, (dust) particles or topography on the substrate are an obstacle to the propagating contact; applying gentle pressure to the stamp helps spread the contact to the rest of the substrate. Hybrid stamps are convenient to mount on printing tools [50] such as modified mask aligners, or on home-built printers using step motors to print substrates (up to 40 cm in lateral dimensions) with curved stamps while controlling the pressure applied to the stamp during printing [67]. Alignment of the stamp to preexisting structures on the substrate is desirable to print cell adhesion molecules on electrodes or to pattern substrates with multiple types of proteins, for example [39, 50]. Other methods already employed in soft lithography might be applicable for printing proteins. One example is printing surfaces and curved surfaces with cylindrical stamps [5, 10].

3.6.6
Characterization of the Printed Patterns

Surface-sensitive techniques such as ellipsometry, contact angle microscopy, and X-ray photoelectron spectroscopy can provide chemical information about surfaces printed with proteins. Patterns are best characterized using: (i) AFM, for which no labeling of the proteins is necessary; (ii) fluorescence and scanning confocal fluorescence microscopy; (iii) scanning electron microscopy, provided that the protein layer attenuates the emission of secondary electrons sufficiently to yield fair contrast [68]; or (iv) time-of-flight secondary ion mass spectroscopy [69]. AFM yields rich data concerning the contrast and resolution of the patterns, as well as the appearance and height of the printed layer, but suffers from the difficulty in localizing small printed areas [16]. Fluorescence microscopy is conversely effective in localizing signals from fluorescently labeled proteins, albeit with much less resolution than AFM. Colocalization of fluorophores having different spectral properties allows the detection of successively different types of proteins forming complex patterns. Optimization of the fluorescent signals is greatly facilitated by à priori knowledge of the geometry of a printed pattern. Detection of unlabeled proteins is possible by immunoassays with detecting Abs either fluorescently labeled or conjugated with a reporter enzyme. In the latter case, the enzymatic conversion of chromogenic precursor into a precipitating fluorescent product keeps the fluorescent signal localized to the printed areas [17]. Staining using electroless deposition also keeps signals local [23]. Microcontact printing proteins onto diffraction gratings [51] or surfaces suited for plasmon resonance [26] offers the exciting capability of following binding events in real time and over sites in parallel.

3.7
Outlook

The possibilities of microcontact printing biomolecules on a variety of surfaces with spatial control and resolution down to single protein molecules is unprecedented. Many fields could benefit in principle from these achievements. Surfaces could be decorated with high-quality patterns of microcontact-printed proteins for diagnostic applications. Very small volumes and quantities of reagents would be necessary for this purpose, using inking strategies based on microfluidic networks. In this case, numerous different types of protein could be patterned next to each other and used for surface ligand assays [70]. Printing proteins with tools that control the pressure during printing – hybrid stamps that are accurate and mechanically stable – and in alignment with predefined structures on the substrate has been demonstrated. The next steps are to build on these concepts, and mass fabricate high-quality arrays of proteins on surfaces such as glass slides or polystyrene surfaces for diagnostic applications. Stamps of various types can be devised. Some incorporate proteins in solution in a hydrogel-based reservoir, some have sites capable of binding target molecules from a complex ink, and others display numerous inked sites of micrometer lateral dimensions. Wherever biomolecules can be used on a surface, they might be placed advantageously by means of μCP. This is clearly the case for the growth of cells on surfaces, for which it becomes possible to construct hybrid architectures with cells connected to parts of electronic devices [53]. Positioning biomolecules on a biosensor surface is equally interesting. Microcontact printing can pattern areas of a sensing element with great precision and contrast, enabling the real-time monitoring of binding events on a surface. Possibly, the interaction between a large number of analytes and multiple printed sites could be screened for multianalyte immunoassays or drug screening. Microcontact printing is also well suited for preparing samples to investigate the biophysical properties of single biomolecules. Arrays of single biomolecules provide the advantage of having multiple sites to study each immobilized molecule with easy localization, without suffering from averaging effects, and with minimal signal degradation (photobleaching). It appears that although µCP was originally developed as a tool for applications in lithography [5], it is such a versatile technique that chemists and biologists have diverted it towards many more purposes. Microcontact printing clearly has unique, impressive features to manipulate and pattern biomolecules on surfaces, and it will be interesting to see how these will translate into firmly established applications for diagnostics, and biology in general.

Acknowledgments

I am very grateful to my colleagues Bruno Michel for his indefectible support of the work, to André Bernard and Jean Philippe Renault for having pioneered and carried out challenging experiments on microcontact printing proteins, and to Sergei Amontov, Helen Berney, Hans Biebuyck, Alexander Bietsch, Hans Rudolf Bosshard, Isabelle Caelen, Dora Fitzli, Matthias Geissler, Bert Hecht, David Juncker, Max Kreiter, Heinz Schmid, Peter Sonderegger, Richard Stutz, Heiko Wolf, and Marc Wolf for their close collaboration on our biopatterning activities.

References

[1] See for example the proceedings on µTAS: *Micro Total Analysis Systems 2002*: Y. Baba, S. Shoji, A. van den Berg (eds), Kluwer Academic Publishers, Dordrecht, The Netherlands.

[2] R. S. Kane, S. Takayama, E. Ostuni, D. E. Ingber, G. M. Whitesides, *Biomaterials* **1999**, *20*, 2363–2376.

[3] (a) D. R. Reyes, D. Iossifidis, P.-A. Auroux, A. Manz, *Anal. Chem.* **2002**, *74*, 2623–2636; (b) P.-A. Auroux, D. Iossifidis, D. R. Reyes, A. Manz, *Anal. Chem.* **2002**, *74*, 2637–2652.

[4] A. Kumar, H. A. Biebuyck, G. M. Whitesides, *Langmuir* **1994**, *10*, 1498–1511.

[5] Y. Xia, G. M. Whitesides, *Angew. Chem., Int. Ed. Engl.* **1998**, *37*, 550–575.

[6] A. Bietsch, B. Michel, *J. Appl. Phys.* **2000**, *88*, 4310–4318.

[7] S. J. Clarson, J. A. Semlyen (eds), *Siloxane Polymers*, PTR Prentice Hall, Englewood Cliffs, NJ, **1993**.

[8] H. Schmid, B. Michel, *Macromolecules* **2000**, *33*, 3042–3049.

[9] T. W. Odom, J. C. Love, D. B. Wolfe, K. E. Paul, G. M. Whitesides, *Langmuir* **2002**, *18*, 5314–5320.

[10] Y. Xia, D. Qin, G. M. Whitesides, *Adv. Mater.* **1996**, *8*, 1015–1017.

[11] A. Bernard, E. Delamarche, H. Schmid, B. Michel, H. R. Bosshard, H. A. Biebuyck, *Langmuir* **1998**, *14*, 2225–2229.

[12] M. A. Markowitz, D. C. Turner, B. D. Martin, B. P. Gaber, *Appl. Biochem. Biotechnol.* **1997**, *68*, 57–68.

[13] A. Bernard, D. Fitzli, P. Sonderegger, E. Delamarche, B. Michel, H. R. Bosshard,

H. A. Biebuyck, *Nature Biotechnol.* **2001**, *19*, 866–869.

[14] B. D. Martin, B. P. Gaber, C. H. Patterson, D. C. Turner, *Langmuir* **1998**, *14*, 3971–3975.

[15] B. D. Martin, S. L. Brandow, W. J. Dressick, T. L. Schull, *Langmuir* **2000**, *16*, 9944–9946.

[16] J. P. Renault, A. Bernard, A. Bietsch, B. Michel, H. R. Bosshard, E. Delamarche, M. Kreiter, B. Hecht, U. P. Wild, *J. Phys. Chem. B* **2003**, *107*, 703–711.

[17] A. Bernard, J.-P. Renault, B. Michel, H. R. Bosshard, E. Delamarche, *Adv. Mater.* **2000**, *12*, 1067–1070.

[18] J. L. Tan, J. Tien, C. S. Chen, *Langmuir* **2002**, *18*, 519–523.

[19] A. Kumar, G. M. Whitesides, *Appl. Phys. Lett.* **1993**, *63*, 2002–2004.

[20] H. Hillborg, U. W. Gedde, *IEEE Trans. Dielectrics and Electrical Insulation* **1999**, *6*, 703–717.

[21] G. S. Ferguson, M. K. Chaudhury, H. A. Biebuyck, G. M. Whitesides, *Macromolecules* **1993**, *26*, 5870–5875.

[22] C. Donzel, M. Geissler, A. Bernard, H. Wolf, B. Michel, J. Hilborn, E. Delamarche, *Adv. Mater.* **2001**, *13*, 1164–1167.

[23] J. P. Renault, A. Bernard, D. Juncker, B. Michel, H. R. Bosshard, E. Delamarche, *Angew. Chem., Int. Ed. Engl.* **2002**, *41*, 2320–2323.

[24] B. R. Young, W. G. Pitt, S. L. Cooper, *J. Colloid Interface Sci.* **1988**, *124*, 28–43.

[25] J. D. Andrade, V. Hlady, *Adv. Polym. Sci.* **1986**, *79*, 1–63.

[26] H. B. Lu, J. Homola, C. T. Campbell, G. G. Nenninger, S. S. Yee, B. D. Ratner, *Sensors and Actuators B* **2001**, *74*, 91–99.

[27] L. A. Kung, L. Kam, J. S. Hovis, S. G. Boxer, *Langmuir* **2000**, *16*, 6773–6776.

[28] D. Stamou, C. Duschl, E. Delamarche, H. Vogel, *Angew. Chem., Int. Ed.* **2003**, *42*, 5580–5583.

29] D. Losic, J. G. Shapter, J. J. Gooding, *Langmuir* **2001**, *17*, 3307–3316.

[30] D. W. Branch, B. C. Wheeler, G. J. Brewer, D. E. Leckband, *IEEE Trans. Biomed. Eng.* **2000**, *47*, 290–300.

[31] M. Nishizawa, K. Takoh, T. Matsue, *Langmuir* **2002**, *18*, 3645–3649.

[32] D. W. Branch, J. M. Corey, J. A. Weyhenmeyer, G. J. Brewer, B. C. Wheeler, *Med. Bio. Engineer. Comput.* **1998**, *36*, 135–141.

[33] J. S. Hovis, S. G. Boxer, *Langmuir* **2000**, *16*, 894–897.

[34] Z. Yang, A. Chilkoti, *Adv. Mater.* **2000**, *12*, 413–417.

[35] J. Lahann, I. S. Choi, J. Lee, K. F. Jensen, R. Langer, *Angew. Chem., Int. Ed. Engl.* **2001**, *40*, 3166–3169.

[36] A. T. A. Jenkins, N. Boden, R. J. Bushby, S. D. Evans, P. F. Knowles, R. E. Miles, S. D. Ogier, H. Schönherr, G. J. Vancso, *J. Am. Chem. Soc.* **1999**, *121*, 5274–5280.

[37] E. Delamarche, H. A. Biebuyck, H. Schmid, B. Michel, *Adv. Mater.* **1997**, *9*, 741–746.

[38] C. D. James, R. C. Davis, L. Kam, H. G. Craighead, M. Isaacson, J. N. Turner, W. Shain, *Langmuir* **1998**, *14*, 741–744.

[39] C. D. James, R. D. Davis, M. Meyer, A. Turner, S. Turner, G. Withers, L. Kam, G. Banker, H. Craighead, M. Isaacson, J. Turner, W. Shain, *IEEE Trans. Biomed. Eng.* **2000**, *47*, 17–21.

[40] L. Kam, W. Shain, J. N. Turner, R. Bizios, *Biomaterials* **2001**, *22*, 1049–1054.

[41] M. Scholl, C. Sprössler, M. Denyer, M. Krause, K. Nakajima, A. Maelicke, W. Knoll, A. Offenhäusser, *J. Neurosci. Methods* **2000**, *104*, 65–75.

[42] J. Hyun, H. Ma, P. Banerjee, J. Cole, K. Gonsalves, A. Chilkoti, *Langmuir* **2002**, *18*, 2975–2979.

[43] P. Harder, M. Grunze, R. Dahint, G. M. Whitesides, P. E. Laibinis, *J. Phys. Chem. B* **1998**, *102*, 426–436.

[44] K. L. Prime, G. M. Whitesides, *J. Am. Chem. Soc.* **1993**, *115*, 10714–10721.

[45] S. R. Sheth, D. Leckband, *Proc. Natl. Acad. Sci. USA* **1997**, *94*, 8399–8404.

[46] E. Delamarche, C. Donzel, F. Kamounah, H. Wolf, M. Geissler, R. Stutz, P. Schmidt-Winkel, B. Michel, H. J. Mathieu, K. Schaumburg, *Langmuir* **2003**, *19*, 8749–8758.

[47] B. Michel, A. Bernard, A. Bietsch, E. Delamarche, M. Geissler, D. Juncker, H. Kind, J.-P. Renault, H. Rothuizen, H. Schmid, P. Schmidt-Winkel, R. Stutz, H. Wolf, *IBM J. Res. Develop.* **2001**, *45*, 697–719.

[48] C. Y. Hui, A. Jagota, Y. Y. Lin, E. J. Kramer, *Langmuir* **2002**, *18*, 1394–1407.

[49] F. Schwesinger, R. Ros, T. Strunz, D. Anselmetti, H.-J. Güntherodt, A. Honegger, L. Jermutus, L. Tiefenauer, A. Plückthun,

Proc. Natl. Acad. Sci. USA **2000**, *97*, 9972–9977.

[50] L. Lauer, S. Ingebrandt, K. Scholl, A. Offenhäusser, *IEEE Trans. Biomed. Eng.* **2001**, *48*, 838–842.

[51] P. M. St. John, R. Davis, N. Cady, J. Czajka, C. A. Batt, H. G. Craighead, *Anal. Chem.* **1998**, *70*, 1108–1111.

[52] J. Lahann, M. Balcells, T. Rodon, J. Lee, I. S. Choi, K. F. Jensen, R. Langer, *Langmuir* **2002**, *18*, 3632–3638.

[53] See also Chapter 5 by A. Offenhäusser and K. Vogt.

[54] S. C. Lin, F. G. Tseng, H. M. Huang, C. Y. Huang, C. C. Chieng, *Fresenius J. Anal. Chem.* **2001**, *371*, 202–208.

[55] L. G. Mendoza, P. McQuary, A. Mongan, R. Gangadharan, S. Brignac, M. Eggers, *BioTechniques* **1999**, *27*, 778–788.

[56] J. B. Delehanty, F. S. Ligler, *Anal. Chem.* **2002**, *74*, 5681–5687.

[57] H. D. Irenowicz, S. Howell, F. E. Regnier, R. Reifenberger, *Langmuir* **2002**, *18*, 5263–5268.

[58] J. Tien, C. M. Nelson, C. S. Chen, *Proc. Natl. Acad. Sci.* **2002**, *99*, 1758–1762.

[59] M. Geissler, A. Bernard, A. Bietsch, H. Schmid, B. Michel, E. Delamarche, *J. Am. Chem. Soc.* **2000**, *122*, 6303–6304.

[60] E. Delamarche, A. Bernard, H. Schmid, B. Michel, H. A. Biebuyck, *Science* **1997**, *276*, 779–781.

[61] D. Juncker, H. Schmid, U. Drechsler, H. Wolf, M. Wolf, B. Michel, N. de Rooij, E. Delamarche, *Anal. Chem.* **2002**, *74*, 6139–6144.

[62] D. Qin, Y. Xia, G. M. Whitesides, *Adv. Mater.* **1996**, *8*, 917–919.

[63] T. W. Odom, V. R. Thalladi, J. C. Love, G. M. Whitesides, *J. Am. Chem. Soc.* **2002**, *124*, 12112–12113.

[64] G. T. Hermanson, A. K. Mallia, P. K. Smith (eds), *Immobilized Affinity Ligand Techniques*, Academic Press Inc., San Diego, CA, **1992**.

[65] Z. Yang, A. M. Belu, A. Liebmann-Vinson, H. Sugg, A. Chilkoti, *Langmuir* **2000**, *16*, 7482–7492.

[66] J. Hyun, Y. Zhu, A. Liebmann-Vinson, Jr., T. P. Beebe, A. Chilkoti, *Langmuir* **2001**, *17*, 6358–6367.

[67] E. Delamarche, J. Vichiconti, S. A. Hall, M. Geissler, W. Graham, B. Michel, R. Nunes, *Langmuir* **2003**, *19*, 6567–6569.

[68] G. P. Lopez, H. A. Biebuyck, R. Haerter, A. Kumar, G. M. Whitesides, *J. Am. Chem. Soc.* **1993**, *115*, 10774–10781.

[69] A. M. Belu, Z. Yang, R. Aslami, A. Chilkoti, *Anal. Chem.* **2001**, *73*, 143–150.

[70] R. P. Ekins, H. Berger, F. W. Chu, P. Finkch, F. Krause, *Nanobiology* **1998**, *4*, 197–220.

4
Cell–Nanostructure Interactions

Joachim P. Spatz

4.1
Introduction

Cell–cell and cell–extracellular matrix (ECM) adhesion is a complex, highly regulated process which plays a crucial role in most fundamental cellular functions including motility, proliferation, differentiation, and apoptosis [1, 2]. Focal adhesions are the primary cellular domains responsible for surface adhesion. These are complex multimolecular assemblies consisting of transmembrane proteins, the integrin receptors, and cytoplasmic proteins, such as vinculin, paxillin, and focal adhesion kinase (for an overview, see Figure 4.1) [3–9]. Cell binding to the ECM results in local accumulation of integrins, cytoplasmic proteins, which form focal adhesion clusters (FAC), and reorganization of the actin cytoskeleton, which generates forces to the underlying substrate by the work of myosin molecular motors [10].

Despite enormous progress and challenging studies in the field of cell adhesion during the past 50 years [3], numerous questions concerning the signaling of focal adhesion which are based on single protein assembly remain unsolved. We have only partial knowledge of the existence of hierarchical and cooperative arrangements and synergetic interactions between focal adhesion proteins. Likewise, we have a very limited understanding of the function of cell adhesion, and of the significance of focal adhesion cluster size, shape, characteristic length scales between proteins in a FAC, and protein assembly dynamics. Future knowledge in this area, when combined with tools that control these processes in focal adhesion, would allow for the tuning of cell adhesion and its associated signaling, with molecular precision.

Integrin clustering into linear objects is clearly observed using either differential interference contrast (DIC) microscopy or fluorescent optical microscopy following immunohistochemical staining of a specific protein involved in FAC formation [11]. However, it is not yet known how the adhesion and signaling of cells is coordinated by integrin clustering, integrin–integrin separation distances and integrin pattern geometries in cell membranes, nor how many integrins are necessarily involved in the formation of stable adhesion [12]. Control of these structural arrangements in the cell membrane is offered by adhesive nanotemplates or substrate topographies, and points clearly to the value of

Nanobiotechnology. Edited by Christof Niemeyer, Chad Mirkin
Copyright © 2004 WILEY-VCH Verlag GmbH & Co. KgaA, Weinheim
ISBN 3-527-30658-7

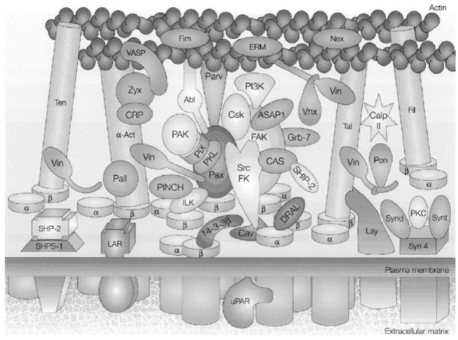

Nature Reviews | Molecular Cell Biology

Figure 4.1 Schematic depicting the complexity of the main molecular domains of cell–matrix adhesions by Geiger [3]. The primary adhesion receptors are heterodimeric (α and β) integrins, represented by orange cylinders. Additional membrane-associated molecules enriched in these adhesions (red) include syndecan-4 (Syn4), layilin (Lay), the phosphatase leukocyte common antigen-related receptor (LAR), SHP-2 substrate-1 (SHPS-1) and the urokinase plasminogen activator receptor (uPAR). Proteins that interact with both integrin and actin, and which function as structural scaffolds of focal adhesions, include α-actinin (α-Act), talin (Tal), tensin (Ten) and filamin (Fil), shown as golden rods. Integrin-associated molecules in blue include: focal adhesion kinase (FAK), paxillin (Pax), integrin-linked kinase (ILK), down-regulated in rhabdomyosarcoma LIM-protein (DRAL), 14-3-3β and caveolin (Cav). Actin-associated proteins (green) include vasodilator-stimulated phosphoprotein (VASP), fimbrin (Fim), ezrin–radixin–moesin proteins (ERM), Abl kinase, nexillin (Nex), parvin/actopaxin (Parv) and vinculin (Vin). Other proteins, many of which might serve as adaptor proteins, are colored purple and include zyxin (Zyx), cysteine-rich protein (CRP), palladin (Pall), PINCH, paxillin kinase linker (PKL), PAK-interacting exchange factor (PIX), vinexin (Vnx), ponsin (Pon), Grb-7, ASAP1, syntenin (Synt), and syndesmos (Synd). Among these are several enzymes, such as SH2-containing phosphatase-2 (SHP-2), SH2-containing inositol 5-phosphatase-2 (SHIP-2), p21-activated kinase (PAK), phosphatidylinositol 3-kinase (PI3K), Src-family kinases (Src FK), carboxy-terminal src kinase (Csk), the protease calpain II (Calp II) protein kinase C (PKC). Enzymes are indicated by lighter shades. For details, see Ref. [3].

nanostructures in an understanding of molecular dimensions and processes in focal adhesion formation.

Integrins are heterodimers formed by the noncovalent association of α and β subunits; the β subunit recognizes the RGD (Arginine-Glycine-Aspatate) motif, a sequence which is present in many ECM proteins. The α and β tails form together a V-shaped flexible structure of an estimated lateral diameter of 80–120 Å [13, 14], with the head ranging between 57–73 Å for $\alpha IIb\beta_3$-integrin, and ~90 × 60 × 45 Å for $\alpha_v\beta_3$-integrin [15, 16]. Ligand binding affinity is influenced by the conformational changes in the receptor caused by the extracellular environment, and also by the interactions with the cytoplasmic proteins [17].

A prominent example where spatial arrangement of RGD ligands in a defined and rigid geometry fulfills important functions is given by the adenovirus. Its capsid contains the penton base protein, a pentamer that promotes virus entry in cells via α_v-integrins. In fact, the penton base protein presents RGD sequences on the tips of a regular stiff pentagon having a side length of ~60 Å [18] (Figure 4.2). Since only the pentameric form mediates integrin-specific adhesion of nonactivated lymphoid cells, while the monomeric does not, the polyvalent binding of several integrins to the RGD sequences is mandatory and assumed to be controlled by the rigidity of this protein [19]. This rigidity of the virus template also allows the adenovirus to escape neutralization by IgG antibodies of the immunosystem directed against RGD integrin receptor sites due to steric hindrance. Thus, the 60 Å separation between RGD sequences demonstrates a minimal distance necessary for integrin heads to bind simultaneously but which is too small for the IgG to neutralize the virus.

The adenovirus is a major example where nature demonstrates regulation of cell interactions with the extracellular site by single ligand pattern of specific geometry, and this length of scale opens challenging opportunities for understanding and tailoring cell func-

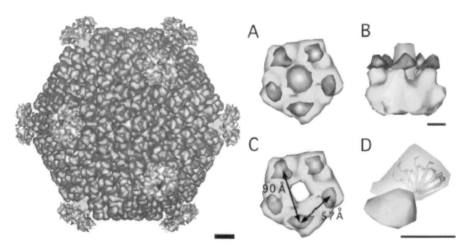

Figure 4.2 Left: A cryoelectron microscopy image reconstruction of the adenovirus binding complex. The penton base capsomers at the icosahedral vertices are shown in yellow, the reconstructed portion of the flexible fibers in green, the remaining capsid density in blue and the Fab density in magenta. The complex is viewed along an icosahedral 3-fold axis. The scale bar is 100 Å. Right: The adenovirus penton base protrusions. (A) Top view of the penton base (yellow) and fiber (green), along with weak protrusion density (red). (B) Side view of the external portion of the penton base with fiber. (C) Top view of the penton base, showing the distances between weak protrusions. (D) An enlargement of a single penton base protrusion with a loop modeled from the crystallographic RGD peptide. The arginine side chain is shown in blue, and the aspartic acid side chain in red. The loops are shaded with a transparency gradient to denote motion. The scale bars are 25 Å [18].

tions with ultrahigh sensitivity. The formation of a regular and stiff ligand template is a basic requirement for mimicking such molecularly defined adhesive "keys" that are set by the spatial distribution of ligands for single integrin occupation. As described in the next section, methods developed from nanotechnological systems already approach fidelity and functionality to provide these requirements for control of cell adhesion on a molecular level.

4.2
Methods

Making use of advanced opportunities from material sciences for the identification, location, and systematic manipulation of molecular components at interfaces identifies great potential in the development of new materials for biophysical and biochemical investigations, and particularly in the field of cell adhesion. In principle, cellular adhesion studies on the nanoscale may be divided into the reaction of cells to variations in substrate topography, or to the presence of a chemical contrast along a substrate. While topography induces surface roughness and such greater adhesive areas, the substrate's chemical contrast points to an opportunity of controlling transmembrane and intracellular molecules, as well as protein distributions.

A group of researchers at Glasgow University demonstrated the response of cells not only to micrometric but also to nanometric scale topography [20–22]. The formation of nanotopography was explored using methods based on polymer demixing; that is, demixing of polystyrene and polybromostyrene, where nanoscale islands of reproducible height were fabricated. The islands were shown to affect cell spreading compared with planar surfaces, where morphological, cytoskeletal, and molecular changes in fibroblast reaction to 13 nm-high islands were observed. It should be noted that this topography length scale is on the order of roughness which is induced by adhesive proteins such as fibronectin or lamin as well as collagen fibers that are dominant within the ECM. The cellular responses were characterized using methods such as scanning electron microscopy, fluorescence microscopy, and gene microarray. The results showed that cells respond to the islands by broad gene up-regulation, notably in areas of cell signaling, proliferation, cytoskeleton, and the production of ECM proteins. Results obtained with microscopy confirmed the microarray findings and highlighted several corresponding points between cytoskeletal, morphological, and genetic observations. Moreover, they also showed a synergy from the formation of focal contacts to the up-regulation of genes required for fibroblast differentiation. These findings indicate that increased cell attachment and spreading is required for up-regulation proliferation and matrix synthesis.

One prerequisite when preparing model systems to demonstrate the role of spatial ligand distributions and ligand concentration in cell adhesion in vitro is the availability of a nonadhesive surface; this permits specific cellular responses to be attributed entirely to the interaction with specific adhesion-mediating ligands. Polyethylene glycol (PEG) or polyethylene oxide (PEO) -based substrates are widely used as biologically inert interfaces, and recent developments have included the grafting of high molecular weight PEG [23] and star-shaped PEG macromolecules to substrates [24, 25], or the use of oligo(ethylene oxide) functionalized self-assembled monolayers (SAMs) [26, 27].

The surface concentration and spatial distribution of cell-adhesive ligands in such a biologically inert PEG or PEO background may be controlled statically by mixing bioactive macrosystems with unsubstituted molecules [28], or dynamically by the electrochemical control of ligand release [29]. The modification of inert polymers with a cell-adhesive motif that often contains the amino acid sequence RGD (which is also found in fibronectin) and its influence on cell adhesion and signaling has been recently reviewed in detail by Kessler et al. [30].

Cell attachment to interfaces depends on many factors, including the affinity and specificity of surface-bound ligands to integrins, the mechanical strength of ligand support and linkage, spacer length, overall ligand concentration, and ligand clustering [30]. As an example, the number of attached cells is clearly correlated to RGD surface density, as indicated by a sigmoidal increase with RGD concentration [31]. This suggests that there is a minimum ligand density for cell response. As a general rule, a higher RGD surface density is related to intense cell spreading, cell survival, and focal contact formation. Ever since the early days of RGD-mediated cell adhesion, discussions have been ongoing on the subject of how many RGD molecules are required to induce not only cell attachment but also cell spreading and focal contact formation. Hubbell produced a benchmark result by showing amounts as low as 1 fmol RGD ligands cm^{-2} to be sufficient for cell spreading, and as low as 10 fmol cm^{-2} to be sufficient for the formation of focal contacts and stress fibers [32]. In these studies, RGD molecules were covalently bound to glycophase glass coverslips via NH_2-terminal primary amines. Smart macromolecular designs of PEG molecules such as PEG-stars allow the control of an average number of RGD ligands per star (one, five, or nine YGRGD peptides per star) as demonstrated by Griffith and Lauffenburger in a series of publications and illustrated in Figure 4.3 [25, 33–35]. Cell adhesion and movement was observed for 1000 ligands μm^{-2}, or more [25]. The macromolecular approach has the advantage that the large and flexible chains may account for different cell-binding activities that are probably caused by local enrichment of ligands at the cell membrane and anchoring compliance. However, it may not account for control of precise ligand clustering, as the ligand template may be not well-ordered, as well as being too flexible. This permits ligands to cluster by pure chemical affinity, or cells to arrange ligands at their convenience [34]. This situation does not apply to the adenovirus, where the control of cell function is achieved by the arrangement of single ligands in patterns on a rigid template.

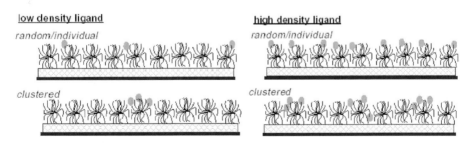

Figure 4.3 Schematic illustration of star polymer as a tether to present ligand (shaded oval) in a manner in which the total average concentration (left to right) and the spatial distribution, from homogeneous to highly clustered (top versus bottom), can be independently varied (from Ref. [25]).

A clear understanding of how the adhesion and signaling of cells depend on the composition, size, and distribution of FACs has long been limited to patterning studies of ligands on submicrometer patches. By using micro contact printing [36], surfaces patterned with adhesive and nonadhesive domains have been prepared at scales down to the micrometer level. These surfaces were then successfully used to control geometrically both cell shape and viability [36]. The results of these studies indicated strongly that cell shape and integrin distribution are able to control the survival/apoptosis of cells, and can switch between these two basic programs of the cells.

Even smaller adhesion pattern were prepared using dip-pen nanolithography, where cell-adhesive patches (retronectin) of 200 nm diameter and 700 nm separation still showed attachment of cells (Figure 4.4) (see also Chapter 19) [37].

The control of defined spacing between adhesive ligands on interfaces at protein length scales between 10 and 200 nm over large surface areas remains a challenge. However, this is the length scale on which protein clustering in focal adhesions occurs when cells adhere to interfaces. The exact spatial control of receptor clustering in the cell membrane on this length scale is demanding for challenging concepts from nanotechnology which offer a rigid nanoadhesive pattern with flexible geometries over extended surface areas at rather low production costs.

A substrate-patterning strategy based on the self-assembly of polystyrene-block-poly (2-vinylpyridine) (PS-b-P2VP) diblock copolymer micelles covers the indicated length scale; that is, diblock copolymer micelle lithography [38–42]. PS-b-P2VP diblock copolymers form reverse micelles in toluene. The core of a micelle consists of associated P2VP blocks which complex $HAuCl_4$ if this is added to the micellar solution. Dipping and retracting a substrate from such a solution results in uniform and extended monomicellar films supported by the substrate. Each micelle contains approximately the same quantity of Au. Treating these films with a gas plasma results in the deposition of highly

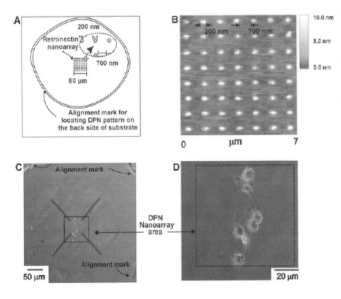

Figure 4.4 Cell adhesion to a pattern of retronectin fabricated by dip-pen nanolithography (DPN) [37].
(A) Diagram describing the cell adhesion experiment on the DPN-generated pattern. The total patterned area is 6400 mm². The alignment marks were generated by scratching a circle into the backside of the Au-coated glass substrate. (B) Topography image (contact mode) of the retronectin protein array. (C) Large-scale optical microscope image showing the localization of cells in the nanopatterned area. (D) Higher-resolution optical image of the nanopatterned area, showing intact cells.

regular Au-nanodots, thereby forming a rather perfect hexagonal pattern on solid-state interfaces such as glass or Si-wafers. A preparation scheme is presented in Figure 4.5A. The scanning electron microscopy (SEM) images in Figure 4.5B–E show Au-dots as white spots which are arranged in pattern on Si-wafers by self-assembly of polystyrene-*block*-poly(2-vinylpyridine(HAuCl$_4$)$_{0.5}$) diblock copolymer micelles, that is PS-*b*-P[2VP(HAuCl$_4$)$_{0.5}$]. The nanoscopic patterns consist of Au-nanodots (3, 5, 6, or 8 nm in diameter) with spacings between dots of 28, 58, 73, and 85 nm respectively adjusted by the molecular weights of PS-b-P2VP and the amount of HAuCl$_4$ added to the micellar solution. A side view of the Au-nanodots on a Si-wafer is shown in the high-resolution electron transmission microscopy image in Figure 4.5F.

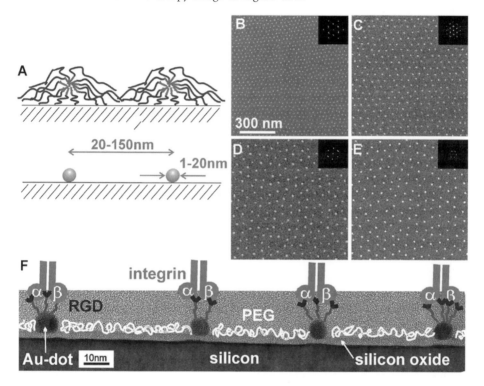

Figure 4.5 Micellar block copolymer lithography and biofunctionalization. (A) Scheme of diblock copolymer micelle lithography. (B–E) Extended Au-nanodot pattern are displayed using scanning electron microscopy [38]. Uniform Au-nanodots (bright spots) of: (B) 3 nm by PS(190)-*b*-P[2VP(HAuCl$_4$)$_{0.5}$](190); (C) 5 nm by PS(500)-*b*-P[2VP(HAuCl$_4$)$_{0.5}$](270); (D) 6 nm by PS(990)-*b*-P[2VP(HAuCl$_4$)$_{0.5}$](385); and (E) 8 nm by PS(1350)-*b*-P[2VP(HAuCl$_4$)$_{0.5}$](400) deposited onto Si-wafers are shown. The number in brackets refers to the number of monomer units in each block which control the separation between Au-dots. These varied between (B) 28, (C) 58, (D) 73 and (E) 85 nm. The Au-dots form extended, nearly perfect hexagonally-close packed pattern as indicated by the Fourier transformed images (inset) which show second-order intensity spots. (F) Biofunctionalization of the Au-nanodots pattern [43]. The Au-dots are presented as side view micrographs using a high-resolution transmission electron microscope. Molecules and Proteins are drawn schematically. Since the Au-dot is sufficiently small, it is most likely that only one integrin transmembrane receptor occupies one dot.

These nanostructures serve as chemical templates for the spatial arrangement of RGD-based ligands, as shown schematically in Figure 4.5F [43]. Biofunctionalization of the interface comprises binding a polyethylene oxide layer to the silicon oxide substrate between the Au-nanodots to avoid any nonspecific adsorption of proteins or parts of a cell membrane. Subsequently, the Au-dots are functionalized by RGD ligands that bind selectively to Au from a solution containing these ligands. In this study, cyclic RGD molecules have been used, that is c(RGDfK)-thiol, as has been synthesized by the group of Kessler (TU Munich). c(RGDfK)-thiols contain the cell-adhesive RGD sequence which is recognized by $\alpha_v\beta_3$-integrin with high affinity [44, 45]. In Figure 4.5F, Au-dots and integrins are drawn approximately to scale, indicating that the size of a Au-nanodot provides dimensions of a chemical anchor point to which, potentially, only one integrin can bind. This is a very valuable tool, as the pattern dimensions and geometries control the assembly of single integrins to form the basis of a focal adhesion cluster. Thus, uniform patterning of extended substrate areas by diblock copolymer micelle lithography provides access to an important length-scale for cell-adhesion studies that is hardly accessible with other techniques.

In Figure 4.6, MC3T3-osteoblasts were seeded on glass and examined after one day using optical phase-contrast microscopy. Only three-fourths of the glass substrate area was patterned with Au-nanodots, with different spacings between the dots. The Au-dots were functionalized by c(RGDfK)-thiols, and the free glass was passivated by PEG. A line of cells marks the borderline of the nano-pattern area (white arrows). The right

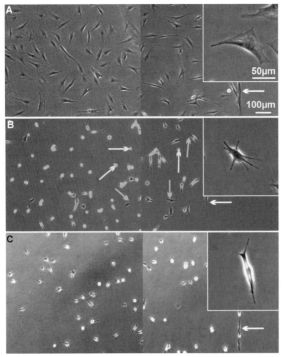

Figure 4.6 Optical phase-contrast microscopy images of MC3T3-osteoblasts on nanopatterns of different spacing [43]: (A) ~58 nm; (B) ~73 nm. Cells mark the borderline of nanostructures. Extending the separation between individual dots from 58 nm to 73 nm causes failure in cell spreading. (C) Cells attach also to Au-nanodots that were not covered by c(RGDfK)-thiols, but cell adhesion and spreading is truly limited, as in the case of 73-nm spacing between dots.

side was entirely passivated against cell adhesion; thus, cell adhesion and attachment is only observed on the left side of the images. When plated on Au-nanodot patterns with various spacing, functionalized by c(RGDfK)-thiols, MC3T3-osteoblasts show different adhesion behavior. It is clear that cells spread very well on the 58-nm (Figure 4.6A) patterns, appearing as they do on uniformly RGD- or fibronectin-coated surfaces (not shown). On the other hand, hardly any cell spreading is observed on substrates with 73-nm spaced nanodots (Figure 4.6B). Quiescent and migrating cells are also visible. Quiescent cells present a rounded shape which causes strong scattering of light, while migrating cells are usually characterized by extended filopodia (see arrows). These observations have been repeated with additional cell types, i.e., REF52-fibroblasts, 3T3-fibroblasts and B16-melanocytes, indicating a universal characteristic cell-adhesion behavior. Figure 4.6C shows MC3T3-osteoblasts on Au-nanodots separated by 58 nm and not conjugated to c(RGDfK)-thiols. Cell spreading on these surfaces is rather poor, and few cells remain attached after gentle rinsing.

The molecular formation of focal contacts and the assembly of actin stress fibers in MC3T3-osteoblasts adhering to these nanopattern substrates were investigated by culturing cells for one day, fixing and staining them against vinculin and actin. Figure 4.7 presents confocal micrographs where c(RGDfK) covered Au-dots had spacings of (A) 58 and (B) 73 nm. It is obvious that the pattern with adhesive c(RGDfK) Au-nanodot separations of 58 nm establishes well-constituted, quite long vinculin clusters (shown as green) and well-defined actin stress fibers (shown as red). The adhesion area of these cells is a factor of ~4 greater. Fairly blurred images of vinculin and actin distribution were obtained when Au-dots were not covered by c(RGDfK)-thiols (not shown), or the separation between the dots was 73 nm (B). It is also of note that only these cells which remained on substrates after fixing and staining could establish either strong or stable adhesion (see Figure 4.6A), or cells which could at least form nonstable adhesions (shown by red arrows in Figure 4.6B). All other cells (yellow arrows in Figure 4.6B) were washed off by the fixing and staining process.

The increase in dot separation distances causes a decrease in global dot density. Therefore, the observed limitation of cell adhesion at increased dot separation could be reasoned either on the global density of c(RGDfK)-thiol covered Au-dots or on the local dot-to-dot distance. In order to address this issue, "micro"-nanostructured interfaces were created as described in Ref. [38]. This technique allows for deposition of a defined number of Au-nanodots in a confined area of the substrate. The surfaces were designed such that the global dot density was 90 dots μm^{-2}, and thus significantly smaller than in all cases of extended Au-dot pattern (280 dots μm^{-2}, with 58 nm and 190 dots μm^{-2} with the 73-nm separated dots). The local dot density was organized in $2 \times 2 \ \mu m^2$ patches of 58-nm spaced dots was 280 dots μm^{-2} (Figure 4.8A). Figure 4.8B shows a bright-field optical micrograph taken 3 hours after plating of MC3T3-osteoblasts on the substrate. Clearly, the cells are confined only to the structured area, and the process of cell spreading was advanced (as shown in the inset). After 24 hours (C), well-spread cells are present in this area, whereas cells located outside the frame (indicated by arrows) are poorly spread. Figure 4.8D shows a confocal fluorescent micrograph of a cell from (C) after immunohistochemical staining for vinculin (green) and actin (red), thereby demonstrating the confinement of focal adhesion to the square pattern and the origin of actin stress fibers from there. In this case, the average focal adhesion length was 2.6 ± 0.9 µm; this value is between the side length of one square pattern and its diagonal. The distribution in focal adhesion length is remarkably narrow and displays the confinement by a square. Cells do not adhere to all squares, but in some areas a separation distance between focal adhesions of 1.5 µm is recognized as shown in the inset of Figure 4.8D. If cultured on pattern uniformly structured with dots separated by 58 nm, these cells form focal adhesion lengths with a mean value of 5.6 ± 2.7 µm (Figure 4.7A).

These adhesion experiments indicate that local dot–dot separation, rather than global dot density, was critical for inducing cell adhesion and focal adhesion assembly. Thus, for example, the dot density located under cells attached to the "micro"-nanostructured

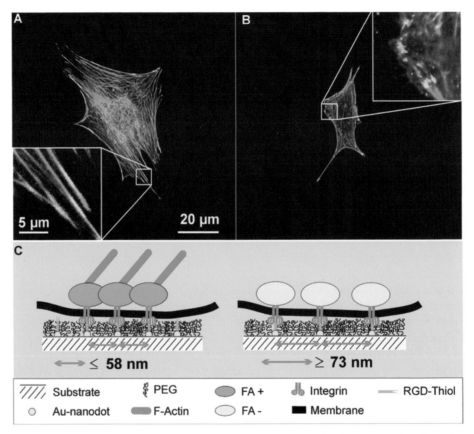

Figure 4.7 A pair of confocal fluorescent optical micrographs of MC3T3-osteoblasts stained for vinculin (green) and actin (red) [43]. Cells interacting with Au-nanodot patterns with Au-dot spacing of (a) 58 nm and (b) 73 nm. (C) Scheme of biofunctionalized nanopattern to control integrin clustering in cell membranes: Au-dots are conjugated with c(RGDfK)-thiols and areas between cell adhesive Au-dots are passivated by PEG against cell adhesion. Therefore, cell adhesion is mediated entirely via c(RGDfK)-covered Au-nanodots. A separation of Au/RGD dots by ≥73 nm causes limited cell attachment and spreading and actin stress fiber formation because of restricted integrin clustering. This is indicated by failure of focal adhesion activation (FA–), whereas distances between dots of ≤58 nm caused focal adhesion activation (FA+).

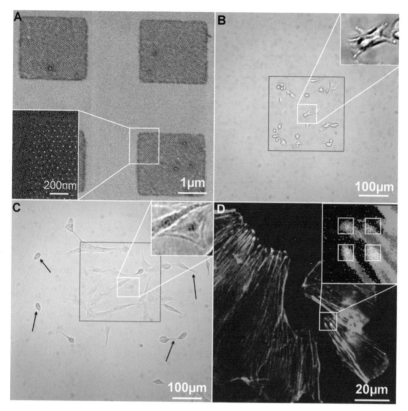

Figure 4.8 MC3T3-osteoblast adhesion on "micro"-nanostructures occupied by c(RGDfK)-thiols [43]. (A) SEM image of "Micro"-nanostructures: SEM-micrograph of 5-nm Au-dots separated by 58 nm in a hexagonally-close packed pattern localized in 2 × 2 μm squares which are separated by 1.5 μm [38]. The bright-field optical micrograph of adhesive MC3T3-osteoblasts on pattern shown in (A) is covering the area in the marked box after (B) 3 h and (C) 24 h of cell culture. (D) Fluorescent optical micrograph of MC3T3-osteoblast showing the location of FA by staining for vinculin (green). FA appear as small strip-like, bright green objects on the pattern in (A). The actin filaments are seen in red.

squares, consisting of 58-nm separated dots is considerable lower than that of dots located underneath cells attached to a substrate, uniformly patterned by dots, separated by ≥ 73 nm. Nevertheless, cells did form focal adhesions on the former surface, but failed to do so on the latter. This is summarized schematically in Figure 4.7C.

4.3
Outlook

Cell attachment to interfaces depends on many factors, such as the affinity and specificity of surface-bound ligands to integrins, the mechanical strength of ligand support and linkage, spacer length, overall ligand concentration, and ligand clustering [30]. This survey of challenging investigations concerned with cell adhesion on nanostructured interfaces concludes that highly intriguing cellular processes are stimulated and controlled by substrate nanotopography and spatial ligand patterning for single integrin receptor occupation. Thus, nanoadhesive patterns offer the unique opportunity to define length scales in multi-molecular complexes within focal adhesions, with unprecedented resolution as small as a single protein. Variations in nanoadhesive site organization, including alterations in ligand template pliability and presentation of small dot clusters, for example pairs or triplets, may shed light on the minimal molecular number of an effective integrin cluster necessary to obtain cell attachment, spreading or migration, and also of the possible pattern-specific features (molecularly defined adhesion "keys") that trigger cell adhesion-based signaling [43].

References

[1] H. M. Blau, D. J. Baltimore, *J. Cell Biol.* **1991**, *112*, 781-783.

[2] E. Ruoslahti, B. Obrink, *Exp. Cell. Res.* **1996**, *227*, 1-11.

[3] B. Geiger, A. Bershadsky, R. Pankov, K. M. Yamada, *Nature Rev. Mol. Cell Biol.* **2001**, *2*, 793.

[4] E. Zamir, B. Geiger, *J. Cell Sci.* **2001**, *114*, 3583; E. Zamir, B. Geiger, *J. Cell Sci.* **2001**, *114*, 3577.

[5] D. R. Critchley, *Curr. Opin. Cell Biol.* **2000**, *12*, 133.

[6] F. G. Giancotti, E. Ruoslahti, *Science* **1999**, *285*, 1028.

[7] R. O. Hynes, *Cell* **1987**, *4*, 549.

[8] S. Miyamoto, S. K. Akiyama, K. M. Yamada, *Science* **1995**, *267*, 883.

[9] S. Levenberg, B. Z. Katz, K. M. Yamada, B. Geiger, *J. Cell Sci.* **1998**, *111*, 347.

[10] N. Q. Balaban, U. S. Schwarz, D. Rivelin, P. Goichberg, G. Tzur, L. Sabanay, D. Mahalu, S. Safran, A. Bershadsky, L. Addadi, B. Geiger, *Nature Cell Biol.* **2001**, *3*, 466.

[11] B. Alberts, D. Bray, J. Lewis, M. Raff, K. Roberts, J. D. Watson, *Molecular Biology of the Cell*, 3rd edn., Garland Publishing, New York, USA, **1994**.

[12] L. A. Lasky, *Nature* **1997**, *390*, 15-17.

[13] J. W. Weisel, C. Nagaswami, G. Vilaire, J. S. Bennett, *J. Biol. Chem.* **1992**, *267*, 16637-16643.

[14] M. V. Nermut, N. M. Green, P. Eason, S. S. Yamada, K. M. Yamada, *EMBO J.* **1988**, *7*, 4093-4099.

[15] J. P. Xiong, T. Stehle, B. Diefenbach, R. Zhang, R. Dunker, D. L. Scott, A. Joachimiak, S. L. Goodman, M. A. Arnaout, *Science* **2001**, *294*, 339.

[16] E. M. Erb, K. Tangemann, B. Bohrmann, B. Müller, J. Engel, *Biochemistry* **1997**, *36*, 7395-7402.

[17] J. P. Xiong, T. Stehle, R. Zhang, A. Joachimiak, M. Frech, S. L. Goodman, M. A. Arnaout, *Science* **2002**, *296*, 151-155.

[18] P. L. Stewart, C. Y. Chiu, S. Huang, T. Muir, Y. Zhao, B. Chait, P. Mathias, G. R. Nemerow, *EMBO J.* **1997**, *16*, 1189-1198.

[19] D. G. Stupack, E. Li, S. A. Silletti, J. A. Kehler, R. L. Geahlen, K. Hahn, G. R. Nemerow, D. A. Cheresh, *J. Cell Biol.* **1999**, *144*, 777-788.

[20] M. J. Dalby, S. J. Yarwood, M. O. Riehle, H. J. H. Johnstone, S. Affrossman, A. S. G. Curtis, *Exp. Cell Res.* **2002**, *276*, 1–9.

[21] M. J. Dalby, M. O. Riehle, H. Johnstone, S. Affrossman, A. S. G. Curtis, *Biomaterials* **2002**, *23*, 2945–2954.

[22] M. J. Dalby, S. Childs, M. O. Riehle, H. J. H. Johnstone, S. Affrossman, A. S. G. Curtis, *Biomaterials* **2003**, *24*, 927–935.

[23] D. L. Elbert, J. A. Hubbell, *Biomacromolecules* **2001**, *2*, 430-441.

[24] G. B. Brown, Spatial Control of Ligand Presentation on Biomaterials Surfaces, in: *Materials Science and Engineering.* MIT, Cambridge, **1999**, pp. 165.

[25] G. Maheshwari, G. Brown, D. A. Lauffen-burger, A. Wells, L. G. Griffith, *J. Cell Sci.* **2000**, *113*, 1677.

[26] M. Mrksich, S. M. Whitesides, *ACS Symp. Ser.* **1997**, *680*, 361.

[27] R. L. C. Wang, H. J. Kreuzer, M. Grunze, *J. Phys. Chem. B* **1997**, *101*, 9767-9773.

[28] C. Roberts, C. S. Chen, M. Mrksich, V. Martichonok, D. E. Ingber, G. M. Whitesides, *J. Am. Chem. Soc.* **1998**, *120*, 6548-6555.

[29] W. S. Yeo, C. D. Hodneland, M. Mrksich, *ChemBioChem* **2001**, *2*, 590-592.

[30] U. Hersel, C. Dahmen, H. Kessler, *Biomaterials* **2003**, *24*, 4385.

[31] M. Kantlehner, P. Schaffner, D. M. J. Fin-singer, A. Jonczyk, B. Diefenbach, B. Nies, G. Holzemann, S. L. Goodman, H. Kessler, *ChemBioChem.* **2000**, *1*, 107-114.

[32] S. P. Massia, J. A. Hubbell, *J. Cell Biol.* **1991**, *114*, 1089-1100.

[33] L. Y. Koo, D. J. Irvine, A. M. Mayes, D. A. Lauffenburger, L. G. Griffith, *J. Cell Sci.* **2002**, *115*, 1423-1433.

[34] D. J. Irvine, A. M. Mayes, L. G. Griffith, *Biomacromolecules* **2001**, *2*, 85-94.

[35] D. J. Irvine, A. V. G. Ruzette, A. M. Mayes, L. G. Griffith, *Biomacromolecules* **2001**, *2*, 545.

[36] C. S. Chen, M. Mrksich, S. Huang, G. M. Whitesides, D. E. Ingber, *Science* **1997**, *276*, 1425.

[37] K. B. Lee, S. J. Park, C. A. Mirkin, J. C. Smith, M. Mrksich, *Science* **2002**, *295*, 1702.

[38] R. Glass, M. Arnold, M. Möller, J. P. Spatz, *Adv. Funct. Mater.* **2003**, *13*, 569; R. Glass, M. Möller, J. P. Spatz, *Nanotechnology* **2003**, *14*, 1153.

[39] J. P. Spatz, S. Mößmer, M. Möller, T. Her-zog, A. Plettl, P. Ziemann, *Langmuir* **2000**, *16*, 407.

[40] J. P. Spatz, S. Mößmer, M. Möller, *Chem. Eur. J.* **1996**, *2*, 1552-1555.

[41] J. P. Spatz, A. Roescher, M. Möller, *Adv. Mater.* **1996**, *8*, 337.

[42] J. P. Spatz, S. Sheiko, M. Möller, *Macromolecules* **1996**, *29*, 3220-3226.

[43] M. Arnold, E. Ada Cavalcanti-Adam, R. Glass, J. Blümmel, W. Eck, M. Kant-lehner, H. Kessler, J. P. Spatz, *Chem Phys Chem*, in press.

[44] M. Pfaff, K. Tangemann, H. Kessler, *J. Biol. Chem.* **1994**, *269*, 20233.

[45] R. Haubner, D. Finsinger, H. Kessler, *Angew. Chem.* **1997**, *109*, 1440.

5
Defined Networks of Neuronal Cells in Vitro

Andreas Offenhäusser and *Angela K. Vogt*

5.1
Introduction

The growth of neurons into networks of controlled geometry is of major interest in the field of cell-based biosensors, neuroelectronic circuits, neurological implants, and pharmaceutical testing, as well as in fundamental biological questions about neuronal interactions. The precise control of the network architecture can be achieved by defined engineering of the surface material properties – this process is called neuronal cell patterning. Within the literature, it is possible to find a wide range of methods for the formation of well-defined structures on surfaces for cell patterning, and these will be reviewed with regard to the formation of neuronal networks. These methods can be mainly divided into two types. On the one hand, cell patterning can be induced by shapes and textures formed in the substrate; this is called topographical patterning. On the other hand, the term "chemical patterning" is used when differences in (bio)chemical adhesion properties are produced on the surface of the substrate. The pattern-inducing effect is not always evident however, as on the one side the interaction of ECM proteins with topographical structures can result in (bio)chemical differences on the substrate, whilst on the other side the (bio)chemical modification of a surface always includes physical effects.

In the second section of the chapter some background information will be provided about the signaling within a biological neuronal network, together with an overview of the history of in-vitro neuronal cell patterning. In the third section, some recent results will be presented from current methods used in neuronal cell patterning, and this will be followed by a discussion of the application of these methods in the field of bioelectronic devices.

Nanobiotechnology. Edited by Christof Niemeyer, Chad Mirkin
Copyright © 2004 WILEY-VCH Verlag GmbH & Co. KgaA, Weinheim
ISBN 3-527-30658-7

5.2
Overview: Background and History

5.2.1
Physiology of Information Processing within Neuronal Networks

Neurons use differences in electrochemical potential to encode information, and messages can be passed on to other neurons through either chemical or electrical connections. Chemical synapses comprise the conversion of an electrical signal – the action potential arriving at an axon terminal – into a chemical signal; that is, the release of a neurotransmitter into the synaptic cleft. The binding of a neurotransmitter to specific postsynaptic receptors triggers the opening of an intrinsic ion channel in the membrane of the postsynaptic cell. The resulting ion flux alters the transmembrane potential and facilitates or suppresses the generation of a new action potential in this cell [1]. The excitatory neurotransmitters acetycholine, glutamate, and serotonin open cation channels (Na^+, K^+, Ca^{2+}) which depolarize the postsynaptic cell, thus facilitating the generation of an action potential [2]. Inhibitory neurotransmitters, such as glycine and γ-aminobutyric acid (GABA), activate anion channels (Cl^-, HCO_3^-) that lead to hyperpolarization, thereby suppressing neuronal firing. At electrical synapses, ion channels connect the cytoplasm of the pre- and postsynaptic cells, and some current from the presynaptic cells also flows through these low-resistance, high-conductance channels. This current can depolarize the postsynaptic cell and, as a consequence, can induce an action potential [3]. The gap junction channels of an electrical synapse thus mediate electrical signal transmission. Rectifying and nonrectifying electrical synapses do not appear to differ in ultrastructure. At both type of synapses, markers such as fluorescent dyes flow readily between the pre- and postsynaptic cells through the junction. The major difference between the two classes of electrical synapses may reside in the extent to which channel gating is sensitive to voltage. However, chemical synapses – in contrast to electrical ones – exhibit plasticity and thus are thought to be responsible for processes such as learning and memory. Therefore, their presence in the system is absolutely crucial for the study of network behavior, plasticity, and activity-dependent changes.

5.2.2
Topographical Patterning

The employment of topographical patterning techniques started in the early 1960s [4] and 1970s. At this time, typically planar substrates with etched or scribed grooves or glass fibers were used as the means to study cellular patterning. With increasing evidence that curvature was the main effect in cell guidance on surfaces, more groups began to examine the effects of varying groove depth, width, and spacing. In the 1980s, lithography was used to microfabricate grooved surfaces, in particular by utilizing anisotropic etching of silicon wafers. For further details about concepts, materials, surface structures, and possible cellular and biomolecular mechanisms for topographically patterning that were presented over the past, the reader is referred to reviews by Curtis and Wilkinson [5, 6] and by Jung et al. [7]. During the late 1980s and the early 1990s, the Glasgow group was starting

Figure 5.1 Left: Two nerve cells (dorsal root ganglion, rat) growing in adjacent 12 μm-wide, 3 μm-deep grooves with superimposed parallel adhesive tracks (laminin) (From S. T. Britland et al., *Exp. Biol. Online* **1996**, *1*, 2). Right: Nerve cells on microelectrode array with topographical guidance structures (From M. Denyer et al., *Cell. Eng.* **1997**, *2*, 122–131).

to study the relative effects of groove depth and pitch, and cell guidance by ultrafine structured quartz and silicon surfaces produced by electron beam lithography [8]. Later, a combination of adhesive stripes and topographic features was studied in regard of cell patterning [9]. These authors showed that for strong adhesive stripes and shallow grooves, the cells aligned along the adhesive stripes. As groove depth increased, the degree of patterning increased along the groove direction. An example of topographical patterning is shown in Figure 5.1. Further discussion of topographical methods for neuronal cell patterning can be found later in this chapter.

5.2.3
Chemical Patterning

The first results of chemical cell patterning were introduced during the mid-1960s when Carter et al. discovered that fibroblasts adhered preferentially to palladium islands evaporated onto a polyacetate surface [10]. In 1975, Letourneau was using this method to study the alignment of chick dorsal root ganglion neurons on palladium regions on polymeric substrates [11]. He could demonstrate that the cells adhered well on the metal regions when evaporated onto tissue culture plastic, but showed only weak adhesion when the palladium was surrounded by polyornithine. These studies showed that differences ('contrast') between adjacent regions are necessary in order to obtain cell patterning. Later, the role and function of extracellular matrix (ECM) proteins were used to improve cell adhesion and growth and promote neurite extension [12–18]. These proteins were adsorbed onto solid surfaces in order to study cell adhesion in vitro. A major advance was the identification of the Arginine-Glycine-Aspartate (RGD) cell adhesive domain present in some ECM proteins [19, 20]. The RGD domain binds specifically to integrin receptors on the outer membrane of the cell. The authors have also used a covalent tethering method of the active peptide sequence to the surface. Later, other recognition subunits in proteins responsible for cell adhesion were identified, including the B1 chain of the laminin [17, 21]. By using only the protein recognition sequences in combination with a spacer molecule to achieve cell adhesion, any issues of protein conformation can be neglected.

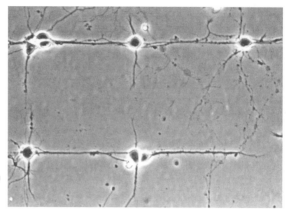

Figure 5.2 Networks of neurons on a pattern of polylysine on glass. The polylysine was patterned through a quartz mask with a pulse from a 193 nm wavelength excimer laser. (From Ref. [27].)

In 1988, Kleinfeld, Kahler and Hockberger used photolithographic techniques for the patterning of silanes on surfaces for the controlled adhesion and growth of neurons [15]. This report probably had the highest impact on the field of neuronal cell patterning. In 1992, the Curtis group photolithographically patterned laminin in lines of 6 μm width for the outgrowth of neurites [22]. At about the same time, the Fromherz group demonstrated that photolithography could be used to pattern ECM proteins for the guided outgrowth of leech neurons in culture [23, 24], while in the mid-1990s the Aebischer group covalently modified polymeric substrates for neuronal cell patterning [25, 26].

Similar to the photolithographic methods, organic thin films can be patterned by using photochemical reactions. The Wheeler group used selective laser ablation to grow rat hippocampal neurons on grids of polylysine with varying line width, intersection distance, and nodal diameter resulting in a very high compliance (Figure 5.2) [27]. Photoablation was also used to pattern ultrathin polymer layers in order to control the adsorption of proteins and the adhesion and spatial orientation of neuronal cells onto surfaces (for example, see Ref. [28]). Further examples of photochemical patterning of neuronal cells will be discussed later.

5.3
Methods

Based on the overview presented, we will now focus in more detail on the methods currently used to produce defined networks of neuronal cells in vitro.

5.3.1
Topographical Patterning

Topographical cell patterning methods have been continuously developed over the past decade. These methods are based on lithography and structuring techniques developed in microelectronics industry, and are used either to pattern resist on the surface of a silicon wafer and selectively etch away the material of interest, or selectively to deposit a layer of material to yield topographical patterns on the substrate [29].

Craighead et al. used micron-size topographical features to influence the pattern formation of neuronal cells [30]. They observed the preferred attachment of astrocytes and neurons to arrays of silicon pillars, although this mechanism is not completely understood. For smaller feature sizes electron beam lithography is used to create feature sizes down to tens of nanometers for applications in neuronal cell patterning as shown by the Cornell and the Glasgow groups [31, 32].

Topographical patterning was also used in a different way: cells and neurites are immobilized in deep structures rather than modulating the cytoskeleton by imposing mechanical restrictions on the plasma membrane. This approach is based on observations that the mechanical forces generated by the cells will move or rearrange the neurites during the culture period. Structures with a high-aspect ratio can be realized by lithography and structuring methods developed for the production of Micro-Electro-Mechanical Systems (MEMS). Maher et al. have used this approach to grow neuronal cells in deep pits on a silicon substrate, and recorded data from them using metal microelectrodes [33]. Griscom et al. explored a three-dimensional (3D) microfluidic array to influence cell placement and neural guidance. The complex 3D high-aspect ratio structures of poly-dimethylsiloxane (PDMS) were made directly on structured silicon wafers and using EPON SU-8 negative photoresist [34].

Recently, Merz and Fromherz have built on the results of such experiments, and have obtained well-defined networks of cultured neurons from the pond snail *Lymnaea stagnalis* by growing them in a microstructured polyester photoresist (SU-8) on a silicon substrate [35]. By applying electrophysiological techniques, they studied pairs of nerve cells that had formed connections, which were identified to be electrical synapses.

5.3.2
Photolithographic Patterning

Photolithographic techniques are well established for mass production of silicon chips with a resolution and alignment precision in the sub-μm range. The pattern in the photoresist, which is generated by light exposure through a mask followed by chemical development, can be transferred into thin films of molecules immobilized on a surface [15]. However, organic solvents and alkaline solutions used in the process may influence the stability of functional molecules. Clark et al. [36] showed that photolithography could be used to pattern laminin, which guided neurite outgrowth similarly well as other methods [14]. Standard photoresist techniques have been adapted to generate micropatterns of proteins on glass by using lift-off and plasma-etching techniques [37, 38].

5.3.3
Photochemical Patterning

Photochemical patterning can be used to pattern self-assembled monolayers (SAM) or thin films of organic molecules by exposing the surface to UV light through either a photomask or a metal mask. Usually, illumination with UV light causes oxidation of the molecules in the exposed areas (for example alkanethiolate oxidize to alkanesulfonate), and this alters the properties of the organic molecules – that is, their solubility. By immersing the patterned substrate in a solution with another organic molecule, the illuminated region can be modified with a second monolayer [39]. This method has been used successfully for the adhesion and growth of rat hippocampal neurons on circuit-like patterns employing μm features [40]. SAMs of silanes on glass have been used in combination with deep-UV photopatterning: trimethoxysilylpropyldiethynenetriamine (DETA) supports cell adhesion and outgrowth; and (tridecafluoro-1,1,2,2-tetrahydrooctyl)-1-1dimethylchlorosilane (13F) is cell repellent [41]. It was shown that networks of neurons grown on DETA patterns against a background of 13F develop normal chemical synapses in culture [42]. Photochemical patterning in combination with versatile crosslinking chemistry and tailored peptides has proved to be very effective in patterning the growth of hippocampal neurons [43]. The patterning was achieved by applying UV-photomasking technique and the chemically attachment of a synthetic peptide derived from a neurite-outgrowth-promoting domain of the B2 chain of laminin. The attachment was carried out by coupling the peptide to an amine-derived glass surface using a heterobifunctional crosslinker.

5.3.4
Microcontact Printing

Microcontact printing (μCP) uses an elastomeric stamp to create patterns of organic molecules on surfaces, and was initially developed by Whitesides group to print patterns of monolayers of alkanethiols onto gold substrates [44]. The method allows the patterning of surfaces with biomolecules, and has been studied extensively in the context of biosensors and high-throughput bioassays (see Chapter 3). Application of this technique to the investigation of cell–substrate interactions [45] has mainly focused on endothelial cell adhesion and the control of neuronal process outgrowth for the creation of defined neuronal networks [46–51], as shown in Figure 5.3.

The procedure starts with a photolithography step to produce the mold (master stamp). The patterning of neuronal networks requires high-relief stamps which can be realized, for example, by photoresists with a high-aspect ratio and a thickness of more than 5 μm. The elastomeric stamp is prepared by casting PDMS against the patterned photoresist. The PDMS stamp is inked with a solution of organic molecules, dried, and placed in contact with a surface. The organic molecules are transferred only at those regions where the stamp contacts the surface. The patterned surface can be in the range of several cm^2 in size, and the features can have an edge resolution in the sub-μm range. When a pattern of cell-attracting components is stamped onto a background material that repels cell adhesion, the attachment and outgrowth of cells – for example neurons – is restricted to the regions where transfer took place, confining the geometry of the forming networks. The cells in such networks have been shown to be interconnected by chemical synapses allowing signal transduction along the predefined pathways (Figure 5.4) [51a].

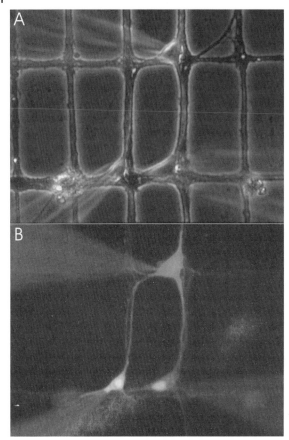

Figure 5.3 Network of rat embryonic cortical neurons grown on a pattern of ECM proteins created by microcontact printing onto polystyrene. (A) Phase-contrast microscopy. (B) The connectivity of the cells is visualized by microinjection of three different fluorescent dyes into three cells in the course of a patch–clamp measurement. (Figures provided courtesy of A. K. Vogt.)

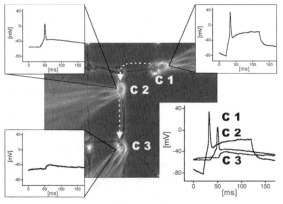

Figure 5.4 Signal transduction in a geometrically confined network of rat embryonic neurons. The cells were cultivated on a pattern of ECM proteins against a background of polystyrene. Synaptic connectivity was observed through patch–clamp measurements. A signal evoked experimentally in cell 1 (C1) traveled through two synapses via cell 2 to cell 3, indicating a simple functional network. (Figures provided courtesy of A. K. Vogt.)

5.4
Outlook

> Neuro from the nerves, the silver paths. Romancer. Neuromancer. (...) "I met Neuro-
> mancer. He talked about your mother. I think he's something like a giant ROM
> construct, for recording personality, only it's full RAM. The constructs think
> they're there, like it's real, but it just goes on forever." (...)
> Case chewed his lower lip and grazed out across the plateaus of the Eastern Seaboard
> Fission Authority, into the infinite neuroelectronic void of the matrix.
>
> William Gibson Neuromancer

Bioelectronic interfacing is a topic that inspires and fascinates not only science-fiction wri-
ters and movie makers. Indeed, the possibilities arising from an interweaving of neuronal
networks with microelectronics – a "marriage of biological systems with technology" – are
probably many more than we can envision to date. Cell-based hybrids as biosensors, neuronal prostheses, neuroelectronic circuits and artificial intelligence are only the first issues
we are aiming at.

The cultivation of neurons on field effect transistors for extracellular stimulation and
signal recording – one of the requirements for such applications – is already possible,
as shown in Figure 5.5. Being able to precisely pattern neuronal networks is another
scientific advance that takes us a step towards these goals. In order to communicate reliably with a silicon chip, cell adhesion must be confined precisely to defined areas on the
chip, as must be the pathways of connectivity [52–54]. However, control over polarity and
synapse formation is required for the controlled design of networks of neuronal cells. The
potential applications for such neuron–chip systems are ambitious – for example, artificial
photoreceptors which can be implanted into an irreversibly damaged eye and are able to
communicate directly with the optical nerve. Other ideas for neuronal prostheses include
the bridging of damaged sections of the spinal cord by neurosilicon chips, or the targeting

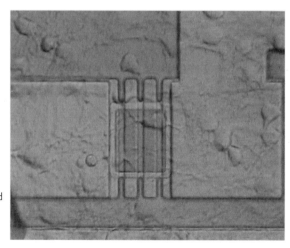

Figure 5.5 Brainstem neurons
cultured for 7 days on a laminin-coated
triple gate structure for extracellular
recordings. (Figure provided courtesy
of S. Ingebrandt.)

of neurite outgrowth after injury, specifically to re-innervate damaged tissue. Apart from medical applications, neuronal networks on silicon are thought to be implementable into neurocomputers to support data processing directly, without the need to unravel the principles underlying neuronal signal transduction.

On the other hand, exactly this unraveling of neuronal information processing is perhaps the most fascinating problem addressed by researchers working with patterned neurons. As simplified systems with defined and manipulatable geometry, these networks may teach us much about the impact of connectivity on the input and output of a network, about the way neurons integrate incoming signals, and how different parameters influence the transmission, routing, and processing of such signals. The rules underlying these actions, which lead to the selective amplification of some signals and concurrent attenuation of others, may be regarded as the neuronal language or code. Deciphering this code is of fundamental interest in diverse scientific disciplines, since it seems to be the basis for many abilities of the brain, such as learning, associative recognition, and memory.

Attempts to model the function of the human brain with computers have been made already. The term "neuronal network" also stands for algorithms that are supposed to enable a computer to learn for example to recognize and associate patterns in the way a person would when they are "fed" with the proper teaching material. The abilities of such a system of course are limited to very particular types of tasks, and they are also rather sensitive with respect to the correct training experiments [55]. Implementing new findings on principles of neuronal signal integration and processing into such algorithms may help us to refine them, ultimately creating computers with abilities that were so far restricted to humans, such as associative memory or creativity.

References

[1] Nicholls, J. C., Martin, A. R., Wallace, B. G., *From Neuron to Brain*, 3rd edition, Sinauer Associates, Sunderland, Mass., **1992**.

[2] Hucho, F., Weise, C., *Angew. Chem. – Int. Edit.* **2001**, *40*, 3101–3116.

[3] Kandel, E., Schwartz, J., Jessell, T., *Principles of Neural Science*; 3rd edition, Prentice-Hall International, Inc., London, **1991**.

[4] Rosenberg, M. D., *Science* **1963**, *139*, 411–412.

[5] Curtis, A., Wilkinson, C., *Biomaterials* **1997**, *18*, 1573–1583.

[6] Curtis, A. S. G., Wilkinson, C. D., *J. Biomater. Sci.-Polymer Ed.* **1998**, *9*, 1313–1329.

[7] Jung, D. R., Kapur, R., Adams, T., Giuliano, K. A., Mrksich, M., Craighead, H. G., Taylor, D. L., *Crit. Rev. Biotechnol.* **2001**, *21*, 111–154.

[8] Clark, P., Connolly, P., Curtis, A. S. G., Dow, J. A. T., Wilkinson, C. D. W., *Development* **1990**, *108*, 635–644.

[9] Britland, S., Morgan, H., Wojiak-Stodart, B., Riehle, M., Curtis, A., Wilkinson, C., *Exp. Cell Res.* **1996**, *228*, 313–325.

[10] Carter, S. B., *Nature* **1965**, *208*, 1183–1187.

[11] Letourneau, P. C., *Dev. Biol.* **1975**, *44*, 92–101.

[12] Manthrope, M., Engvall, E., Ruoslahti, E., Longo, F. M., Davis, G. E., Varon, S., *J. Cell Biol.* **1983**, *97*, 1882–1890.

[13] Ruegg, U. T., Hefti, F., *Neurosci. Lett.* **1984**, *49*, 319–324.

[14] Hammarback, J. A., Palm, S. L., Furcht, L. T., Letourneau, P. C., *J. Neurosci. Res.* **1985**, *13*, 213–220.

[15] Kleinfeld, D., Kahler, K. H., Hockberger, P. E., *J. Neurosci.* **1988**, *8*, 4098–4120.

[16] Kleinman, H. K., Ogle, R. C., Cannon, F. B., Little, C. D., Sweeney, T. M., *Proc. Natl. Acad. Sci. USA* **1988**, *85*, 1282–1286.

[17] Kleinman, H. K., Sephel, G. C., Tashiro, K.-I., Weeks, B. S., Burrows, B. A., Adler, S. H., Yamada, Y., Martin, G. R., *Ann. N. Y. Acad. Sci.* **1990**, *580*, 302–310.

[18] Lewandowska, K., Balachander, N., Sukenik, C. N., Culp, L. A., *J. Cell. Physiol.* **1989**, *141*, 334–345.

[19] Ruoslahti, E., Pierschbacher, M. D., *Cell* **1986**, *44*, 517–518.

[20] Ruoslahti, E., Pierschbacher, M. D., *Science* **1987**, *238*, 491–497.

[21] Graf, J., Iwamoto, Y., Sasaki, M., Martin, G. R., Kleinman, H. K., Robey, F. A., Yamada, Y., *Cell* **1987**, *48*, 989–996.

[22] Curtis, A. S. G., Breckenridge, L., Connolly, P., Dow, J. A. T., Wildinson, C. D. W., Wilson, R., *Med. Biol. Eng. Comput.* **1992**, *30*, CE33–CE36.

[23] Fromherz, P., Schaden, H., Vetter, T., *Neurosci. Lett.* **1991**, *129*, 77–80.

[24] Fromherz, P., Schaden, H., *Eur. J. Neurosci.* **1994**, *6*, 1500–1504.

[25] Valentini, R. F., Vargo, T. G., Gardella, J. A., Aebischer, P., *J. Biomater. Sci.-Polymer Ed.* **1993**, *5*, 13–36.

[26] Ranieri, J. P., Bellamkonda, R., Bekos, E. J., Gardella, J. A., Mathieu, H. J., Ruiz, L., Aebischer, P., *Int. J. Dev. Neurosci.* **1994**, *12*, 725–735.

[27] Corey, J. M., Wheeler, B. C., Brewer, G. J., *J. Neurosci. Res.* **1991**, *30*, 300–307.

[28] Bohanon, T., Elender, G., Knoll, W., Koberle, P., Lee, J. S., Offenhäusser, A., Ringsdorf, H., Sackmann, E., Simon, J., Tovar, G., Winnik, F. M., *J. Biomater. Sci.-Polymer Ed.* **1996**, *8*, 19–39.

[29] Sze, S. M., *Semiconductor devices: Physics and Technology*, John Wiley Sons, Inc., New York, **1985**.

[30] Craighead, H. G., Turner, S. W., Davis, R. C., James, C., Perez, A. M., John, P. M. S., Isaacson, M. S., Kam, L., Shain, W., Turner, J. N., Banker, G., *Biomed. Microdev.* **1998**, *1*, 49–64.

[31] Turner, S., Kam, L., Isaacson, M., Craighead, H. G., Shain, W., Turner, J., *J. Vac. Sci. Technol. B* **1997**, *15*, 2848–2854.

[32] Wilkinson, C. D. W., Curtis, A. S. G., Crossan, J., *J. Vac. Sci. Technol. B* **1998**, *16*, 3132–3136.

[33] Maher, M. P., Dvorak-Carbone, H., Pine, J., Wright, J. A., Tai, Y. C., *Med. Biol. Eng. Comput.* **1999**, *37*, 110–118.

[34] Griscom, L., Degenaar, P., LePioufle, B., Tamiya, E., Fujita, H., *Jpn. J. Appl. Phys.* **2001**, *40*, 5485–5490.

[35] Merz, M., Fromherz, P., *Adv. Mater.* **2002**, *14*, 141–144.

[36] Clark, P., Britland, S., Connolly, P., *J. Cell Sci.* **1993**, *105*, 203–212.

[37] Tai, H. C., Buettner, H. M., *Biotechnol. Prog.* **1998**, *14*, 364–370.

[38] Sorribas, H., Padeste, C., Tiefenauer, L., *Biomaterials* **2002**, *23*, 893–900.

[39] Dulcey, C. S., Georger, J. H., Krauthamer, V., Stenger, D. A., Fare, T. L., Calvert, J. M., *Science* **1991**, *252*, 551–554.

[40] Ravenscroft, M. S., Bateman, K. E., Shaffer, K. M., Schessler, H. M., Jung, D. R., Schneider, T. W., Montgomery, C. B., Custer, T. L., Schaffner, A. E., Liu, Q. Y., Li, Y. X., Barker, J. L., Hickman, J. J., *J. Am. Chem. Soc.* **1998**, *120*, 12169–12177.

[41] Stenger, D. A., Pike, C. J., Hickman, J. J., Cotman, C. W., *Brain Res.* **1993**, *630*, 136–147.

[42] Ma, W., Liu, Q. Y., Jung, D., Manos, P., Pancrazio, J. J., Schaffner, A. E., Barker, J. L., Stenger, D. A., *Dev. Brain Res.* **1998**, *111*, 231–243.

[43] Matsuzawa, M., Liesi, P., Knoll, W., *J. Neurosci. Methods* **1996**, *69*, 189–196.

[44] Kumar, A., Whitesides, G. M., *Appl. Phys. Lett.* **1993**, *63*, 2002–2004.

[45] Singhvi, R., Kumar, A., Lopez, G. P., Stephanopoulos, G. N., Wang, D. I. C., Whitesides, G. M., Ingber, D. E., *Science* **1994**, *264*, 696–698.

[46] Branch, D. W., Corey, J. M., Weyhenmeyer, J. A., Brewer, G. J., Wheeler, B. C., *Med. Biol. Eng. Comput.* **1998**, *36*, 135–141.

[47] Branch, D. W., Wheeler, B. C., Brewer, G. J., Leckband, D. E., *IEEE Trans. Biomed. Eng.* **2000**, *47*, 290–300.

[48] James, C. D., Davis, R. C., Kam, L., Craighead, H. G., Isaacson, M., Turner, J. N., Shain, W., *Langmuir* **1998**, *14*, 741–744.

[49] Scholl, M., Sprossler, C., Denyer, M., Krause, M., Nakajima, K., Maelicke, A., Knoll, W., Offenhausser, A., *J. Neurosci. Methods* **2000**, *104*, 65–75.

[50] Yeung, C. K., Lauer, L., Offenhausser, A., Knoll, W., *Neurosci. Lett.* **2001**, *301*, 147–150.

[51] Kam, L., Shain, W., Turner, J. N., Bizios, R., *Biomaterials* **2001**, *22*, 1049–1054.

[51a] Vogt, A. K., Lauer, L., Knoll, W. Oftenhäusser, A., *Biotechnol. Prog.* **2003**, *19*, 1562–1568.

[52] Zeck, G., Fromherz, P., *Proc. Natl. Acad. Sci. USA* **2001**, *98*, 10457–10462.

[53] Chang, J. C., Brewer, G. J., Wheeler, B. C., *Biosens. Bioelectron.* **2001**, *16*, 527–533.

[54] Lauer, L., Ingebrandt, S., Scholl, M., Offenhausser, A., *IEEE Trans. Biomed. Eng.* **2001**, *48*, 838–842.

[55] Gasteiger, J., Zupan, J., *Angew. Chem. – Int. Edit.* **1993**, *32*, 503–527.

6
S-Layers

Uwe B. Sleytr, Eva-Maria Egelseer, Dietmar Pum, and *Bernhard Schuster*

6.1
Overview

The fabrication of supramolecular structures and devices requires molecules that are capable of interlocking in a predictable, well-defined manner. Thus, molecular self- assembly systems which exploit the molecular-scale manufacturing precision of biological systems are prime candidates for supramolecular engineering. Although self-assembly of molecules is an ubiquitous strategy of morphogenesis in nature, in molecular nanotechnology these unique features of molecules are not yet fully exploited for the functionalization of surfaces and interfaces and for hierarchical self-assembly systems as required for the production of biomimetic membranes and encapsulating systems.

Crystalline bacterial cell-surface layers (S-layers) have been optimized during billions of years of biological evolution as one of the simplest biological membranes [1–3]. S-layers are composed of a single protein or glycoprotein species endowed with the ability to assemble into monomolecular arrays on the supporting cell envelope component of prokaryotic organisms (bacteria and archaea). The wealth of information accumulated on the structure, chemistry, assembly, genetics, and function of S-layers has led to a broad spectrum of applications for life and material sciences [3–7].

Abbreviations

Bet v1   major birch pollen allergen
BLMs   bilayer lipid membranes
cAB   camel antibody sequence recognizing lysozyme
CdS   cadmium sulfide
DNA   deoxyribonucleic acid
H_2S hydrogen sulfide
IgE   immunoglobulin E
IL-8   interleukin 8
MFMs   microfiltration membranes
MPL   main phospholipid of *Thermoplasma acidophilum*

Nanobiotechnology. Edited by Christof Niemeyer, Chad Mirkin
Copyright © 2004 WILEY-VCH Verlag GmbH & Co. KgaA, Weinheim
ISBN 3-527-30658-7

rSbpA   recombinant S-layer protein of *Bacillus sphaericus* CCM 2177
SbpA   S-layer protein of *Bacillus sphaericus* CCM 2177
SbsB   S-layer protein of *Geobacillus stearothermophilus* PV72/p2
SbsC   S-layer protein of *Geobacillus stearothermophilus* ATCC 12980
SCWP   secondary cell wall polymer
S-layer   surface layer
SLH   S-layer homology domain
SPR   surface plasmon resonance
SUMs   S-layer ultrafiltration membranes
t-PA   tissue type plasminogen activator

6.1.1
Chemistry and Structure

With few exceptions, S-layers are composed of a single homogeneous protein or glycoprotein species with molecular weights ranging from 40 to 200 kDa (Table 6.1). The results of amino acid analysis of various S-layer proteins and the secondary structure estimated by protein sequence data and circular dichroism measurements on S-layer proteins are summarized in Table 6.1. Few posttranslational modifications are known to occur in S-layer proteins, including cleavage of amino- or carboxy-terminal fragments, phosphorylation, and glycosylation of amino acid residues (Table 6.1) [3, 7]. The latter is a remarkable characteristic of many archaeal and some bacterial S-layer proteins, and in this way the glycan chains and linkages differ significantly from those of eukaryotes [3, 8–10].

Electron microscopy studies on the mass distribution of the lattices were generally per-

Table 6.1 Properties of S-layers

- The relative molecular mass of constituent subunits in the range of 40 kDa to 200 kDa
- These are weakly acidic proteins (pI ~4–6), except *Methanothermus fervidus* (pI = 8.4) and lactobacilli (pI > 9.5)
- Large amounts of glutamic acid, aspartic acid (~15 mol.%) and hydrophobic amino acids (~40–60 mol.%), and a high lysine content (~10 mol.%)
- Hydrophilic and hydrophobic amino acids do not form extended clusters
- No or low content of sulfur-containing amino acids
- In most S-layer proteins, ~20% of the amino acids are organized as α-helix, and about 40% occur as β-sheets
- Aperiodic foldings and β-turn content may vary between 5 and 45%
- S-layer lattices can have oblique (p1, p2), square (p4), or hexagonal (p3, p6) symmetry
- The center-to-center spacing of the morphological unit can range from 3 nm to 35 nm
- The lattices are generally 5 nm to 20 nm thick (in archaea, up to ~70 nm)
- S-layer lattices exhibit pores of identical size and morphology
- The pore sizes range from approximately 2 nm to 8 nm
- In many S-layers, two or even more distinct classes of pores are present
- The pores can occupy 30–70% of the surface area
- The outer surface is generally less corrugated than the inner surface
- Posttranslational modifications of S-proteins include: (i) cleavage of N- or C-terminal fragments; (ii) glycosylation; and (iii) phosphorylation of amino acid residues.

formed on negatively stained preparations or unstained, thin, frozen foils (Figure 6.1a). Two- and three-dimensional analysis, including computer image enhancement, revealed structural information down to a range of 0.5–1.5 nm (Figure 6.1b) [11–14]. High-resolution images of the surface topography of S-layers were also obtained using underwater atomic force microscopy (Figure 6.1c) [3, 15–17]. A common feature of S-layers is their smooth outer surface and more corrugated inner surface.

The proteinaceous subunits of S-layers can be aligned in lattices with oblique, square, or hexagonal symmetry (Figure 6.1d) with center-to-center spacing of the morphological units of between 3 and 35 nm. Hexagonal lattice symmetry is predominant among archaea [18, 19]. S-layers are very porous membranes, with pores occupying between 30 and 70 % of their surface area (see Table 6.1). Since S-layers are in most cases assemblies of identical subunits, they exhibit pores of identical size and morphology. However, in many protein lattices two or more distinct classes of pores with diameters in the range of 2 to 8 nm have been identified [19–21].

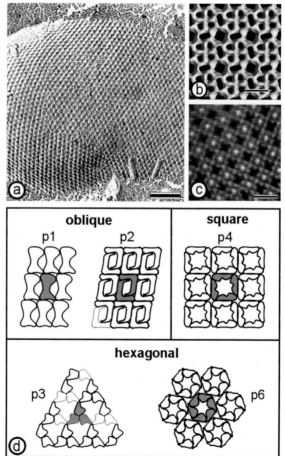

Figure 6.1 (a) Freeze-etching preparation of whole cells of *Thermoanaerobacter thermohydrosulfuricus* L111-69 revealing a hexagonally ordered array. Scale bar = 100nm. (b) Three-dimensional model of the S-layer of *Bacillus stearothermophilus* NRS 2004/3a/V2 exhibiting oblique lattice symmetry. The protein meshwork shows one square-shaped, two elongated, and four small pores per morphological unit. (c) Computer image reconstruction of scanning force microscopic images of the topography of the square S-layer lattice from *Bacillus sphaericus* CCM 2177. The images were taken under water. The surface corrugation corresponding to a gray scale tram black to white is 1.8 nm. Scale bars in (b) and (c) = 10 nm. (d) Schematic drawing of the different S-layer lattice types. The regular arrays exhibit either oblique (p1, p2), square (p4), or hexagonal lattice symmetry (p3, p6). The morphological units are composed of one, two, three, four, or six identical subunits. (Reproduced from Ref. [3], with permission from Wiley-VCH.)

In both Gram-positive bacteria and archaea, the lattice assembles on the surface of the wall matrix (e. g., peptidoglycan or pseudomurein), whereas in Gram-negative bacteria the S-layer is attached to components of the outer membrane (e. g., lipopolysaccharides). In most archaea the S-layer represents the exclusive cell-wall component external to the cytoplasmic membrane.

6.1.2
Genetics and Secondary Cell-Wall Polymers

During the past decade, numerous S-layer genes from organisms of quite different taxonomic affiliations have been cloned and sequenced [1, 7, 22, 23]. Considering the frequently highly competitive situation of closely related organisms in their natural habitats, it is obvious that the S-layer surface must contribute to diversification rather than to conservation. This can be achieved by S-layer variation leading to the expression of different types of S-layer genes, or to the recombination of partial coding sequences. S-layer variation was studied in detail for *Campylobacter fetus*, an important pathogen for humans and ungulates [24, 25], but was also observed for nonpathogens such as *Geobacillus stearothermophilus* [26–28]. Although it was proposed for several years that sequence identities among S-layer proteins are extremely rare, or do not even exist, it is now apparent that high sequence identities are limited to the N-terminal region that is responsible for anchoring the protein to the cell surface by binding to an accessory secondary cell-wall polymer (SCWP), and which is covalently linked to the peptidoglycan backbone. In this context, three repeats of S-layer homology (SLH) motifs, consisting of 50–60 amino acids each [29], have been identified at the N-terminal part of many S-layer proteins [22]. If present, SLH motifs are involved in SCWP-mediated anchoring of the S-layer protein to the peptidoglycan layer [22, 30–37]. During the past few years, a considerable amount of information on the chemical composition and structure of SCWPs from different organisms has been accumulated [8, 30, 33, 38–40], indicating a highly specific lectin-type recognition mechanism between the S-layer protein and a distinct type of SCWP. In a recent study, the interaction of the S-layer protein SbsB of *G. stearothermophilus* PV72/p2 and the corresponding SCWP was assessed by surface plasmon resonance (SPR) biosensor technology [41]. By using two truncated forms consisting either of the three SLH motifs or the residual part of SbsB, the exclusive and complete responsibility of a functional domain formed by the three SLH motifs of the S-layer protein SbsB for SCWP recognition was clearly confirmed. The interaction proved to be highly specific for the carbohydrate component, and strong evidence for glycan pyruvylation was provided [41]. In contrast to most S-layer proteins of Gram-positive bacteria, those of *G. stearothermophilus* wild-type strains [34, 42] and *Lactobacillus* [31, 43] do not possess SLH-motifs. Nevertheless, the N-terminal part of *G. stearothermophilus* wild-type strains is highly conserved and recognizes a net negatively charged SCWP as the proper binding site [31, 34]. The production of different truncated forms of the S-layer protein SbsC of *G. stearothermophilus* ATCC 12980 confirmed that the N-terminal part is exclusively responsible for cell-wall binding, but this positively charged segment is not involved in the self-assembly process [35] and seems to fold independently of the remainder of the protein sequence.

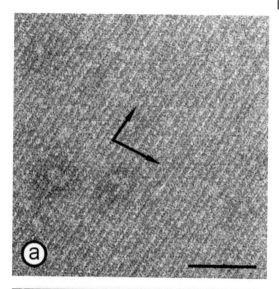

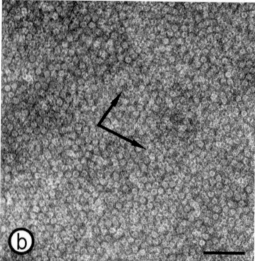

Figure 6.2 Cell wall fragments carrying a chimeric S-layer formed by the fusion protein BS1(S1)$_3$ (a) were capable of binding biotinylated ferritin (b). At BS1(S1)$_3$, one core streptavidin is fused to the C-terminus of the S-layer protein SbsB of *Geobacillus stearothermophilus* PV72/p2. The proteins were refolded to heterotetramers consisting of one chain of fusion protein and three chains of streptavidin. (a) Self-assembly was enabled by the specific interaction between an accessory cell-wall polymer that is part of the cell wall of *G. stearothermophilus* PV72/p2, and the SLH-domain of the fusion protein. (b) Bound biotinylated ferritin reflected the underlying S-layer lattice. The preparations were negatively stained with uranyl acetate for TEM. The arrows indicate the base vectors of the oblique p1 lattice; scale bars = 100 nm. (c) The cartoon shows the orientation of BS1(S1)$_3$ after SLH-enabled self-assembly with the streptavidin carrying outer face of the S-layer exposed. (Reproduced from Ref. [45]; copyright (2002) National Academy of Sciences, USA.)

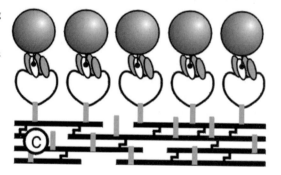

In order to determine at which amino acid positions of the S-layer proteins foreign peptide sequences could be fused without interfering with the self-assembly and recrystallization properties, the structure–function relationship of distinct segments of different S-layer proteins had to be elucidated. In the case of the S-layer protein SbpA of *Bacillus sphaericus* CCM 2177, it could be demonstrated that the C-terminal end of the full-length form of recombinant rSbpA (rSbpA$_{31-1268}$) was only available to a limited extent, but was fully accessible in the C-terminally truncated form rSbpA$_{31-1068}$ [37]. Based on these results, the C-terminally truncated form was exploited as base form for the construction of further S-layer fusions proteins, incorporating either the major birch pollen allergen Bet v1 (rSbpA$_{31-1068}$ /Bet v1) or a camel antibody sequence recognizing lysozyme as an epitope (rSbpA$_{31-1068}$ /cAB) [37, 44]. Owing to the versatile applications of the streptavidin–biotin interaction as a biomolecular coupling system, minimum-sized core-streptavidin (118 amino acids) was fused either to N-terminal positions of the S-layer protein SbsB or attached to the C-terminus of this S-layer protein (Figure 6.2) [45]. The fusion proteins and core-streptavidin were produced independently in *Escherichia coli*, isolated and refolded to heterotetramers consisting of one chain of fusion protein and three chains of streptavidin. As determined by a fluorescence titration method, the biotin binding capacity of the heterotetramers was 80 % in comparison to homotetrameric streptavidin, indicating that at least three of the four core streptavidin residues were accessible and active. Due to the ability of the heterotetramers to recrystallize in suspension, on liposomes, and on silicon wafers, this chimeric S-layer can be used as self-assembling nanopatterned molecular affinity matrix to arrange biotinylated compounds on a surface (Figure 6.2) [45].

6.1.3
Assembly

A complete solubilization of S-layers composed of native or recombinant proteins into their constituent subunits can generally be achieved with high concentrations of hydrogen bond-breaking agents (e. g., guanidine hydrochloride). In summarizing the results from different disintegration procedures, it was concluded that: (i) in general, bacterial S-layer proteins are not covalently linked to each other or the supporting cell wall component; (ii) different combinations of weak bonds (hydrophobic bonds, ionic bonds, and hydrogen bonds) are responsible for the structural integrity of S-layers; and (iii) bonds holding the S-layer subunits together are stronger than those binding the S-layer lattices to the underlying envelope layer or membrane [5, 6, 46, 47].

6.1.3.1 **Self-Assembly in Suspension**
S-layers isolated from a broad spectrum of prokaryotic organism have shown the inherent ability to reassemble into two-dimensional arrays after removal of the disrupting agent used in the dissolution procedure (Figure 6.3). High-resolution electron microscopical studies in combination with digital image processing have shown that crystal growth is initiated simultaneously at many randomly distributed nucleation points and proceeds in-plane until the crystalline domains meet, thus leading to a closed, coherent mosaic of individual several micrometer large S-layer domains [48–50]. Most important for ap-

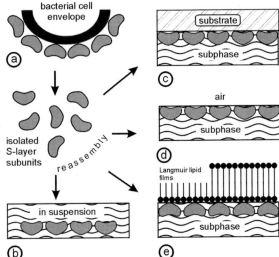

Figure 6.3 (a) Schematic illustration of the recrystallization of isolated S-layer subunits into crystalline arrays. The self assembly process can occur in suspension (b), on solid supports (c), at the air/water interface (d), and on Langmuir lipid films (e). (Reproduced from Ref. [3], with permission from Wiley-VCH.)

plied S-layer research, the formation of these self-assembled arrays is only determined by the amino-acid sequence of the polypeptide chains and, consequently, the tertiary structure of the S-layer protein species [51, 52]. The self-assembly products may have the form of flat sheets, open-ended cylinders or closed vesicles [46, 53, 54]. The shape and size of the self-assembly products depends strongly on the environmental parameters during crystallization such as temperature, pH, ion composition, and/or ionic strength.

6.1.3.2 Recrystallization at Solid Supports

Reassembly of isolated S-layer proteins into larger crystalline arrays can be also induced on solid surfaces. In particular, the recrystallization of S-layer proteins on technologically relevant substrates such as silicon wafers (Figure 6.4), carbon-, platinum- or gold electrodes and on synthetic polymers already revealed a broad application potential for the crystalline arrays in micro- and nanotechnology [14, 48, 55, 56]. The formation of coherent crystalline arrays depends strongly on the S-layer protein species, the environmental conditions of the bulk phase and, in particular, on the surface properties of the substrate.

6.1.3.3 Recrystallization at the Air/Water Interface and on Langmuir Lipid Films

Reassembly of isolated S-layer subunits at the air/water interface and on Langmuir–Blodgett lipid films (see below) has proven to be an easy and reproducible way to generate coherent S-layer lattices on a large scale. In accordance with S-layers recrystallized on solid surfaces the orientation of the protein arrays at liquid interfaces is determined by the anisotropy in the physico-chemical surface properties of the protein lattice. Electron microscopical examinations revealed that recrystallized S-layers were oriented with their outer charge neutral, more hydrophobic face against the air/water interface and with their negatively charged, more hydrophilic inner face against charge neutral, charged or zwitterionic headgroups of phospho- or tetraether lipid films [57]. As with S-layer lattices recrys-

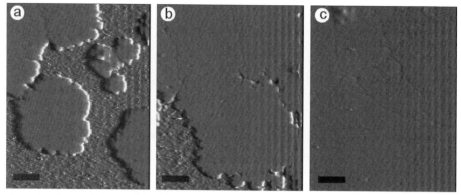

Figure 6.4 Recrystallization of the S-layer protein SbpA of *Bacillus sphaericus* CCM 2177 on a hydrophilic silicon wafer. The atomic force microscopical images show that crystal growth is initiated simultaneously at many randomly distributed nucleation points (a) and proceeds in-plane until the crystalline domains meet (b), thus leading to a closed, coherent mosaic of individual several micrometer large S-layer domains (c). Scale bars = 0.5 μm; Z-range = 12 nm. (Figure courtesy of E. Györvary and O. Stein.)

tallized on solid surfaces, S-layer protein monolayers consist of a closed mosaic of individual monocrystalline domains.

6.2
Methods

6.2.1
Diagnostics

Studies on the structure, morphogenesis, genetics, and function of S-layers revealed that these isoporous monomolecular arrays have a considerable application potential in biotechnology, molecular nanotechnology, and biomimetics. The repetitive features of S-layers have led to their applications in the production of S-layer ultrafiltration membranes (SUMs), as supports for a defined covalent attachment of functional molecules (e. g., enzymes, antibodies, antigens, protein A, biotin, and avidin) as required for affinity and enzyme membranes, in the development of solid-phase immunoassays, or in biosensors [3, 7, 22, 58, 59].

In dipstick-style solid-phase immunoassays, the respective monoclonal antibody was covalently bound to the carbodiimide-activated carboxylic acid groups of the S-layer lattice [60]. Proof of principle was demonstrated for different types of SUM-based dipsticks. For example, for the diagnosis of type I allergies (determination of IgE in whole blood or serum against the major birch pollen allergen Bet v1), for quantification of tissue type plasminogen activator (t-PA) in patients' whole blood or plasma for monitoring t-PA levels during the course of thrombolytic therapies after myocardial infarction, or for determination of interleukin 8 (IL-8) in the supernatants of human umbilical vein endothelial cells induced with lipopolysaccharides [7, 61, 62].

Alternative or complementary to existing S-layer technologies, genetic approaches are currently used for the construction of chimeric S-layer fusion proteins incorporating biologically active sequences without hindering the self-assembly of S-layer subunits into regular arrays on surfaces and in suspension. In the chimeric S-layer proteins $rSbsC_{31-920}$/Bet v1 and $rSbpA_{31-1068}$/Bet v1 carrying the major birch pollen allergen Bet v1 at the C-terminal end, the surface location and functionality of the fused allergen was demonstrated by binding Bet v1-specific IgE [37, 63]. These fusion proteins can be used for building up arrays for diagnostic test systems to determine the concentration of Bet v1-specific IgE in patients' whole blood, plasma, or serum samples [62]. In order to build up functional monomolecular S-layer protein lattices on solid supports (e. g., gold, silicon, or glass), the surface must be functionalized with covalently attached chemically modified SCWP, to which the S-layer fusion proteins bind with their N-terminal part, leaving the C-terminal part with the fused functional sequence exposed to the ambient environment. Owing to the versatile applications of the streptavidin–biotin interaction as a biomolecular coupling system, S-layer-streptavidin fusion proteins were constructed [45]. The two-dimensional protein lattices displayed streptavidin in defined repetitive spacing, and proved to be capable of binding biotin and also biotinylated functional molecules (see Figure 6.2). Thus, the chimeric S-layer can be seen as a feasible tool to arrange different biotinylated targets (e. g., proteins, allergens, antibodies, or oligonucleotides) on a surface which will find application in protein, allergy, or DNA-chip technology. Furthermore, chimeric S-layers recrystallized on solid supports with a defined orientation are also expected to be a key element in the rational design of highly integrated diagnostic devices (Lab-on-Chip). Another application potential can be seen in the development of label-free detection systems [44]. In the SPR or surface acoustic wave technique, specific binding of functional molecules (e. g., proteins or antibodies) to the sensor chip functionalized with an oriented chimeric S-layer can be visualized directly by a mass increase on the chip without the need for any labeled compound.

To conclude, such supramolecular biomimetic structures consisting of a functional S-layer fusion protein recrystallized in defined orientation on SCWP-coated solid supports allow the development of new label-free detection systems as required for biochip technology.

6.2.2
Lipid Chips

Since it became evident that typically free-standing bilayer lipid membranes (BLMs) survive for only minutes to hours and are very sensitive toward vibration and mechanical shocks [64–66], stabilization of BLMs is imperatively necessary to utilize the function of cell membrane components for practical applications (e. g., as lipid chips). S-layer proteins can be exploited as supporting structures for BLMs (Figure 6.5) since they stabilize the lipid film and largely retain their physical features (e. g., thickness, fluidity) [57].

In the following section the most promising methods to attach lipid membranes on porous or solid supports in order to generate attractive lipid chips and membrane protein-based devices are described. In general, lipid membranes attached to a porous support combine the advantage of possessing an essentially unlimited ionic reservoir on each side of the lipid membrane and of easy manual handling. A new strategy is the application of an SUM with the S-layer as stabilizing and biochemical layer between the BLM and the porous support. SUMs are isoporous structures with very sharp molecular exclusion limits and were manufactured by depositing S-layer-carrying cell wall fragments under high pressure on commercial microfiltration membranes (MFMs) with an average pore size of approximately 0.4 μm [67, 68]. After deposition, the S-layer lattices are chemically crosslinked to form a coherent smooth surface ideally suited for depositing lipid membranes.

Composite SUM-supported bilayers (Figure 6.5C) are tight structures with breakdown voltages well above 500 mV during their whole life-time of ~8 hours [69]. For a comparison, lipid membranes on a plain nylon MFM revealed a life-time of about 3 hours, and ruptured at breakdown voltages of ~210 mV. Specific capacitance measurements and reconstitution experiments revealed functional lipid membranes on the SUM as the pore-forming protein α-hemolysin could be reconstituted to form lytic channels. For the first time, the opening and closing behavior of even single α-hemolysin pores (see also

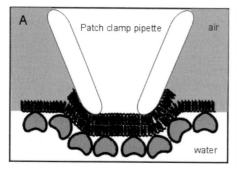
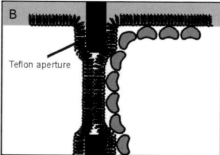
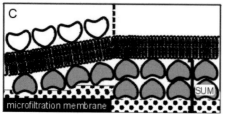
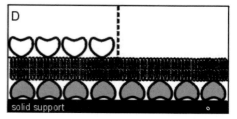

Figure 6.5 Schematic illustrations of various S-layer-supported lipid membranes. (A) Bilayer lipid membranes (BLMs) have been generated across an aperture of a patch–clamp pipette using the Tip-Dip method, and a closed S-layer has been recrystallized from the aqueous subphase. (B) A folded membrane has been generated to span a Teflon aperture using the method of Montal and Mueller [71]. Subsequently, S-layer protein can be injected into one or both compartments (not shown), whereby the protein self-assembles to form closely attached S-layer lattices on the BLMs. (C) On an S-layer ultrafiltration membrane (SUM) a BLM can be generated by a modified Langmuir–Blodgett (LB) technique. As a further option, a closed S-layer lattice can be attached on the external side of the SUM-supported BLM (left part). (D) Solid supports can be covered by a closed S-layer lattice, and subsequently BLMs can be generated using combinations of the LB and Langmuir–Schaefer techniques, and vesicle fusion. As shown in (C), a closed S-layer lattice can be recrystallized on the external side of the solid supported BLM (left part).

Chapter 7) could be measured with membranes generated on a porous support [69]. The main phospholipid of *Thermoplasma acidophilum* (MPL), a membrane-spanning tetraether lipid, has also been transferred on an SUM using a modified Langmuir–Blodgett technique [70, 71]. Again, SUM-supported MPL-membranes allowed reconstitution of functional molecules, as proven by measurements on single gramicidin pores. Recrystallization of an additional monomolecular S-layer protein lattice on the lipid-faced side of SUM-supported MPL membranes increased the lifetime significantly to 21.2 ± 3.1 hours [70].

Solid-supported membranes (Figure 6.5D) were developed in order to overcome the fragility of free-standing BLMs, and also to enable biofunctionalization of inorganic solids (e. g., semiconductors, gold-covered surfaces) for the use in sensing devices such as lipid chips [72, 73]. Various types of solid-supported lipid membranes often show considerable drawbacks as there is a limited ionic reservoir at the side facing the solid support, the membranes often appear to be leaky (noninsulating), and large domains, protruding from the membrane, may become denatured by the inorganic support [57, 74–78]. Again, S-layer proteins have been studied to elucidate their potential as stabilizing and separating ultrathin layer, which maintains also the structural and dynamic properties of the lipid membranes. Silicon substrates have been covered by a closed S-layer lattice and bilayers were deposited by the Langmuir–Blodgett technique [79–81]. Lateral diffusion of fluorescently labeled lipid molecules in both layers have been investigated by fluorescence recovery after photobleaching studies [82]. In comparison with hybrid lipid bilayers (lipid monolayer on alkylsilanes) and lipid bilayers on dextran, the mobility of lipids was highest in S-layer-supported bilayers. Most importantly, the S-layer cover could prevent the formation of cracks and other inhomogenities in the bilayer [82]. These results have demonstrated that the biomimetic approach of copying the supramolecular architecture of archaeal cell envelopes opens new possibilities for exploiting functional lipid membranes at meso- and macroscopic scale. Moreover, this technology has the potential to initiate a broad spectrum of lipid chips applicable for sensor technology, diagnostics, electronic or optical devices, and high-throughput screening for drug discovery.

6.2.3
S-Layers as Templates for the Formation of Regularly Arranged Nanoparticles

The reproducible formation of nanoparticle arrays in large scale with predefined lattice spacing and symmetries remains a challenge in the development of future generations of molecular electronic devices (see also Chapter 19). This is particularly true for the realization of self-assembly and bottom-up approaches, as these strategies acquire the highest efficiency in a fabrication process. Biomolecular templating has proven to be very attractive, as the self-assembly of molecules into monomolecular arrays is an intrinsic property of many biological molecules and has already grown into a scientific and engineering discipline crossing the boundaries of several established fields (see also Chapters 16 and 17).

The first approach in using S-layers as templates in the generation of perfectly ordered nanoparticle arrays was developed by Douglas and coworkers [55]. S-layer fragments of *Sulfolobus acidocaldarius* were deposited on a smooth carbon surface and metal coated by evaporation of a ~1 nm-thick tantalum/tungsten film. Subsequently, this protein–

metal heterostructure was ion milled, leading to 15 nm-sized holes hexagonally arranged according to the center-to-center spacing of the S-layer of 22 nm. Later on, this approach was further optimized using fragments of the same S-layer species on a smooth graphite surface and titanium oxide for the metal coating [56]. After oxidation in air and fast-atom beam milling at normal incidence, a thin (~3.5 nm) metallic nanoporous mask with pores in the 10 nm range was obtained. The same group used low-energy electron-enhanced etching to pattern the surface properties of a silicon substrate through the regularly arranged pores of the S-layer [83]. After etching and removal of the S-layer, the patterned surface was oxidized in an oxygen plasma, leading to a nanometric array of etched holes (18 nm diameter) which served as nucleation sites in the formation of an ordered array of nanometric titanium metal clusters. In a similar approach using argon ion etching in the final step, the S-layer of *Deinococcus radiodurans* was used as a nanometric template for patterning ferromagnetic films [84]. Uniform hexagonal patterns of 10 nm-wide dots and lattice spacing of 18 nm were fabricated from 2.5 nm-thick sputter-coated Co, FeCo, Fe, FeNi, and NiFe films.

More recently, a synthesis pathway for the fabrication of nanoparticles by wet chemical processes and S-layers as nanometric templates was developed [85–88]. In this approach, self-assembled S-layer structures were exposed to a metal–salt solution (e. g., $[AuCl_4]^-$, $[PtCl_4]^{2-}$), followed by slow reaction with a reducing agent such as hydrogen sulfide (H_2S). Nanoparticle superlattices were formed according to the lattice spacing and symmetry of the underlying S-layer. Furthermore, since the precipitation of the metals was confined to the pores of the S-layer, the nanoparticles also resembled the morphology of the pores. The first example exploiting this technique was the precipitation of cadmium sulfide (CdS) on S-layer lattices composed of SbsB and SbpA [85]. After incubation of the S-layer self-assembly products with a $CdCl_2$ solution for several hours, the hydrated samples were exposed towards H_2S for at least one or two days. The generated CdS nanoparticles were 4–5 nm in size, and their superlattice resembled the oblique lattice symmetry of SbsB (a = 9.4 nm, b = 7.4 nm, γ = 80°), or the square lattice symmetry of SbpA (a = b = 13.1 nm, γ = 90°), respectively. In a similar approach, a superlattice of 4–5 nm-sized gold particles was formed by using SbpA (with previously induced thiol groups) as a template for the precipitation of a tetrachloroauric (III) acid solution [86] (Figure 6.6a). Gold nanoparticles were formed either by reduction of the metal salt with H_2S or under the electron beam in a transmission electron microscope. The latter approach is technologically important as it allows those areas where nanoparticles are formed to be defined. As determined by electron diffraction, the gold nanoparticles were crystalline but their ensemble was not crystallographically aligned. The wet chemical approach was used in the formation of Pd- (salt: $PdCl_2$), Ni- ($NiSO_4$), Pt- ($KPtCl_6$), Pb- ($Pb(NO_3)_2$) and Fe- ($KFe(CN)_6$) nanoparticle arrays (unpublished results), and for producing platinum nanoparticles on the S-layer of *Sporosarcina ureae* [87, 88].

Unfortunately, wet chemical methods do not allow varying size or composition of nanoparticles in the fabrication process. Thus, the binding of preformed nanoparticles into regular arrays on S-layers has significant advantages in the development of nanoscale electronic devices. Based on the studies of binding biomolecules (e. g. enzymes or antibodies) onto S-layers, it has already been demonstrated that gold or CdSe nanoparticles can be electrostatically bound in regular arrangements on S-layers [89–91] (Figure 6.6b). The

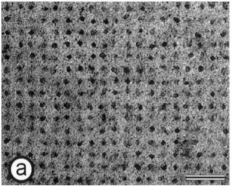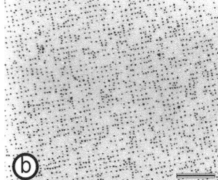

Figure 6.6 (a) Electron microscopical image of gold nanoparticles (mean diameter 4.5 nm) obtained using wet chemistry. An S-layer with square lattice symmetry served as template in the precipitation of the metal salt. The gold nanoparticles were formed in the pore region of the protein meshwork under the electron beam. Scale bar = 50 nm. (b) Electron microscopical image of preformed gold nanoparticles (mean diameter 4 nm) regularly bound on the surface of an S-layer with square lattice symmetry. Electrostatic interactions between the surface of the nanoparticles and functional domains on the S-layer are responsible for the binding. Scale bar = 100 nm. (Reproduced from Ref. [91], with permission from Elsevier.)

nanoparticles were either negatively charged due to surface citrate ions or positively charged due to surface coating with poly-l-lysine.

In summary, these experiments have clearly shown that S-layers are perfectly suited to control the formation of nanoparticle arrays, either by direct precipitation from the vapor or liquid phase, or by binding preformed nanoparticles. The S-layer approach provides for the first time a biologically based fabrication technology for the self-assembly of molecular electronic or optic devices.

6.3
Outlook

At present, most applications developed for using S-layers depend on the in vitro self-assembly capabilities of native S-layer proteins in suspension, on the surface of solids (e. g., silicon wafers, metals, polymers), Langmuir-lipid films, and liposomes. Once the regular arrays have been formed, a broad spectrum of very precise chemical modifications can be applied for tailoring the physico-chemical properties of S-layers and for a defined binding of differently sized functional molecules. In particular, the possibility of immobilizing or growing other materials (e. g., silicon oxide, metals) on top of recrystallized S-layer lattices with most accurate spatial controlled architecture opens up many new possibilities in nanofabrication and supramolecular engineering [6, 7].

An important line of development for the specific tuning of structural and functional features concerns the genetic manipulation of S-layer proteins. Recent studies have clearly demonstrated that truncated S-layer proteins incorporating specific functional domains of other proteins maintain the self-assembly capability into regular arrays [5, 35]. This approach can lead to new isoporous ultrafiltration membranes, affinity structures, enzyme membranes, ion- and metal particle-selective binding matrices, microcarriers, biosensors, diagnostics, biocompatible surfaces, and vaccines [37, 44, 45, 63, 92].

Moreover, biomimetic approaches copying the supramolecular principle of virus envelopes such as S-layer-coated liposomes will provide new strategies for drug targeting and drug delivery. Preliminary studies have also provided strong evidence that S-layers have a great potential as patterning elements for non-life science applications (e.g., nonlinear optics and molecular electronics) [90].

References

[1] U. B. Sleytr, T. J. Beveridge, *Trends Microbiol.* **1999**, *7*, 253–260.

[2] U. B. Sleytr, P. Messner, D. Pum, M. Sára, *Mol. Microbiol.* **1993**, *10*, 911–916.

[3] U. B. Sleytr, P. Messner, D. Pum, M. Sára, *Angew. Chem. Int. Ed.* **1999**, *38*, 1034–1054.

[4] U. B. Sleytr, M. Sára, *Trends Biotechnol.* **1997**, *15*, 20–26.

[5] U. B. Sleytr, M. Sára, D. Pum, B. Schuster, *Prog. Surf. Sci.* **2001**, *68*, 231–278.

[6] U. B. Sleytr, M. Sára, D. Pum, B. Schuster, in: M. Rosoff (ed.), *Nano-Surface Chemistry*, Marcel Dekker, New York , Basel, **2001**, pp. 333–389.

[7] U. B. Sleytr, M. Sára, D. Pum, B. Schuster, P. Messner, C. Schäffer, in: A. Steinbüchel, S. Fahnestock (eds), *Biopolymers*, Vol. 7, Wiley-VCH, Weinheim, Germany, **2003**, pp. 285–338.

[8] C. Schäffer, P. Messner, in: W. Herz, H. Falk, G. W. Kirby (eds), *Prokaryotic Glycoproteins*, Springer, Wien, New York, **2003**, pp. 51–124.

[9] P. Messner, G. Allmaier, C. Schäffer, T. Wugeditsch, S. Lortal, H. König, R. Niemetz, M. Dorner, *FEMS Microbiol. Rev.* **1997**, *20*, 25–46.

[10] M. Sumper, F. T. Wieland, in: J. Montreuil, J. F. G. Vliegenthart, H. Schachter (eds), *Glycoproteins*, Elsevier, Amsterdam, **1995**, pp. 455–473.

[11] W. Baumeister, G. Lembcke, R. Dürr, B. Phipps, in: J. R. Fryer, D. L. Dorset (eds), *Electron Crystallography of Organic Molecules*, Kluwer, Dordrecht, **1990**, pp. 283–296.

[12] W. Baumeister, G. Lembcke, *J. Bioenerg. Biomembr.* **1992**, *24*, 567–575.

[13] T. J. Beveridge, *Curr. Opin. Struct. Biol.* **1994**, *4*, 204–212.

[14] D. Pum, U. B. Sleytr, *Trends Biotechnol.* **1999**, *17*, 8–12.

[15] F. Ohnesorge, W. M. Heckl, W. Häberle, D. Pum, M. Sára, H. Schindler, K. Schilcher, A. Kiener, D. P. E. Smith, U. B. Sleytr, G. Binnig, *Ultramicroscopy* **1992**, *42*, 1236–1242.

[16] S. Karrasch, R. Hegerl, J. Hoh, W. Baumeister, A. Engel, *Proc. Natl. Acad. Sci. USA* **1994**, *91*, 836–838.

[17] D. J. Müller, W. Baumeister, A. Engel, *J. Bacteriol.* **1996**, *178*, 3025–3030.

[18] H. König, *Can. J. Microbiol.* **1988**, *34*, 395–406.

[19] U. B. Sleytr, P. Messner, D. Pum, M. Sára, Appendix, in: U. B. Sleytr, P. Messner, D. Pum, M. Sára (eds), *Crystalline Bacterial Cell Surface Proteins*. Landes/Academic Press, Austin, TX, **1996**, pp. 211–225.

[20] W. Baumeister, I. Wildhaber, B. M. Phipps, *Can. J. Microbiol.* **1989**, *35*, 215–227.

[21] S. Hovmöller, A. Sjögren, D. N. Wang, *Prog. Biophys. Mol. Biol.* **1988**, *51*, 131–163.

[22] M. Sára, U. B. Sleytr, *J. Bacteriol.* **2000**, *182*, 859–868.

[23] E. Akca, H. Claus, N. Schultz, G. Karbach, B. Schlott, T. Debaerdemaeker, J. P. Declercq, H. König, *Extremophiles* **2002**, *6*, 351–358.

[24] J. Dworkin, M. J. Blaser, *Mol. Microbiol.* **1997**, *26*, 433–440.

[25] S. A. Thompson, M. J. Blaser, in: I. Nachamkin, M. J. Blaser (eds), *Campylobacter*, ASM Press, Washington, DC, **2000**, pp. 321–347.

[26] M. Sára, B. Kuen, H. F. Mayer, F. Mandl, K. C. Schuster, U. B. Sleytr, *J. Bacteriol.* **1996**, *178*, 2108–2117.

[27] E. M. Egelseer, T. Danhorn, M. Pleschberger, C. Hotzy, U. B. Sleytr, M. Sára, *Arch. Microbiol.* **2001**, *177*, 70–80.

[28] H. C. Scholz, E. Riedmann, A. Witte, W. Lubitz, B. Kuen, *J. Bacteriol.* **2001**, *183*, 1672–1679.

[29] A. Lupas, H. Engelhardt, J. Peters, U. Santarius, S. Volker, W. Baumeister, *J. Bacteriol.* **1994**, *176*, 1224–1233.

[30] W. Ries, C. Hotzy, I. Schocher, U. B. Sleytr, M. Sára, *J. Bacteriol.* **1997**, *179*, 3892–3898.

[31] E. M. Egelseer, K. Leitner, M. Jarosch, C. Hotzy, S. Zayni, U. B. Sleytr, M. Sára, *J. Bacteriol.* **1998**, *180*, 1488–1495.

[32] M. Sára, C. Dekitsch, H. F. Mayer, E. M. Egelseer, U. B. Sleytr, *J. Bacteriol.* **1998**, *180*, 4146–4153.

[33] N. Ilk, P. Kosma, M. Puchberger, E. M. Egelseer, H. F. Mayer, U. B. Sleytr, M. Sára, *J. Bacteriol.* **1999**, *181*, 7643–7646.

[34] M. Jarosch, E. M. Egelseer, D. Mattanovich, U. B. Sleytr, M. Sára, *Microbiology.* **2000**, *146*, 273–281.

[35] M. Jarosch, E. M. Egelseer, C. Huber, D. Moll, D. Mattanovich, U. B. Sleytr, M. Sára, *Microbiology.* **2001**, *147*, 1353–1363.

[36] M. Sára, *Trends Microbiol.* **2001**, *9*, 47–49.

[37] N. Ilk, C. Völlenkle, E. M. Egelseer, A. Breitwieser, U. B. Sleytr, M. Sára, *Appl. Environ. Microbiol.* **2002**, *68*, 3251–3260.

[38] C. Schäffer, H. Kählig, R. Christian, G. Schulz, S. Zayni, P. Messner, *Microbiology.* **1999**, *145*, 1575–1583.

[39] C. Schäffer, N. Müller, P. K. Mandal, R. Christian, S. Zayni, P. Messner, *Glycoconjug. J.* **2000**, *17*, 681–690.

[40] C. Steindl, C. Schäffer, T. Wugeditsch, M. Graninger, I. Matecko, N. Müller, P. Messner, *Biochem. J.* **2002**, *368*, 483–494.

[41] C. Mader, D. Moll, C. Hotzy, C. Huber, U. B. Sleytr, M. Sára, *Biochemistry* submitted.

[42] B. Kuen, U. B. Sleytr, W. Lubitz, *Gene.* **1994**, *145*, 115–120.

[43] H. J. Boot, C. P. Kolen, P. H. Pouwels, *Mol. Microbiol.* **1996**, *21*, 799–809.

[44] M. Pleschberger, A. Neubauer, E. M. Egelseer, S. Weigert, B. Lindner, U. B. Sleytr, S. Muyldermans, M. Sára, *Bioconjug. Chem.* **2003**, *14*, 440–448.

[45] D. Moll, C. Huber, B. Schlegel, D. Pum, U. B. Sleytr, M. Sára, *Proc. Natl. Acad. Sci. USA* **2002**, *99*, 14646–14651.

[46] U. B. Sleytr, P. Messner, in: H. Plattner (ed.), *Electron Microscopy of Subcellular Dynamics*, CRC Press, Boca Raton, Florida, **1989**, pp. 13–31.

[47] U. B. Sleytr, P. Messner, *Annu. Rev. Microbiol.* **1983**, *37*, 311–339.

[48] D. Pum, U. B. Sleytr, in: Sleytr, U. B., P. Messner, D. Pum, M. Sára (eds), *Crystalline Bacterial Cell Surface Layer Proteins (S-Layers)*, Academic Press, R. G. Landes Company, Austin, USA, **1996**, pp.175–209.

[49] D. Pum, U. B. Sleytr, *Supramol. Science* **1995**, *2*, 193–197.

[50] D. Pum, M. Weinhandl, C. Hödl, U. B. Sleytr, *J. Bacteriol.* **1993**, *175*, 2762–2766.

[51] U. B. Sleytr, *Int. Rev. Cytol.* **1978**, *53*, 1–64.

[52] U. B. Sleytr, *Nature* **1975**, *257*, 400–402.

[53] P. Messner, D. Pum, U. B. Sleytr, *J. Ultrastruct. Mol. Struct. Res.* **1986**, *97*, 73–88.

[54] U. B. Sleytr, R. Plohberger, in: W. Baumeister, W. Vogell (eds), *Electron Microscopy at Molecular Dimensions*, Springer-Verlag, Berlin, Heidelberg, New York, **1980**, pp. 36–47.

[55] K. Douglas, N. A. Clark, *Appl. Phys. Lett.* **1986**, *48*, 676–678.

[56] K. Douglas, G. Devaud, N. A. Clark, *Science* **1992**, *257*, 642–644.

[57] B. Schuster, U. B. Sleytr, *Rev. Mol. Biotechnol.* **2000**, *74*, 233–254.

[58] M. Sára, U. B. Sleytr, *Appl. Microbiol. Biotechnol.* **1989**, *30*, 184–189.

[59] U. B. Sleytr, M. Sára, D. Pum, in: A. Ciferri (ed.), *Supramolecular Polymers*, Marcel Dekker, Inc., New York, Basel, **2000**, pp. 177–213.

[60] A. Breitwieser, S. Küpcü, S. Howorka, S. Weigert, C. Langer, K. Hoffmann-Sommergruber, O. Scheiner, U. B. Sleytr, M. Sára, *Biotechniques* **1996**, *21*, 918–925.

[61] U. B. Sleytr, H. Bayley, M. Sára, A. Breitwieser, S. Küpcü, C. Mader, S. Weigert, F. M. Unger, P. Messner, B. Jahn-Schmid, B. Schuster, D. Pum, K. Douglas, N. A. Clark, J. T. Moore, T. A. Winningham, S. Levy, I. Frithsen, J. Pankovc, P. Beale, H. P. Gillis, D. A. Choutov, K. P. Martin, *FEMS Microbiol. Rev.* **1997**, *20*, 151–175.

[62] A. Breitwieser, C. Mader, I. Schocher, K. Hoffmann-Sommergruber, W. Aberer,

O. Scheiner, U. B. Sleytr, M. Sára, *Allergy* **1998**, *53*, 786–793.

[63] A. Breitwieser, E. M. Egelseer, D. Moll, N. Ilk, C. Hotzy, B. Bohle, C. Ebner, U. B. Sleytr, M. Sára, *Protein Eng.* **2002**, *15*, 243–249.

[64] T. H. Tien, A. L. Ottova, *J. Membr. Sci.* **2001**, *189*, 83–117.

[65] B. Raguse, V. Braach-Maksvytis, B. A. Cornell, L. G. King, P. D. J. Osman, R. J. Pace, L. Wieczorek, *Langmuir* **1998**, *14*, 648–659.

[66] M. Zviman, H. T. Tien, *Biosens. Bioelectron.* **1991**, *6*, 37–42.

[67] M. Sára, U. B. Sleytr, *J. Bacteriol.* **1987**, *169*, 2804–2809.

[68] S. Weigert, M. Sára, *J. Membrane Sci.* **1995**, *106*, 147–159.

[69] B. Schuster, D. Pum, M. Sára, O. Braha, H. Bayley, U. B. Sleytr, *Langmuir* **2001**, *17*, 499–503.

[70] B. Schuster, S. Weigert, D. Pum, M. Sára, U. B. Sleytr, *Langmuir* **2003**, *19*, in press.

[71] M. Montal, P. Mueller, *Proc. Natl. Acad. Sci. USA* **1972**, *69*, 3561–3566.

[72] E. Sackmann, M. Tanaka, *Trends Biotechnol.* **2000**, *18*, 58–64.

[73] B. A. Cornell, V. L. Braach-Maksvytis, L. G. King, P. D. Osman, B. Raguse, L. Wieczorek, R. J. Pace, *Nature* **1997**, *387*, 580–583.

[74] W. Knoll, C. W. Frank, C. Heibel, R. Naumann, A. Offenhäusser, J. Rühe, E. K. Schmidt, W. W. Shen, A. Sinner, *Rev. Mol. Biotechnol.* **2000**, *74*, 137–158.

[75] D. P. Nikolelis, T. Hianik, U. J. Krull, *Electroanalysis* **1999**, *11*, 7–15.

[76] S. Heyse, T. Stora, E. Schmid, J. H. Lakely, H. Vogel, *Biochim. Biophys. Acta* **1998**, *1376*, 319–338.

[77] A. L. Plant, *Langmuir* **1993**, *9*, 2764–2767.

[78] E. Kalb, S. Frey, L. K. Tamm, *Biochim. Biophys. Acta* **1992**, *1103*, 307–316.

[79] A. Zasadzinski, R. Viswanathan, L. Madson, J. Garnaes, K. D. Schwartz, *Science* **1994**, *263*, 1726–1733.

[80] I. Langmuir, V. J. Schaefer, *J. Am. Chem. Soc.* **1937**, *59*, 1406–1417.

[81] K. J. Blodgett, *J. Am. Chem. Soc.* **1935**, *57*, 1007–1022.

[82] E. Györvary, B. Wetzer, U. B. Sleytr, A. Sinner, A. Offenhäusser, W. Knoll, *Langmuir* **1999**, *15*, 1337–1347.

[83] T. A. Winningham, H. P. Gillis, D. A. Choutov, K. P. Martin, J. T. Moore, K. Douglas, *Surf. Sci.* **1998**, *406*, 221–228.

[84] M. Panhorst, H. Brückl, B. Kiefer, G. Reiss, U. Santarius, R., Guckenberger, *J. Vac. Sci. Technol. B* **2001**, *19*, 722–724.

[85] W. Shenton, D. Pum, U. B. Sleytr, S. Mann, *Nature* **1997**, *389*, 585–587.

[86] S. Dieluweit, D. Pum, U. B. Sleytr, *Supramol. Science* **1998**, *5*, 15–19.

[87] M. Mertig, R. Kirsch, W. Pompe, H. Engelhardt, *Eur. Phys. J.* **1999**, *9*, 45–48.

[88] W. Pompe, M. Mertig, R. Kirsch, R. Wahl, L. C. Ciachi, J. Richter, R. Seidel, H. Vinzelberg, *Z. Metallkd.* **1999**, *90*, 1085–1091.

[89] S. R. Hall, W. Shenton, H. Engelhardt, S. Mann, *Chem. Phys. Chem.* **2001**, *3*, 184–186.

[90] E. Györvary, A. Schroedter, D. V. Talapin, H. Weller, D. Pum, U. B. Sleytr, *J. Nanosci. Nanotechnol.* submitted.

[91] U. B. Sleytr, E. Györvary, D. Pum, *Prog. Organic Coatings* **2003**, in press.

[92] V. Weber, S. Weigert, M. Sára, U. B. Sleytr, D. Falkenhagen, *Therapeutic Apheresis* **2001**, *5*, 433–438.

7
Engineered Nanopores

Hagan Bayley, Orit Braha, Stephen Cheley, and Li-Qun Gu

7.1
Overview

Engineered nanopores have potential applications in many areas of technology, including separation science, sensing, and drug delivery. In this chapter, we focus on recent developments regarding pores that span membranes and have diameters of 10 nm or less. We emphasize pores into which new properties have been engineered, highlighting our own recent investigations into α-hemolysin.

7.1.1
What is a Nanopore?

Nanopores occur in nature, and in the biological literature they are simply known as pores when they are more than 1 nm or so in diameter, or channels when they are narrower. For certain applications in technology, biological pores are being engineered directly. In other cases, new types of pores are being constructed that are based on biological structures [1]. The biological pores that concern us here are transmembrane proteins. Cells also contain soluble pores, such as chaperonins and enzymes that handle nucleic acids, which may be of interest in the future in the context of engineering catalysis into nanopores. Transmembrane proteins fall into two main classes: helix bundles, and β barrels [2]. In general, helix bundles constitute various classes of receptors and channels. Natural examples of these proteins are relatively difficult to engineer: (i) because the helices are usually in close contact and alterations to the side chains disrupt the structure of the protein; and (ii) because they have evolved highly specialized functions, which tend to be difficult to change. Nevertheless, some of the earliest attempts to build membrane proteins de novo involved transmembrane helices. For example, bundles of individual synthetic helices were conceived by the groups of DeGrado and Lear [3], and template-assembled synthetic proteins (TASP) were used initially by Montal, Vogel and colleagues (Figure 7.1a) [4]. Some progress has been made on de novo single-chain bundles [5, 6], but as yet this work lacks a solid structural basis. By contrast, β barrels are open structures, permitting dramatic alterations of natural scaffolds [7]. So far, the β–barrel fold has been found in all the outer membrane

Nanobiotechnology. Edited by Christof Niemeyer, Chad Mirkin
Copyright © 2004 WILEY-VCH Verlag GmbH & Co. KgaA, Weinheim
ISBN 3-527-30658-7

proteins of Gram-negative bacteria that have been examined, as well as in certain pore-forming toxins. Importantly, the side chains that project into a β barrel do not interact strongly with one another, and can be altered to manipulate the properties of the pores. Recently, progress has been made on barrel-like structures formed from helices [8, 9], and we can expect more work in this area in the future.

Among the natural channels that have been engineered as nanopores, the most prominent have been porins, proteins that control the permeability of the bacterial outer membrane, and α–hemolysin (αHL), a pore-forming toxin secreted by *Staphylococcus aureus*. Work on both these systems has been aided by high-resolution crystal structures (Figure 7.1b,c). At this point, the engineering of porins has been limited, but they do constitute a promising system [10–12]. The αHL pore is a heptameric structure (Figure 7.1c) [13]. The transmembrane β barrel is topped by a cap domain that contains a large internal cavity. Both the cap and barrel have been engineered by various means (see below). Our laboratory has developed methods for producing pure heteromeric pores, most usefully of the form WT_6ENG_1 with just one altered subunit. The engineered subunit (ENG) is chemically or genetically tagged to confer an altered charge. The wild-type (WT) and engineered subunits are then co-assembled and the desired heteromer isolated by preparative sodium dodecyl sulfate-polyacrylamide gel electrophoresis (SDS-PAGE) [14]. αHL belongs to a family of toxins, which includes the leukocidins. The latter form pores comprising two

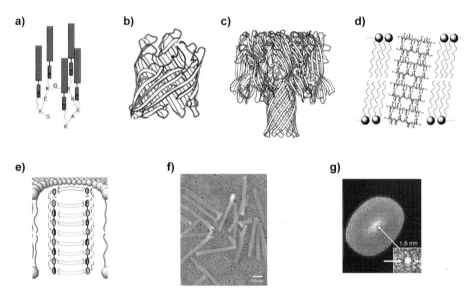

Figure 7.1 Examples of various classes of nanopore. (a) Helix bundle in the form of a template-assembled synthetic protein (TASP) [116]. (b) A single subunit of the trimeric porin OmpF (inner diameter ~2 nm) [117]. (c) The α-hemolysin pore (inner diameter of barrel ~2 nm) [13]. (d) Nanotube formed from stacked cyclic peptides (inner diameter ~0.75 nm) [16]. (e) Synthetic barrel based on octiphenyl staves with attached peptides (inner diameter ~1.5 nm) [20]. (f) Electron micrograph of silica nanotubes made in a porous alumina template (inner diameter, e.g., ~20 nm). The template was subsequently dissolved [1]. (g) Electron micrograph of a nanopore in silicon nitride sculpted with an Ar ion beam (inner diameter ~2 nm) [27].

different, but related, polypeptide chains. We have shown that one leukocidin pore is an octamer, rather than a heptamer, containing four subunits of each class [15]. The fact that both heptamers and octamers exist suggests that it may be possible to control diameter in this family of pores by engineering the subunit–subunit interface.

Designed pores containing β structure have been developed by the groups of Ghadiri and Matile. The Ghadiri group pioneered this area by introducing stacked cyclic peptides containing alternating D- and L-amino acids, which form nanotubes that span lipid bilayers (Figure 7.1d) [16]. The structures resemble gramicidin, a naturally occurring peptide antibiotic that forms transmembrane pores from two head-to-head β spirals. The nanotubes can be stabilized by stapling them together in a metathesis reaction between olefinic amino acid side chains [17]. The diameter of the nanotubes can be adjusted by altering the ring size of the peptides [18], and β-amino acids can form similar structures [19]. Matile's group has developed synthetic barrels from octiphenyl staves to which all L-amino acid peptides are attached (Figure 7.1e) [20]. In principle, the structure of the octiphenyl barrels is highly manipulatable. The diameter can be changed by altering the length of the attached peptides, and the number of staves (which depends upon the steric and electrostatic interactions between the peptides) [21]. The length of the barrel can be controlled by adjusting the length of the stave [22]. Finally, heteromeric pores containing two types of staves have been assembled [21]. While limited structural information has been obtained on Ghadiri's nanotubes [23], the structures of the octiphenyl barrels are based on model building and a substantial body of indirect evidence.

Several groups have made nanopores from materials with no connection to biology [1]. For example, commercial track-etched polymer membranes are made by bombarding thin films with nuclear fission particles. The damage tracks are then etched, with base for example in the case of polycarbonate, to form uniform pores as small as 10 nm in diameter, which may be several μm in length. To make smaller pores of far shorter effective length, asymmetric etching techniques have been developed to generate cone-shaped structures [24]. For example, poly(ethylene terephthalate) through which a single high-energy Au ion had passed was treated with concentrated sodium hydroxide from one side, with a stop solution of formic acid on the other side. The membrane was in an electrical cell so that breakthrough could be registered and etching stopped at that point. Further, the applied potential was such that hydroxide ion moved away from the etch site after breakthrough (electro-stopping). Individual pores formed in this way exhibit many of the properties of biological pores, such as ion selectivity, and open and closed states [25].

The chemically etched pores will probably be difficult to make and derivatize in a reproducible way. Fortunately, Martin's group has developed sophisticated approaches for generating nanotubules with modifiable internal surfaces within the pores of track-etched polycarbonate filters. For example, a gold film is deposited on the surface of the pores by electroless plating [26]. This serves both to control the diameter of the nanotubules, which can be made as small as 1 nm, and to provide a reactive surface (see below).

Individual nanopores have also been made by using a low-energy beam of Ar ions to "sculpt" a hole in a thin layer of silicon nitride [27]. Surprisingly, it was demonstrated that the beam can close as well as open a hole, depending on the beam flux and duty cycle and the sample temperature. Because the size of the hole could be monitored during sculpting by counting the transmitted Ar ions, it was possible to produce nanopores of known diameter by feedback control of the beam, and in this way diameters below 1.8 nm were obtained (Figure 7.1g).

7.1.2
Engineering Nanopores

Protein pores are engineered by using mutagenesis and targeted chemical modification. The primary interest lies in engineering the interior (lumen) of the pores. Mutagenesis can be used to introduce any of the twenty natural amino acids, which provide a variety of side chains of differing size, shape, polarity, and reactivity. Non-natural amino acid mutagenesis can also be used to provide side chains that do not occur in nature, such as ketones, alkenes, and azides. Originally, non-natural amino acids were introduced by using preloaded suppressor tRNAs in an in-vitro translation system, a technology that is sufficiently demanding to prevent its routine use. Currently, *Escherichia coli* strains are under development that provide the mutant suppressor tRNAs and the engineered synthetases required to load them. Indeed, it is even possible to have the bacteria make an amino acid ! [28]. In targeted chemical modification, an amino acid side chain (usually cysteine) is selectively modified with a reagent. This again adds to the diversity of functionalities that can be incorporated into a protein. Given a high-resolution structure, these approaches to protein engineering allow modifications to be placed at well-defined locations. If more than one modification is made, the geometrical relationship between the new side chains is known.

Our group recently implemented a new concept in protein engineering – that of noncovalent modification. Specifically, cyclodextrins and other host molecules (adapters) were allowed to bind to positions within the lumen of the transmembrane barrel of the αHL pore that could be defined by mutagenesis (Figure 7.2) [29–32]. The adapters remain capable of binding guests and therefore noncovalent modification can introduce new binding sites within the pore. The vast literature on host–guest interactions supports the enormous potential of this approach.

Designed pores containing β structure, such as those of Ghadiri and Matile, can be engineered using chemical synthesis with a variety of amino acids, and of course there is no requirement for using the natural side chains. The side chains in Ghadiri's nanotubes all project outwards, which makes it difficult to engineer function, although progress has been made by using functionalized peptides to cap the structures [33]. In Matile's synthetic barrels, just as in a natural β barrel, the side chains of the amino acids project alternately outwards and into the lumen, permitting the introduction of functional groups within the pore. However, although it is in principle possible, the exquisite precision of protein engineering, in which individual changes are made at specific positions, has not yet been achieved with the synthetic barrels. At present, for example, 32 or 64 histidine residues are introduced at one time.

As expected, the internal surfaces of Martin's gold-plated nanotubules are amenable to derivatization with a variety of thiols that vary in size, shape, and polarity [34]. Interestingly, the surfaces of the gold-plated nanotubules can also be charged electrostatically by using an applied potential [35]. Recently, the idea of electromodulation has been de-

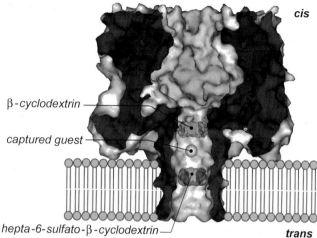

cis

β-*cyclodextrin*

captured guest

hepta-6-sulfato-β-cyclodextrin

trans

Figure 7.2 α-Hemolysin pore modified with noncovalent adapters. The illustration is based on the superposition of structures of αHL and cyclodextrins. The positions of the cyclodextrins were surmised from mutagenesis experiments. In the structure shown, sites for two different cyclodextrins were engineered. An organic molecule can be trapped in the cavity between the two sites [31].

monstrated in which electrostatic charging is used to attract an amphiphilic surfactant to the wall of the nanotubule and thereby render the surface hydrophobic [36]. By using silane chemistry, proteins have been attached to the internal surface of another class of nanotubules made of silica in an alumina template (see Figure 7.1f) [37]. Asymmetrically etched poly(ethylene terephthalate) can also be chemically modified. For example, the free carboxyl groups can be neutralized by methylation with diazomethane [25] and additional chemistry that would alter surface properties can readily be conceived. Nevertheless, in these cases, the inability to control the number of modified sites or organize them in space remains a severe drawback.

7.1.3
What Can a Nanopore Do?

In Arthur Karlin's succinct words: "Ion channels open, conduct ions selectively, and close" (Figure 7.3) [38]. The nanopores we are considering also conduct ions, and even though the diameters of the pores are larger than the typical ion channel, they do exhibit at least modest ion selectivity. Importantly, because they are large they can also transport large molecules across a membrane or bilayer. Natural pores, such as gap junctions, can gate (open and close in response to a stimulus), but so far little attention has been given to this issue with engineered nanopores. More important has been the generation of sites within the pores at which blockers can bind (Figure 7.3d), by analogy with drugs such as verapamil, a Ca-channel blocker. These properties of pores can be elaborated and enhanced by various engineering techniques, and we describe selected examples here. Emphasis is given to investigations on the αHL pore, for which such studies are relatively advanced.

The ion selectivity of porins [39] and the αHL pore [32] can be altered by the replacement of residues within the lumen. In general, positively charged residues produce anion selectivity, and negatively charged residues cation selectivity. Similarly, in the case

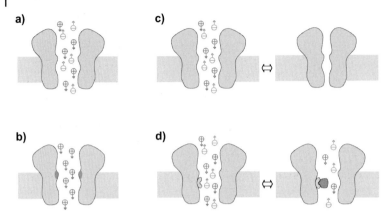

Figure 7.3 Basic properties of channels and pores. (a) Ion conduction; (b) ion selectivity; (c) gating; (d) channel block.

of Matile's artificial barrels, the amino acid side chains that project into the lumen can be used to control selectivity [40]. Ghadiri's nanotubes remain cation-selective in the presence of positively charged cyclic peptide caps, but the selectivity is substantially reduced [33]. When the neutral adapter molecule β-cyclodextrin is lodged within the lumen of the αHL pore, the weak anion selectivity of the pore is enhanced. By contrast, when the anionic hepta-6-sulfato-β-cyclodextrin is used, the pore becomes cation-selective [41]. Ion selectivity can also be controlled with a tightly adsorbed but noncovalently bound internal surface layer. Matile and colleagues have shown that phosphate tightly bound to positively charged residues in the lumen of artificial barrels renders them cation-selective, while Mg^{2+} ions bound to internal carboxylates render the barrels anion-selective [40]. Underivatized Au nanotubes are unselective in KF solutions, but cation-selective in KCl and KBr because of adsorption of anions to the Au surface [35]. The selectivity of the Au nanotubes can also be controlled by direct electrostatic charging of the metal surface [35].

There is still a great deal to learn about the transport of larger molecules through nanopores. For rigid molecules there is molecular mass cut-off, although transport will be reduced before the cut-off is reached because of an increased barrier to entry (hard-sphere model) and interactions with the walls during transport. In one example, glucose was shown to pass through nanotubes made from cyclic peptides of ten amino acids, but not through those of eight amino acids [18]. Au nanotubules can also mediate size separation [26]. Several examples of facilitated transport have been demonstrated with various derivatized inorganic nanotubules. In the case of small organic molecules, a high degree of selectivity has been obtained after reacting the walls of Au nanotubules with various alkane thiols. For example, nanotubules with a low diameter of 1.5 nm that had been reacted with hexadecanethiol transported toluene more than 100 times faster than pyridine in aqueous solution [34]. It seems likely that the derivatized nanotubules do not fill with water, and that the toluene diffuses through an internal hydrophobic phase. Electromodulation of the hydrophobicity of the interior of Au nanotubules is a recent innovation [36]. The enantiomers of chiral molecules have been separated in nanotubules [37, 42]. For example, a racemate can be passed though silica nanotubules within which a stereospecific antibody has been attached [37]. It is likely that this separation works by a facilitated transport mechanism in which the permeant molecules partition strongly into nanotubules containing a high density of cognate binding sites. For facilitated transport to occur, these sites must exchange the permeant molecules rapidly. In the case of the αHL pore, the focus has been on the transport of macromolecules. The cut-off for a rigid spherical molecule based on the width of the narrowest constriction in the lumen is less than 1000 Da, but this has not been tested in rigorous experiments. Elongated polymers of much higher mass such as single-stranded DNA [43] and poly(ethylene glycol) (PEG) [44] can pass through the αHL pore. For a highly flexible molecule such as PEG there is a high entropic cost to entry into the pore, which greatly reduces the rate of transport [45]. A great deal of experimental [46, 47] and theoretical effort ([48] and references therein) continues to be devoted to this aspect of nanopore research because of the interest in making sensors for various polymers.

α-Helical ion channels are gated (opened and closed) by voltage, ligands, or mechanical force. β-Barrel pores gate naturally by mechanisms that are poorly understood and depend upon variables such as the pH and applied potential. Attempts are being made to engineer voltage gating into porins. For example, positively charged peptide loops have been inserted into turns of a porin. The applied potential pushes the charged insert into the barrel at one polarity and removes it at the other [11], which is akin to the so-called ball-and-chain inactivation of K^+ channels [49]. Although weakly asymmetric conduction has been achieved with the porin, more work is required to obtain clear-cut gating. Certain derivatives of gramicidin also appear to gate in a ball-and-chain fashion [50]. There are several other attractive possibilities for controlling the activity of pores. For example, attempts have been made to open and close pores with light, and in such an attempt Woolley and colleagues attached an azobenzene near the mouth of the gramicidin pore [51]. But again, all-or-none gating has not yet been achieved. Interestingly, control of the access of organic molecules to cavities within mesoporous silica was recently achieved through reversible photochemical dimerization of covalently attached coumarins [52].

A variety of blocker sites have been engineered into the αHL pore (Figure 7.4), in this case with a view towards sensor technology (see below). The simplest sites involve mutagenesis with natural amino acids. For example, the introduction of histidine residues at sites within the lumen of the αHL pore allows block by various divalent metal cations

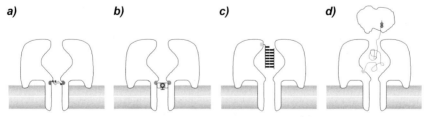

Figure 7.4 Blocker and related sites in the αHL pore. (a) Genetically engineered site for phosphate anions [54]. (b) Site for organic molecules formed with a noncovalent adapter [29]. (c) Covalently attached oligonucleotide for duplex formation [57]. (d) Ligand attached to a long PEG chain for the capture of protein analytes [59].

[14, 53], while the introduction of arginine residues permits the binding of anionic phosphate esters [54] (Figure 7.4a). Sites for organic molecules can be formed from noncovalent adapters (Figure 7.4b) [29–31, 55], although in principle other forms of engineering might be used. Interestingly, the cap domain of the αHL pore, which contains a 45 Å-diameter cavity, is large enough to contain a short covalently attached oligonucleotide, an 8-mer for example (Figure 7.4c). The tethered oligonucleotide can act as a binding site for complementary strands [56–58]. Most proteins are too large to be accommodated within the lumen of the αHL pore. Nevertheless, the attachment of a protein to an external ligand can be transmitted to the interior and hence affect the conductance of the pore. This has been achieved by attaching the ligand to the end of a molecular fishing line, namely a long PEG chain [59] (Figure 7.4d). When the protein is bound, the mobility of the part of the chain that lies within the pore is altered, which produces a characteristic change in the electrical signal during bilayer recording.

In addition to the fundamental properties of pores discussed above, there are additional features that might be engineered. For example, many membrane proteins form two-dimensional crystals, which have contributed to structural studies. Engineered arrays of protein nanopores might be exploited in the same way as bacterial S-layers. S-layers are porous, two-dimensional arrays of a single protein species that envelope a variety of bacteria and are being explored for various applications including ultrafiltration, the geometrically defined immobilization of macromolecules, and acting as templates for the formation of arrays of nanocrystals [60].

7.1.4
What are the Potential Applications of Nanopores?

There are many exciting prospects for the utilization of engineered nanopores, and in some cases practical applications are emerging. For example, as described above, Martin and colleagues have effected separations with various modified inorganic nanopores. The separations have been based on several properties including molecular mass [26], charge [35], hydrophobicity [34, 36], and even stereochemistry [37, 42]. In principle, similar separations might be carried out with engineered protein nanopores, but with the exception of S-layers with which size separation has been accomplished [60], proteins have not yet been exploited for this purpose.

Another area of application of protein nanopores is in cell permeabilization. Because of its importance, the utility of reversible plasma membrane permeabilization in cell and tissue preservation has received the most attention. A mutant αHL pore, H5, which can be closed with divalent metal ions, can be used to introduce small molecules into cells [61]. Once the cells are loaded, efflux can be prevented by Zn(II) ions at the micromolar concentrations normally found in plasma. This procedure has been used to introduce sugars such as sucrose and trehalose into cells (Figure 7.5) [62]. Cells treated in this way have far higher survival rates after cryopreservation [63] and desiccation [64, 65] than those that have not been treated.

Various therapeutic approaches might also be mediated by nanopores. First, nanopores might be used in direct attacks on infectious or malignant cells. For example, Ghadiri's group has demonstrated that selected cyclic peptides have a rapid bactericidal action on

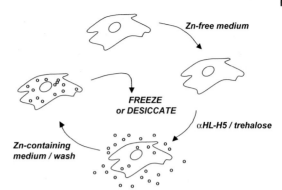

Figure 7.5 Loading mammalian cells with trehalose by using αHL-H5 pores, which can be blocked completely by Zn(II) ions [63].

both Gram-negative and Gram-positive species [66]. They suggest that the nanotubes formed from the peptides act by assembling at target membranes and forming "carpets", thereby rendering them leaky, rather than by acting as individual pores. Malignant mammalian cells might repair themselves after permeabilization, and it has been suggested that the effect of pore formation might be enhanced in the presence of cytotoxic molecules that would not normally enter target cells. Selectivity might be achieved by directing pore-forming proteins to target cells with antibodies or lectins. Additional selectivity might be achieved by activation of the pore-forming protein at the target cell surface. Along these lines, αHL polypeptides that can be activated by a tumor protease, cathepsin B, have been obtained by screening a library of inactive two-chain αHL mutants containing a combinatorial cassette encoding thousands of potential protease recognition sites [67].

Nanopores have excellent prospects in sensor technology. For example, the ionic current flowing through a pore in a transmembrane potential and the modulation of the current by analytes that act, for instance, as blockers can be monitored. In this way, sensors based on either many pores (macroscopic currents) or single pores are being devised. Here, we focus on single pores and the reader is referred to reviews, where key papers on macroscopic approaches can be located [68–70]. Single pores are used in stochastic sensing where individual interactions between the pores and analyte molecules are detected [47, 70, 71]. In stochastic sensing, the frequency of occurrence of individual binding events reveals the concentration of an analyte, while the analyte can be identified from the signature of the binding events (the extent and duration of current block, for example) (Figure 7.6). As indicated in the previous section, a variety of blocker sites have been engineered into the αHL pore, and they can be used to detect a wide variety of analytes ranging from small cations and anions, to organic molecules, to macromolecules including DNA and proteins. Polymers can be detected by nanopores in a second mode in which their transit through the pores is registered [46, 47]. This was first demonstrated with single-stranded DNA [43], but there is no reason why the procedure cannot be adapted to other polymers. Here, the nanopore is acting as more than a simple Coulter counter, because limited information about the structure of the polymer is revealed by the changes in current during transit and the transit velocity. It has been proposed that DNA might be sequenced by this approach, but this seems to be an unlikely proposition in its present

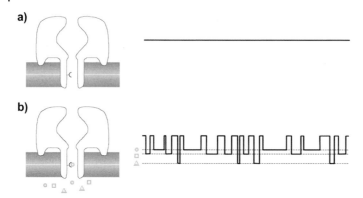

Figure 7.6 Basis of stochastic sensing [71]. Individual interactions between single pores and analyte molecules are detected. (a) Engineered nanopore in the absence of analyte. A flat current trace is seen. (b) Engineered nanopore in the presence of analytes. Individual binding events are manifested as transient reductions in current. The frequency of occurrence of binding events reveals the concentration of an analyte. Further, an analyte can be identified from the signatures of the events.

formulation [47]. However, protein engineering has not been applied to this idea in earnest, and it might be of help.

Finally, engineered nanopores have applications in basic science. For example, the ability to introduce molecules into cells is an important tool in cell biology, and pore-forming proteins are one of the devices that have been used for this purpose [72]. Obviously, the size of the pore determines what goes in and, just as importantly, what comes out. For example, αHL has often been used to bring small molecules (up to ~1000 Da) into cells, while streptolysin O (which forms pores ~35 nm in diameter) can be used to transfer macromolecules of up to 100 kDa. Prolonged permeabilization of cells is lethal, and so a means of closing the pores is required. In some cases, wild-type lesions can be eliminated. For example, resealing of the pores formed by streptolysin O occurs by an unknown mechanism that is promoted by Ca^{2+} ions [73]. Engineered pores can also be used and αHL-H5, which can be closed with divalent metal ions [61], has found several application in the laboratory. For example, Palczewski and colleagues used αHL-H5 to load nucleotides into retinal rod cells [74].

In our own laboratory, it is becoming clear that the αHL pore can be used as a nanoreactor for the investigation of both noncovalent and, most recently, covalent interactions at the single molecule level. Single molecule studies are capable of revealing information that is obscured in ensemble measurements [75]. These studies stem from our attempts to make sensor elements from the αHL pore (see above). In the case of sensors, it does not matter whether the kinetics seen in bulk solution are maintained inside or close to the pore. However, our studies suggest that several binary noncovalent interactions occur with similar kinetics and thus possess similar K_d values to those observed in bulk solvent. These include host–guest interactions [29], DNA duplex formation [56], and protein–ligand interactions [59]. While we can say that association and dissociation rate constants are within a factor of ten of those in bulk solvent, more detailed comparisons will be required to ascertain just how close they are. One partner in each of these interactions was a small molecule. Where a macromolecule must assume a compact state to enter the pore, overall K_d values are increased [45, 76]. In recent studies, we have investigated a variety of covalent chemistries within the αHL pore at the single molecule level, and again the observed kinetic constants are close to those seen in bulk solvent [77, 78].

7.1.5
Keeping Nanopores Happy

One issue of great importance in sensor technology is the search for more robust platforms for existing nanopores (Figure 7.7), or simply more robust nanopores [71]. This is an area that is gaining from several energing aspects of nanotechnology. The lipid bilayers used in the laboratory are usually formed across apertures in polymer films, such as Teflon. These bilayers are mechanically unstable, and a great deal of work is aimed at obtaining a storable and shippable alternative [79–82]. One opportunity lies with improved apertures, and several based on silicon substrates have been reported. For example, Peterman and colleagues recently showed that bilayers can be formed across 25-μm apertures in silicon nitride made by established fabrication technology [83] (Figure 7.7a). The capacitance of the device was reduced by oxidation of exposed silicon substrate and the use of a polyimide surface coating. The bilayers were more robust than those on Teflon apertures, as judged by their ability to withstand high potentials. Another promising area is the use of porous supports, such as nanoporous glass, agarose, etc. One interesting example, with which steady improvements are being made, is the use of S-layers [84], the porous two-dimensional protein crystals from bacterial envelopes (Figure 7.7b). Recent investigations have demonstrated that S-layer-supported bilayers have an increased resistance to hydrostatic pressure [85]. The exploration of bilayers on solid supports continues. In general, these bilayers have been too leaky for single-channel current recording. Recently, tethered supported bilayers on ultrasmooth Au surfaces have been obtained with specific capacitance and impedance values rivaling those of planar lipid bilayers [86] (Figure 7.7c). Therefore, if sufficiently small surface areas are used, it should be possible to examine single nanopores in such systems [87]. However, a second serious issue is the high resistance of the volume between the bilayer and the support, which is a consequence of the modest reservoir of mobile ions, at least with tethered bilayers [88, 89]. It may also be possible to replace lipid bilayers completely. For example, polymerizable block copolymers appear to be a promising substitute [90]. In short, progress has been made, but we do not yet have a bilayer that can be slipped into a briefcase.

Figure 7.7 Different classes of robust bilayer. (a) Bilayer across a small aperture formed in silicon nitride [83]. (b) Bilayer on a porous support comprising a bacterial S-layer [84]. (c) Tethered supported bilayer on a smooth gold surface [86].

7.2
Methods

In this chapter we have covered a wide variety of types of nanopores, and it would be presumptuous of us to discuss the methodology involved in areas where we do not work directly. Instead, we present a summary of the methods that have been crucial for the progress that has been made with the αHL pore.

7.2.1
Protein Production

The monomeric form of αHL and its mutants are obtained in abundance and high purity by expression in *Staphylococcus* [91] or *E. coli* [92]. When necessary, the purifications can be aided with C-terminal His tags. Conveniently, small amounts of αHL are available by in-vitro transcription and translation, which produces around 50 μg mL^{-1} [92]. For single-channel recording in the laboratory, we use only nanograms of protein. Other applications likewise require modest amounts; for example, a 1 m^2, two-dimensional crystal of αHL pores would contain only 3.5 mg of protein. The in-vitro production technology is readily scaled up to provide milligrams and even grams [93].

While assembly into heptamers can be facilitated by receptors on target cells, such as rabbit red blood cells [94], the αHL pore self-assembles on artificial lipid bilayers [95]. The heptameric pore is robust and remains stable in the denaturing detergent SDS at up to 65 °C [96]. This stability is the basis of the separation by SDS-PAGE of heterohep-tamers, pores containing various combinations of engineered and WT subunits. The availability of pure heteroheptameric species has been a key aspect of our work on the αHL pore. To achieve the electrophoretic separation, the mutant subunits are tagged; originally, the tag was a targeted chemical modification with a disulfonate near the C terminus [14], but more recently we have used a genetically encoded oligoaspartate tail [57].

7.2.2
Protein Engineering

With the benefit of high-resolution structural information, protein engineering [97] allows substitute amino acid side chains or chemical modifications to be placed in a protein at defined sites. Engineering studies on membrane proteins have been limited compared to other systems, largely because the proteins can be tricky to handle and because, until quite recently, there has been a lack of structural information. As noted earlier, de-novo design is a difficult task, and very few studies have been conducted on genetically encoded designed membrane proteins. "Redesign" is more readily accomplished, and since 1996 it has been aided in the case of αHL by the availability of a 1.9 Å crystal structure of the heptameric pore. More recently, structural studies of LukF, a related toxin, have provided information about the monomeric form of αHL [98, 99]. Therefore, the impact of structural changes on both the pore and its ability to assemble can be appraised. The αHL pore is a "blank slate" in terms of engineering, and the goal has been to introduce function rather than change function – which has been a difficult task when attempted in proteins more specialized than αHL. In αHL, most desired changes are made on the internal surface of the protein, which is an unusual challenge. Nevertheless, the open structure of the αHL lumen is more readily manipulated than a narrow channel. For example, we have replaced the transmembrane β barrel with a retro (reversed) amino acid sequence and obtained a functional pore [7].

In practice, redesign is done both by rational approaches (e. g., the examination of molecular models) and hit-and-miss methods (e. g., the screening of libraries of mutants). The approaches are often combined. For example, structural studies may suggest a region for intensive mutagenesis. This approach has been applied to the transmembrane region of αHL by using a semi-synthetic gene in which selected residues are encoded by a replaceable cassette [7]. Nowadays, most forms of site-directed mutagenesis are greatly accelerated by tools such as the polymerase chain reaction and ligation-free recombination of DNA fragments. Targeted chemical modification is an invaluable adjunct to direct mutagenesis. In general, modifications are carried out at specific introduced cysteine residues, and in the case of αHL all manner of additions have been made, including the attachment of small molecules with and without poly(ethylene glycol) linkers, peptides, and oligonucleotides [56, 59]. Finally, as noted above, it is possible to make heteromeric αHL pores, which are most commonly molecules with six unaltered subunits and one engineered subunit.

7.2.3
Electrical Recording

While ensembles of pores can be useful in many circumstances, one notable capability is the means to examine function in intricate detail by single-channel recording. Single-channel recording is also the basis of stochastic sensing. Both monomeric and heptameric αHL pores can assemble or insert into lipid bilayers in a single orientation. Unlike many other pores, the αHL pore remains open indefinitely at moderate transmembrane potentials. Under typical conditions (in 1 M NaCl), the pore has a high conductance of 700 pS, and changes of a few percent in the current carried by a single pore can be measured under optimal conditions. Many measurements can be carried out on a timescale of tens of microseconds. Single-channel recording allows the determination of the basic properties of a nanopore with confidence, including conductance values, ion selectivity, stability, and uniformity. Further, the noncovalent and covalent interactions of exogenous molecules with nanopores can be examined with exquisite sensitivity.

7.2.4
Other Systems

Analogous methods to those described above will have to be developed for other nanopore systems, if they are to reach the state of sophistication that, in the case of the αHL pore, has been aided by structural studies, protein engineering, and single-channel recording. It is not at all clear how the structures of many of the alternative systems might be examined at atomic resolution. Nor is it clear how they might be modified at discrete sites with defined stoichiometry. If the current passing through single nanopores is to be recorded, they will have to be assembled into systems that do not suffer from current leaks and are of low intrinsic capacitance.

7.3
Outlook

As suggested by the preceding discussion, many improvements might be made to existing nanopore technology.

7.3.1
Rugged Pores

There are limits to the stability of proteins, and therefore several alternative approaches have been taken to the preparation of nanopores, as described earlier. In the long term, whichever techniques are used must be amenable to mass production and compatible with array technology (see below). Of course, a major advantage of protein nanopores is the ability to carry out functionalization with precision by replacing amino acid side chains or carrying out targeted chemical modification. While alternative nanopores, made from polymers, carbon nanotubes, inorganic materials, etc., can be randomly derivatized, it is not yet clear how defined patterns of functionalization or single-site alterations might be made. One compromise would be to immobilize a single adapter molecule (as we have used with the αHL pore) at the entrance of a mechanically and thermally stable nanopore, such as an Au nanotubule.

7.3.2
Supported Bilayers

To accommodate protein nanopores, it is clear that additional work is needed on supported bilayers and related areas. The formation of defect-free bilayers appears to require ultrasmooth surfaces [86]. "Edge effects" – that is, leaks at the perimeter of the bilayers – must also be tackled. In addition, improved electrical properties demand a substantial reservoir of electrolyte between the bilayer and the support surface [89]. Finally, a reasonable fraction of the incorporated nanopores should be active. Pores with an appreciable extramembranous domain that is located between the bilayer and the underlying surface, might be denatured by the support unless precautions such as polymer cushions are used. In this area, a near-term goal is to observe single channels in supported bilayers [87].

7.3.3
Membrane Arrays

Arrays of lipid bilayers would be useful for screening the properties of membrane proteins and for monitoring their interactions with other molecules (Figure 7.8). The same technology could be used for array-based sensors. Several assays for pore-forming molecules

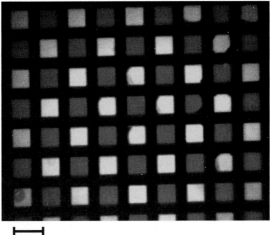

Figure 7.8 Membrane array on a glass surface. The pattern was formed by photolithography of a glass slide coated with photoresist. Each corral contains a bilayer doped with either 1 % NBD-phosphatidyl ethanolamine or 1 % Texas Red-phosphatidyl ethanolamine. A false-color image was generated from epifluorescence images recorded through a home-built macroscope. (E. T. Castellana and P. S. Cremer, Texas AM University, unpublished data.)

800 μm

have been developed in which liposomes are used in microtiter formats [100–102]. When they are more fully developed, addressable bilayer arrays containing pores will be far more versatile. Several procedures for generating bilayer arrays have been developed, including the formation of corrals by scratching off or blotting the bilayer, or by patterning the support in various ways, for example by depositing the bilayer with a poly-dimethylsiloxane (PDMS) stamp or by making corrals with various barriers including metals, metal oxides, and proteins [103]. Recently, barriers have been formed by the polymerization of diacetylenic lipids [104]. All corrals can be filled with a single type of bilayer by exposure to a vesicle suspension, or individual corrals can be filled with a micropipette [105, 106] or by flow patterning [103]. Two goals of further development are to address each corral electrically and to wash each one individually by using a microfluidic system. It should also become possible to arrange other classes of pores, such as those fabricated with ion beams, in addressable formats.

7.3.4
Alternative Protein Pores

Advances in protein folding and structural biology will drive the search for alternative protein pores. While research on αHL has been extremely fruitful, the pore is a heptamer and therefore intricate to handle. A single-chain, monomeric pore would be ideal, and we have begun to explore the monomeric porin OmpG [107, 108]. It is likely that yet more sophisticated engineering techniques will be applied to protein nanopores than have been in the past. For example, multistep syntheses on the lumen wall might be used to produce complex host molecules within a pore. Because highly stable bilayers with desirable properties have been difficult to obtain, the possibility of directly housing a protein nanopore inside a mechanically stable pore of slightly greater diameter should be considered. The surfaces of both the protein nanopore and the recipient pore (e. g., an Au nanotubule) could be tailored for compatibility (Figure 7.9).

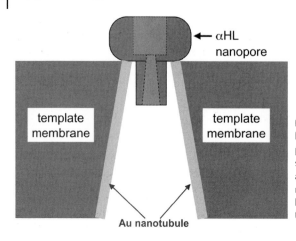

Figure 7.9 Possible means for stabilizing a single protein pore. The nanopore (here the αHL pore) is placed inside a mechanically stable pore (here a Au nanotubule). The surface of the nanopore and the αHL pore are tailored for compatibility. (C. R. Martin, University of Florida.)

7.3.5
Pores with New Attributes and Applications

Several advanced applications of nanopore engineering are being initiated. For example, the αHL pore can be used as a "nanoreactor" for investigating both noncovalent and covalent interactions. The study of noncovalent interactions elaborates upon the work on sensors. Single-channel current traces contain kinetic information. For example, for a simple binary system, such as a cyclodextrin–guest interaction, k_{on}, k_{off} and K_d can be readily obtained. However, while the values are close to those in bulk solution, further work is required to determine whether there are significant differences. Considerations include steric effects, transmembrane and local potentials, solvent confinement, and electroosmosis. In pursuit of ideas for sensing reactive molecules, we have recently begun to explore covalent interactions at the single molecule level with αHL as a nanoreactor, and related considerations come into play here. In addition, it might soon be possible to build various catalytic activities into the pores. This has a precedent in nature, as several enzymes contain a series of active sites located inside an internal channel [109]. Substrates enter the tunnel at one end and products leave at the other. These channels span up to 100 Å, which is the length of the αHL pore. Esterase and ribonuclease activity in a synthetic barrel lined with histidine residues has been observed by Matile and colleagues [110, 111]. By using the variety of substitutions available to protein engineering, and the ability to place them with known geometry, it may be possible to improve turnover and couple catalysis to transmembrane transport.

7.3.6
Theory

With a continually expanding output of experimental data, much of which is purely descriptive, there is a need for theory and computation in studies of nanopores. Of course, the applications of calculations and simulations to membrane channels and pores is well developed, though far from perfected. By combining aspects of results from structural studies, molecular dynamics simulations, Brownian dynamics and electrodiffusion theory, estimates of fundamental properties of channels and pores have been obtained, such as conductance values, the shape of current–voltage curves, and the magnitude of ion selectivity [112, 113]. The topics covered in this chapter suggest additional areas for exploration such as the properties of water [114] and polymers [48] under confinement, and electroosmosis in pores of nanometer dimensions [115, 118]. It should go without saying that all of this must be coupled to experimentation.

Acknowledgments

The authors thank several colleagues for help with the figures, and acknowledge grants from the U. S. Department of Energy, the Multidisciplinary University Research Initiative (ONR 1999), the National Institutes of Health, the Office of Naval Research and DARPA (MOLDICE program). H.B. is the holder of a Royal Society-Wolfson Research Merit Award.

References

[1] C. R. Martin, P. Kohli, *Nature Rev. Drug Disc.* **2003**, *2*, 29–37.

[2] G. E. Schulz, *Biochim. Biophys. Acta* **2002**, *1565*, 308–317.

[3] K. S. Åkerfeldt, J. D. Lear, Z. R. Wasserman, L. A. Chung, W. F. DeGrado, *Acc. Chem. Res.* **1993**, *26*, 191–197.

[4] H. Bayley, *Curr. Opin. Biotechnol.* **1999**, *10*, 94–103.

[5] S. Lee, T. Kiyota, T. Kunitake, E. Matsumoto, S. Yamashita, K. Anzai, G. Sugihara, *Biochemistry* **1997**, *36*, 3782–3791.

[6] E. Matsumoto, T. Kiyota, S. Lee, G. Sugihara, S. Yamashita, H. Meno, Y. Aso, H. Sakamoto, H. M. Ellerby, *Biopolymers* **2001**, *56*, 96–108.

[7] S. Cheley, O. Braha, X. Lu, S. Conlan, H. Bayley, *Protein Sci.* **1999**, *8*, 1257–1267.

[8] A. J. Wallace, T. J. Stillman, A. Atkins, S. J. Jamieson, P. A. Bullough, J. Green, P. J. Artymiuk, *Cell* **2000**, *100*, 265–276.

[9] B. North, C. M. Summa, G. Ghirlanda, W. F. DeGrado, *J. Mol. Biol.* **2001**, *311*, 1081–1090.

[10] G. E. Schulz, *Curr. Opin. Struct. Biol.* **1996**, *6*, 485–490.

[11] M. Bannwarth, G. E. Schulz, *Prot. Eng.* **2002**, *15*, 799–804.

[12] S. Terrettaz, W.-P. Ulrich, H. Vogel, Q. Hong, L. G. Dover, J. H. Lakey, *Prot. Sci.* **2002**, *11*, 1917–1925.

[13] L. Song, M. R. Hobaugh, C. Shustak, S. Cheley, H. Bayley, J. E. Gouaux, *Science* **1996**, *274*, 1859–1865.

[14] O. Braha, B. Walker, S. Cheley, J. J. Kasianowicz, L. Song, J. E. Gouaux, H. Bayley, *Chem. Biol.* **1997**, *4*, 497–505.

[15] G. Miles, L. Movileanu, H. Bayley, *Protein Sci.* **2002**, *11*, 894–902.

[16] J. D. Hartgerink, T. D. Clark, M. R. Ghadiri, *Chem. Eur. J.* **1998**, *4*, 1367–1372.

[17] T. D. Clark, M. R. Ghadiri, *J. Am. Chem. Soc.* **1995**, *117*, 12364–12365.

[18] J. R. Granja, M. R. Ghadiri, *J. Am. Chem. Soc.* **1994**, *116*, 10785–10786.

[19] T. D. Clark, L. K. Buehler, M. R. Ghadiri, *J. Am. Chem. Soc.* **1998**, *120*, 651–656.

[20] S. Matile, *Chem. Rec.* **2001**, *1*, 162–172.

[21] G. Das, N. Sakai, S. Matile, *Chirality* **2002**, *14*, 18–24.

[22] G. Das, S. Matile, *Chirality* **2001**, *13*, 170–176.

[23] M. R. Ghadiri, J. R. Granja, R. A. Milligan, D. E. McRee, N. Khazanovich, *Nature* **1993**, *366*, 324–327.

[24] P. Y. Apel, Y. E. Korchev, Z. Siwy, R. Spohr, M. Yoshida, *Nucl. Intr. Meth. Physics Res. B* **2001**, *184*, 337–346.

[25] C. A. Pasternak, G. M. Alder, P. Y. Apel, C. L. Bashford, D. T. Edmonds, Y. E. Korchev, A. A. Lev, G. Lowe, M. Milovanovich, C. W. Pitt, T. K. Rostovtseva,

N. I. Zhitariuk, *Radiation Measurements* **1995**, *25*, 675–683.

[26] K. B. Jirage, J. C. Hulteen, C. R. Martin, *Science* **1997**, *278*, 655–658.

[27] J. Li, D. Stein, C. McMullan, D. Branton, M. J. Aziz, J. A. Golovchenko, *Nature* **2001**, *412*, 166–169.

[28] R. A. Mehl, J. C. Anderson, S. W. Santoro, L. Wang, A. B. Martin, D. S. King, D. M. Horn, P. G. Schultz, *J. Am. Chem. Soc.* **2003**, *125*, 935–939.

[29] L.-Q. Gu, O. Braha, S. Conlan, S. Cheley, H. Bayley, *Nature* **1999**, *398*, 686–690.

[30] J. Sanchez-Quesada, M. R. Ghadiri, H. Bayley, O. Braha, *J. Am. Chem. Soc.* **2000**, *122*, 11758–11766.

[31] L.-Q. Gu, S. Cheley, H. Bayley, *Science* **2001**, *291*, 636–640.

[32] L.-Q. Gu, S. Cheley, H. Bayley, *J. Gen. Physiol.* **2001**, *118*, 481–494.

[33] J. Sánchez-Quesada, M. P. Isler, M. R. Ghadiri, *J. Am. Chem. Soc.* **2002**, *124*, 10004–10005.

[34] J. C. Hulteen, K. B. Jirage, C. R. Martin, *J. Am. Chem. Soc.* **1998**, *120*, 6603–6604.

[35] M. Nishizawa, V. P. Menon, C. R. Martin, *Science* **1995**, *268*, 700–702.

[36] S. B. Lee, C. R. Martin, *J. Am. Chem. Soc.* **2002**, *124*, 11850–11851.

[37] S. B. Lee, D. T. Mitchell, L. Trofin, T. K. Nevanen, H. Söderlund, C. R. Martin, *Science* **2002**, *296*, 2198–2200.

[38] J. M. Pascual, A. Karlin, *J. Gen. Physiol.* **1998**, *111*, 717–739.

[39] K. Saxena, V. Drosou, E. Maier, R. Benz, B. Ludwig, *Biochemistry* **1999**, *38*, 2206–2212.

[40] N. Sakai, N. Sordé, G. Das, P. Perrottet, D. Gerard, S. Matile, *Org. Biomol. Chem.* **2003**, *1*, 1226–1231.

[41] L.-Q. Gu, M. Dalla Serra, J. B. Vincent, G. Vigh, S. Cheley, O. Braha, H. Bayley, *Proc. Natl. Acad. Sci. USA* **2000**, *97*, 3959–3964.

[42] B. B. Lakshmi, C. R. Martin, *Nature* **1997**, *388*, 758–760.

[43] J. J. Kasianowicz, E. Brandin, D. Branton, D. W. Deamer, *Proc. Natl. Acad. Sci. USA* **1996**, *93*, 13770–13773.

[44] L. Movileanu, S. Cheley, S. Howorka, O. Braha, H. Bayley, *J. Gen. Physiol.* **2001**, *117*, 239–251.

[45] L. Movileanu, H. Bayley, *Proc. Natl. Acad. Sci. USA* **2001**, *98*, 10137–10141.

[46] S. M. Bezrukov, *J. Membr. Biol.* **2000**, *174*, 1–13.

[47] H. Bayley, C. R. Martin, *Chem. Rev.* **2000**, *100*, 2575–2594.

[48] M. Muthukumar, *J. Chem. Phys.* **2003**, *118*, 5174–5184.

[49] M. Zhou, J. H. Morais-Cabral, S. Mann, R. MacKinnon, *Nature* **2001**, *411*, 643–644.

[50] G. A. Woolley, V. Zunic, J. Karanicolas, A. S. I. Jaikaran, A. V. Starostin, *Biophys. J.* **1997**, *73*, 2465–2475.

[51] L. Lien, D. C. J. Jaikaran, Z. H. Zhang, G. A. Woolley, *J. Am. Chem. Soc.* **1996**, *118*, 12222–12223.

[52] N. K. Mal, M. Fujiwara, Y. Tanaka, *Nature* **2003**, *421*, 350–353.

[53] O. Braha, L.-Q. Gu, L. Zhou, X. Lu, S. Cheley, H. Bayley, *Nature Biotechnol.* **2000**, *17*, 1005–1007.

[54] S. Cheley, L.-Q. Gu, H. Bayley, *Chem. Biol.* **2002**, *9*, 829–838.

[55] L.-Q. Gu, H. Bayley, *Biophys. J.* **2000**, *79*, 1967–1975.

[56] S. Howorka, L. Movileanu, O. Braha, H. Bayley, *Proc. Natl. Acad. Sci. USA* **2001**, *98*, 12996–13001.

[57] S. Howorka, S. Cheley, H. Bayley, *Nature Biotechnol.* **2001**, *19*, 636–639.

[58] S. Howorka, H. Bayley, *Biophys. J.* **2002**, *83*, 3202–3210.

[59] L. Movileanu, S. Howorka, O. Braha, H. Bayley, *Nature Biotechnol.* **2000**, *18*, 1091–1095.

[60] M. Sára, U. Sleytr, *J. Bacteriol.* **2000**, *182*, 859–868.

[61] M. J. Russo, H. Bayley, M. Toner, *Nature Biotechnol.* **1997**, *15*, 278–282.

[62] J. P. Acker, X. Lu, V. Young, H. Cheley, H. Bayley, A. Fowler, M. Toner, *Biotechnol. Bioeng.* **2003**, *82*, 525–532.

[63] A. Eroglu, M. J. Russo, R. Bieganski, A. Fowler, S. Cheley, H. Bayley, M. Toner, *Nature Biotechnol.* **2000**, *18*, 163–167.

[64] T. Chen, J. P. Acker, A. Eroglu, S. Cheley, H. Bayley, A. Fowler, M. Toner, *Cryobiology* **2001**, *43*, 168–181.

[65] J. P. Acker, A. Fowler, B. Lauman, S. Cheley, M. Toner, *Cell Preserv. Technol.* **2002**, *1*, 129–140.

[66] S. Fernandez-Lopez, H.-S. Kim, E. C. Choi, M. Delgado, J. R. Granja, A. Khasanov, A. Kraehenbuehl, G. Long, D. A. Weinberger, K. M. Wilcoxen, M. R. Ghadiri, *Nature* **2001**, *412*, 452–455.

[67] R. G. Panchal, E. Cusack, S. Cheley, H. Bayley, *Nature Biotechnol.* **1996**, *14*, 852–856.

[68] H. A. Fishman, D. R. Greenwald, R. N. Zare, *Annu. Rev. Biophys. Biomol. Struct.* **1998**, *27*, 165–198.

[69] C. Ziegler, W. Göpel, *Curr. Opin. Chem. Biol.* **1998**, *2*, 585–591.

[70] H. Bayley, O. Braha, L.-Q. Gu, *Adv. Mater.* **2000**, *12*, 139–142.

[71] H. Bayley, P. S. Cremer, *Nature* **2001**, *413*, 226–230.

[72] S. Bhakdi, U. Weller, I. Walev, E. Martin, D. Jonas, M. Palmer, *Med. Microbiol. Immunol.* **1993**, *182*, 167–175.

[73] I. Walev, S. C. Bhakdi, F. Hofmann, N. Djonder, A. Valeva, K. Aktories, S. Bhakdi, *Proc. Natl. Acad. Sci. USA* **2001**, *98*, 3185–3190.

[74] A. E. Otto-Bruc, R. N. Fariss, J. P. Van Hooser, K. Palczewski, *Proc. Natl. Acad. Sci. USA* **1998**, *95*, 15014–15019.

[75] X. S. Xie, H. P. Lu, *J. Biol. Chem.* **1999**, *274*, 15967–15970.

[76] L. Movileanu, S. Cheley, H. Bayley, *Biophys. J.* **2003**, *85*, 897–910.

[77] T. Luchian, S.-H. Shin, H. Bayley, *Angew. Chem. Int. Ed.* **2003**, *42*, 1926–1929.

[78] S.-H. Shin, T. Luchian, S. Cheley, O. Braha, H. Bayley, *Angew. Chem. Int. Ed.* **2002**, *41*, 3707–3709.

[79] F. S. Ligler, T. L. Fare, K. D. Seib, B. See, J. W. Smuda, A. Singh, P. Ahl, M. E. Ayers, A. Dalziel, P. Yager, *Medical Instrumentation* **1988**, *22*, 247–256.

[80] E. Sackmann, M. Tanaka, *Trends Biotechnol.* **2000**, *18*, 58–64.

[81] E.-K. Sinner, W. Knoll, *Curr. Opin. Chem. Biol.* **2001**, *5*, 705–711.

[82] J. T. Groves, *Curr. Opin. Drug Discov. Devel.* **2002**, *5*, 606–612.

[83] M. C. Peterman, J. M. Ziebarth, O. Braha, H. Bayley, H. A. Fishman, D. A. Bloom, *Biomed. Microdevices* **2002**, *4*, 231–236.

[84] B. Schuster, D. Pum, O. Braha, H. Bayley, U. B. Sleytr, *Biochim. Biophys. Acta* **1998**, *1370*, 280–288.

[85] B. Schuster, U. B. Sleytr, *Biochim. Biophys. Acta* **2002**, *1563*, 29–34.

[86] S. M. Schiller, R. Naumann, K. Lovejoy, H. Kunz, W. Knoll, *Angew. Chem. Int. Ed. Engl.* **2003**, *42*, 208–211.

[87] G. Wiegand, K. R. Neumaier, E. Sackmann, *Rev. Sci. Instrum.* **2000**, *71*, 2309–2320.

[88] B. A. Cornell, V. L. B. Braach-Maksvytis, L. G. King, P. D. J. Osman, B. Raguse, L. Wieczorek, R. J. Pace, *Nature* **1997**, *387*, 580–583.

[89] G. Krishna, J. Schulte, B. A. Cornell, R. J. Pace, P. D. Osman, *Langmuir* **2003**, *19*, 2294–2305.

[90] W. Meier, C. Nardin, M. Winterhalter, *Angew. Chem. Int. Ed. Engl.* **2000**, *39*, 4599–4602.

[91] A. Valeva, A. Weisser, B. Walker, M. Kehoe, H. Bayley, S. Bhakdi, M. Palmer, *EMBO J.* **1996**, *15*, 1857–1864.

[92] B. J. Walker, M. Krishnasastry, L. Zorn, J. J. Kasianowicz, H. Bayley, *J. Biol. Chem.* **1992**, *267*, 10902–10909.

[93] T. Lamla, W. Stiege, V. A. Erdmann, *Mol. Cell. Proteomics* **2002**, *1.6*, 466–472.

[94] A. Hildebrand, M. Pohl, S. Bhakdi, *J. Biol. Chem.* **1991**, *266*, 17195–17200.

[95] G. Menestrina, *J. Membr. Biol.* **1986**, *90*, 177–190.

[96] B. Walker, H. Bayley, *Protein Eng.* **1995**, *8*, 491–495.

[97] W. F. ed. DeGrado, *Chem. Rev.* **2001**, *101*, special issue.

[98] R. Olson, H. Nariya, K. Yokota, Y. Kamio, E. Gouaux, *Nature Struct. Biol.* **1999**, *6*, 134–140.

[99] J.-D. Pédelacq, L. Maveyraud, G. Prévost, L. Baba-Moussa, A. González, E. Courcelle, W. Shepard, H. Monteil, J.-P. Samama, L. Mourey, *Structure* **1999**, *7*, 277–288.

[100] S. Kolusheva, L. Boyer, R. Jelinek, *Nature Biotechnol.* **2000**, *18*, 225–227.

[101] J. M. Rausch, W. C. Wimley, *Anal. Biochem.* **2001**, *293*, 258–263.

[102] G. Das, P. Talukdar, S. Matile, *Science* **2002**, *298*, 1600–1602.

[103] J. T. Groves, S. G. Boxer, *Acc. Chem. Res.* **2002**, *35*, 149–157.

[104] K. Morigaki, T. Baumgart, A. Offenhäusser, W. Knoll, *Angew. Chem. Int. Ed. Engl.* **2001**, *40*, 172–174.

[105] P. S. Cremer, T. Yang, *J. Am. Chem. Soc.* **1999**, *121*, 8130–8131.

[106] T. Yang, E. E. Simanek, P. S. Cremer, *Anal. Chem.* **2000**, *72*, 2587–2589.

[107] S. Conlan, Y. Zhang, S. Cheley, H. Bayley, *Biochemistry* **2000**, *39*, 11845–11854.

[108] M. Behlau, D. J. Mills, H. Quader, W. Kühlbrandt, J. Vonck, *J. Mol. Biol.* **2001**, *305*, 71–77.

[109] X. Huang, H. M. Holden, F. M. Raushel, *Annu. Rev. Biochem.* **2001**, *70*, 149–180.

[110] B. Baumeister, N. Sakai, S. Matile, *Org. Lett.* **2001**, *3*, 4229–4232.

[111] B. Baumeister, S. Matile, *Macromolecules* **2002**, *35*, 1549–1555.

[112] M. S. P. Sansom, I. H. Shrivastava, K. M. Ranatunga, G. R. Smith, *Trends Biochem. Sci.* **2000**, *25*, 368–374.

[113] B. Roux, *Curr. Opin. Struct. Biol.* **2002**, *12*, 182–189.

[114] U. Raviv, P. Laurat, J. Klein, *Nature* **2001**, *413*, 51–54.

[115] A. T. Conlisk, J. McFerran, Z. Zheng, D. Hansford, *Anal. Chem.* **2002**, *74*, 2139–2150.

[116] G. G. Kochendoerfer, J. M. Tack, S. Cressman, *Bioconjug. Chem.* **2002**, *13*, 474–480.

[117] S. W. Cowan, R. M. Garavito, J. N. Jansonius, J. A. Jenkins, R. Karlsson, N. König, E. F. Pai, R. A. Pauptit, P. J. Rizkallah, J. P. Rosenbusch, G. Rummel, T. Schirmer, *Structure* **1995**, *3*, 1041–1050

[118] L.-Q. Gu, S. Cheley, H. Bayley, *Proc. Natl. Acad. Sci. USA*, in press.

8
Genetic Approaches to Programmed Assembly

Stanley Brown

8.1
Introduction

In biological systems, proteins are the predominant catalysts, motors, pumps and channels, they form many rigid and flexible structures, and also act as scaffolds in assembly
processes. They can be expected to also provide these functions in contrived nanosystems.
The first step in applying proteins to nanoassembly is the isolation of peptides/proteins
which are able to adhere to the surface of specific materials. Moreover, many peptides
able to bind a material have the secondary trait of modulating the formation of that
material. Since chimeric binding peptides having dissimilar specificities can easily be
produced by recombinant DNA techniques, we are rapidly approaching an era of
programmed formation and assembly of materials at the nanometer scale.

A frequently employed strategy for the isolation of binding peptides is phage display. In
display technologies, vast random populations of peptides are prepared, each peptide
physically joined to the genes which encodes it. Peptides able to adhere to a target surface
are selected by adhesion of the composite entity to that target. The recovery of individual
members of a population having a defined property such as adhesion is a genetic experiment. Hence, carefully devised genetic searches are critical to realize the full potential of
display technology. The power of this approach should not be surprising, since the genetic
analysis of phage and bacteria formed the foundation of molecular biology.

This chapter will introduce the various systems of display technology focusing on the
analysis of both the enriched populations and recovered peptides. It will also review
some of the protein structures available for controlling the spatial orientation of the recovered peptides.

8.2
Order from Chaos

All display strategies start with large populations of partially random peptides. Peptides
having a desired binding property are recovered from the population. Figure 8.1 uses
an *Escherichia coli* cell-surface display [1, 2] (reviewed in Ref. [3]) to illustrate the recovery

Nanobiotechnology. Edited by Christof Niemeyer, Chad Mirkin
Copyright © 2004 WILEY-VCH Verlag GmbH & Co. KgaA, Weinheim
ISBN 3-527-30658-7

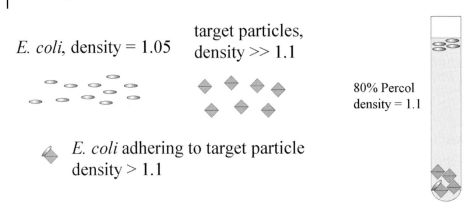

Figure 8.1 Enrichment cycle. A population of bacteria displaying random peptides on their outer surfaces is mixed with the target powder, allowing the bacteria to adhere. The density is raised with Percol, and the suspension is centrifuged. Bacteria adhering to the target powder sediment with it, while bacteria failing to adhere remain suspended and are discarded with the supernatant. The target powder with adhering bacteria is transferred to a bacteriological growth medium and the bacteria are permitted to multiply. This represents a single enrichment cycle that can be repeated many times.

process. Peptides encoded by recombinant genes containing partially random DNA are synthesized within bacteria. Each bacterium contains a different partially random gene and thus synthesizes a different peptide. The recombinant gene is designed so the peptide becomes anchored to the outer surface of the bacterium that encodes it. The anchoring can be by fusion to an integral outer membrane protein (Figure 8.2) or by fusion to appendages such as fimbrae [4]. If a peptide binds to a specific material, it will cause the bacterium to adhere to that material. The bacterial population is mixed with the target material, after which the target material with any adhering bacteria is recovered and transferred to a bacteriological growth medium. The suspension is incubated and the bacteria then multiply. This constitutes one enrichment cycle, and this may be repeated many times. Since E. coli are typically 1–2 µm long and 0.5 µm in diameter, the target particles must be large enough to change the density of the bacterium–particle aggregate. The target material can also be in the form of a sheet or plate and removed from the bacterial suspension after the candidates are permitted to adhere. If the target material is fluorescent, enrichment by binding can be combined or replaced with fluorescence activated cell sorting (FACS) [5].

With phage display, the peptides are exposed on the surface of the bacterial virus (phage) encoding them. The filamentous phage M13 and fd are probably the most familiar platform for displaying peptides. The wild-type phage are approximately 1 µm long and 6 nm in diameter [6]. Proteins at both ends [7, 8] and along the length [6, 9] of each virion can accommodate the insertion of foreign peptides (Figure 8.2). The enrichment process with phage display is similar to that with cell-surface display, except that the phage bound to the target material are released and amplified by infecting the bacterial host. There is abundant literature reviewing phage display, including a monograph [10] and a manufacturer's web site [11].

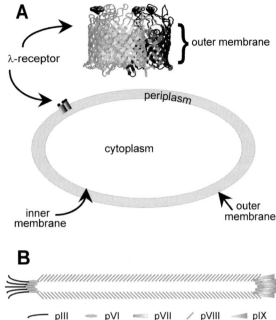

Figure 8.2 Cell-surface and phage display. (A) Bacterial cell-surface display based on the λ-receptor [1]. The upper portion is a ribbon diagram of the λ-receptor [46] homotrimer as integrated in the outer membrane. The site where foreign peptides are inserted is shown as space-filling atoms in black. Cell components are not drawn to scale. Many engineered λ-receptors are displayed on the surface of induced cells. (B) The single-stranded DNA phage, M13 or fd. All coat proteins except pVI have been used to display peptides. Components are not drawn to scale.

A third method is also used to display peptides physically linked to the genes encoding them, ribosome or mRNA display [12, 13] (reviewed in Ref. [14]). Each type of display system has inherent advantages and limitations (Table 8.1).

Some systems of cell-surface display appear to only be able to detect tight-binding peptides (dissociation constants <<1 nM) [15–17]. Phage display does not suffer from this artifact, and tight-binding peptides can be isolated by the use of off-rate selections [18]. In this strategy, the phage are first allowed to adhere to the target material in the form of a large piece such as a plate. After the unbound phage are washed away, the plate is

Table 8.1 Display methods

Method	Advantages	Disadvantages
Ribosomes/ mRNA	Very large populations Resistance to many solvents	In-vitro reactions Sensitive to inhibitors
Phage	Vast literature Many successes with commercially available libraries Off-rate selections	Must be released from surface free of inhibitors and must be infective
Cells	Do not have to be released from the target prior to growth High-valency display permits isolation by FACS	Selections limited to physiological conditions

incubated with a large excess of the target material in the form of small particles. Phage dissociating from the initial target rapidly bind to the small particles. After a period of incubation, the initial target retains those phage that dissociate very slowly.

8.3
Monitoring Enrichment

The enrichment process is Darwinian. The most obvious selective parameter is adhesion to the target material, although the population is, by design, genetically diverse. The recovery of bacteria is influenced by efficiency of release from the target material, growth rate and length of time required for physiological recovery after return to the growth medium. Similarly, recovery of phage is influenced by the efficiency of phage infection, phage production, and phage stability. Consequently, whether the ultimate trait desired is adhesion to a specific material or a secondary trait such as modifying the crystallization or precipitation of that material (discussed below), individual survivors of the enrichment

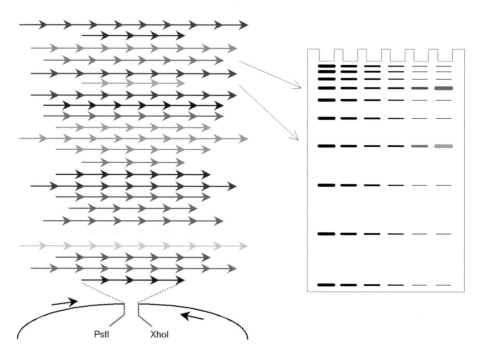

Figure 8.3 Monitoring population complexity. (A) A population of random genes. In this example, the genes encode repeating polypeptides, but the method is applicable to any population of genes of varying length [21, 22]. Here, the repeating oligonucleotides are inserted between *Pst*I and *Xho*I sites in the display vector. The population of genes after each cycle of enrichment is subjected to PCR amplification of the inserts plus some flanking DNA by use of oligonucleotide primers complementary to vector DNA sequences. The distribution of insert sizes is monitored by gel electrophoresis. The lanes depict the initial distribution of insert sizes and the distribution following increasing numbers of enrichment cycles.

must be characterized. Thus, a relevant parameter in determining the end point of the enrichment is the complexity of the enriched population, how many candidate peptides can the investigator reasonably expect to test.

It would be unusual to initiate a search using a population with fewer than millions of different members. The complexity of a population reduced to not many more than a dozen different members can easily be evaluated by sequencing the DNA of individual clones. The complexity of populations reduced to hundreds or perhaps a few thousand different members can be analyzed by the presence or absence of restriction enzyme recognition sites [19, 20]. Quite often, the experimenter would be prepared to examine dozens of survivors of an enrichment, especially in a search for a secondary trait such as altered crystallization. Thus, it would be convenient to monitor the complexity of the enriched populations and note when the complexity traverses the desired range.

Although most random populations vary in sequence but not in length, the utility of monitoring population complexity can be illustrated with populations that vary in both sequence and length. The examples in Figure 8.3 use repeating polypeptides, but the

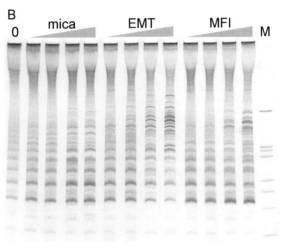

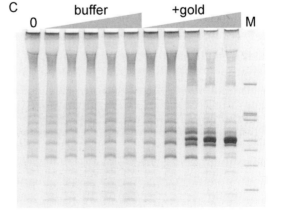

Figure 8.3 Monitoring population complexity. (B) PCR analysis of populations enriched by binding to three aluminum silicates, mica, and EMT and MFI zeolites [17]. 'M' are size standards or 'markers', pBR322 DNA digested with HinFI and HindIII. '0' is the original population. The bands appear diffuse because of the large number of different sequences comprising each size class. For each enrichment, the survivors of 10, 11, 12, and 13 cycles are examined. (C) PCR analysis of populations enriched by binding to gold powder or a mock enrichment with buffer alone (no gold added). 'M' and '0' are the same as in (B). For each enrichment, the survivors of two, three, four, five, and six cycles are examined.

method is suitable for any population that varies in both sequence and length [21, 22]. The strategy is depicted in panel A of Figure 8.3. The population contains inserts of varying size, and within each size class are inserts of many different sequences. A consequence of this dual variation is that a given sequence is present only in a certain size insert. This allows the sequence distribution to be displayed as the distribution of size classes. As a subset of peptides becomes enriched, the size of the inserts encoding them becomes more prominent. When the insert region of the population is amplified by PCR and separated by gel electrophoresis, we see the bands representing the enriched genes become pronounced. The results of this analysis can be seen in enrichments for binding to aluminum silicates [17] (Figure 8.3B). The bands produced from the initial population are diffuse because there are many different sequences in each size class. As enrichment proceeds, some bands become more prominent but they also become sharper as they arise from fewer different sequences. For both zeolites, the pattern continues to change with the number of enrichment cycles, but both enriched populations retain many different members. The pattern ceases to change after the second to last cycle of enrichment for binding to mica. Thus, further enrichment for mica-binding would not improve the likelihood of finding *bona fide* mica binders. Figure 8.3C shows the same analysis applied to an enrichment for gold binding. The gold-binding enrichment started with the same

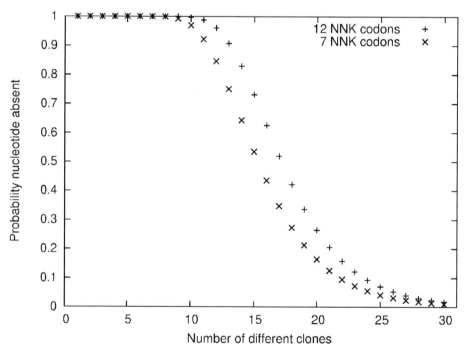

Figure 8.4 Probability of missing nucleotide. Shown is the probability that at least one 'N' position would be missing at least one of the four nucleotides among the indicated number of random sequences (clones). Commercially available libraries that have yielded inorganic surface-binding and -modifying peptides contain seven and 12 NNK codons.

population used for the aluminum silicate-binding enrichments. In the gold-binding enrichments, the population rapidly reduces to few members. In fact, more candidates can easily be tested than appear to remain after six cycles of enrichment. After four cycles of enrichment, although the population is dramatically reduced when compared to the initial library, the enriched population appears sufficiently complex to justify testing a number of individual survivors.

The complexity of populations varying in sequence can be probed by examining the nucleotide distribution. At each initially random position in the sequence, the probability of observing the absence of one of the four nucleotides increases as the population complexity declines. The expected probability for observing an absent nucleotide is shown in Figure 8.4. Thus, sequencing the DNA of a population can monitor its distribution without sequencing individual members of the population.

8.4
Quantification of Binding and Criteria for Specificity

Two aspects comprise this subject. The first is the affinity of the peptide for its target, and many authors report an equilibrium binding constant or the force necessary to separate the peptide from its target. The second is the specificity – an indication of how poor is the affinity of the peptide to undesired targets. A knowledge of these two values will aid in determining the effective concentrations of components in the assembly process. As in all areas of science, when considering values reported in the literature, one must pay close attention to experimental design. In most cases, binding of the peptide itself was measured, at least indirectly. However, cell extracts and phage lysates are complex mixtures. Evidence that a phenomenon is due to the recovered peptide rather than another constituent, including the product of the display vector, may require the incorporation of subtle controls.

8.5
Unselected Traits and Control of Crystallization/Reactivity

A clear distinction should be made between selections and screens. As discussed above, enriching a phage library for those that bind to a target is a selection, only the "fittest" survive. Testing individual phage from the enriched population for binding is a screen, as we identify both those that bind and those that do not. However, the selected trait is only one of many possible traits that can be screened. For example, as binding a substrate is the first step in catalysis, peptides from an appropriately enriched population can be tested for either stimulating or inhibiting a reaction. Similarly, peptides can be tested for diverting a reaction to new products. Binding peptides isolated in such a manner have been shown to stimulate the formation of ZnS nanocrystals [23] (Figure 8.5), and alter the shape of growing gold [24] and silver [25] crystals.

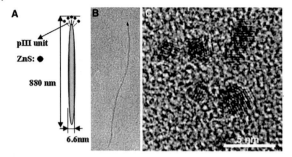

Figure 8.5 Characterization of the dilute A7-ZnS suspension using TEM. (A) Schematic diagram of the individual A7 phage and ZnS nanocrystals. The pIII peptide unit and the ZnS nanocrystal bound to A7 phage are not drawn to scale. (B) TEM image of an individual A7 phage (880 nm in length) and ZnS nanocrystals, stained with 2% uranyl acetate. (C) High-resolution TEM image of 0.01% A7-ZnS suspension, showing lattice fringe images of five wurtzite ZnS nanocrystals. The d spacing of the nanocrystals was 0.22 nm, corresponding to (102) plane. (Reproduced from Ref. [23].)

8.6
Dominant Traits, Interpretation of Gain-of-Function Mutants

Most peptides isolated in a search for binding peptides have a property not displayed by the phage; they are classical gain-of-function mutants. Therefore, although the peptide may mediate some function seen elsewhere such as in a natural biological system, we cannot *à priori* declare the mechanism by which the peptide acts to be the same as in the other system. This may seem a little abstract, but a clear example comes from our work. We isolated proteins that altered the shape of growing gold crystals [24]. These proteins probably act by altering the local environment at the surface of the growing crystal by a mechanism of acid catalysis rather than by providing a template for crystallization. This says that the control of crystal shape by proteins in natural biological systems can be carried out by altering the local environment, not that it necessarily is by altering the local environment. Many mechanisms are likely to be observed among proteins or peptides isolated from searches as discussed here, though many of them may not be commonly used by natural biological systems. Nonetheless, searches as discussed here are likely to identify new possible mechanisms that can be considered when investigating biological processes.

8.7
Interpretation and Requirement for Consensus Sequences

In the display strategies discussed here, the peptides are encoded by genes. The nature of the genetic code biases the initial population. In the natural, 64-member code, methionine and tryptophan are each encoded by only one nucleotide triplet, but arginine, leucine and serine are each encoded by six triplets and there are three triplets encoding stop. Although a 48-member code can eliminate two of the stop codons in populations where the orientation of the random DNA cannot be controlled [21], nearly all populations in which the orientation of the random DNA is controlled use a 32-member code. The triplets NNS or NNK can encode all 20 amino acids and retain only one stop codon. With a 32-member code, arginine, leucine and serine are each encoded by only three triplets, reducing the bias of the amino acid distribution. The frequency of encoded amino acids in the initial population also influences the interpretation of peptide sequences recovered from a search. Obviously, if the sequences are random, we expect to recover amino acids encoded by three codons more frequently than amino acids encoded by one codon.

If a common amino acid sequence is found in several peptides which bind the target, the analysis is simple (Table 8.2). However, in many cases either too few peptides were examined or the target has too many different features that are recognized by proteins for a consensus sequence to be identified. Although an overall amino acid composition of the recovered peptides may be expected [17, 26, 27], reflecting the charge or hydrophobicity of the target surface, if the binding is specific, we would expect only a few amino acids to constitute the binding site which contacts the recognized surface features. If only a small portion of the amino acid sequence is conserved, the overall amino acid composition is unlikely to vary from random in a statistically significant manner [26]. A fascinating counter-example derives from an analysis of silica-precipitating peptides. The four most efficient silica-precipitating peptides isolated from a phage display library contain at least five histidines [28]. The probability of at least five histidine codons appearing among 12 random expressed NNK codons is 0.004. Recovering such an amino acid distribution four times is even more significant, suggesting the histidines contribute to the precipitation of silica. Equally interesting is the absence of histidines in a diatom peptide that precipitates silica which can be explained by the extensive posttranslational modification of the diatom peptide [29]. This last observation should not discourage comparing recovered peptide sequences with sequences of naturally occurring proteins in public databases. A comparison of a ZnO-binding peptide with the SWISS-PROT data base found 15 of the 24 amino acids to be identical with a putative Zn-binding protein [22].

Table 8.2 Examples of consensus sequences

Target	Sequence
gold crystal shape	GASL–SEKL [24]
chromium-binding[a]	QHQK [15]
iron oxide-binding[b]	RR(S/T)-(R/K)HH [21, 45][c]
	RSK-R [21, 45]

[a] Surface oxidation not monitored
[b] Two forms of iron oxide
[c] Serine and threonine (S, T) are both hydroxy amino acids.
Arginine and lysine (R, K) are both basic amino acids.

8.8
Sizes of Proteins and Peptides

Proteins are macromolecules, and planned experiments may be influenced by the size of the peptides or proteins used. Core proteins occupy approximately 1.2 nm^3 per kilodalton [30]. This means that the average 12-amino-acid peptide, if spherical, would have a diameter of approximately 1.4 nm, excluding the hydration shell. If two heterologous binding peptides are fused to associate two different types of particles, the location and space occupied by the protein must be considered. A second value to consider is the maximum length of a fully extended peptide chain, about 3.6 Å per amino acid [31]. A flexible hinge peptide frequently used to fuse two peptide chains is G$_4$S (gly-gly-gly-gly-ser) [32] and would be expected to have a maximum length of 1.8 nm.

8.9
Mix and Match, Fusion Proteins, and Context-Dependence

Unless we can design and prepare bifunctional or multifunctional binding proteins, our ability to employ peptides in nanoassembly will be limited. Although it is tempting to think of proteins/peptides as modular, this is often not the case [33]. In many cases, peptides are isolated as part of a much larger protein or structure, and they may depend on the associated protein for folding or function. As with block copolymers, the peptide is tethered to another peptide with certain solubility properties. Each peptide or block can influence the structure and thus the properties of the other. Therefore, unless the peptides have been demonstrated to have a binding property independent of the display structure, fusing two peptides to prepare a bifunctional reagent may not always be successful. One source of confusion about the likelihood of successful fusions of bacterial proteins may arise from the early gene fusion work of Beckwith's laboratory (reviewed in Ref. [34]). Here, the fusions of proteins to β-galactosidase were functional because the authors selected function – that is, they were able to detect only those fusions which retained the functions of the constituents. This is not to say that peptides are unlikely to retain properties when fused to various other peptides – it only suggests that the properties be verified in the newly made fusion and occasionally a number of candidate peptides may have to be tested.

8.10
Mix and Match, Connecting Structures

What types of nanostructures can we expect to assemble with the aid of engineered proteins (for an introduction, see Ref. [35])? How should peptides be fused to create such engineered proteins? In some cases, we may want the peptides joined by a connector of a controlled length. Should the connector be flexible or rigid? We may want the peptides held in a fixed orientation relative to each other. Additionally, we may want many peptides in a large array. Numerous frameworks addressing the above problems are available among proteins of known structure. The survey below provides only a superficial starting point for consideration.

Structures on the µm scale can be based on M13 [23] (see Figure 8.2B), and the length of the phage can be varied over a several fold range by varying the length of the phage genome [36]. Multi-subunit proteins have been assembled by juxtaposing them as hybrids of pVII and pIX [8]. On a smaller scale, the flexible G_4S [32] linker described above reaches a maximum length of approximately 1.8 nm. The flexibility and length of linker peptides can be varied enormously [37]. The protein chains of antibodies can be fused directly to control the orientation of the two binding sites [38]. A rigid connector is the β-roll of *Pseudomonas aeruginosa* alkaline protease, which has the interesting property of spacing calcium cations approximately every 4.7 Å along its length [39]. Another rigid structure is the tail fiber of phage T4, which can accommodate the insertion of peptides in P37 [40].

The insertion of the protein streptavidin in a bacterial S-layer protein displays streptavidin in a regular, rigid two-dimensional array [41] (see Chapter 6). Streptavidin binds biotin tightly, and reactive derivatives of biotin allow it to be conjugated to many materials; alternately, it can be conjugated enzymatically to proteins containing a biotin-acceptor peptide [16, 42, 43].

8.11
Outlook

The isolation of peptides/proteins which are capable of adhering to the surface of specific materials has great potential in nanotechnology. Moreover, many peptides have the secondary trait of modulating the formation of the material they bind. Since chimeric binding peptides having dissimilar specificities can easily be produced by recombinant DNA techniques, we can expect to use peptides/proteins not only for the programmed formation and assembly of particles but also for the controlled localization of catalysts.

Assembly processes can be sequential, and subsets of the assembly process can take place in different reactions. Thus, proteins/peptides must only distinguish the constituents of the reaction in which they participate – they need not distinguish components of the eventual nanomachine that are absent or hidden in the reaction they mediate. One strategy for the programmed self-assembly of nanomachines could emulate solid-phase synthesis. The assembling machine or substructure is retained on a solid support as components are introduced, and the excess is removed with the liquid phase. Surface-binding proteins can provide, in addition to integral constituents of the machines, a simple and effective mechanism for immobilizing the assembling structure.

The control of nanoparticle formation may require carefully controlled conditions to be identified after an extensive survey of reaction conditions and constituents (for a recent report, see Ref. [44]). Binding peptides as discussed above may provide additional tools to the chemist in surveys of possible reaction constituents.

References

[1] A. Charbit, J. C. Boulain, A. Ryter, M. Hofnung, *EMBO J.* **1986**, *5*, 3029–3037.

[2] R. Freudl, S. MacIntire, M. Degen, U. Henning, *J. Mol. Biol.* **1986**, *188*, 491–494.

[3] G. Georgiou, C. Stathopoulos, P. S. Daugherty, A. R. Nyak, B. L. Iverson, R. Curtis, *Nature Biotechnol.* **1997**, *15*, 29–34.

[4] L. Hedegaard, P. Klemm, *Gene* **1989**, *85*, 115–124.

[5] W. Chen, G. Georgiou, *Biotechnol. Bioeng.* **2002**, *79*, 496–503.

[6] P. Malik, T. D. Terry, L. R. Gowda, A. Langara, S. A. Petukhov, M. F. Symmons, L. C. Welsh, D. A. Marvin, R. N. Perham, *J. Mol. Biol.* **1996**, *260*, 9–21.

[7] G. P. Smith, *Science* **1985**, *228*, 1315–1317.

[8] C. Gao, S. Mao, C.-H. L. Lo, W. Wirsching, R. A. Lerner, K. D. Janda, *Proc. Natl. Acad. Sci. USA* **1999**, *96*, 6025–6030.

[9] F. Felici, L. Castagnoli, A. Musacchio, R. Jappelli, G. Cesareni, *J. Mol. Biol.* **1991**, *222*, 301–310.

[10] B. K. Kay, J. Winter, J. McCafferty, *Phage display of peptides and proteins, a laboratory manual*, Academic Press, San Diego, **1996**.

[11] New England Biolabs, http://www.neb.com/neb/products/drug_discovery/phd.html

[12] L. C. Mattheakis, R. R. Bhatt, W. J. Dower, *Proc. Natl. Acad. Sci. USA* **1994**, *91*, 9022–9026.

[13] G. Cho, A. D. Keefe, R. Liu, D. S. Wilson, J. W. Szostak, *J. Mol. Biol.* **2000**, *297*, 309–319.

[14] W. J. Dower, L. C. Mattheakis, *Curr. Opin. Chem. Biol.* **2002**, *6*, 390–398.

[15] S. Brown, *Nature Biotechnol.* **1997**, *15*, 269–272.

[16] S. Brown, *Nano Lett.* **2001**, *1*, 391–394.

[17] S. Nygaard, R. Wendelbo, S. Brown, *Adv. Mater.* **2002**, *14*, 1853–1856.

[18] R. E. Hawkins, S. J. Russel, G. Winter, *J. Mol. Biol.* **1992**, *226*, 889–896.

[19] J. D. Marks, H. R. Hoogenboom, T. P. Bonnert, J. McCafferty, A. D. Griffiths, G. Winter, *J. Mol. Biol.* **1991**, *222*, 581–597.

[20] R. B. Christian, R. N. Zuckerman, J. M. Kerr, L. Wang, B. A. Malcolm, *J. Mol. Biol.* **1992**, *227*, 711–718.

[21] S. Brown, *Proc. Natl. Acad. Sci. USA* **1992**, *89*, 8651–8655.

[22] K. Kjærgaard, J. K. Sørensen, M. A. Schembri, P. Klemm, *Appl. Environ. Microbiol.* **2000**, *66*, 10–14.

[23] S.-W. Lee, C. Mao, C. E. Flynn, A. M. Belcher, *Science* **2002**, *296*, 892–895.

[24] S. Brown, M. Sarikaya, E. Johnson, *J. Mol. Biol.* **2000**, *299*, 725–735.

[25] R. R. Naik, S. J. Stringer, G. Agarwal, S. E. Jones, M. O. Stone, *Nature Mater.* **2002**, *1*, 169–172.

[26] N. B. Adey, A. H. Mataragnon, J. E. Rider, J. M. Carter, B. K. Kay, *Gene* **1995**, *156*, 27–31.
From libraries expressing random peptides as part of M13 pIII, the authors isolated phage that adhere to polystyrene and polyvinyl chloride. We also find polymer-binding peptides can be isolated but have observed they were often toxic if permitted to accumulate in contact with the cytoplasmic membrane (G. Barbarella and S. B., unpublished observation). We imagine their hydrophobic nature can disturb the membrane. The toxicity rendered their characterization difficult.

[27] S. R. Whaley, D. S. English, E. L. Hu, P. F. Barbara, A. M. Belcher, *Nature* **2000**, *405*, 665–668.

[28] R. R. Naik, L. L. Brott, S. J. Clarson, M. O. Stone, *J. Nanosci. Nanotechnol.* **2002**, *2*, 95–100.

[29] N. Krüger, S. Lorenz, E. Brunner, M. Sumper, *Science* **2002**, *298*, 584–586.

[30] J. Tsai, R. Taylor, C. Chothia, M. Gerstein, *J. Mol. Biol.* **1999**, *290*, 253–266.

[31] L. Pauling, R. B. Corey, H. R. Branson, *Proc. Natl. Acad. Sci. USA* **1951**, *37*, 205–211.

[32] J. S. Huston, D. Levinson, M. Mudgett-Hunter, M. S. Tai, J. Novotny, M. N. Margolies, R. J. Ridge, R. E. Bruccoleri, E. Haber, R. Crea, H. Oppermann, *Proc. Natl. Acad. Sci. USA* **1988**, *85*, 5879–5883.

[33] T. Matsuura, K. Miyai, S. Trakulnaleamesai, T. Yomo, Y. Shima, S. Miki, K. Yama-

moto, I. Urabe, *Nature Biotechnol.* **1999**, *17*, 58–61.

[34] P. Bassford, J. Beckwith, M. Berman, E. Brickman, M. Casadaban, L. Guarente, I. Saint-Girons, A. Sarthy, M. Schwartz, H. Shuman, T. Silhavy, in: J. H. Miller, W. S. Resnikoff (eds), *The Operon*, Cold Spring Harbor Laboratory, Cold Spring Harbor, New York, **1978**, pp. 245–261.

[35] C. M. Niemeyer, *Angew. Chem. Int. Ed.* **2001**, *40*, 4128–4158.

[36] W. M. Barnes, M. Bevan, *Nucleic Acids Res.* **1983**, *11*, 349–368.

[37] C. R. Robinson, R. T. Sauer, *Proc. Natl. Acad. Sci. USA* **1998**, *95*, 5929–5934.

[38] P. Holliger, T. Prospero, G. Winter, *Proc. Natl. Acad. Sci., USA* **1993**, *90*, 6444–6448.

[39] U. Baumann, S. Wu, K. M. Flaherty, D. B. McKay, *EMBO J.* **1993**, *12*, 3357–3364.

[40] P. Hyman, R. Valluzzi, E. Goldberg, *Proc. Natl. Acad. Sci. USA* **2002**, *99*, 8488–8493.

[41] D. Moll, C. Huber, B. Schlegel, D. Pum, U. B. Sleytr, M. Sára, *Proc. Natl. Acad. Sci. USA* **2002**, *99*, 14646–14651.

[42] J. E. Cronan, K. E. Reed, *Methods Enzymol.* **2000**, *326*, 440–458.

[43] D. Beckett, E. Kovaleva, P. J. Schatz, *Protein Sci.* **1999**, *8*, 921–929.

[44] Y. Sun, Y. Xia, *Science* **2002**, *298*, 2176–2179.

[45] C. F. Barbas, J. S. Rosenblum, R. A. Lerner, *Proc. Natl. Acad. Sci. USA* **1993**, *90*, 6385–6389.

[46] T. Schirmer, T. A. Keller, Y. F. Wang, J. P. Rosenbusch, *Science* **1995**, *267*, 512–514.

9
Microbial Nanoparticle Production

Murali Sastry, Absar Ahmad, M. Islam Khan, and *Rajiv Kumar*

9.1
Overview

Inorganic materials in the form of hard tissues are an integral part of most multicellular biological systems. Hard tissues are generally biocomposites containing structural bioma-cromolecules and some 60 different kinds of minerals that perform a variety of vital structural, mechanical, and physiological functions [1]. Unicellular organisms such as bacteria and algae also are capable of synthesizing inorganic materials, both intra- and extracellularly [2]. Examples of such organisms include magnetotactic bacteria which synthesize magnetite particles [3–5] (see also Chapter 10), diatoms and radiolarians that synthesize siliceous materials [6, 7], and S-layer bacteria that synthesize gypsum and calcium carbonate as surface layers [8]. These bioinorganic materials can be extremely complex both in structure and function, and also exhibit exquisite hierarchical ordering from the nanometer to macroscopic length scales which has not even remotely been achieved in laboratory-based syntheses. While the study of inorganic structures in biological systems would impact both the physical sciences (geology, mineralogy, physics, chemistry) and biological sciences (zoology, microbiology, evolution, physiology, cellular biology), one area where there is perhaps the greatest potential for application is that of materials science, particularly nanomaterials. Indeed, one branch of materials science where considerable development has already taken place is in the design and crystal growth of minerals such as calcium carbonate, hydroxyapatites, and gypsum by the use of biomimetic methods [9].

An important aspect of nanotechnology concerns the development of experimental procedures for the reproducible synthesis of nanomaterials of controlled size, polydispersity, chemical composition, and shape. Though solution-based chemical methods enjoy a long history dating back to the pioneering work of Faraday on the synthesis of aqueous gold colloids [10], increasing pressure to develop green chemistry, eco-friendly methods for nanomaterial synthesis has resulted in researchers turning to biological organisms for inspiration. It is interesting to note that while biotechnological applications such as remediation of toxic metals have long employed microorganisms such as bacteria [11, 12] and yeast [13], the detoxification process occurring by reduction of the metal ions or by formation of insoluble complexes with the metal ion (e. g., metal sulfides) in the form

Nanobiotechnology. Edited by Christof Niemeyer, Chad Mirkin
Copyright © 2004 WILEY-VCH Verlag GmbH & Co. KgaA, Weinheim
ISBN 3-527-30658-7

of nanoparticles, the possibility of using such microorganisms in the deliberate synthesis of nanomaterials is a recent phenomenon. An amalgamation of curiosity, environmental compulsions, and conviction that nature has evolved the best processes for synthesis of inorganic materials on nano- and macro-length scales has contributed to the development of a relatively new and largely unexplored area of research based on the use of microbes in the biosynthesis of nanomaterials.

Microbes are classically defined as microscopic organisms, the term most often having been used to signify bacteria. In this chapter, we will use the general definition to include in addition to bacteria, actinomycetes (both prokaryotes) and algae, yeasts, and fungi (eukaryotes). Some of the earliest reports on the accumulation of inorganic particles in microbes can be traced to the work of Zumberg, Sigleo and Nagy (gold in Precambrian algal blooms) [14], Hosea and coworkers (gold in algal cells) [15], Beveridge and co-workers (gold in bacteria) [16], Aiking and co-workers (CdS in bacteria) [11], Reese and co-workers (CdS in yeast) [17, 18], Temple and LeRoux (ZnS in sulfate-reducing bacteria) [19], and Blakemore, Maratea and Wolfe (magnetite in bacteria) [20]. More recent and detailed investigations into the use of microbes in the deliberate synthesis of nanoparticles of different chemical compositions include bacteria for silver [21–24], gold [24, 25–27] CdS [28–30] ZnS (sphalerite) [31], magnetite [3–5, 32], iron sulfide [33, 34], yeast for PbS [35] and CdS [36], and algae for gold [37]. In all these studies, the nanoparticles are formed intracellularly, but may be released into solution by suitable treatment of the biomass. Recently, we have shown that fungi when challenged with aqueous metal ions lead to the formation of nanoparticles both intra- and extracellularly [38–42]. Different genera of fungi have been identified for the extracellular synthesis of gold [38], silver [39], and CdS quantum dots [40], as well as the intracellular growth of nanocrystals of gold [41] and silver [42]. Extremophilic actinomycetes such as *Thermomonospora* sp. have also been used to synthesize fairly monodisperse gold nanoparticles extracellularly [43]. Yacaman and co-workers have demonstrated the growth of gold nanoparticles in sprouts, roots and stems of live alfalfa plants [44, 45].

Among the different microbes studied for the biosynthesis of nanoparticles, bacteria have received the most attention [11, 16, 19–34]. In a series of papers, Tanja Klaus and co-workers showed that the metal-resistant bacterium, *Pseudomonas stutzeri* AG259 (originally isolated from a silver mine), when challenged with high concentrations of silver ions during culturing resulted in the intracellular formation of silver nanoparticles of variable shape [21–23]. This is illustrated in the transmission electron microscopy (TEM) image of a *Pseudomonas stutzeri* AG259 cell with a number of silver particles located intracellularly (Figure 9.1). The particles were crystalline, were often observed to form at the poles of the bacteria, and were not particularly monodisperse, ranging in size from a few nm to 200 nm [21]. Most of the nanoparticles were found to be composed of elemental silver, while occasionally the formation of Ag_2S was observed [21]. The exact mechanism leading to the formation of intracellular silver nanoparticles in *P. stutzeri* AG259 is yet to be elucidated. Biofilms of metal nanoparticles embedded in a biological matrix may have important applications in the synthesis of eco-friendly and economically viable cermet materials for optically functional thin film coatings [23]. Jorger, Klaus and Granqvist showed that heat treatment of the Ag nano-bacteria biomass yielded hard coatings of a cermet that was resistant to mechanical scratching with a knife and whose optical proper-

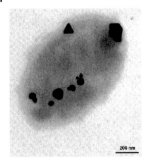

Figure 9.1 Silver-based crystals with different morphology, size and chemical composition produced by *P. stutzeri* AG259. Triangular, hexagonal and spheroid Ag-containing particles are accumulated at different cellular binding sites in the periplasmic space of the bacterial cell. (Reprinted with permission from Ref. [23]; © 2000 Wiley-VCH).

ties could be tailored by varying the silver loading factor [23]. The cermet material was composed primarily of graphitic carbon and up to 5% by weight (of the dry biomass) of silver.

In an interesting recent study, Nair and Pradeep have demonstrated that bacteria not normally exposed to large concentrations of metal ions also may be used to grow nanoparticles [24]. These authors showed that *Lactobacillus* strains present in buttermilk, when exposed to silver and gold ions, resulted in the large-scale production of nanoparticles within the bacterial cells [24]. The exposure of lactic acid bacteria present in the whey of buttermilk to mixtures of gold and silver ions could also be used to grow nanoparticles of alloys of gold and silver [24]. The UV-visible spectra of the bacterial colloids after exposure to pure silver and gold ions as well as a mixture of the two ions, are shown in Figure 9.2. The surface plasmon vibrations from the silver and gold bacterial colloids occur at 439

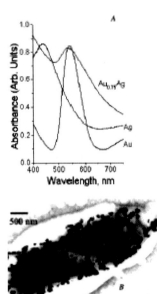

Figure 9.2 (A) Comparison of the UV/visible absorption spectra of bacterial colloids of pure Au and Ag with an alloy colloid of starting composition $Au_{0.75}Ag$. The peak maxima are 547, 439, and 537 nm for Au, Ag and $Au_{0.75}Ag$, respectively. Note that there is no peak due to Ag colloid in the alloy. The spectra have been moved vertically as there is a shift in baseline from sample to sample. (B) TEM of a bacterium with alloy crystallites. [111] Zone axis was seen in the electron diffraction; smaller crystallites were also seen outside the bacterium. (Reprinted with permission from Ref. [24]; © 2002 American Chemical Society).

and 547 nm respectively, while for the mixed alloy case it is centered at 537 nm. In the case of bacteria exposed to a mixture of the metal ions, the fact that the plasmon vibration wavelength is within the range defined by pure silver and gold nanoparticles, together with the absence of a distinct vibration corresponding to pure silver, was argued by Nair and Pradeep to indicate the formation of an alloy of the composition $Au_{0.75}Ag$, and not a core-shell structure [24]. By using a series of time-dependent UV-visible spectroscopy and TEM measurements, Nair and Pradeep concluded that the nucleation of the silver and gold nanoparticles occurs outside the bacterium (presumably on the cell surface through sugars and enzymes in the cell wall), following which the metal nuclei are transported into the cell where they aggregate and grow to larger-sized particles. The presence of noble metal nanocenters is known to enhance Raman spectroscopic signatures [46], and this feature was used by the authors to probe the internal chemical environment in the bacteria [24].

There is much interest in the development of protocols for the synthesis of semiconductor nanoparticles such as CdS for application as quantum-dot fluorescent biomarkers in cell labeling [47]. Simple variation of the particle size enables tailoring of the band gap and, consequently, the color of the quantum dots during UV-light irradiation [47]. Bacteria have been used with considerable success in the synthesis of CdS nanoparticles [28–30]. Holmes and co-workers have shown that exposure of the bacterium *Klebsiella aerogenes* to Cd^{2+} ions resulted in the intracellular formation of CdS nanoparticles in the size range 20–200 nm [29]. They also showed that the composition of the nanoparticles formed was a strong function of buffered growth medium for the bacterium. In an interesting extension of the bacteria-based methodology for the growth of magnetic nanoparticles,

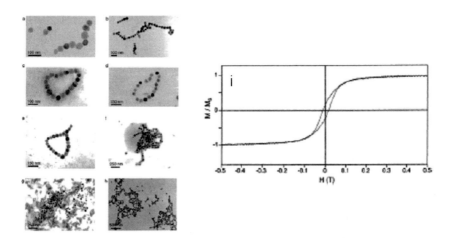

Figure 9.3 Representative selection of cluster morphologies for magnetite (Fe_3O_4) colloids extracted from cells imaged by transmission electron micrographs. Note the magnetic flux closure rings in images (c), (d), and (e). The tendency to form string-like aggregates can be clearly seen in for example, images (b) and (f). (i) Normalized magnetization M/M_S (M_S is the measured saturation magnetization at high field of dried magnetite particles extracted from cells) versus the applied magnetic field H. (Reprinted with permission from Ref. [4]; © 2002 American Chemical Society).

Roh and co-workers showed that metals such as Co, Cr, and Ni may be substituted into magnetite crystals biosynthesized in the thermophilic iron-reducing bacterium *Thermoanaerobacter ethanolicus* (TOR-39) [5]. This procedure led to the formation of octahedral-shaped magnetite nanoparticles in large quantities that co-existed with a poorly crystalline magnetite phase near the surface of the cells. A more fundamental investigation into the assembly of single-domain magnetite particles harvested from the bacterium *Magnetospirillum magnetotacticum* into folded-chain and flux-closure ring morphologies was carried out by Philipse and Maas [4]. The TEM images in Figure 9.3a–h show the magnetite particles extracted from the bacterial biomass by sonication. The particles are ~4.7 nm in diameter and predominantly organized in the form of rings (Figure 9.3c–e) and, more infrequently, as linear superstructures. The magnetite crystals are single domains with large magnetic moments that, when constrained to lie on a two-dimensional surface, are responsible for the head-to-tail assembly. The circular structures were explained by the authors to be flux-closure rings of in-plane dipoles. In conventional ferrofluids, the magnetic moments of the particles are much smaller than that observed for biogenic magnetite and therefore, such linear and ring-like structures have not been observed. Figure 9.3i shows data obtained from magnetization measurements of dried magnetite particles harvested from the bacterial cells. Based on these measurements and the magnitude of the remnant magnetization/coercive field, it was established that the biogenic magnetite nanoparticles are not superparamagnetic [4].

It has long been recognized that the exposure of yeasts such as *Candida glabrata* and *Schizosaccharomyces pombe* to Cd^{2+} ions leads to the intracellular formation of CdS quantum dots [17, 18]. In this particular case, the biochemical process resulting in the nanoparticle formation is well understood [17, 18]. Yeast cells exposed to Cd^{2+} ions produce metal-chelating peptides (glutathiones),and this is accompanied by an increase in the intracellular sulfide concentration and the formation of nanocrystalline CdS. The biogenic CdS quantum dots are capped and stabilized by the peptides, glutathione and its derivative phytochelatins with the general structure $(\gamma\text{-Glu-Cys})_n\text{Gly}$ [17, 18]. Based on an extensive screening program, Kowshik and co-workers have identified the yeast, *Torulopsis* sp. as being capable of intracellular synthesis of nanoscale PbS crystallites when exposed to aqueous Pb^{2+} ions [35]. The PbS nanoparticles were extracted from the biomass by freeze–thawing, and analyzed using a variety of techniques. A blue shift in the absorption edge suggested that the particles were indeed in the quantum size regime. A HRTEM image of the PbS nanoparticles harvested from the *Torulopsis* sp. biomass is shown in Figure 9.4, and shows clearly that the particles are quite spherical

a)

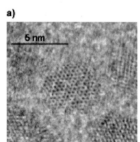

b)

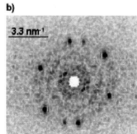

Figure 9.4 (a) An HRTEM image showing the near-spherical PbS nanocrystallites; (b) their diffraction pattern. (Reprinted with permission from Ref. [35]; © 2002 Wiley-VCH).

in shape and range in size from 4 to 8 nm. The particles are crystalline and exhibit a well-developed electron diffraction pattern (Figure 9.4b) with evidence for mixed cubic and hexagonal phases in the particles. Ultimately, biogenic nanoparticles would have to compete with chemically synthesized nanoparticles in terms of performance in devices. As a step in this direction, Kowshik et al. have shown that CdS quantum dots synthesized intracellularly in *Schizosaccharomyces pombe* yeast cells exhibit ideal diode characteristics [36]. Biogenic CdS nanoparticles in the size range 1–1.5 nm were used in the fabrication of a heterojunction with poly(*p*-phenylenevinylene). Such a diode exhibited 75 mA cm^{-2} current in the forward bias mode at 10 V, while breakdown occurred at 15 V in the reverse direction.

The use of fungi in the synthesis of nanoparticles is a relatively recent addition to the list of microbes discussed above. A detailed screening process involving approximately 200 genera of fungi resulted in two genera which, when challenged with aqueous metal ions such as AuCl$_4^-$ and Ag^{2+}, yielded large quantities of metal nanoparticles either extracellularly (*Fusarium oxysporum*) [38, 39] or intracellularly (*Verticillium* sp.) [41, 42]. The inset of Figure 9.5B shows flasks containing the *Verticillium* sp. biomass before (flask on top) and after exposure to 10^{-4} M HAuCl$_4$ solution for 72 hours [41]. The appearance of a distinctive purple color in the fungal biomass indicates formation of gold nanoparticles and can clearly be seen in the UV-visible absorption spectrum recorded from the gold-loaded biomass as a resonance at ~540 nm (Figure 9.5A, curve 2). This resonance is clearly missing in the biomass before exposure to gold ions (Figure 9.5A, curve 1) and in the filtrate (dotted line, Figure 9.5A) after reaction of *Verticillium* with the gold ions. The gold ions are thus reduced intracellularly, further evidence of which is provided by TEM analysis of thin sections of the cells after formation of gold nanoparticles (Figure 9.5C and D). In the low-magnification TEM image (Figure 9.5C), a number of nearly spherical gold nanoparticles are seen very close to the surface of the cells. At higher magnification (Figure 9.5D), the nanoparticles ranging in size from 5 nm to 200 nm with an average size of 20 $\pm$ 8 nm are clearly seen populating both the cell wall and cytoplasmic membrane of the fungus. Furthermore, the gold nanoparticles are crystalline, as can be seen from the powder X-ray diffraction pattern recorded from the biofilm (Figure 9.5B). The Bragg reflections are characteristic of face-centered cubic (fcc) gold structure. The reduction of the gold ions is expected to be due to reaction with enzymes present in the cell walls of the mycelia [41]. Exposure of *Verticillium* sp. to silver ions resulted in a similar intracellular growth of silver nanoparticles [42].

From the application point of view, extracellular synthesis of nanoparticles would be more important. We have observed that exposure of the fungus *Fusarium oxysporum* to aqueous gold and silver ions leads to the formation of fairly monodisperse nanoparticles in solution [38, 39]. Even more exciting was the finding that exposure of *Fusarium oxysporum* to aqueous CdSO$_4$ solution yielded CdS quantum dots extracellularly [40]. Figure 9.6A shows the CdS nanoparticles formed after reaction of 10^{-4} M CdSO$_4$ solution with the *Fusarium oxysporum* biomass for 12 days. The particles are reasonably monodisperse, and range in size from 5 to 20 nm. X-ray diffraction analysis of a film of the particles formed on a Si (111) wafer clearly showed that the particles were nanocrystalline with Bragg reflections characteristic of hexagonal CdS (Figure 9.6B). Reaction of the fungal biomass with aqueous CdNO$_3$ solution for an extended period of time did not yield CdS

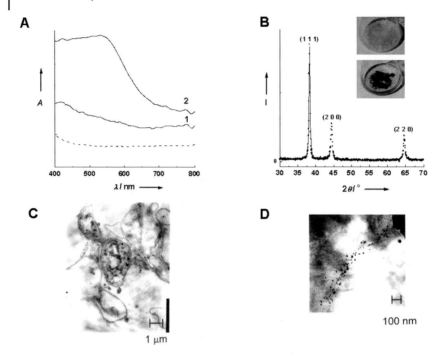

Figure 9.5 (A) UV/Visible spectra recorded from biofilms of the *Verticillium* sp. fungal cells before (curve 1) and after (curve 2) exposure to 10^{-4} M aqueous HAuCl₄ solution for 72 hours. The spectrum recorded from the HAuCl₄ solution after immersion of the fungal cells for 72 hours is shown for comparison (dashed line). (B) X-ray diffraction pattern recorded from an Au nano-*Verticillium* biofilm formed on a Si (111) wafer. The principal Bragg reflections are identified. The inset shows pictures of the *Verticillium* fungal cells after removal from the culture medium (flask on top) and after exposure to 10^{-4} M aqueous solution of HAuCl₄ for 72 hours (flask at bottom). (C, D) TEM images at different magnifications of thin sections of stained *Verticillium* cells after reaction with AuCl₄⁻ ions for 72 hours. (Reprinted with permission from Ref. [41]; © 2001 Wiley-VCH).

nanoparticles, indicating the possibility of release of a sulfate reductase enzyme into solution. The inset of Figure 9.6B shows the polyacrylamide gel electrophoresis (PAGE) results of the aqueous extract exposed to the fungal biomass for 12 days. The electrophoresis measurements indicate the presence of at least four protein bands in the extract. Reaction of the protein extract after dialysis (using a dialysis bag with 3kDa molecular weight cutoff) with CdSO₄ solution did not yield CdS nanoparticles. However, addition of ATP and NADH to the dialysate restored the CdS formation capability of the protein extract. It is believed that the same proteins are also responsible for the reduction of gold and silver ions. The gold, silver, and CdS nanoparticles were stable in solution for many months due to stabilization by surface-bound proteins [38, 40]. The development of a rational nanoparticle biosynthesis procedure using specific enzymes secreted by fungi in both the intra- and extracellular synthesis of nanoparticles has many attractive associated features. Plant pathogenic fungi produce copious quantities of enzymes, are usually nonpathogenic to humans, and are easily cultured in the laboratory.

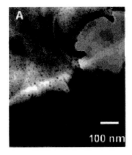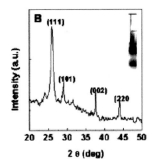

Figure 9.6 (A) Bright-field TEM image of CdS nanoparticles formed by reaction of CdSO$_4$ with the fungal biomass for 12 days. (B) X-ray diffraction pattern recorded from the CdS nanoparticle film deposited on a Si (111) wafer. The inset shows the native gel electrophoresis of aqueous protein extract obtained from *Fusarium oxysporum* mycelia; 10% (w/v) polyacrylamide slab gel, pH 4.3 (Reprinted with permission from Ref. [40]; © 2002 American Chemical Society).

9.2
Outlook

A case for the serious investigation of microorganisms such as bacteria, algae, yeasts, actinomycetes, and fungi as possible inorganic nanofactories has been made. A number of issues from the nanotechnology and microbiology points of view require to be addressed before such a biosynthesis procedure can compete with existing physical and chemical synthesis protocols. The elucidation of biochemical pathways leading to metal ion reduction or formation of insoluble complexes in the different classes of microbes is essential in order to develop a rational microbial nanoparticle synthesis procedure. Likewise, an understanding of the surface chemistry of the biogenic nanoparticles (i. e., the nature of capping surfactants/peptides/proteins) would be equally important. This would then lead to the possibility of genetically engineering microbes to overexpress specific reducing molecules and capping agents, thereby controlling not only the size of the nanoparticles but also their shape. The rational use of constrained environments within cells such as the periplasmic space and cytoplasmic vesicular compartments (e. g., magnetosomes) to modulate nanoparticle size and shape is an exciting possibility. The range of chemical compositions of nanoparticles currently accessible by microbial methods is currently extremely limited and confined to metals, some metal sulfides and iron oxide. Extension of the protocols to enable reliable synthesis of nanocrystals of other oxides (TiO$_2$, ZrO$_2$, etc.) and nitrides, carbides, etc. could make microbial synthesis a commercially viable proposition.

Equally intriguing are questions related to the metal ion reduction/reaction process in cellular metabolism, and whether the nanoparticles formed as byproducts of the reduction process have any role to play in cellular activity (e. g., magnetite in magnetotactic bacteria). Plant organisms (e. g., fungi) are not normally exposed to high concentrations of metal ions such as Cd^{2+}, AuCl$_4^-$ and Ag^{2+}. The fact that, when challenged, they secrete enzymes that are capable of metal ion reduction – and indeed conversion of sulfates to sulfides – suggests that evolutionary processes may be at play.

Acknowledgments

The authors would like to thank graduate students Priyabrata Mukherjee, Satyajyoti Senapati, and Deendayal Mandal for their enthusiastic contributions to much of the experimental work.

References

[1] H. A. Lowenstam, *Science* **1981**, *211*, 1126.

[2] K. Simkiss, K. M. Wilbur, *Biomineralization*, Academic Press, New York, **1989**.

[3] R. B. Frankel, R. P. Blakemore (eds). *Iron Biominerals*, Plenum Press, New York, **1991**.

[4] A. P. Philipse, D. Maas, *Langmuir* **2002**, *18*, 9977.

[5] Y. Roh, R. J. Lauf, A. D. McMillan, C. Zhang, C. J. Rawn, J. Bai, T. J. Phelps, *Solid State Commun.* **2001**, *118*, 529.

[6] N. Kröger, R. Deutzmann, M. Sumper, *Science* **1999**, *286*, 1129.

[7] J. Parkinson, R. Gordon, *Trends Biotech.* **1999**, *17*, 190.

[8] S. Schultz-Lam, G. Harauz, T. J. Beveridge, *J. Bacteriol.* **1992**, *174*, 7971.

[9] S. Mann, *Biomineralization. Principles and Concepts in Bioinorganic Materials Chemistry*, Oxford University Press, Oxford, **2001**.

[10] M. Faraday, *Philos. Trans. R. Soc. London* **1857**, *147*, 145.

[11] H. Aiking, K. Kok, H. van Heerikhuizen, J. van't Riet, *Appl. Environ. Microbiol.* **1982**, *44*, 938.

[12] J. R. Stephen, S. J. Maenaughton, *Curr. Opin. Biotechnol.* **1999**, *10*, 230.

[13] R. K. Mehra, D. R. Winge, *J. Cell. Biochem.* **1991**, *45*, 30.

[14] J. E. Zumberg, A. C. Sieglo, B. Nagy, *Miner. Sci. Eng.* **1978**, *10*, 223.

[15] M. Hosea, B. Greene, R. McPherson, M. Heinzl, M. D. Alexander, D. W. Darnall, *Inorg. Chim. Acta* **1986**, *123*, 161.

[16] T. J. Beveridge, R. J. Doyle, *Metal Ions and Bacteria*, J. Wiley Sons, New York, **1989**.

[17] R. N. Reese, D. R. Winge, *J. Biol. Chem.* **1988**, *263*, 12832.

[18] C. T. Dameron, R. N. Reese, R. K. Mehra, A. R. Kortan, P. J. Carroll, M. L. Steigerwald, L. E. Brus, D. R. Winge, *Nature* **1989**, *338*, 596.

[19] K. L. Temple, N. LeRoux, *Econ. Geol.* **1964**, *59*, 647.

[20] R. P. Blakemore, D. Maratea, R. S. Wolfe, *J. Bacteriol.* **1979**, *140*, 720.

[21] T. Klaus, R. Joerger, E. Olsson, C.-G. Granqvist, *Proc. Natl. Acad. Sci. USA* **1999**, *96*, 13611.

[22] T. Klaus-Joerger, R. Joerger, E. Olsson, C.-G. Granqvist, *Trends Biotech.* **2001**, *19*, 15.

[23] R. Joerger, T. Klaus, C.-G. Granqvist, *Adv. Mater.* **2000**, *12*, 407.

[24] B. Nair, T. Pradeep, *Cryst. Growth Des.* **2002**, *2*, 293.

[25] T. J. Beveridge, R. G. E. Murray, *J. Bacteriol.* **1980**, *141*, 876.

[26] G. Southam, T. J. Beveridge, *Geochim. Cosmochim. Acta* **1996**, *60*, 4369.

[27] D. Fortin, T. J. Beveridge, in: *Biomineralization. From Biology to Biotechnology and Medical Applications*, E. Baeuerien (ed.), Wiley-VCH, Weinheim, **2000**, p. 7.

[28] D. P. Cunningham, L. L. Lundie, *Appl. Environ. Microbiol.* **1993**, *59*, 7.

[29] J. D. Holmes, P. R. Smith, R. Evans-Gowing, D. J. Richardson, D. A. Russell, J. R. Sodeau, *Arch. Microbiol.* **1995**, *163*, 143.

[30] P. R. Smith, J. D. Holmes, D. J. Richardson, D. A. Russell, J. R. Sodeau, *J. Chem. Soc., Faraday Trans.* **1998**, *94*, 1235.

[31] M. Labrenz, G. K. Druschel, T. Thomsen-Ebert, B. Gilbert, S. A. Welch, K. M. Kemner, G. A. Logan, R. E. Summons, G. De Stasio, P. L. Bond, B. Lai, S. D. Kelly, J. F. Banfield, *Science* **2000**, *290*, 1744.

[32] D. R. Lovley, J. F. Stolz, G. L. Nord, E. J. P. Phillips, *Nature* **1987**, *330*, 252.

[33] J. H. P. Watson, D. C. Ellwood, A. K. Sper, J. Charnock, *J. Magn. Magn. Mater.* **1999**, *203*, 69.

[34] J. H. P. Watson, B. A. Cressey, A. P. Roberts, D. C. Ellwood, J. M. Charnock, *J. Magn. Magn. Mater.* **2000**, *214*, 13.

[35] M. Kowshik, W. Vogel, J. Urban, S. K. Kulkarni, K. M. Paknikar, *Adv. Mater.* **2002**, *14*, 815.

[36] M. Kowshik, N. Deshmukh, W. Vogel, J. Urban, S. K. Kulkarni, K. M. Paknikar, *Biotech. Bioeng.* **2002**, *78*, 583.

[37] M. G. Robinson, L. N. Brown, D. Beverley, *Biofouling* **1997**, *11*, 59.

[38] P. Mukherjee, S. Senapati, D. Mandal, A. Ahmad, M. I. Khan, R. Kumar, M. Sastry, *ChemBioChem.* **2002**, *3*, 461.

[39] A. Ahmad, P. Mukherjee, S. Senapati, D. Mandal, M. I. Khan, R. Kumar, M. Sastry, *Coll. Surf. B.* **2003**, *28*, 313.

[40] A. Ahmad, P. Mukherjee, D. Mandal, S. Senapati, M. I. Khan, R. Kumar, M. Sastry, *J. Am. Chem. Soc.* **2002**, *124*, 12108.

[41] P. Mukherjee, A. Ahmad, D. Mandal, S. Senapati, S. R. Sainkar, M. I. Khan, R. Ramani, R. Parischa, P. V. Ajaykumar, M. Alam, M. Sastry, R. Kumar, *Angew. Chem. Int. Ed.* **2001**, *40*, 3585.

[42] P. Mukherjee, A. Ahmad, D. Mandal, S. Senapati, S. R. Sainkar, M. I. Khan, R. Parischa, P. V. Ajayakumar, M. Alam, R. Kumar, M. Sastry, *Nano Lett.* **2001**, *1*, 515.

[43] A. Ahmad, S. Senapati, M. I. Khan, R. Kumar and M. Sastry, *Langmuir* **2003**, *19*, 3550.

[44] J. L. Gardea-Torresdey, J. G. Parsons, E. Gomez, J. Peralta-Videa, H. E. Troiani, P. Santiago, M. J. Yacaman, *Nano Lett.* **2001**, *2*, 374.

[45] J. Gardea-Torresdey, E. Gomez, J. Peralta-Videa, J. Parsons, H. Troiani, M. Jose-Yacaman. *Langmuir* **2003**, *19*, 1357.

[46] Y. C. Cao, R. Jin, C. A. Mirkin, *Science* **2002**, *297*, 1536.

[47] W. C. W. Chan, D. J. Maxwell, X. Gao, R. E. Bailey, M. Han, S. Nie, *Curr. Opin. Biotech.* **2002**, *13*, 40.

10
Magnetosomes: Nanoscale Magnetic Iron Minerals in Bacteria

Richard B. Frankel and *Dennis A. Bazylinski*

10.1
Introduction

10.1.1
Magnetotactic Bacteria

Magnetotaxis is the orientation and navigation along magnetic field lines by motile, aquatic, bacteria [1, 2]. Magnetotactic bacteria are generally found in chemically stratified water columns or sediments where they occur predominantly in or below the microaerobic redox transition zone, between the aerobic zone of upper waters or sediments and the anaerobic regions of the habitat [3]. They are a diverse group of microorganisms with respect to morphology, physiology, and phylogeny [4, 5]. Commonly observed morphotypes include coccoid to ovoid cells, rods, vibrios, and spirilla of various dimensions, and multicellular forms. All known magnetotactic bacteria are motile by means of flagella, and possess a cell-wall structure characteristic of Gram-negative bacteria. The arrangement of flagella varies between species/strains: cells with polar or bipolar single flagella and others with bundles of flagella have been observed.

Magnetotactic bacteria are difficult to isolate and grow in pure culture. Most cultured strains belong to the genus *Magnetospirillum* (American Type Culture Collection, Washington, DC). Several other freshwater magnetotactic spirilla in pure culture have not yet been completely described [6]. Other species of cultured magnetotactic bacteria include a number of incompletely characterized organisms: the marine vibrios, strains MV-1 and MV-2; a marine coccus, strain MC-1; and a marine spirillum, strain MV-4 [7]. There is also an anaerobic, sulfate-reducing, rod-shaped bacterium, *Desulfovibrio magneticus*, strain RS-1 [8]. These cultured organisms, except strain RS-1, are obligate or facultative microaerophiles and all are chemoorganoheterotrophic, although the marine strains also grow chemolithoautotrophically.

Nanobiotechnology. Edited by Christof Niemeyer, Chad Mirkin
Copyright © 2004 WILEY-VCH Verlag GmbH & Co. KgaA, Weinheim
ISBN 3-527-30658-7

10.1.2
Magnetosomes

All magnetotactic bacteria contain magnetosomes, which are intracellular structures comprising magnetic iron mineral crystals enveloped by a membrane vesicle [9–11]. The magnetosome membrane (MM) is presumably a structural entity that anchors the crystals at particular locations in the cell [12], as well as the locus of biological control over the nucleation and growth of the magnetosome crystals. The MM in the genus *Magnetospirillum* is a lipid bilayer consisting of neutral lipids, free fatty acids, glycolipids, sulfolipids, and phospholipids [10, 11]. It is often located adjacent to the cytoplasmic membrane, although there is no clear microscopic evidence for direct connections between the two. Empty and partially-filled vesicles have been reported in iron-starved cells, suggesting that magnetosome vesicles are formed prior to the deposition of the mineral crystals [10].

The magnetosome magnetic mineral phase consists of magnetite, Fe_3O_4, or greigite, Fe_3S_4. Each magnetotactic species or strain exclusively produces either magnetite or greigite magnetosomes, except for one marine organism that produces magnetosomes of both kinds [3]. The magnetosome crystals are typically of order 35 to 120 nm in length, although crystals with lengths of ~200 nm are known [12]. In most magnetotactic bacteria, the magnetosomes are organized in one or more straight chains of various lengths parallel to the long axis of the cell [13], as shown in Figure 10.1. Dispersed aggregates or clus-

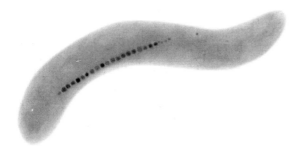

Figure 10.1 Transmission electron micrograph (TEM) of a cell of *Magnetospirillum magnetotacticum* showing the intracellular chain of magnetosomes. The chain of magnetosomes is approximately 1100 nm long. Excluding the smallest crystals at its ends, it contains 19 magnetite crystals that have an average diameter and an average separation of ~45 and ~9.5 nm, respectively.

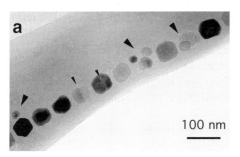
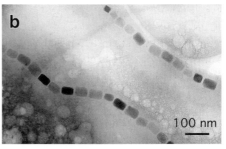

Figure 10.2 Higher magnification TEMs of magnetosome magnetite crystals in (a) *M. magnetotacticum*, and (b) cultured strain MV-1. Arrows in (a) indicate crystal twinning and anomalously small crystals.

ters of magnetosomes occur in some magnetotactic bacteria, usually at one side of the cell, which often corresponds to the site of flagellar insertion [14, 15]. The narrow size range and consistent morphologies of the magnetosome crystals in each species or strain [16], and the consistent crystallographic orientation of the magnetosomes in chains [17], as shown in Figure 10.2, are clear indications that the magnetotactic bacteria exert a high degree of control over magnetosome formation and arrangement.

10.1.3
Cellular Magnetic Dipole and Magnetotaxis

The size-range and linear arrangement of magnetosomes within a magnetotactic bacterium are highly significant for the magnetic properties of the cell [18]. The magnetosome crystals are within the permanent single-magnetic-domain (SD) size-range for both magnetite and greigite, and are thus uniformly magnetized with the maximum magnetic dipole moment per unit volume [18, 19]. Magnetic crystals larger than SD size are nonuniformly magnetized because of formation of magnetic domains or vortex configurations; this has the effect of significantly reducing their magnetic dipole moments. On the other hand, very small SD particles are superparamagnetic (SPM). Although SPM particles are uniformly magnetized, their magnetic dipole moments are not constant because of spontaneous, thermally induced, reversals which produce a time-averaged moment of zero. Therefore, magnetotactic bacteria produce magnetosomes with the optimum particle size for the maximum, permanent, magnetic dipole moment per magnetosome.

The arrangement of the SD magnetosomes in chains maximizes the dipole moment of the cell because magnetic interactions between the magnetosomes in a chain cause each magnetosome moment to orient spontaneously and in parallel with others along the chain axis, minimizing the magnetostatic energy. Thus the total dipole moment of the chain, **M**, is the algebraic sum of the moments of the individual magnetosomes in the chain [18]. However, this is true only because the magnetosomes are physically constrained by the magnetosome membranes in the chain configuration. If free to float in the cytoplasm, magnetosomes would likely clump, resulting in a much smaller net dipole moment than in the chain. For organisms such as *Magnetospirillum magnetotacticum*, the remanent moment is the maximum possible moment of the chain [20].

Magnetotaxis results from the passive orientation of a swimming magnetotactic bacterium along the local magnetic field by the torque exerted by the field **B** (e. g., the geomagnetic field) on the cellular dipole moment **M** [18]. A chain of 10–20 magnetosomes, each of dimension 50 nm, would be sufficient for the orientation of a magnetotactic bacterium in the geomagnetic field at ambient temperature. Since the chain of particles is fixed within the cell [21], the entire cell is oriented by the torque exerted on the magnetic dipole by the magnetic field. If the magnetic field is decreased, the time-averaged orientation of the cell along the field is decreased and the migration rate of the cell along the magnetic field direction is decreased, even though the forward-swimming speed of the cell is unchanged. Thus, magnetotactic bacteria essentially behave like self-propelled magnetic dipoles.

The potential energy (E) of the cellular magnetic dipole moment in the magnetic field is given by:

$$E_\Theta = - MB \cos\Theta \tag{1}$$

where Θ is the angle between **M** and **B**. Thermal energy at ambient temperatures will tend to cause misalignment of the swimming bacterium. In thermal equilibrium at temperature T, the probability of the moment having energy E_Θ is proportional to the Boltzmann factor $\exp(-E_\Theta/kT)$, where k is Boltzmann's constant. The thermally averaged projection of the dipole moment on the magnetic field $<\cos\Theta>$ can be determined from the Langevin theory of paramagnetism and is given by the Langevin function $L(\alpha)$ [18]:

$$<\cos\Theta> = L(\alpha) = \coth(\alpha) - 1/\alpha \tag{2}$$

where $\alpha = MB/kT$. $L(\alpha) = 0$ for $\alpha = 0$ and asymptotically approaches 1 as α approaches ∞. In particular, $L = 0.9$ when $\alpha = 10$.

Experimental determination of the average dipole moment per cell of *M. magnetotacticum* by electron holography gave a value of 5×10^{-16} Am^2 [20]. In the geomagnetic field of ca. 50 µT at room temperature, $L(\alpha)$ is greater than 0.8, meaning that the migration rate of cells along the local direction of the geomagnetic field would be 80% of their forward-swimming speed. If the number of magnetosomes, and hence **M**, is low, then the alignment of the cell and its migration along the field lines is inefficient. On the other hand, increasing the number of magnetosomes beyond a certain value will not significantly improve the alignment of the cell in the field because of the asymptotic approach of $L(\alpha)$ to 1 for large α. Magnetotactic bacteria control the biomineralization process to produce a sufficient number of magnetosomes of optimal size for efficient magnetic navigation in the geomagnetic field.

10.1.4
Magneto-Aerotaxis

Like most other free-swimming bacteria, magnetotactic bacteria propel themselves through the water by rotating their helical flagella [22]. Because of their magnetosomes, magnetotactic bacteria are passively oriented and actively migrate along the local magnetic field **B**, which in natural environments is the geomagnetic field. When distinct morphotypes of magnetotactic bacteria, isolated and grown in pure culture, were studied in oxygen concentration gradients using thin, flattened capillaries (Vitrocom, Inc.), it became clear that magnetotaxis and aerotaxis work together in these bacteria [22]. The behavior observed in these strains has been referred to as "magneto-aerotaxis", and two different magneto-aerotactic mechanisms – termed polar and axial – are found in different bacterial species. For both polar and axial magnetotactic bacteria, the cellular magnetic dipole remains oriented along the local magnetic field, but the direction of migration along the magnetic field lines is determined by the sense of flagellar rotation, which in turn is controlled by aerotactic receptors. Thus, a magnetotactic bacterium is essentially a self-propelled magnetic dipole with an oxygen sensor. Magnetotaxis effectively turns a three-dimensional search problem to find the optimal oxygen concentration into a one-dimensional search problem by using the magnetic field.

10.1.5
Magnetite Crystals in Magnetosomes

High-resolution transmission electron microscopy, selected-area electron-diffraction studies and electron holography have revealed that the magnetite crystals within magnetotactic bacteria are of relatively high structural perfection [16] and have been used to determine their idealized morphologies [17, 23]. The morphologies are all derived from combinations of {111}, {110}, and {100} forms (a form refers to the equivalent symmetry related lattice planes of the crystal structure) with some distortions [16]. These include cuboctahedral ([100] + [111]), and elongated, nonequidimensional prismatic (Figure 10.3). The cubocta-hedral crystal morphology preserves the symmetry of the face-centered cubic spinel structure – that is, all symmetry-related crystal faces are equally developed. In the elongated and prismatic morphologies, some symmetry-related faces are unequally developed. This implies anisotropic growth conditions, for example, an anisotropic ion flux into the magnetosome membrane. It is thought that magnetite forms from a precursor, an amorphous iron oxide phase [17].

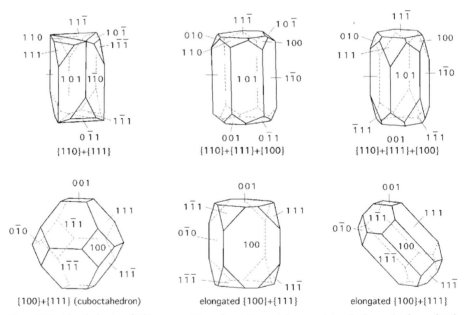

Figure 10.3 Morphologies of Fd3m magnetite and greigite crystals comprising the low index forms {100}, {111}, and {110}. Anisotropic growth causes symmetry breaking in all but the cuboctahedron (lower left).

10.1.6
Greigite Crystals in Magnetosomes

Whereas virtually all freshwater, magnetotactic bacteria have been found to synthesize magnetite magnetosomes, many marine, estuarine, and salt-marsh species produce iron sulfide magnetosomes consisting primarily of the magnetic iron sulfide, greigite (Fe_3S_4) [24–26]. While none of these organisms is currently available in pure culture, recognized greigite-producing magnetotactic bacteria include a multicellular, magnetotactic prokaryote [27] and a variety of relatively large, rod-shaped bacteria. The greigite crystals in their magnetosomes are thought to form from nonmagnetic precursors including mackinawite (tetragonal FeS) and possibly a sphalerite-type cubic FeS [26]. Like magnetite crystals in magnetosomes, the morphologies of the greigite crystals also appear to be species- and/or strain-specific [28]. As noted above, there is one reported instance of a marine bacterium that contains magnetite and greigite magnetosomes co-organized within the same magnetosome chain [3].

10.1.7
Biochemistry and Gene Expression in Magnetosome Formation

Knowledge of the biochemical and genetic controls on magnetite production is essential to understanding how the magnetotactic bacteria produce magnetosomes and organize them in chains. Although progress has been made by several laboratories, and the genomes of two magnetotactic bacteria (*M. magnetotacticum* and strain MC-1) [29] have been sequenced, the overall process is not well understood and is a focal point for future work.

Magnetosome mineral formation must begin with transport of Fe into the cell and deposition within the MM vesicles to form a saturated Fe solution [17, 30, 31] (Figure 10.4). Manipulation of the redox conditions within the MM vesicles so that [Fe(III)]/[Fe(II)] was ~2, corresponding to a redox potential of about –100 mV at elevated pH, would make magnetite the most stable Fe-oxide phase [32]. The MM could also provide sites for nucleation and growth of the magnetite crystals. Interactions of the MM with the faces of the growing crystal could affect crystal morphology [17].

Schüler and Baeuerlein [33] found that initially Fe-starved cells became magnetic (i. e., formed magnetosomes) within ~10 minutes after incubation in 10 μM Fe(III). Since Fe can amount to ca. 2–3 % of the dry weight of a magnetotactic bacterium [33, 34], an efficient Fe uptake system is required. An important question in the magnetite synthesis

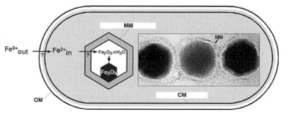

Figure 10.4 Model for magnetosome mineral formation in the magnetosome membrane vesicles in magnetotactic bacteria. CM, cytoplasmic membrane; MM, magnetosome membrane; OM, outer membrane. (Adapted from a figure kindly provided by D. Schüler.)

process concerns whether Fe(III) or Fe(II) is transported into magnetosome membrane vesicles. Cells of *Magnetospirillum gryphiswaldense* mainly take up Fe(III) via a high-affinity system [33]. Research has shown that magnetotactic bacterial cells have Fe(III) reductase activity [35], and a Fe(III) reductase has been isolated from *M. magnetotacticum*. [36]. Although this protein appears to be mainly cytoplasmic, it may be bound on the inner side of the cytoplasmic membrane and could participate in the reduction of Fe(III) as it enters the cell [36].

Since the MM is thought to be of paramount importance in magnetosome mineral formation, researchers have focused on the role MM proteins – that is, proteins which occur in the magnetosome membrane but not in the soluble (periplasmic and cytoplasmic) fraction, nor the cytoplasmic or outer membranes, in magnetosome synthesis [7, 31, 36, 37]. Two different approaches have been used: (i) N-terminal amino acid sequencing of the MM proteins, followed by "reverse genetics" to obtain the gene sequences for these proteins; and (ii) performing biochemical protein comparison of mutants that do not produce magnetosomes with wild-type strains, then again using "reverse genetics" to determine gene sequences.

The *mam* (mam: magnetosome membrane) genes [11] appear to be conserved in a large gene cluster within several magnetotactic bacteria (*Magnetospirillum* species and strain MC-1) and may be involved in magnetite biomineralization. Grünberg et al. [11] cloned and sequenced some of the *mam* genes in *M. gryphiswaldense* that were assigned to two different genomic regions. The proteins resulting from these gene sequences exhibited the following homologies: MamA to tetratricopeptide repeat proteins; MamB to cation diffusion facilitators; and MamE to HtrA-like serine proteases. The gene sequences of MamC and MamD showed no homology to existing proteins. A gene cluster containing *MamA* and *MamB* was also found in *M. magnetotacticum* and strain MC-1 which also contained genes that showed no homology with known genes or proteins in established databases [11]. Definitive functions have not yet been ascribed to these proteins.

The *MagA* gene found in *M. magneticum* strain AMB-1 [38] encodes a protein that is a proton-gradient-driving H/Fe(II) antiporter. The proton-driving pump is situated on the cytoplasmic membrane. *MagA* has been expressed in *Escherichia coli*, and membrane vesicles prepared from these cells that contained the *magA* gene product took up Fe, though only when ATP was supplied [38], indicating that Fe uptake was energy-dependent. It was also shown, using a *magA-luc* fusion protein, that magA is a membrane protein localized in the cell membrane and possibly the magnetosome membrane [39].

Three other genes that encode MM specific proteins, *mms6*, *mms16*, and *mpsA*, have been obtained from *Magnetospirillum* strain AMB-1 [40, 41]. These genes were also found in the genome of *M. magnetotacticum*. MpsA exhibits homology to an acyl-CoA transferase, while Mms16 shows GTPase activity and is possibly involved in MM vesicle formation by invagination and budding from the cytoplasmic membrane [42]. Mms6, the most abundant of the three, is apparently bound to magnetite and may function in regulation of crystal growth [42].

10.1.8
Applications of Magnetosomes

Magnetosomes have been exploited in a number of applications [30, 44, 46]. Commercial applications include the immobilization of enzymes for use in biosensors, the formation of magnetic antibodies in various fluoroimmunoassays and the quantification of IgG, the detection and separation of various cell types, and the transfer of genes into cells. Magnetosomes also show promise as MRI contrast enhancement agents. Whole magnetotactic bacteria have been used for cell separations, as oxygen biosensors, and in studies of magnetic domains in meteorites and terrestrial rocks.

10.2
Research Methods

Research on magnetosomes in magnetotactic bacteria primarily involves conventional, well-known, microbiological and molecular biological methods. Details on culturing magnetotactic bacteria, extracting magnetosomes and analyzing MM proteins and genes are given in a number of references (e. g., [6, 10, 11, 43−45]).

10.3
Conclusion and Future Research Directions

Magnetotactic bacteria have solved the problem of constructing an internal, permanent, magnetic dipole that is sufficiently robust so that a cell will be aligned along the geomagnetic field as it swims, yet be no longer than the length of the cell (ca. 1–2 µm). The solution involves a hierarchical structure – the magnetosome chain – and a mineralization process in which the mineral type, grain size and placement in the cell are all controlled by the cell. Primary control is presumably exerted by the magnetosome membrane through MM proteins. The roles of the MM proteins and the details of the magnetosome mineralization process are the most important issues to be elucidated.

Directions for future research include the development of genetic systems in order to move genes in and out of magnetotactic bacteria, and to prepare knock-out mutants of strains in order to determine the functions of the MM proteins in magnetosome mineralization. It would be useful to develop convenient assays for the MM proteins. It would also be useful to develop methods for determining the mineral intermediates of magnetite and greigite mineralization and quantifying them over time. Finally, there are a number of bacteria with magnetite magnetosomes, but no bacteria with greigite magnetosomes, available in pure culture. Methods need to be devised for culturing the latter, in order to compare the biomineralization processes for greigite and magnetite.

References

[1] R. P. Blakemore, *Science* **1975**, *190*, 377–379.

[2] R. P. Blakemore, *Annu. Rev. Microbiol.* **1982**, *36*, 217–238.

[3] D. A. Bazylinski, R. B. Frankel, B. R. Heywood, S. Mann, J. W. King, P. L. Donaghay, A. K. Hanson, *Appl. Environ. Microbiol.* **1995**, *61*, 232–3239.

[4] S. Spring, D. A. Bazylinski, in: *The Prokaryotes*. published on the web at http://www.springer-ny.com/, Springer, New York, **2000**.

[5] R. Amann, R. Rossello-Mora, D. Schüler, in *Biomineralization: From Biology to Biotechnology and Medical Application*, in: E. Baeuerlein (ed.), Wiley-VCH, Weinheim, **2000**, pp. 47–60.

[6] D. Schüler, S. Spring, D. A. Bazylinski, *Syst. Appl. Microbiol.* **1999**, *22*, 466–471.

[7] D. A. Bazylinski, R. B. Frankel, in: *Environmental Microbe–Metal Interactions*, D. R. Lovley (ed.), ASM Press, Washington DC, **2000**, pp. 109–144.

[8] T. Sakaguchi, J. G. Burgess, T. Matunaga, *Nature* **1993**, *365*, 47–49.

[9] D. L. Balkwill, D. Maratea, R. P. Blakemore, *J. Bacteriol.* **1980**, *141*, 1399–1408.

[10] Y. A. Gorby, T. J. Beveridge, R. P. Blakemore, *J. Bacteriol.* **1988**, *170*, 834–841.

[11] K. Grünberg, C. Wawer, B. M. Tebo, D. Schüler, *Appl. Environ. Microbiol.* **2001**, *67*, 4573–4582.

[12] M. Farina, B. Kachar, U. Lins, R. Broderick, H. Lins de Barros, *J. Microsc.*, **1994**, *173*, 1–8.

[13] R. P. Blakemore, R. B. Frankel, *Sci. Am.* **1981**, *245*, 58–65.

[14] (a) K. M. Towe, T. T. Moench, *Earth Planet. Sci. Lett.*, **1981**, *52*, 213–220. (b) T. T. Moench, *Antonie van Leeuwenhoek* **1988**, *54*, 483–496.

[15] B. L. Cox, R. Popa, D. A. Bazylinski, B. Lanoil, S. Douglas, A. Belz, D. L. Engler, K. H. Nealson, *Gemicrobiol. J.* **2002**, *19*, 387–406.

[16] B. Devouard, M. Pósfai, X. Hua, D. A. Bazylinski, R. B. Frankel, P. R. Buseck, *Am. Mineral.* **1998**, *83*, 1387–1398.

[17] S. Mann, R. B. Frankel, in *Biomineralization: Chemical and Biochemical Perspectives*, in: S. Mann, J. Webb, R. J. P. Williams (eds), VCH, Weinheim, **1989**, pp. 389–426.

[18] R. B. Frankel, *Annu. Rev. Biophys. Bioeng.* **1984**, *13*, 85–103.

[19] B. M. Moskowitz, *Rev. Geophys. Supp.* **1995**, *33*, 1234–128.

[20] R. E. Dunin-Borkowski, M. R. McCartney, R. B. Frankel, D. A. Bazylinski, M. Posfai, P. R. Buseck, *Science* **1998**, *282*, 2868–1870.

[21] S. Ofer, I. Nowick, R. Bauminger, G. C. Papaefthymiou, R. B. Frankel, R. P. Blakemore, *Biophys J.* **1984**, *46*, 57–64.

[22] R. B. Frankel, D. A. Bazylinski, M. S. Johnson, B. L. Taylor, *Biophys. J.* **1997**, *73*, 994–1000.

[23] S. Mann, N. H. C. Sparks, R. G. Board, *Adv. Microb. Physiol.* **1990**, *31*, 125–181.

[24] S. Mann, N. H. C. Sparks, R. B. Frankel, D. A. Bazylinski, H. W. Jannasch, *Nature* **1990**, *343*, 258–261.

[25] D. A. Bazylinski, R. B. Frankel, A. J. Garratt-Reed, S. Mann, in: *Iron Biominerals*, R. B. Frankel, R. P. Blakemore (eds), Plenum, New York, **1990**, pp. 239–255.

[26] (a) M. Pósfai, P. R. Buseck, D. A. Bazylinski, R. B. Frankel, *Science* **1998**, *280*, 880–883; (b) M. Pósfai, P. R. Buseck, D. A. Bazylinski, R. B. Frankel, *Am. Mineral.* **1998**, *83*, 1469–1481.

[27] E. F. DeLong, R. B. Frankel, D. A. Bazylinski, *Science* **1993**, *259*, 803–806.

[28] B. R. Heywood, S. Mann, R. B. Frankel, *Mat. Res. Soc. Symp. Proc.* **1991**, *218*, 93–108.

[29] Joint Genomics Institute, http://www.jgi.doe.gov/JGI_microbial/html/

[30] D. Schüler, *J. Mol. Microb. Biotechnol.* **1999**, *1*, 79–86.

[31] E. Baeuerlein, in: *Biomineralization: From Biology to Biotechnology and Medical Application*, E. Baeuerlein 9ed., Wiley-VCH, Weinheim, **2000**, pp. 61–79.

[32] R. M. Garrels, C. L. Christ, *Solutions Minerals and Equilibria*, Harper and Row, New York, **1965**.

[33] (a) D. Schüler, E. Baeuerlein, *J. Bacteriol.* **1998**, *180*, 159–162. (b) D. Schüler, E. Baeuerlein, *Arch. Microbiol.* **1996**, *166*, 301–307.

[34] D. A. Bazylinski, R. B. Frankel, H. W. Jannasch, *Nature* **1988**, *334*, 518–519.

[35] (a) L. C. Paoletti, R. P. Blakemore, *Curr. Microbiol.* **1988**, *17*, 339–342. (b) W. F. Guerin, R. P. Blakemore, *Appl. Environ. Microbiol.* **1992**, *58*, 1102–1109.

[36] Y. Fukumori, in: *Biomineralization: From Biology to Biotechnology and Medical Application*, E. Baeuerlein (ed.), Wiley-VCH, Weinheim, **2000**, pp. 93–107.

[37] D. Schüler, in: *Biomineralization: From Biology to Biotechnology and Medical Application*, E. Baeuerlein (ed.), Wiley-VCH, Weinheim, **2000**, pp. 109–118.

[38] C. Nakamura, J. G. Burgess, K. Sode, T. Matsunaga, *J. Biol. Chem.* **1995**, *270*, 28392–28396.

[39] C. Nakamura, T. Kikuchi, J. G. Burgess, T. Matsunaga, *J. Biochem.* **1995**, *118*, 23–27.

[40] T. Matsunaga, N. Tsujimura, Y. Okamura, H. Takeyama, *Biochem. Biophys. Res. Commun.* **2000**, *268*, 932.

[41] Y. Okamura, H. Takeyama, T. Matsunaga, *J. Biol. Chem.* **2001**, *276*, 48183–48188.

[42] T. Matsunaga and Y. Okamura, private communication.

[43] D. Schultheiss, D. Schüler, *Arch. Microbiol.*, **2003**, *179*, 89–94.

[44] (a) T. Matsunaga, T. Sakaguchi, *J. Biosci. Bioeng.*, **2000**, *90*, 1–13. (b) T. Matsunaga, T. Sakaguchi, in: *Biomineralization: From Biology to Biotechnology and Medical Application*, E. Baeuerlein (ed.), Wiley-VCH, **2000**, pp. 119–135

[45] D. A. Bazylinski, A. J. Garratt-Reed, R. B. Frankel, *Microsc. Res. Tech.*, **1994**, *27*, 389–401.

[46] R. C. Reszka, in: *Biomineralization: From Biology to Biotechnology and Medical Application*, E. Baeuerlein (ed.), Wiley-VCH, Weinheim, **2000**, pp. 81–91.

11

Bacteriorhodopsin and its Potential in Technical Applications

Norbert Hampp and *Dieter Oesterhelt*

11.1
Introduction

In biotechnology, the production of biological macromolecules for technical processes is state-of-the-art. The biosynthetic capabilities of cells go far beyond those of organic chemistry. Materials such as functional biomolecules, enzymes, antibodies, and hormones are indispensable in the food industry, cleaning, medical diagnosis, pharmacy, and therapy, and consequently many companies supplying systems such as DNA chips and readers, and high-throughput screening platforms were established to meet demands. The biotechnology industry is booming – indeed, it is very likely to become the most important high-tech industry within the next few decades.

Why is "bio" the coming technology? The advantages and the need for enduring miniaturizations are an increasing challenge as long as conventional lithographic methods are employed. The utilization of self-organization principles and bioengineering of functional biological structures seems to be a promising alternative approach. Re-engineering of biomolecules in order to realize technically desirable functions has become possible.

However, there remains another problem – namely, the communication between classical microsystems (in particular electronic systems) and the nanoscaled biomolecules. This interface remains the major challenge in the realization of "cross-technology" products.

Last – but not least – there is the problem of stability, as biomaterials are less stable than organic and semiconductor structures. This is of course a problem today, but in future technologies, where repair mechanisms like those in living organisms may be implemented, this obstacle may be overcome. Today, we have the need to seek methods for stabilizing the structures of biomaterials before they can be considered for technical applications.

Bacteriorhodopsin has been studied over the past two decades as a material for technical applications. Its stability is adequate, it has several technically interesting functions, tools for both its modification and production in technical quantities have been developed, and it offers various interface principles, whether optical, electrical, or chemical.

In the following sections, the activities for technical applications of bacteriorhodopsin will be reviewed and future developments will be discussed.

Nanobiotechnology. Edited by Christof Niemeyer, Chad Mirkin
Copyright © 2004 WILEY-VCH Verlag GmbH & Co. KgaA, Weinheim
ISBN 3-527-30658-7

11.2
Overview: The Molecular Properties of Bacteriorhodopsin

In the first section of this chapter, the organisms in which bacteriorhodopsin is found are described, after which the processes used to modify this protein and to produce it in large quantities will be outlined. Finally, the current technical applications of bacteriorhodopsin will be summarized.

11.2.1
Haloarchaea and their Retinal Proteins

Archaea form, together with Bacteria and Eucarya, the three domains in life. Archaea are unicellular organisms thriving in a variety of habitats. Most of the archaea so far isolated and cultivated prefer extreme environments, although recent investigations have revealed archaea to be a standard component in biomasses of terrestrial and marine environments [1, 2]. Archaea are split into the crenarchaeotal and the euryarchaeotal branches. Members of the crenarchaeotal family comprise the full temperature range of life, from habitats at $-1.8\,°C$ in the Antarctic to hyperthermal springs where *Pyrolobus fumarii* holds the world record in optimal temperatures for growth at $113\,°C$ [3]. Typical of the euryarchaeota are the methanogenic archaea. These occur ubiquitously in locations where organic matter is decomposed under strictly anaerobic conditions, such as in aquatic sediments, in marshes, or in the rumen of herbivores, and they contribute heavily to biogenic methane production. A second branch of the euroarchaeota comprises extreme halophilic organisms, the so-called halobacteria; these are ubiquitous on earth wherever salt occurs in solute concentrations close to saturation (Figure 11.1).

Archaea and their proteins won commercial interest for the unusual physico-chemical conditions under which they live and reproduce. The best example is bioleaching (biohydrometallurgy) of ore rubble, which was first carried out during the 1950s for copper and is now used to produce a variety of metals [4]. Heat-stable enzymes are valuable molecular biological tools; examples include heat-stable catalysts for the polymerase chain reaction, or use in washing powders. Further applications of archaeal proteins which are halophilic or especially resistant to extreme pH values might become apparent in the near future. A high potential for applications in nanotechnology lies in the fabrication of devices on the basis of archaeal surface-layer proteins [5]. These readily form two-dimensional crystals, and actually occur on the cell surface as natural crystals with pores of precisely defined size.

To date, six genera of halophilic archaea have been identified [6], and retinal proteins occur ubiquitously among these. In two genera, the extreme halophilicity is connected to a second extreme condition of life, alkalophilism, and these archaea are found in large masses in salty natron lakes. Halophilic archaea in general grow on organic substrates and have optimized their bioenergetics during evolution. With a sufficient supply of organic nutrients and oxygen, they respire in standard fashion. However, it must be noted that oxygen solubility in saturated salt solution is about five times lower than in water. Under the anaerobic conditions that commonly occur in their habitats, these organisms have acquired three alternative means for energy conversion: either they use nitrate,

Figure 11.1 *Halobacterium salinarum* is found in nature in concentrated salt solutions as they occur in salines. A purple color is caused by bacteriorhodopsin, and this is the key protein of the photo- synthetic capabilities of *H. salinarum*. The proton pathway with the amino acids involved and the lysine-bound retinylidene residue are shown in the structural model of bacteriorhodopsin (foreground).

dimethylsulfoxide or trimethylaminoxide as end electron acceptors [7, 8] or they ferment arginine to carbamoyl phosphate for the production of ATP [9–11]. The third choice is a very powerful system, and the second route to photosynthesis in nature. This system does not use chlorophyll-based reaction centers (as do bacteria and plants) but rather relies on retinal as a photon absorber and retinal-containing proteins as energy transducers [12]. The light-driven proton pump bacteriorhodopsin drives a proton circuit across the cell membrane, and ATP is produced via photophosphorylation as the chemical energy source for cell growth. The process is supported by the chloride pump halorhodopsin, which converts light energy via chloride transport into electrochemical energy used for the maintenance of osmotic balance.

Halobacteria in nature grow very slowly, due to the limited supply of organic nutrients in hypersaline lakes. (The reader is referred to Refs. [7, 8] for a detailed description of their ecophysiology and the natural cycle of their "blooms" in salt lakes.) An example of such massive growth density is shown in Figure 11.1; an indication of heavy growth is the presence of an intense reddish to purple color. These blooms may occur only once each year (or even less frequently), but under optimization of nutrient supply in the laboratory generation times are usually in the order of hours. For biotechnological use, these two facts have to be considered: (i) a salt concentration of 25 % is required to prepare peptone media, and this may be detrimental even to stainless steel fermentors; and (ii) peptone media are usually expensive to produce. These two obstacles are respectively overcome by using salt-resistant fermenting units and low-priced yeast extracts to replace the peptone. Another point to consider is the genetic stability of strains. Here, extensive progress has been made in the availability of stable standard strains used in biotechnology. In particular, strains which produce bacteriorhodopsin can be maintained stably by the use of phototrophic selection procedures [13].

Halophilic archaea, among them the genus Halobacteria, are unique in the sense that they are the only group of archaea that contains retinal proteins. While originally thought to occur only in higher animals capable of vision, retinal proteins recently have also been found in unicellular plant organisms [14], in fungi [15], in bacteria [16] and, most diverse in function, in halophilic archaea [12]. While all other groups seem to harbor either one of the two principal functions of retinal proteins – that is, a sensory function (eucaryotes) or a presumed energy-converting function (bacteria) – only halophilic archaea have developed a set of four retinal proteins, two of which serve a sensory function, while two convert light energy to chemical energy.

Retinal, or vitamin-A aldehyde, originates from β-carotene by oxidative cleavage in the center of the molecule. The aldehyde in the free state is a chemically labile molecule with five conjugated double bonds. It is oxygen-sensitive and shows light-induced isomerization around all double bonds. Light and oxygen together (photooxidation) destroy the free retinal easily. All known proteins containing retinal protect the molecule against photooxidation and select specific photoisomerization reactions, e.g., from 11-*cis* retinal to all-*trans* retinal in visual pigments and all-*trans* retinal to 13-*cis* retinal in the haloarchaeal proteins. All retinal proteins known are intrinsic membrane proteins and possess a transmembrane helical topography. Retinal always binds to the ε-amino group of a lysine residue of the seventh transmembrane helix, and a protonated Schiff base results; this becomes embedded in a cage of amino acids, which in turn drastically modifies the spectroscopic, chemical, and photochemical properties. For example visual pigments cover the

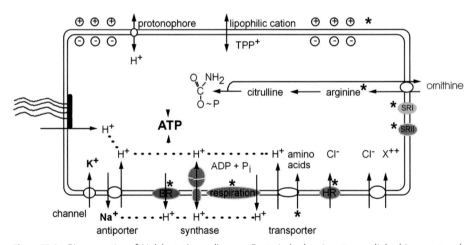

Figure 11.2 Bioenergetics of *Halobacterium salinarum*. Bacteriorhodopsin acts as a light-driven, outward-directed proton pump. The generated proton gradient over the cell membrane drives a membrane-bound ATPase. These two proteins together form the simplest photosynthetic system known.

color range of the entire visible spectrum, and bacteriorhodopsin is chemically more stable than most proteins. Moreover, its light fastness is greater than that of organic dye molecules. Four molecular structures of retinal proteins are presently known: the visual pigment rhodopsin [17]; bacteriorhodopsin [e.g., 18], halorhodopsin [19]; and sensory rhodopsin II [e.g., 20].

Halobacterium salinarum, for example, makes extensive physiological use of retinal proteins (Figure 11.2, see p. 149). Bacteriorhodopsin drives the above-mentioned photosynthetic process for the production of ATP in light, and is also coupled to a high-capacity energy storage system in the form of a molar potassium gradient [21]. In order to maintain osmotic balance during growth – when a volume increase occurs under conditions of isoosmolar conditions, both inside and outside the cell – halorhodopsin is used as a light-driven anion pump [22]. This transports chloride ions into the cells, against the existing electrochemical potential, and allows a net salt accumulation to occur during the volume increase. This is a second way of avoiding the extensive use of respiratory energy at the expense of organic nutrients, by using light energy instead. Finally, two other retinal proteins occur in the cell which monitor the intensity and wavelength of the environmental illumination. These two photoreceptors – sensory rhodopsin I and II – receive orange light as an attractive stimulant, and blue light and near UV-light as repulsive stimulants (for a review, see Ref. [23]). Both photoreceptors signal to a two-component system of the cell, thereby regulating the frequency of the flagellar motor's changes in rotational sense for directing halobacterial cells into an environment of optimal conditions. These photoreceptors are only two among a total of 18 receptors, all of which convert chemical and physical signals of the environment to direct the cell into areas of optimal growth.

The structures of all four halobacterial retinal proteins are very similar, and the molecular structures of two – bacteriorhodospin and halorhodopsin – have been elucidated to the atomic level. Moreover, based on details of the sequence alignment of the two ion pumps and the two sensors, together with the details of some two dozen other structures of archaeal retinal proteins, a tree has been created for this unique family of proteins [24]. Of particular interest in biophysical and biochemical terms is that the functions of the proton pump, the chloride pump, and the sensors are largely inter-convertible, either by varying the physical conditions under which the molecules operate, or by introducing minor genetic modifications. As an example, a point mutation in bacteriorhodopsin will convert this proton pump into a chloride pump [25, 26].

11.2.2
Structure and Function of Bacteriorhodopsin

Bacteriorhodopsin is by far the best-studied archaeal retinal protein. It is naturally overproduced in the cells under conditions of illumination and limited aeration of a growing cell culture, but constitutive overproducers have been isolated. These cell lines produce up to 300 000 copies per cell, covering about 80 % of the cell surface as patches of two-dimensional natural crystals of bacteriorhodopsin with specific lipidic molecular species forming the so-called purple membrane [18]. Being a paradigm of light-driven proton pumps and seven-transmembrane helical proteins, the intense studies on this molecule over the past three decades by dozens of laboratories have produced a large body of knowledge on its biophysics, biochemistry, and molecular biology. Thus, the biotechnology of this "myoglobin" of membrane proteins is well founded on detailed knowledge about this molecule (for a recent review, see Ref. [27]).

Like the other archaeal retinal proteins, bacteriorhodopsin is an intrinsic membrane protein with the common seven-transmembrane helix topology (see Figure 11.1) and an approximate molecular weight of 26 kDa. The seven helices are arranged in two arcs: an inner arc with helices B, C, and D; and an outer arc with helices E, F, G, and A. A transmembrane pore is formed mainly between helices B, C, F, and G. The retinal is bound to Lys216 in helix G as a protonated Schiff base, which interrupts the pore and separates an extracellular (EC) half channel (Figure 11.1, downward oriented) from a cytoplasmic (CP) half channel (Figure 11.1, upward oriented).

The retinal side chain in the binding pocket is closely packed between four tryptophan residues, and the positively charged Schiff base interacts electrostatically with the protein environment. The chromophore is defined as the retinylidene moiety and the side chains in contact with it. Its color is tuned by electrostatic interactions specifically with the Schiff base and a complex counterion consisting of several amino acids. Light absorption causes photoisomerization of the retinal (all-*trans* to 13-*cis* and vice versa) and energy storage which includes ion-affinity shifts (H^+ in bacteriorhodopsin and Cl^- in halorhodopsin) of the Schiff base and internal binding sites as well as conformational changes as central elements of the catalytic cycle. This cycle may formally be represented as a sequence of six steps which are indispensable for transport: the all-*trans* to 13-*cis* photoisomerization and its thermal reversal (isomerization, I); a reversible change in accessibility (switch, S) of the Schiff base for ions in the EC and CP channel respectively; and ion transfer (T) reactions to and from the Schiff base active center (Figure 11.3). Specifically in wild-type bacteriorhodopsin the order of these elementary steps is photoisomerization, proton transfer from the Schiff base to the acceptor aspartic acid (D) 85 in the EC channel, change in accessibility of the Schiff base from EC to CP, proton transfer from the donor aspartic acid (D) 96 in the CP channel to the Schiff base and thermal reisomerization, followed by reset of the Schiff base accessibility from CP to EC. The two proton transfer steps are intimately linked to cooperative changes in the EC and CP channels where hydrogen networks exist [28], and proton conduction over the total distance of about 48 Å is very likely based on a Grotthuss-type mechanism [29]. Structural key players in the EC channel are, besides D85, the arginine residue 82 and the two glutamic acid residues 194 and 204. They are connected by a water bridge to form a dyad which, by all likelihood, is the proton-release unit on the EC surface [18]. Other water molecules in the EC channel also play an important role [30, 31] in addition to amino acid side chains. One water molecule is located between D85 and the Schiff base, and disappears in the structure of the key intermediate M [32]. Three more water molecules below D85 are interconnected with side chains to form an extended hydrogen network as suggested by early experiments [33].

The CP channel from which the Schiff base later in the catalytic cycle receives the proton back is functionally dominated by the protonated D96. The distance between D96 and the nitrogen is 15 Å, and the hydrophobic nature of the space between them could provide an insulating layer against the membrane potential of 280 mV in the ground state. One water molecule occurs in contact with D96, and one water molecule is located between D96 and the Schiff base. Again, changes of this part of the molecule are observed in

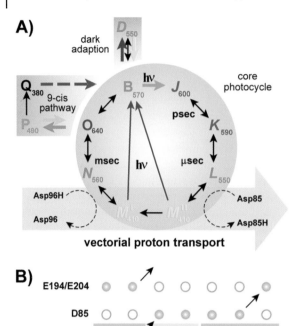

Figure 11.3 Photocycle of bacterior-hodopsin. (A) Upon absorption of a photon, the initial B-state of bacterior-hodopsin is converted photochemically to the J-state from where a series of thermal steps leads back to the initial state. The proton transport is intimately coupled to the photocycle, which is observed as a sequence of intermedi-ates which are represented by the common single-letter code with their absorption maxima given as subscripts. In the dark, bacteriorhodopsin relaxes thermally to the D-state which has 13-*cis* configuration. The resulting mixture of B- and D-states is called dark-adapted bacteriorhodopsin. From the O-state, a photochemical conversion of all-*trans* to 9-*cis* retinal is possible which is not thermally reisomerized to the initial state. (B) The proton transport and related retinal configurations as well as accessibility of the nitrogen in the Schiff-base linkage between retinal and Lys216 are indicated. This se-quence represents several of the mo-lecular changes involved in the proton transport.

the key intermediate M [32]. At the proton entrance of the CP channel, aspartic acid D38 plays a role in the refeeding mechanism of protons [34].

The most dramatic effect of mutations on the structure and function of bacteriorhodop-sin have the two carboxylates at positions 85 and 96. The removal of the proton acceptor D85 prevents the deprotonation of the Schiff base and almost completely annihilates the proton transport function (for a detailed description, see Ref. [35]). The lack of the proton donor D96 slows the catalytic cycle by a factor of up to several hundred, depending on temperature, humidity, azide concentration, etc. [36]. The reason is that the lack of the in-ternal proton donor renders the rate of reprotonation of the Schiff base depending on ex-ternal pH. Another important feature of the M-intermediate is its capacity to absorb blue light, and by this to reconvert photochemically into the initial state. Thus, in mutants lack-ing D96 the life-time of the M-intermediate and therefore the speed of color changes can be regulated either by physico-chemical parameters such as pH, temperature, and humid-ity, or alternatively by simultaneous application of two photons of different quality, for example green and blue photons.

The three most important features, which eventually lead to biotechnological use of bacteriorhodopsin on the basis of its catalytic cycle are:

1. The color changes, which can be used for any type of information processing and storage process.
2. The photoelectric events which are due to the changing geometry of the Schiff base upon photoisomerization and the movement of the proton. Such electric changes occur from the picosecond to the millisecond time regime.
3. The pH change between the inside and the outside of bacteriorhodopsin-containing membrane systems as the net result of proton translocation.

So far, the color changes are under most intensive investigation because of the velocity of light-triggered reactions in bacteriorhodopsin and the possibility of regulating the speed of the color changes over a very wide range.

11.2.3
Genetic Modification of Bacteriorhodopsin

The molecular biology and genetics of halophilic archaea are developing at a slow, but constant, pace. The development of a transformation system was a major breakthrough, as this opened the door to site-specific mutagenesis and gene deletion or replacement [37, 38]. The basic process consists of a protoplast (spheroblast) preparation by incubation of cells with EDTA to remove magnesium ions. These are necessary for the integrity of the surface layer formed by the glycoprotein in the cell wall of archaea. Spheroblasts are incubated with a mixture of vector DNA carrying an antibiotic resistance and polyethyleneglycol. After a curing period in complex medium the cells are plated with antibiotic to select transformed clones [39].

Not many antibiotics have been reported as efficient agents against halophiles, and only two resistance genes have been cloned and inserted into suitable transformation vectors. Mevinolin inhibits the β-hydroxymethylglutaryl CoA (HMG) reductase and thus prevents isoprenoid synthesis of archaea. As lipids of halophilic archaea are diphytanoyl ethers of substituted glycerol, mevinolin completely blocks growth [40]. Novobiocin prevents DNA replication by inhibiting Gyrase B. The two selection marker genes used are *hmg* and *gyrB*; these were isolated from *Haloferax volcanii* and should have highly homologous genes in most other halophilic archaea [41].

The creation of site-specific retinal protein mutants requires appropriate host strains and vectors. The very stable strain *H. (halobium) salinarum* L33 [42] was introduced as a host for bacteriorhodopsin muteins [39]. It was found as a spontaneous mutation, carries an insertion element in the bop gene, and does not revert to a wild-type phenotype under any condition tested. Further strains were selected after either spontaneous or induced mutation which lack one or more of the retinal proteins, and many more strains with various phenotypes bearing favorable properties for investigations on the bioenergetics and signal transduction of halophiles have been reported. These are not covered in this chapter. More recently, systematic deletions of retinal protein genes have been described; one such example is the strain SNOB (S9 without *bop*) which is derived from the bacteriorhodopsin-overproducing strain S9 and has the *bop* gene deleted [43]. In consecutive rounds of a deletion procedure which uses the same antibiotic for selection repeatedly, all four retinal protein genes have been deleted to reach the strain NAOMI (now all opsins missing; M. Otsuka and D. Oesterhelt, unpublished results). This strain has the advantage that complementation with opsin genes allows the combination of their physiological function at will.

Vectors have been introduced first on the basis of a replicative element and a mevinolin-resistant determinant from a halophilic cell, together with a replicon and an ampicillin resistance from *E. coli* to serve as shuttle vectors [44]. In many laboratories, a plethora of vectors was subsequently designed with a size smaller than 10 kb, and without halobacterial origin of replication (suicide vectors) to enforce homologous recombination and therefore stable integration into the halobacterial genome.

For interruption of a functional *bop* gene, the resistance gene used for selection is usually inserted into the coding sequence and remains stably associated with it permanently. Site-specific mutated genes can be favorably introduced into wild-type (S9) or mutated (L33) background in the following way. The resistance gene is placed next to the mutant *bop* gene on the suicide vector DNA. After transformation, antibiotic-resistant clones are selected which must result from a single crossover genetic event leading to the integration of vector DNA into the chromosome. These clones are allowed to grow without selection pressure, and are then replica plated with and without antibiotic. Among several genetic results is one which removes the vector DNA with the resistance gene and the original version of *bop* gene and leaves the mutated *bop* gene in the correct place stably integrated. These clones are found at various frequencies (usually one among several hundred) as the phenotype which does not grow on the antibiotic-containing plate but on its antibiotic-free counterpart. Besides genetic stability, the procedure provides the additional advantage of repeated use of the same antibiotic in further rounds of genetic alterations.

Mutations of bacteriorhodopsin have been produced by the hundred and, once published, are available from the various laboratories. While the genetic background into which the mutated bop genes are introduced is often different, the mutagenesis procedure presently is usually the overlapping PCR method which replaced the preceding approach of gapped duplex DNA and is easily and quickly applied.

Recently, a β-galactosidase gene as a reporter gene has been shown to act as an indicator gene for promoter strength and for blue-white selection procedures [45, 46, 75].

In conclusion, the production of bacteriorhodopsin muteins has become a routine method, but the maintenance of stable strains overproducing these muteins requires much care. In addition, the maximal level of mutein production can be very different, is unpredictable, and whether a given mutein will produce the crystalline arrangement of the purple membrane is also in doubt.

11.2.4
Biotechnological Production of Bacteriorhodopsins

The biotechnology of bacteriorhodopsin is based on simplicity of its isolation and chemical and photochemical stability when in the form of purple membranes. As mentioned, bacteriorhodopsin forms 2D crystals in vivo, and these can be isolated as purple membranes. The isolation is facilitated by two facts: (i) halobacterial cells are unstable in water and cell constituents are released upon lysis; (ii) the cell membrane is fragmented, and for unknown reasons the crystalline patches of the purple membrane are set free as fragments of largest size and highest buoyant density. These two specific features are used in the isolation procedure, either in a combination of sedimentation and isopycnic gradient centrifugation leading to a product of highest purity, or by a filtration procedure. The purification step yields a product which is 95–100% pure depending on the conditions and the mutein under consideration. Although these methods have not yet been exposed to economic competition, it is expected that the biotechnological production of purple membranes might in the future represent a competitive biomaterial in information technology. Certain limits of this material should be mentioned, however. As a membrane protein, bacteriorhodopsin cannot be produced via inclusion bodies as although the refolding and reconstitution of the active chromoprotein into membranes is possible in principle, it is certainly not feasible on an economic basis. Halobacterial cells do not form invaginations of their cell membrane like phototrophic bacteria (e. g., *Rhodopseudomonas*), which concentrate their reaction centers for photosynthesis in units called chromatophores. Although the purple membrane may occupy about 80% of the cell membrane area, the amount of purple membranes obtained from cells is comparably low. It is needless to point out that the secretion of the integral membrane protein bacteriorhodopsin from cells into the medium has no biochemical basis.

In practical terms, the current production of bacteriorhodopsin is at a level of 25 g m^{-3} nutrient broth. A 25-g quantity of bacteriorhodopsin would produce 50 L of suspension with an optical density of 1 (at a layer thickness of 1 cm this would allow the passage of only 10% of light; i. e. it is an intense color). The molecular biology or production of bacteriorhodopsin variants has been established over the years, and at present the genetics is well known, including the sequence of the entire genome [47]. One problem in the molecular genetics of the halophilic archaea, especially *H. salinarum*, is that of insertion elements, as this sometimes causes instability of strains. However, with increasing knowledge on transformation procedures, vector construction and the creation of new antibiotics for the cells, these problems may be removed, at least for industrial purposes. It should be mentioned again, however, that cells producing bacteriorhodopsin variants do not always overproduce bacteriorhodopsin as do wild-type cells.

11.3
Overview: Technical Applications of Bacteriorhodopsin

The remarkable physico-chemical properties of bacteriorhodopsin and its numerous technically attractive molecular functions have led to many potential technical uses for this material [48–50]. In most of these applications, bacteriorhodopsin is used in the purple membrane (PM) form because its deliberation from the crystalline package reduces both the chemical and thermodynamic stability of bacteriorhodopsin to a significant degree.

Bacteriorhodopsin comprises three basic molecular functions which may be used in technical applications: photoelectric, photochromic, and proton transport properties. The molecular mechanisms responsible for each of these properties have been already discussed, and here we will focus briefly on the related applications.

11.3.1
Photoelectric Applications

The use of bacteriorhodopsin as a molecular level photoelectric conversion element is one of the fields where technical applications of the material have been examined. Upon illumination, a photovoltage up to 250 mV per single bacteriorhodopsin layer is generated, and this may be used either as an indicator or control element for various applications.

Triggered by the absorption of a photon, the bacteriorhodopsin molecule undergoes a series of very rapid molecular changes, one of which is the generation of a photovoltage caused by changes in the orientation of molecular dipole moments that are triggered by the isomerization of the retinal on the femtosecond scale (Figure 11.4A). The proton released through the outer proton half channel may be either transferred to the outer medium or conducted along the surface of the PM [51]. The proton conductivity along the PM in *H. salinarum* cells supports delocalization of protons on the surface and proton conduction to the membrane-bound ATPase molecules. In most of the photoelectric applications of PM, the water content of the films is reduced, and this in turn causes proton conduction along the membrane surface. A single PM sheet contains several thousands of unidirectionally oriented BR molecules. Upon illumination, a number of bacteriorhodopsin molecules proportional to the intensity of light will be excited to accomplish a proton transport process. As all of the bacteriorhodopsin molecules in a single PM patch are oriented in the same direction, the voltage generated over a single membrane is independent of the number of active molecules (Figure 11.4B), but the proton current generated is proportional to the light intensity.

The photovoltage generated can be easily measured by embedding the PM layer between two transparent electrodes. The light-triggered photovoltage induces a compensating polarization voltage in the outer electrodes, and this in turn may be detected as a high impedance voltage signal (Figure 11.4C). In a perfectly capacitive coupled system of this type, an induced photovoltage with the characteristics shown in Figure 11.4C is observed. A voltage is induced only during the light intensity change. The polarity of the signal is different for the OFF $\Rightarrow$ ON and the ON $\Rightarrow$ OFF transition. This is called the "differential responsivity" of PM layers.

There are two major issues to be solved for an attractive use of bacteriorhodopsin in photoelectric processing. The first is that a high degree of orientation of the PMs is obtained because counter-oriented PMs cancel out each others' photoelectric effects. The second is that a coupling of the light-dependent proton-motive force of bacteriorhodopsin to the electromotive forces used in conventional electronics needs to be achieved.

11.3.1.1 **Preparation of Oriented PM Layers**
The photoelectric signal of PM is high enough so that a single layer only of PM is needed for applications where the PM acts as a photoelectric indicator molecule. The Langmuir–Blodgett technique is often used for the preparation of such devices. The main problem is to obtain a high degree of orientation of the PM patches. The PM patches are thin (5 nm), large (up to micrometers), and flexible, and for this reason a mechanical orientation is difficult to obtain.

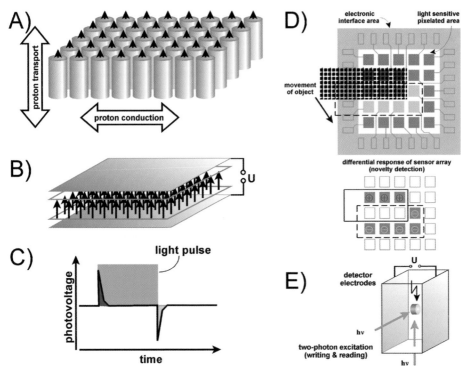

Figure 11.4 Photoelectric properties of bacteriorhodopsin. (A) Absorption of a photon by bacteriorhodopsin causes molecular changes which lead on a pico- to nanosecond timescale to the generation of a photovoltage over the molecule. The bacteriorhodopsin molecules in a single purple membrane (PM) patch are unidirectionally oriented. The light-driven vectorial proton transport of all the bacteriorhodopsin molecules is switched in parallel. This means that, over a single PM patch, it is not the voltage but the proton current which is proportional to the light intensity. The protons transported through the bacteriorhodopsin molecules can either be released to the outer medium or move along the surface of the PM patch due to proton conduction. (B) Oriented bacteriorhodopsin molecules, each represented by an arrow, sandwiched in a capacitor structure can be used as a photoelectric indicator cell. Upon illumination, a charge separation over the bacteriorhodopsin layer is generated which induces a proportional charging of the outer electrode layers. No electric conduction between the bacteriorhodopsin layer and the electrodes is required. The electric field of the charge distribution induced in the electrodes compensates the electric field caused by bacteriorhodopsin. (C) Depending on the type of outer circuitry, either the induced voltage or the induced charge motion can be measured. In the latter case, a signal is recorded which corresponds to the first derivative of the temporal change of the light. (D) In a pixelated structure which is coated with oriented bacteriorhodopsin, photovoltages are measured only in those spots where a change in the light intensity occurs; hence, this is called novelty filtering. (E) In a volume (e. g., in 3-D data storage) the detection of the photovoltage in an outer capacitor structure was considered for readout. If the bacteriorhodopsin in the point of excitation is in the B-state, the absorption of a photon will lead to a photo-induced voltage, but not in the M-state.

The physical properties of the two sides of the PM are definitively different, but the differences are not so large that orientation (e. g., in an electric field) achieves more than a preorientation. Any preorientation obtained needs to be made permanent, for example, by covalent crosslinking or by polymer embedding of the PM patches.

The only reliable method described so far is the coating of a surface with monoclonal antibodies against bacteriorhodopsin [52]. Since the antibodies selectively react with the cytoplasmic or extracellular side of the PM, a highly oriented monolayer of PM can be obtained using this method, although unfortunately it is restricted to a single PM layer.

11.3.1.2 Interfacing the Proton-Motive Force

The other problem in photoelectric applications is that under light exposure bacteriorhodopsin transports protons, but not electrons. The interface between the biological component bacteriorhodopsin and its proton-motive force and the electromotive forces required for conventional electronics needs to be considered in the design of devices. Interfacing the proton-motive force of BR with the electron-conducting outer electrodes requires an electrolyte layer which couples both "worlds". The balancing between the electronic circuitry and the photoelectric properties of a PM layer is crucial, and is the reason why results from different laboratories are so incomparable. The more than complete review published by Hong [53] is recommended for the reader who seeks much more detail on this subject.

11.3.1.3 Application Examples

Ultrafast photodetection was one of the earliest proposals for the use of bacteriorhodopsin in a technical application [54, 55]. Two-dimensional photoelectric arrays, as artificial retinas or as control elements for a liquid crystal spatial light modulator, were developed later. Last, but not least, the photoelectric properties of bacteriorhodopsin can be used for indicator purposes in 3-D memories.

Artificial retinas

Most devices which make use of the photoelectric properties of bacteriorhodopsin are called "artificial retinas" [56–58], the reason being that they offer certain preprocessing features known from the retina, including edge detection and novelty filtering. The physical background is called the "differential response" of PM layers. A device (see Figure 11.4D) comprising an electrode array which is covered with one or more oriented layers of PM may be used for novelty filtering. Each of the electrodes is connected to an amplifier electronics. First, assume that the dark rectangular structure shown in Figure 11.4D prevents a set of electrodes from being exposed to light. Then, if this structure is moved over the light-sensitive sensor area (see arrow), it causes a voltage to be induced in each of the pixel electrodes, with a sign proportional to the light change. Because only those electrodes "fire" where a change of the illumination occurs, this type of sensor is called an "artificial retina", and this type of preprocessing is called "novelty filtering". Due to the differential response of bacteriorhodopsin, the polarity of the electrode signal carries the information whether a pixel was switched to "ON" or to "OFF", and in turn the direction of the movement of the object can be derived.

Electro-optically controlled spatial light modulators

Another use of the artificial retina described above may be in optically addressed spatial light modulators (SLMs). The amplifier electronics which detects the photovoltages induced in the electrode pixels may be connected to a SLM device. In particular, liquid crystal (LC) -based SLMs are state-of-the-art. In this case, the electrodes of the bacteriorhodopsin-based artificial retina are connected one-to-one to the pixels of a LC-SLM. An advanced version omits the wiring and amplifiers. The LC-layer of the SLM is controlled directly by the artificial retina device [59, 60].

Readout in 3-D Memories

Another application of the photoelectric properties of PM was investigated for the readout of volume storage units with bacteriorhodopsin. The basis is a cube of oriented PM patches in either the purple initial state or the yellow M state. Upon illumination of a PM patch in the initial state (which may be addressed by actinic light), a photovoltage signal is induced. A PM patch in the M state would not respond to the actinic light, but two electrodes on the outer surfaces of the bacteriorhodopsin cube could detect the photovoltage generated, and by this method the 3-D distribution of the photochemical states of PM patches could be read out. This is a prerequisite for a 3-D memory based on bacteriorhodopsin (Figure 11.4E).

11.3.2
Photochromic Applications

Most applications currently investigated utilize the photochromic properties of bacteriorhodopsin [49, 61]. During the photocycle, bacteriorhodopsin cycles through a pair of spectroscopically distinguishable intermediates (see Figure 11.3A), all of which have an absorption maximum which is different from that of the initial B state. However, in most applications the photochromism of bacteriorhodopsin is used in connection with the purple to yellow absorption change which is related to M state formation.

Upon acidification, a blue membrane is formed which has a significantly different photocycle. The formation of a 9-*cis* retinal-containing state is observed, and this is thermally stable. In contrast, 13-*cis* retinal is isomerized by the bacteriorhodopsin molecule to all-*trans* retinal at room temperature, and the isomerization of 9-*cis* retinal is not catalyzed. This pathway opens the route to long-term storage materials based on bacteriorhodopsin.

11.3.2.1   **Photochromic Properties of Bacteriorhodopsin**

Isomerization from all-*trans* to 13-*cis* is the first occurrence after the photochemical excitation of bacteriorhodopsin, and this causes significant transient shifts in the absorption spectrum. In addition to the isomerization change, deprotonation of the chromophoric group is observed. In the L to M transition, a proton from the Schiff base nitrogen group is transferred to Asp85, and this deprotonation causes a drastic blue shift of the absorption to 410 nm. The photochromism of bacteriorhodopsin is dominated by the intermediate which has the longest life-time; hence, this forms a "bottleneck" in the photocycle.

The retinylidene residue inside bacteriorhodopsin which forms part of the photochromic group in the molecule is strongly anisotropic. A PM layer may be considered as a crystalline arrangement of chromophoric groups which are oriented angles of 120° between them (Figure 11.5A). However, in most applications where a statistic number of PM patches is used, the angular chromophore distribution appears anisotropic. Due to the anisotropy of the retinylidene groups, excitation of a random distribution of PM patches with polarized light causes the chromophores to become oriented in parallel to the actinic light polarization such that a preferentially converted (and in turn photoinduced) anisotropy is obtained (Figure 11.5B). In solutions containing PM this is masked by diffusion, but in bacteriorhodopsin-films where the PM patches are fixed, the photoinduced anistropy can be easily observed and utilized.

Today, three types of photochromic changes in bacteriorhodopsin have been described which enable different applications. The first is the photochromic shift between the B and M states (Figure 11.5C), and this is used mainly for optical processing tasks. The second is photoerasable data storage using 9-*cis*-containing states of blue membrane or suitably modified BR-variants (Figure 11.5D). However, the very low quantum efficiency for recording of far below 1 % is a major limitation. And last, but not least, permanent photochromic changes obtained through two-photon absorption in bacteriorhodopsin are suitable for long-term data storage (Figure 11.5E).

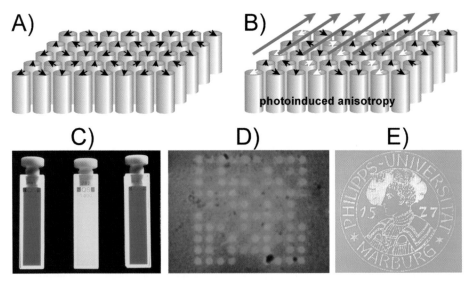

Figure 11.5 Photochromism of bacteriorhodopsin and its application. (A) The retinylidene residues are strongly anisotropic. (B) Upon illumination with polarized light, the retinylidene residues which are in parallel to the electric field vector of the actinic light are preferentially excited and isomerized. (C) Transient photochromic change of bacteriorhodopsin between the initial purple state and the yellowish M-state (middle). (D) The photochemical formation of 9-*cis* retinal may be utilized for photochromic long-term storage. (E) Permanent storage of information in bacteriorhodopsin.

11.3.2.2 Preparation of Bacteriorhodopsin Films

Optical films are prepared from bacteriorhodopsin by polymer embedding. Optically clear, water-soluble polymers are suitable for this purpose (e. g., polyvinylalcohol, gelatin). The film formation is usually carried out by mixing the polymers with PMs and additives in aqueous solution, this being cast on a glass support. The water is generally removed by drying in air, but the films may also be sealed with a second glass plate.

11.3.2.3 Interfacing the Photochromic Changes

The reason why photochromic applications of bacteriorhodopsin are more developed than others is because of the ease with which an interface can be implemented between the bacteriorhodopsin films and any type of optical system. The bacteriorhodopsin film is completely sealed, the only interface being the light which transports energy and information simultaneously.

11.3.2.4 Application Examples

Many applications based on the photochromism of bacteriorhodopsin have been suggested, and some are described briefly here as an indication of the wide range of potential uses.

Photochromic color classifier

As the different rhodopsins in the eye enable color perception, the use of bacteriorhodopsins with different absorption maxima would allow a biomimetic system for color perception to be set up. The photoelectric response of three sensor elements coated with three different bacteriorhodopsin types (i. e., wild-type and two containing retinal analogues) are combined and coupled to a simple neural network for color recognition properties. This functions quite reliably, though it is unclear whether it has any technical advantages over conventional color sensors. Two limitations can be identified in this system. First, the absorption maxima of the three BR types used today are relatively similar, and they do not span the visual wavelength range as well as the human rhodopsins, notably in the blue region. Second, the conventional systems which typically comprise three different color filters and semiconductor light-sensitive elements reliably supply the required information. At present, it is difficult to identify any advantages of the bacteriorhodopsin-based systems over conventional systems [62].

Photochromic inks

Another development is the preparation of photochromic inks. These inks differ from the polymer films for optical recording by their rheological properties. Depending on the method of application (e. g., screen printing, offset printing), the viscosity and surface tension must be considered. The basic photochromic properties are quite similar to those of the optical films, but auxiliaries in the compositions adjust the required application dependent properties. A major problem is to identify suitable compositions which do not interfere with the photochemical properties of the bacteriorhodopsin embedded [63].

Electrochromic inks

The main color shift in bacteriorhodopsin is due not to a primary photochemical reaction but to the protonation change of specific groups, in particular the Schiff base linkage and the Asp85 residue. As protons are charged particles, their removal from the binding position by electric fields should be possible. Indeed, this can be demonstrated, though the speed and efficiency of the decoloration/coloration process is quite low. Nonetheless, the basic principle was successfully demonstrated [64], indicating a potential development of electrochromic paper using bacteriorhodopsin.

Photochromic photographic film

It is well known that a permanent bleaching of bacteriorhodopsin may be achieved with hydroxylamine. The chemical reaction of hydroxylamine with the retinal binding site occurs in an intermediate state only, and no reaction of BR in the B-state is observed. Due to this finding, which dates back to the early retinal extraction experiments, it is possible to fabricate nonreversible optical films from bacteriorhodopsin [65]. These films behave quite similarly to photographic films, except that the compounds needed for the chemical development are already contained in the film. The reason why this process has not been used technically is that, after image formation, the nonreacted hydroxylamine must be removed from the film, or fading of the contrast will occur when stored in light.

Long-term photorewriteable storage of information

Much interest has centered on photorewriteable optical storage with bacteriorhodopsin. For this purpose, the conventional all-*trans* to 13-*cis* conversion is not suitable because of the retinal reisomerization at room temperature in bacteriorhodopsin. In blue membrane, a photochemical conversion from all-*trans* to 9-*cis* retinal, which appears pink in bacteriorhodopsin, may be induced by high light intensities. 9-*cis* retinal is thermally stable and requires photochemical excitation for reconversion of all-*trans*. The main disadvantage of blue membrane is that its formation from PM requires either acidification or the removal of divalent cations, as both cause destabilization of the bacteriorhodopsin molecule. Aggregation of the PM patches is also seen. The requirement is for bacteriorhodopsin variants which have a similar photochemistry but at ambient pH value and without removal of cations. BR variants with alterations in position 85 show such desired properties, and one of the first reported for this purpose was D85N. Another approach is to use wild-type BR at neutral pH values and to switch the material photochemically to the 9-*cis* state from the so-called O-state [66]. This scheme was named branched photocycle memory [67].

Neural networks

Due to the fact that its absorption state may be shifted with blue and yellow light in different directions, bacteriorhodopsin is also a suitable material for neural networks. The output from an absorptive bacteriorhodopsin-cell is used to control the absorption state of another such cell. This has been demonstrated in principle [68], but the bacteriorhodopsin has a "fan-out" of much less than one and for this reason is not really a suitable material for the implementation of optical neural networks. ("Fan-out" is a term from electronics which characterizes the signal power output of a device compared to the required signal power input the same device requires.) A "fan-out" of 10 means that the output terminal of the device supplies enough power that 10 input terminals of identical devices could be supplied with enough energy to signal them the input state reliably. In photochromic devices where no amplification occurs, the "fan-out" is generally less than 1. Without external amplification, it is almost impossible to set up control loops.

3-D information storage

The use of bacteriorhodopsin in 3-D information storage has been investigated for some time. Bacteriorhodopsin shows an astonishingly high two-photon absorption cross-section of the initial B state. This is used in a two-photon absorption set-up to address the absorption state of bacteriorhodopsin in three dimensions [69]. The recording process is quite well handled, but the readout is an intrinsic problem. The advantage of such a memory device is its tolerance towards electromagnetic radiation.

Nonlinear optical filtering

The strongly nonlinear optical response of bacteriorhodopsin towards the incident light intensity has given rise to many applications which use the nonlinear response for image processing purposes (e.g., edge enhancement, noise reduction). The response curve of the bacteriorhodopsin can be tuned over several orders of magnitude by changing the lifetime of the M-state. This may be accomplished by changing the pH, as well as using modified bacteriorhodopsins.

Holographic pattern recognition and interferometry

The use of bacteriorhodopsin as a short-term memory has been tested in several applications. The most challenging was the construction of a real-time holographic pattern recognition system which operates at video frame rate. In this system, the holographic comparison of images allows similarities between objects used for identification purposes for complex images to be quantified [70, 71]. In the bacteriorhodopsin-film, holograms are recorded, read-out and erased at video frame rate. In the early 1990s, when the system was first developed, it was unchallenged by computer systems, but as computing power has advanced the bacteriorhodopsin-film method has been abandoned.

Typical photochromic applications include a holographic real-time correlator (Figure 11.6A) and a holographic camera for nondestructive testing (Figure 11.6B). Examples of photochromic inks made from bacteriorhodopsin are shown in Figure 11.6C, with the initial colored (purple) and bleached, yellowish inks in the foreground and background, respectively. An ID card sample with a bacteriorhodopsin-based optical storage in the purple-colored strip is shown in Figure 11.6D.

11.3.3
Applications in Energy Conversion

The use of bacteriorhodopsin as a light energy-converting device seems to be first choice as with regard to technical applications of the molecule [72, 73]. All applications of this type have as the key element in common a bacteriorhodopsin-driven charge separation

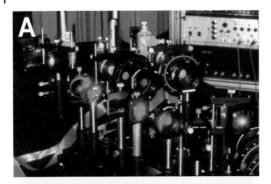

Figure 11.6 Photochromic applications of bacteriorhodopsin. (A) Holographic correlator; (B) holographic camera for interferometric testing; (C) photochromic inks for security applications; (D) optical storage.

over the PM. Although this is the basic function of bacteriorhodopsin embedded in PM, until now it has not been possible to prepare artificial PMs with technically relevant dimensions. Methods devised to overcome this problem have been unable to provide PMs capable of producing passive proton transport, and research is continuing in this area.

11.4
Methods

The methods required for this type of research span from biochemistry and bioengineering to various printing and optical techniques. All have been well documented, with the methods of cultivating halobacteria and isolating PMs being common to all applications.

A comprehensive laboratory manual on Archaea was produced in 1995 [40]. For specific conditions of phototrophic growth [13], laboratory protocols of cultivation and media composition, the reader is referred to Ref. [40] (pages 13–21 and 225–230), while for the isolation of PMs from halobacteria, the reader is referred to Ref. [74]. The genetic modification of halobacteria is described in Refs. [75, 76].

The very first publication on the discovery of the bacteriorhodopsin [77], as well as the latest volume on its applications [78], should also be mentioned in this context.

Information on specific processing steps for the wide range of applications is taken best from the relevant patent applications, and a summary of patents related to applications of bacteriorhodopsin may be found in Ref. [50].

11.5
Outlook

Bacteriorhodopsin is today the biological photochromic material for which technical applications in optical information processing are much more developed than for any other biomaterial. The entire sequence, from an analysis of molecular function to its controlled modification and the development of suitable applications, has been demonstrated with this molecule. Ideas of utilizing the evolutionary optimized functions of biological molecules in technical processes by using genetic engineering to produce tailor-made modified versions of natural molecules with improved technical properties have been demonstrated with bacteriorhodopsin for the first time. Biomaterials as blueprints for technical materials with nanoscale functions form the basis of the concept of nanobionics – and bacteriorhodopsin was the first such example.

References

[1] E. DeLong, *Science* **1998**, *280*, 542–543.

[2] E. F. DeLong, K. Y. Wu, B. B. Prezelin, R. V. M. Jovine, *Nature* **1994**, *371*, 695–697.

[3] E. Blöchl, R. Rachel, S. Burggraf, D. Hafenbradl, H. W. Jannasch, K. O. Stetter, *Extremophiles* **1997**, *1*, 14–21.

[4] R. Amils, A. Ballester (eds), *Biohydrometallurgy and the environment toward the mining of the 21st century.* Proc. Intern. Biohydrometallurgy Symp. IBS 99. Elsevier, Amsterdam, Lausanne, New York, Oxford, Shannon, Singapore, Tokyo, **1999**.

[5] U. B. Sleytr, B. Schuster, D. Pum, *IEEE Eng. Med. Biol.* **2003**, *22*, 140–150.

[6] W. D. Grant, H. Larsen, in: *Bergey's Manual of Systematic Bacteriology,* N. Pfennig (ed.), Williams Wilkins, Baltimore, **1989**, pp. 2216–2233.

[7] A. Oren, *FEMS Microbiol. Rev.* **1994**, *13*, 415–440.

[8] A. Oren, H. G. Trüper, *FEMS Microbiol. Lett.* **1990**, *70*, 33.

[9] R. Hartmann, H.-D. Sickinger, D. Oesterhelt. *Proc. Natl. Acad. Sci. USA* **1980**, *77*, 3821–3825.

[10] A. Ruepp, H. Müller, F. Lottspeich, J. Soppa, *J. Bacteriol.* **1995**, *177*, 1129–1136.

[11] L. I. Hochstein, F. Lang, *Arch. Biochem. Biophys.* **1991**, *288*, 80.

[12] D. Oesterhelt, *Curr. Opinion Struct. Biol.* **1998**, *8*, 489–500.

[13] D. Oesterhelt, G. Krippahl, *Ann. Microbiol. Inst. Pasteur* **1983**, *134B*, 137–150.

[14] K. W. Foster, J. Saranak, N. Patel, G. Zarilli, M. Okabe, T. Kline, K. Nakanishi, *Nature* **1984**, *311*, 756–759.

[15] J. A. Bieszke, E. N. Spudich, K. L. Scott, K. A. Borkovich, J. L. Spudich, *Biochemistry* **1999**, *38*, 14138–14145.

[16] O. Béjà, L. Aravind, E. V. Koonin, M. T. Suzuki, A. Hadd, L. P. Nguyen, S. B. Jovanovich, C. M. Gates, R. A. Feldman, J. L. Spudich, E. N. Spudich, E. F. DeLong, *Science* **2000**, *289*, 1902–1906.

[17] K. Palczewski, T. Kumasaka, T. Hori, C. A. Behnke, H. Motosha, B. A. Fox, I. L. Trong, D. C. Teller, T. Okada, R. E. Stenkamp, M. Yamamoto, M. Miyano, *Science* **2000**, *289*, 739–745.

[18] L.-O. Essen, R. Siegert, W. D. Lehmann, D. Oesterhelt, *Proc. Natl. Acad. Sci. USA* **1998**, *95*, 11673–11678.

[19] M. Kolbe, H. Besir, L.-O. Essen, D. Oesterhelt, *Science* **2000**, *288*, 1390–1396.

[20] V. I. Gordelly, J. Labahn, R. Moukhametlanoc, R. Efremov, J. Granzin, R. Schlesinger, G. Büldt, T. Savopol, A. J. Scheldig, J. P. Klaro, M. Engelhardt, *Nature* **2002**, *439*, 464–465.

[21] G. Wagner, R. Hartmann, D. Oesterhelt, *Eur. J. Biochem.* **1978**, *89*, 169–179.

[22] D. Oesterhelt, *Israel J. Chem.* **1995**, *35*, 475–494.

[23] W. Marwan, D. Oesterhelt. *ASM-News* **2000**, *66*, 83–90.

[24] K. Ihara, T. Umemura, I. Katagiri, T. Kitajima-Ihara, Y. Sugiyama, Y. Kimura, Y. Mukohata, *J. Mol. Biol.* **1999**, *285*, 163–174.

[25] J. Sasaki, L. S. Brown, Y. S. Chon, H. Kandori, A. Maeda, R. Needleman, J. K. Lanyi, *Science* **1995**, *269*, 73–75.

[26] J. Tittor, U. Haupts, Ch. Haupts, D. Oesterhelt, A. Becker, E. Bamberg, *J. Mol. Biol.* **1997**, *271*, 405–416.

[27] U. Haupts, J. Tittor, D. Oesterhelt, *Annu. Rev. Biophys. Biomol. Struct.* **1999**, *28*, 367–399.

[28] R. Rammelsberg, G. Huhn, M. Lübben, K. Gerwert, *Biochemistry* **1998**, *37*, 5001–5009.

[29] M. Wikström, *Curr. Opin. Struct. Biol.* **1998**, *8*, 480–488.

[30] N. A. Dencher, H. J. Sass, G. Büldt, *Biochim. Biophys. Acta* **2000**, *1460*, 192–203.

[31] J. Heberle, *Biochim. Biophys. Acta* **2000**, *1458*, 135–147.

[32] H. Luecke, B. Schobert, H.-T. Richter, J.-P. Cartailler, J. K. Lanyi, *Science* **1999**, *286*, 255–260.

[33] J. le Coutre, J. Tittor, D. Oesterhelt, K. Gerwert, *Proc. Natl. Acad. Sci. USA* **1995**, *92*, 4962–4966.

[34] J. Riesle, D. Oesterhelt, N. A. Dencher, J. Heberle, *Biochemistry* **1996**, *35*, 6635–6643.

[35] U. Haupts, J. Tittor, E. Bamberg, D. Oesterhelt, *Biochemistry* **1997**, *36*, 2–7.

[36] A. Miller, D. Oesterhelt, *Biochim. Biophys. Acta* **1990**, *1020*, 57–64.

[37] R. L. Charlebois, W. L. Lam, S. W. Cline, W. F. Doolittle, *Proc. Natl. Acad. Sci. USA* **1987**, *84*, 8530–8534.

[38] S. W. Cline, W. L. Lam, R. L. Charlebois, L. S. Schalkwyk, W. F. Doolittle, *Can. J. Microbiol.* **1989**, *35*, 148–152.

[39] B. Ni, M. Chang, A. Duschl, J. Lanyi, R. Needleman, *Gene* **1990**, *90*, 169–172.

[40] F. T. Robb, A. R. Place, K. R. Sowers, H. J. Schreier, S. DasSarma, E. M. Fleischmann (eds), *Archaea. A Laboratory Manual.* Cold Spring Harbor Laboratory Press, **1995**.

[41] M. L. Holmes, S. D. Nuttall, M. L. Dyall-Smith, *J. Bacteriol.* **1991**, *173*, 3807–3813.

[42] G. Wagner, D. Oesterhelt, G. Krippahl, J. K. Lanyi, *FEBS Lett.* **1981**, *131*, 341–345.

[43] M. Pfeiffer, T. Rink, K. Gerwert, D. Oesterhelt, H.-J. Steinhoff, *J. Mol. Biol.* **1999**, *287*, 163–171.

[44] W. L. Lam, W. F. Doolittle, *Proc. Natl. Acad. Sci. USA* **1989**, *86*, 5478–5482.

[45] D. Gregor, F. Pfeifer, MICROBIOL-SGM 2001, *147*, 1745–1754.

[46] N. Patenge, A. Haase, H. Bolhuis, D. Oesterhelt, *Mol. Microbiol.* **2000**, *36*, 105–113.

[47] www.halo@mpg.de

[48] R. R. Birge, *Annu. Rev. Phys. Chem.* **1990**, *41*, 683–733.

[49] N. Hampp, *Chem. Rev.* **2000**, *100*, 1755–1776.

[50] N. Hampp, *Appl. Microbiol. Biotechnol.* **2000**, *53*, 633–639.

[51] B. Gabriel, J. Teissie, *Proc. Natl. Acad. Sci. USA* **1996**, *93*, 14521–14525.

[52] K. Koyama, N. Yamaguchi, T. Miyasaka, *Science* **1994**, *265*, 762–765.

[53] F. T. Hong, *Prog. Surf. Sci.* **1999**, *62*, 1–237.

[54] H. W. Trissl, M. Montal, *Nature* **1977**, *266*, 655–657.

[55] F. T. Hong, *BioSystems* **1995**, *35*, 117–121.

[56] T. Miyasaka, K. Koyama, I. Itoh, *Kagaku to Kogyo (Tokyo)* **1993**, *46*, 247–249.

[57] Z. Chen, R. R. Birge, *Trends Biotechnol.* **1993**, *11*, 292–300.

[58] T. Miyasaka, K. Koyama, I. Itoh, *Science* **1992**, *255*, 342–344.

[59] C. Ritzel, K. Meerholz, C. Bräuchle, D. Oesterhelt, *Mol. Cryst. Liq. Cryst. Sci. Technol. A* **1998**, *315*, 443–448.

[60] C. Ritzel, K. Meerholz, C. Bräuchle, D. Oesterhelt, *Proc. SPIE-Int. Soc. Opt. Eng.* **1998**, *3475*, 49–55.

[61] N. Hampp, C. Bräuchle, D. Oesterhelt, *Biophys J.* **1990**, *58*, 83–93.

[62] M. Frydrch, P. Silfsten, S. Parkkinen, J. Parkinnen, T. Jaaskelainen, *Biosystems* **2000**, *54*, 131–140.

[63] N. Hampp, T. Fischer, M. Neebe, *Proc. SPIE* **2002**, *4677*, 121–128.

[64] P. Kolodner, E. Lukashev, Y.-C. Ching, I. Rousso, *Proc. Natl. Acad. Sci. USA* **1996**, *53*, 6012–6015.

[65] N. N. Vsevolodov, T. V. Dyukova, *Trends Biotechnol.* **1994**, *12*, 81–88.

[66] A. Popp, M. Wolperdinger, N. Hampp, C. Bräuchle, D. Oesterhelt, *Biophys. J.* **1993**, *65*, 1449–1459.

[67] N. B. Gillespie, K. J. Wise, L. Ren, J. A. Stuart, D. L. Marcy, Q. Li, L. Ramos, K. Jordan, L. Ren, R. R. Birge, *J. Phys. Chem. B* **2002**, *106*, 13352–13361.

[68] D. Haronian, A. Lewis, *Appl. Opt.* **1991**, *30*, 597–608.

[69] R. R. Birge, R. B. Gross, M. B. Masthay, J. A. Stuart, J. R. Tallent, C. F. Zhang, *Mol. Cryst. Liq. Cryst. Sci. Technol.* **1992**, *B3*, 133–148.

[70] R. Thoma, N. Hampp, *Opt. Lett.* **1994**, *19*, 1364–1366.

[71] R. Thoma, M. Dratz, N. Hampp, *Opt. Eng.* **1995**, *34*, 1345–1351.

[72] M. Eisenbach, C. Weissmann, G. Tanny, S. R. Caplan, *FEBS Lett.* **1977**, *81*, 77–80.

[73] D. Oesterhelt, *FEBS Lett.* **1976**, *64*, 20–22.

[74] Oesterhelt D., Stoeckenius W. *Methods Enzymol.* **1974**, *31*, 667–678.

[75] M. Dyall-Smith, *The Halohandbook: Protocols for halobacterial genetics*, **2001**. http://www.microbiol.unimelb.edu.au/staff/mds/HaloHandbook/index.html

[76] K. J. Wise, N. B. Gillespie, J. A. Stuart, M. P. Krebs, R. R. Birge, *Trends Biotechnol.* **2002**, *20*, 387–394.

[77] D. Oesterhelt, W. Stoeckenius, *Nature New Biol.* **1971**, *233*, 149–152.

[78] N. Vsevolodov, *Biomolecular Electronics*, Birkhäuser, Boston, Basle, Berlin, **1998**.

12
Polymer Nanocontainers

Alexandra Graff, Samantha M. Benito, Corinne Verbert, and *Wolfgang Meier*

12.1
Introduction

In recent years, several very efficient and elegant methods have been developed to prepare hollow polymer particles – the so-called "polymer nanocontainers". The investigation of these systems is a highly active field of research in which numerous new publications appear every month. Due to their high stability and tunable properties, such polymer nanocontainers are believed to have a high potential for applications in biotechnology, such as confined reaction vessels, protective shells for enzymes, or as 'traps' for the selective recovery of biotransformation or polymerase chain reaction products. However, while currently most applications are just beginning to emerge or still only visions, polymer nanocontainers have successfully entered the biomedical field, where they have promoted major interest as drug delivery devices. Here, we will attempt to provide an overview of the existing container systems, discuss their potential for applications in these fields, and outline the technological problems that must be overcome.

12.2
Overview

12.2.1
From Liposomes in Biotechnology to Polymer Nanocontainers in Therapy

The increasing interest in new types of polymer nanocontainers originates from the pioneering studies on lipid vesicles or liposomes that were conducted in the early 1960s. Vesicles are spherically closed lipid bilayers that result from the naturally occurring self-assembly process of amphiphilic molecules. During the past few decades, various methods have been developed for their controlled preparation in the laboratory. In the meantime, liposomes have established a clear position in modern technology. Initially, liposomes served mainly as model systems to study biological membranes, but during the 1970s they were introduced as transport vehicles for drugs. Nowadays, they find also an increasing interest in mathematics and theoretical physics (e. g., topology of two-dimen-

Nanobiotechnology. Edited by Christof Niemeyer, Chad Mirkin
Copyright © 2004 WILEY-VCH Verlag GmbH & Co. KgaA, Weinheim
ISBN 3-527-30658-7

sional surfaces floating in a three-dimensional continuum). In general, liposomes are very important as model systems in biophysics (properties of cell membranes and channels), chemistry (catalysis, energy conversion and photosynthesis), colloid science (stability and thermodynamics of finite systems), biochemistry (function of membrane proteins), and biology (excretion, cell function, trafficking and signaling, gene delivery and function) [1].

It must be emphasized that liposome technology also had considerable impact on the development of new applications – that is, as controlled delivery devices for drugs (antifungals, anticancer agents, vaccines), nonviral gene delivery vectors, cosmetic formulations (skin-care products, shampoo), and diagnostic tools. Over the years, a variety of basic research investigations has led to improvements in their formulation, mainly to increase their stability and interaction characteristics (e. g., 'stealth' liposomes).

Most biotechnological applications of liposomes are based on the compartmentalization that they offer. Ma et al., for example, prepared vesicles from 2,4-tricosadiynoic acid (TCDA) as the lipid matrix and dioctadecyl glyceryl ether-β-glycoside as a receptor to detect *Escherichia coli* [2, 3]. These glycolipid functionalized vesicles are effective colorimetric biosensors which, due to the diacetylene groups, appear blue. Their binding to bacteria creates mechanical stress inside the vesicular membranes, and this induces a change of the effective conjugation length. As a result, the vesicle dispersion turns red. Oberholzer et al. designed liposomal DNA amplification by PCR and minimal cell bioreactors to express proteins [4, 5]. In particular, they demonstrated that DNA replication or ribosomal synthesis of polypeptides can be carried out inside the compartment offered by the aqueous pool of the liposomes. As will be seen later in the chapter, this artificial cell concept has also proved to be of interest in relation to polymer nanocontainers, while future diagnostic applications of these systems appear inevitable.

A major problem with liposomes, however, is that, due to their inherent colloidal and biological instability, they have very short lifetimes and are rapidly cleared from the bloodstream [1], and this in turn considerably limits their potential applications. Hence, enormous efforts have been undertaken during the past few years to design polymeric nanocontainers [6–9] of greater stability. The different container systems that have resulted from these activities will be discussed separately in the following sections.

12.2.2
Dendrimers

Dendrimers are highly branched polymers with radial symmetry and uniform size, which adopt a globular shape in solution [10–12]. Dendritic macromolecules or starburst dendrimers consist of three different structural or topological units that result from an iterative reaction sequence: a central core from which the repetitive branching units extend/emanate radially to finish in the outer layer of end-groups. With each generation – that is, the layers formed in each reaction step – the density is increased due to the geometric growth at each branching point [13]. This dense outer shell gave rise to the earliest concept of dendritic boxes [14]. Here, molecules have been encapsulated during the synthesis of the dendrimer and were retained within the central part of the macromolecules [13, 14]. The release of the molecules could be facilitated by an appropriate modification of the external groups of the dendrimer [13].

However, it is questionable whether dendrimers can be regarded as true nanocontainer systems: the central core groups of the molecules are of crucial importance for their integrity, and it is still an area of discussion whether the end groups of the molecules really form a dense outer layer or if they fold towards the interior, thus producing a dense core. According to Zimmerman [15], the surface will be identified with the end groups, the internal groups with the core, and the repeating units that interconnect both.

Synthetic design affords dendrimers with tailored structures. In general, the repeating units in the interior determine the solubilization properties towards guest molecules, while the functional terminal groups influence the solubility of the dendrimer itself in a given solvent. Particularly interesting examples are, in this context, amphiphilic dendrimers, in which the interior is comprised of hydrophobic moieties and the external groups consist of hydrophilic units. Generally, amphiphilic dendrimers and the more irregular amphiphilic hyper-branched polymers can be regarded as "unimolecular micelles" [16–22]. While classical micelles formed by low molar mass amphiphiles show low stability toward dilution due to the noncovalent interactions responsible for their formation, dendritic unimolecular micelles retain their cohesion regardless of concentration since they are static entities which are covalently linked in a globular fashion.

Similar to conventional micelles, amphiphilic dendrimers can also selectively solubilize hydrophobic guest molecules within their core. Recently, this has been successfully demonstrated for a hydrophobic drug, indomethacin [12]. However, the encapsulation efficiency of these molecules seems to be rather limited.

Interestingly, dendrimers can also be used to prepare real hollow structures by selectively crosslinking their outer shell and degrading the original core region [23].

Although several dendrimers are now commercially available, the preparation of these macromolecules (in particular the synthetic conversion approach) still requires costly and tedious procedures, posing a limiting factor for large-scale applications. Nevertheless, the high stability and the possibility of introducing a rich variety of peripheral functional groups (e.g., receptors or antibodies) make these systems highly interesting as model systems for the targeted delivery of drugs [24–27].

12.2.3
Layer by Layer (LbL) Deposition

A convenient way to produce polymer capsules is to exploit the well-known polyelectrolyte self-assembly at charged surfaces. This chemistry uses a series of layer-by-layer (LbL) deposition steps of oppositely charged polyelectrolytes [28, 29]. The driving force behind the LbL method at each step of the assembly is the electrostatic attraction between the added polymer and the surface. One starts with colloidal particles carrying surface charges, for example, a negative surface charge. Polyelectrolyte molecules having the opposite charge (e.g., polycations) are readily adsorbed due to electrostatic interactions with the surface. Usually, not all of the ionic groups of the adsorbed polyelectrolyte are consumed by the electrostatic interactions. As a result, the original surface charge is usually overcompensated by the adsorbed polymer. Hence, the surface charge of the coated particle changes its sign and is then available for the adsorption of a polyelectrolyte of again opposite

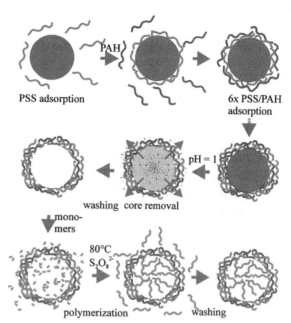

Figure 12.1 Schematic representation of the procedure for preparing hollow spheres using layer-by-layer deposition of oppositely charged polyelectrolytes on colloidal particles and subsequent encapsulation of polymers in a "ship in a bottle" fashion. PSS: sodium poly-styrene sulfonate; PAH: poly(allyl-amine)hydrochloride. (Reproduced from Ref. [81], with permission.)

charge (i. e., a polyanion). As shown diagrammatically in Figure 12.1, such sequential deposition produces ordered polyelectrolyte multilayers.

The size and shape of the resulting core-shell particles is determined by the template colloidal particle, and the formation of particles with diameters ranging from 0.2 to 10 μm has been reported. The thickness of the layered shell is determined by the number of polyelectrolyte layers, and can be adjusted accurately in the nanometer range [30]. Until now, a variety of charged substances, such as synthetic polyelectrolytes, biopolymers, lipids, and inorganic particles have been incorporated as layer constituents to build the multilayer shell on colloidal particles [28–32]. As templates for this approach, mainly colloids consisting of polystyrene latexes or melamine formaldehyde particles [33] have been used, but gold [34] and proteins [35] have also been tested.

Following the complete deposition of a predefined number of layers, the colloidal core can be dissolved and removed. Decomposition products are expelled through the shell wall and removed by several centrifugation and washing cycles [28]. The polyelectrolyte layer shells preserve their hollow sphere morphology and are shape-persistent. It has been shown that small dye molecules can readily permeate such layered polyelectrolyte shells, while larger-sized polymers with molecular weights larger than 4000 Da obviously do not [36].

Biocompatibility and biodegradability are two key parameters in designing biorelated systems. Biopolymers like alginate and polylysine can also be used in a similar way to yield biocompatible nanocapsules [37]. Interestingly, uncharged hydrophobic compounds could be also encapsulated using the LbL technique. A core formed by uncharged low molecular-weight microcrystalline substances (pyrene and fluorescein diacetate) was, in a first step, dispersed in water via micellization with amphiphilic substances such as ionic surfactants, phospholipids, or amphiphilic polyelectrolytes. Subsequently, a LbL procedure of depositing layers of polyelectrolytes rendered stabilized core shell particles. The release of the encapsulated substances, followed via the intrinsic fluorescence of the core forming material, was triggered by the addition of ethanol, which is a good solvent for the microcrystalline core [38]. Unfortunately the conditions in which the release is achieved are not physiological, which in turn prevents the use of this system as an in-vivo drug delivery system.

Wang et al. designed biologically active polymer microcontainers [39]. Using the LBL assembly, they functionalized luminescent polymer containers with anti-immunoglobulin G which rendered them biospecific via their IgG partners. These quantum dot-tagged beads open new opportunities in a range of biotechnological applications. Indeed, these quantum dots exhibit higher photo-bleaching threshold, quantum yield, and chemical stability than their organic fluorophore analogs. Furthermore, their spectral properties can be fine-tuned by controlling their size and, similar to planar LbL luminescent films [40], crosslinked luminescent core-shell particles and hollow capsules could also be used as light-emitting devices. This approach shows much promise in the area of sensors and, particularly, biosensing [41] (see also Chapter 22).

Similar to liposomes, the concept of artificial cells has been also applied to polyelectrolyte microcapsules [42]. Tiourina et al. used hollow microcapsules fabricated by stepwise adsorption of polyelectrolytes and phospholipids as so-called artificial cells. This model biosystem has high permeability for ions. Additionally, ion-channel-forming peptides such as gramicidin and valinomycin were incorporated into the lipid–polymer composite shell of the microcapsules. The resulting membrane potential, which is one of the most important cell parameters, was comparable to that of biological cells.

Nevertheless, it must be expected that the long-term stability of these capsules will depend sensitively on the surrounding environment of the particles. Especially in biological fluids (e. g., blood plasma) or in media of high ionic strength, which may screen the ionic interactions responsible for maintaining their integrity, the long-term stability of such polyelectrolyte shells may be rather limited. However, these problems may be overcome by enhancing the stability of the polyelectrolyte shells using an additional crosslinking polymerization step [43].

12.2.4
Block Copolymer Self-Assembly

Similar to conventional low molar-mass amphiphiles, amphiphilic block copolymers (which are polymers consisting of at least two chemically different parts, hydrophobic versus hydrophilic or rod versus coil) may self-assemble into various lyotropic mesophases [44, 45]. In particular, nanocontainers formed by self-assembled amphiphilic block copolymers have received increasing attention during recent years due to their potential for encapsulating large quantities of guest molecules within their central cavity. Additionally, block copolymer chemistry allows the introduction of a wide variety of different block structures, and this may lead to a plethora of new artificial membrane structures that are inaccessible to conventional lipids. Although the formation of self-assembled block copolymer superstructures follows the same underlying principles as that of low molar-mass amphiphiles, they are considerably more stable due to their larger size, slower dynamics, and inherent steric stabilization. Depending on their block length ratio, the critical aggregation concentration (c.a.c.) of these polymers can be shifted to extremely low values, which in turn makes their superstructures resistant against dilution – an essential requirement for medical applications [46]. Here, we will focus mainly on nanoparticles (micelles, vesicles) formed by such polymers. It must be emphasized however, that a plethora of other nanostructures emerges from block copolymer self-assembly.

12.2.4.1 Shell Cross-linked Knedel's (SCKs)

The term shell cross-linked knedels (SCKs) was first introduced in 1996 [47] to describe a special type of nanoparticles having core-shell morphology. These systems are formed by aggregation of amphiphilic di- and triblock copolymers into micelles [48–50]. An intramicellar crosslinking of the corona-forming blocks leads to the highly stable so-called SCKs, with sizes ranging from 50 to 250 nm [50]. In a second step, the backbone of the core-forming blocks can be cleaved and the low molar-mass degradation products extracted, thus leaving behind nanocages formed by a crosslinked polymer shell (Figure 12.2). Generally, the properties of the shell (i.e., swelling behavior and interactions with the surrounding medium) are governed by the chemical constitution and composition of the original block copolymers.

Recently, poly(ε-caprolactone)-block-poly(acrylic acid)-block-poly(acrylamide) SCKs with a biodegradable core were prepared [50]. Due to the mild conditions necessary to degrade their cores, these systems were expected to have superior properties for biological or medical application. More recently, this concept has also been extended to a so-called "block-copolymer-free" strategy for preparing micelles and hollow spheres under less "drastic conditions". One particularly interesting "block-copolymer free strategy" has

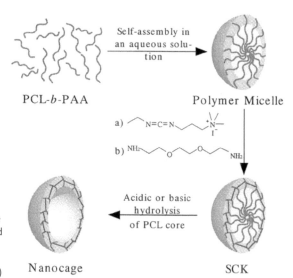

Figure 12.2 General procedure for the preparation of hollow shell cross-linked knedels (SCKs) nanocages from amphiphilic diblock copolymers. (Reproduced from Ref. [50], with permission.)

been described by Liu et al. [51], who used solely hydrogen bonds to interconnect the core and the shell of the micelles instead of using covalently attached block copolymer parts.

12.2.4.2 Block Copolymer Nanocontainers

Micellar structures have been designed for several applications. For example, micelles formed from amphiphilic di- or triblock copolymers have been explored for the solubilization of hydrophobic drugs [52]. In aqueous solution, the hydrophobic blocks form the micellar core while the hydrophilic ones build the corona. The core serves as a microenvironment for the lipophilic drugs, while the outer shell serves as a stabilizing interface between the hydrophobic core and the external medium. Kabanov et al. used Pluronic™ triblock copolymer micelles as delivery vehicles for drug targeting across the blood–brain barrier [53, 54]. Kataoka's group developed micelles formed from copolymers containing a poly(amino acid) core forming block as a delivery system for anti-cancer drugs [55–58].

Block copolymers nanostructures can be used as bile sorbents, which are a possible alternative to commercially available resins that have the side effect of targeting the coronary heart diseases related to elevated cholesterol levels [59]. Recently, the solubilization and release of benzo[a]pyrene and cell tracker CM-DiI in and from micelles consisting of a nontoxic and biodegradable polycaprolactone core surrounded by a poly(ethylene oxide) nontoxic and nonimmunogenic corona have been investigated in terms of loading efficiency, partition coefficient, and release profile [60].

For certain hydrophilic to hydrophobic block length ratios and molecular weight distribution, amphiphilic block copolymers form vesicular structures spontaneously in dilute aqueous solution. Similar to conventional liposomes, these block copolymer nanocontainers may find potential applications in the biotechnology area due to their ability to solubilize molecules in their inner aqueous pool. Initial reports about the controlled direct formation of block copolymer hollow sphere morphologies in aqueous media have been described only very recently, and potential applications are just beginning to emerge.

For example, Meier's group recently described the spontaneous formation of vesicles resulting from the self-assembly of a poly(2-methyloxazoline)-block-poly(dimethylsiloxane)-block-poly(2-methyloxazoline) (PMOXA-PDMS-PMOXA) triblock copolymer [61]. This polymer was additionally modified with reactive methacrylate groups at the ends of the hydrophilic blocks. A free radical polymerization of these methacrylate end groups in the vesicular aggregates led to the formation of shape-persistent polymer nanocontainers, with diameters ranging from 50 to 250 nm [61]. More recently, amphiphilic ABC triblock copolymers, with two different water-soluble blocks A and C (A = poly(2-methyl oxazoline), PMOXA; B = poly(dimethyl siloxane), PDMS; C = polyethylene oxide, PEO) have also been synthesized; these form similar polymer nanospheres, but with superior properties inherent to the asymmetry of their membrane [62]. Interestingly, these polymeric hollow nanospheres combine an extremely high mechanical stability with high flexibility provided by the hydrophobic PDMS middle blocks [61, 63, 64].

12.3
Polymer Nanocontainers with Controlled Permeability

One of the main advantages of polymer nanocontainers is their enormous stability that could provide, for example, an unchanging environment for the encapsulated molecules. However, with regard to potential applications, the stability and low permeability of the polymer walls may be major drawbacks as they prevent the effective loading of preformed containers or the controlled release of encapsulated material.

Recently, several very promising means of overcoming these problems have been introduced, and these will be described in the following sections.

12.3.1
Block Copolymer Protein Hybrid Systems

A new type of hybrid material has emerged from the combination of biological molecules and block copolymers. In one approach, a new class of biologically "active" super-amphiphiles composed of a block copolymer and an enzyme has been designed. This giant amphiphile consists of an enzyme head group and a single covalently connected hydrophobic polymeric tail. This hybrid material was obtained by the coupling of maleimide-functionalized polystyrene to a reduced lipase [65, 66]. Interestingly, the lipase remained functional in the self-assembled superstructures of these 'superamphiphiles'.

A similar pH-sensitive hybrid material was recently presented by Kukula et al. [67] and Chécot et al. [68]. Both groups described the formation of polymer vesicles or "peptosomes" by the self-assembly of poly(butadiene)-block-poly(L-glutamate) in dilute aqueous solution. Poly(L-glutamate) performs a pH-dependent helix-coil transition that does not alter the vesicle morphology. Due to their hydrophilic polypeptide chains, these new copolymer vesicles seem to be particularly suited for biological applications, and they may provide an interesting new bridge between the world of synthetic polymers and biological systems.

Another completely new approach is to reconstitute integral membrane proteins into block copolymer membranes. In Nature, membrane or membrane-associated proteins are responsible for various key functions such as biological signaling pathways or transport across membranes. Many of these membrane proteins possess important pharmacological properties and biotechnological potential.

It is clear that membrane-like superstructures formed by appropriate amphiphilic block copolymers closely resemble typical biological membranes. Actually, it has been shown that membrane proteins could be successfully reconstituted in such artificial polymer membranes. Surprisingly, these proteins remain functional despite the two- to threefold larger thickness of the block copolymer membrane that does not match the hydrophobic–hydrophilic pattern of natural membrane proteins. It seems that this requires a high flexibility of the hydrophobic blocks of the polymers that allows them to adapt to the specific geometric and dynamic requirements of membrane proteins. Under certain conditions (i. e., polymerizable groups at the very ends of the hydrophilic blocks), the proteins survive even a subsequent polymerization of the block copolymer matrix [63, 69]. For instance, the outer membrane protein, OmpF (a channel protein extracted from the

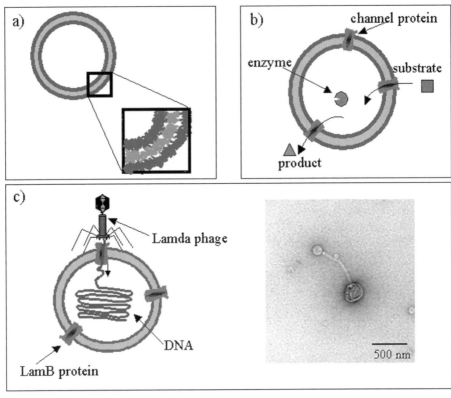

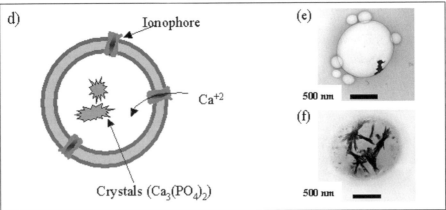

Figure 12.3 (a) Schematic view of a ABA triblock copolymer vesicle and magnification of the structure of its membrane, showing the constituting polymer chains. (b) Representation of a BioNanoreactor with encapsulated β-lactamase and inserted membrane channel proteins to facilitate diffusion of subtrates and products in and out of the nanoreactor. (c) Model of viral DNA encapsulation via phage binding and injection into nanocontainers, and transmission electron micrograph (TEM) showing the binding of a phage onto an ABA triblock copolymer vesicle. (d) Schematic representation of an ABA nanocontainer with incorporated ionophores in its membrane used as biomineralization device. TEMs showing (e) calcium phosphate crystals after 1 h and (f) after 24 h.

outer cell wall of Gram-negative bacteria) has been used to control the permeability of block copolymer nanocontainers (Figure 12.3b). Encapsulated enzymes inside such "nanoreactors" showed full activity and were considerably stabilized against proteolysis and self-denaturation [70]. Moreover, it has been shown that a controlled transmembrane potential could be used to induce a reversible gating transition of the proteins. Since only the open channels allow an exchange of substrates and products between the container's interior and the surrounding medium, such gating activates or deactivates the nanoreactors. In general, these systems have major potential for applications in pharmacy, diagnostics, or biotechnology. For example, suitably engineered channels could be used as prefilters to increase the selectivity of an encapsulated enzyme, or as selective gates to trap biotransformation products inside such nanocontainers, and this would allow a more convenient purification. Moreover, it has been shown recently that membrane receptors can also be incorporated into the walls of such polymer nanocontainers. Interestingly, access to the proteins could be controlled to a certain degree via the length of the hydrophilic blocks of the underlying amphiphilic block copolymers. For longer hydrophilic chains, they are "hidden" below a hydrophilic polymer layer so that larger ligands had no access to them. Such receptors bearing channels provide, for example, an elegant method to load polymer nanocontainers with DNA (Figure 12.3c) [71]. In particular, the small size, the electroneutrality and the low immunogenicity and toxicity of such DNA-loaded nanocontainers renders them highly interesting as new vectors for gene therapy.

Moreover, recent developments have indicated that these receptor-bearing polymer nanocontainers may be of particular interest as biosensors. One major advantage of these systems is that an entire detection and signaling cascade can be incorporated into a single nanocontainer. Block copolymer nanocontainers can be regarded as miniaturized artificial cells [72, 73] which allow massive miniaturization and parallelization (Figure 12.4). In addition, due to their high mechanical and (bio-)chemical stability, the polymer containers provide a constant environment for encapsulated analytic molecules, this being of crucial importance for technical applications where storage of the systems over extended periods of time is required.

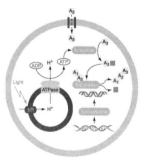

Figure 12.4 Hypothetical representation of a cell-like structure in which DNA is transcribed to RNA and translated to the protein via an encapsulated transcription and translation systems. Amino acids (A_1, A_2, A_3, etc.) are introduced into the compartment via transmembrane channel proteins and later activated (solid square) by ATP. A light- driven bio-energetic system composed of bacteriorhodopsin (bR) and ATP synthase (ATPase) is able to synthetize the needed ATP from ADP and phosphate. (Reproduced from Ref. [73], with permission.)

12.3.2
Stimuli-responsive Nanocapsules

As described above, stimuli-responsive peptides and proteins incorporated into the walls of polymer nanocontainers can be used as "switches" to control molecular exchange across polymer membranes. However, entirely synthetic polymer nanocontainers may also undergo reversible, stimuli-dependent swelling transitions. Such systems can be regarded as mimetics of virion cages, which show a structural transition that leads to the opening of gated pores within the virus shell upon pH changes [74]. Such stimuli-responsive nanocontainers could be obtained by core-shell emulsion polymerization [74]. Here, a two-step polymerization led to crosslinked poly(acrylic acid) hollow spheres that undergo a pH-induced swelling transition. With rising pH, the carboxylic acid groups of the polymer particles of the systems dissociate increasingly, thus leading to a high negative charge density along the polymer backbone. As a result these nanocontainers could increase their diameters by up to a factor of 10, depending on the respective pH and ionic strength. Hence, these containers retained encapsulated material at low pH and released it in "one shot" at high pH.

In the same way, pH-responsive dendrimers can be synthesized, albeit using rather harsh conditions [13]. In this context, dendrimers based on polypropyleneimine [75, 76] are of particular interest due to the potential protonation of their amine residues upon decreasing the pH. It has been shown that in these systems, both anions (e. g., oxoanions such as pertechnate) [76] and hydrophobic substances (pyrene) [77] can be encapsulated and released in a pH-dependent manner.

Not only dendrimers but also the more irregular hyperbranched systems can be tailored to have stimuli-responsive behavior. Krämer et al. showed that modifying the terminal groups of both hyperbranched polyglycerol and polyethyleneimine with acetal/ketals and imines, respectively, together with a hydrophobic outer shell could afford reverse micellar analogs with pH-responsive characteristics [78]. Interestingly, hydrophilic compounds were encapsulated as well as an antitumor drug (mercaptopurine). All of these were released spontaneously upon acidification of the media.

Micron-sized capsules, when prepared via the LbL deposition of weak polyelectrolytes, open at pH values <6 and close at values >8 [79]. The encapsulation of macromolecules in preformed hollow polyelectrolyte capsules was possible by loading them at low pH, whilst a subsequent pH increase captures the material inside the microcapsules. Another possibility is offered by polymerization of hydrophilic monomers in the void volume of similar polyelectrolyte capsules (see Figure 12.1). Here, in contrast to the final polymer, the monomers easily permeate the shells of these systems [80, 81]. As a result, the macromolecules formed during polymerization are entrapped within the capsules. It has been shown that, by encapsulating appropriate polyelectrolytes, it was possible to change the pH by about 2 units inside the containers compared with the surrounding medium. This potentially enables the systems to be used as nanoreactors, for example to carry out acid-catalyzed reactions within the capsules. Moreover, under certain experimental conditions the polymerization of encapsulated monomers takes place mainly within the capsule walls. This leads to an interesting way of modifying the ionic selectivity of the shell, as anionic substances did not translocate the functionalized walls, while cationic probes did. Therefore, a tunable control of the permeability was achieved [82].

As they protect sensitive drugs from proteolytic degradation, pH-responsive microparticles have been proposed for the oral delivery of insulin [83]. The insulin-containing particles retain the substance at low pH in the stomach until they reach a higher pH in the intestine. This delivery system consists of insulin-containing microparticles of cross-linked copolymers of poly(methacrylic acid)-graft-poly(ethylene glycol). The pH sensitivity is due to the reversible formation of interpolymer complexes stabilized by hydrogen bonding between the carboxylic acid protons and the ether groups on the grafted chains. However, due to electrostatic interactions which maintain the integrity of the layered shell, encapsulation and release of some substrates may prove to be limited, as might be the use of such capsules in high ionic strength media such as the biological milieu.

Block copolymer self-assembly appears to be more suitable in this respect. Recently, the group of Okano has designed thermoresponsive polymeric micelles consisting of AB block copolymers of PIPAAm (poly(N-isopropylacrylamide)) blocks and PBMA (poly(butyl methacrylate)) or PSt (polystyrene) blocks capable of encapsulating the hydrophobic drug, adriamycin. PIPAAm-PBMA micelles released the drug only above the reversible thermo-responsive phase transition of PIPAAm [84].

It must be pointed out that this polymer chemistry allows investigations to be made of the integration of temperature, chemical composition, light-sensitive or targeting moieties to these systems, which in turn show great potential for use in sensor technology or diagnostics.

12.4
Nanoparticle Films

One interesting aspect of nanotechnology concerns the formation of nanoparticle layers on a solid support. In the so-called "bottom up" approach, these thin films are formed by surface-modified nanoparticles. In this situation, the attractive electrostatic interactions between charged nanoparticles and functionalized surfaces are frequently exploited [85]. In this context, variations in the pH of a solution can be used to control the degree of ionization of the particle surfaces, which then allows modulation of the electrostatic interactions between nanoparticles and the immobilizing surface.

The uniformly sized dendritic macromolecules are considered as particular promising building blocks for such functionalized surfaces. The large number of end groups at the periphery of a dendrimer and the relative ease of their tailoring leads to a plethora of pathways for surface recognition. The high density of end groups also allows collective processes to occur which could be used as amplification cascades leading to a detectable signal. This may be of value in the development of (bio)sensors [15], nanochip-based release devices, or gene sequencers. Recently, a biosensor has been developed on the basis of SCKs that had been surface-functionalized to promote cell binding [86] via conjugation between the SCK nanoparticles and a biologically active peptide sequence.

12.5
Biomaterials and Gene Therapy

The field of biomaterials focuses on the design of "intelligent" materials – that is, which can respond to their surrounding environment to improve their integration and function. Due to their biocompatibility and responsiveness, the polymer nanoreactors described above may be viewed as a typical example of such materials. The incorporation and controlled release of polypeptide growth factors that are inherent to biological function regulation (e. g., tissue regeneration) could be envisaged in this respect.

As with drug delivery, nonviral gene delivery utilizes a site approach to either increase or decrease the expression of a specific gene by using DNA, RNA, oligonucleotides, or antisense sequences. The design of an optimized vector first requires identification of the desired therapy pathway – that is, cellular uptake either *in vivo* or *in vitro* or directed to a specific tissue.

Gene therapy currently suffers from a lack of safe and efficient carriers. Genetically engineered viruses have a high efficiency, but suffer from a limited genome size when inserting dedicated genes. In addition, safety issues emerging from the virus production itself and their potential immunogenicity and mutagenicity have recently led to the development of various nonviral systems. One approach which has been widely investigated

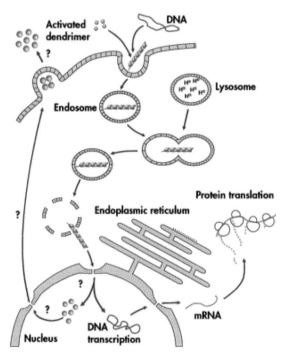

Figure 12.5 Model mechanism of DNA transfection using active dendrimer-mediated uptake. (Reproduced from Ref. [88], with permission.)

is the complexation of DNA with cationic lipids, polycationic polymers, and dendrimers [87, 88]. However, *in vivo* (Figure 12.5) these vectors are affected by interactions with environmental components (e. g., serum proteins) and show only moderate transfection efficiency. Encapsulation in liposomes subsequent to precondensation reduces serum inhibition and enhances the transfection efficiency [88]. However, the poor stability of liposomes in the bloodstream is well known, and therefore polymer vectors which allow receptor-mediated gene delivery offer greater promise. Gene delivery using dendrimers as vehicles, and a comparison with the classic techniques of gene transfer, has been reviewed by Dennig and Duncan [89].

Both liposomes and biocompatible block copolymer nanocontainers, due to their limited blood clearance and drainage into the lymphatic system (in case of tissue injection), enable the genetic material to be protected against the action of endonucleases. In addition, block copolymer chemistry would allow the preparation of nanocontainers with the potential to encapsulate large quantities of guest molecules within their central cavity, and which would also allow crossing of the endothelial barrier. Eventually, block copolymer chemistry might allow the introduction of a wide variety of moieties, cell targeting, endocytosis, and nuclear uptake by the utilization of specific targeting.

Moreover, biocompatible and electrically neutral vectors based on amphiphilic block copolymers could also be prepared which reduce the repulsion between negatively charged plasmid DNA and negatively charged cell membranes, thus facilitating cellular uptake.

12.6
Outlook

It must be emphasized that this overview of the current state of the art is not complete, and that the systems and applications described should be regarded as representative examples only. The possibility of incorporating additional design criteria (e. g., temperature sensitivity, targeting moieties, special surface characteristics) makes polymer nanocontainers highly versatile systems which can be optimized with respect to any desired application. Of particular interest for future developments is the possibility of incorporating biological functions into these synthetic structures. In this context, it is interesting to note that Nature provides many specific, unspecific, or ligand-gated channels (that can also be genetically modified) and other membrane proteins, which can be reconstituted in the polymer walls of the containers. Preliminary investigations in our laboratory show that this provides not only a unique tool to control permeation across the nanocontainer shells, but also their use as molecular motor-driven nanomachines or as nanometer-sized batteries as power supplies [90]. Moreover, by interconnecting different nanoreactors (e. g., containing otherwise incompatible enzymes) it is possible to prepare nanofactory arrays capable of performing multistep syntheses. These systems may be of major interest as self-regulating drug delivery devices or as sensors containing an integrated amplification module for the measured signal.

In general, we believe that the principle of combining the high diversity of polymer chemistry with the functionality of natural proteins and peptides will have many future applications in areas such as drug delivery, sensor technology, energy conversion, diagnostics, and catalysis.

References

[1] D. D. Lasic, *Liposomes: From Physics to Applications*, Elsevier, **1993**.

[2] Z. Ma, J. Li, L. Jiang, *Langmuir* **2000**, *16*, 7801–7804.

[3] Q. Ma, E. E. Remsen, T. Kowalewski, J. Schaefer, K. L. Wooley, *Nano Lett.* **2001**, *1*, 651–655.

[4] T. Oberholzer, M. Albrizio, P. L. Luisi, *Chem. Biol.* **1995**, *2*, 677–682.

[5] T. Oberholzer, K. H. Nierhaus, P. L. Luisi, *Biochem. Biophys. Res. Commun.* **1999**, *261*, 238–241.

[6] C. Nardin, W. Meier, *Chimia* **2001**, *55*, 142–146.

[7] F. Caruso, X. Shi, R. A. Caruso, A. Susha, *Adv. Mater.* **2001**, *13*, 740–744.

[8] A. V. Kabanov, V. Y. Alakhov, *Crit. Rev. Ther. Drug* **2002**, *19*, 1–72.

[9] G. Ibarz, L. Dahne, E. Donath, H. Mohwald, *Adv. Mater.* **2001**, *13*, 1324–1327.

[10] D. K. Smith, F. Diederich, *Top. Curr. Chem.* **2000**, *210*, 183–227.

[11] A. W. Bosman, H. M. Janssen, E. W. Meijer, *Chem. Rev.* **1999**, *99*, 1665–1688.

[12] J. M. J. Frechet, *Proc. Natl. Acad. Sci. USA* **2002**, *99*, 4782–4787.

[13] J. F. G. A. Jansen, E. W. Meijer, E. M. M. de Brabander van den Berg, *J. Am. Chem. Soc.* **1995**, *117*, 4417–4418.

[14] J. F. G. A. Jansen, E. M. M. de Brabander van den Berg, E. W. Meijer, *Science* **1994**, *266*, 1226–1229.

[15] S. C. Zimmerman, L. J. Lawless, *Top. Curr. Chem.* **2001**, *217*, 95–120.

[16] A. Sunder, M. Kramer, R. Hanselmann, R. Mulhaupt, H. Frey, *Angew. Chem. Int. Ed.* **1999**, *38*, 3552–3555.

[17] C. J. Hawker, K. L. Wooley, J. M. J. Frechet, *J. Chem. Soc., Perkin Trans. 1: Org. Bio-Org. Chem. (1972-1999)*, **1993**, 1287–1297.

[18] G. R. Newkome, C. N. Moorefield, G. R. Baker, M. J. Saunders, S. H. Grossman, *Angew. Chem.* **1991**, *103*, 1207–1209 (see also *Angew. Chem., Int. Ed. Engl.*, **1991**, *1230* (1209), 1178–1180).

[19] S. Mattei, P. Seiler, F. Diederich, V. Gramlich, *Helv. Chim. Acta* **1995**, *78*, 1904–1912.

[20] S. Stevelmans, J. C. M. v. Hest, J. F. G. A. Jansen, D. A. F. J. Van Boxtel, E. M. M. de Berg, E. W. Meijer, *J. Am. Chem. Soc.* **1996**, *118*, 7398–7399.

[21] M. Liu, K. Kono, J. M. J. Frechet, *J. Controll. Rel.* **2000**, *65*, 121–131.

[22] M. Liu, K. Kono, J. M. J. Frechet, *J. Polymer Sci. A: Polymer Chem.* **1999**, *37*, 3492–3503.

[23] M. S. Wendland, S. C. Zimmerman, *J. Am. Chem. Soc.* **1999**, *121*, 1389–1390.

[24] R. Esfand, D. A. Tomalia, *Drug Discovery Today* **2001**, *6*, 427–436.

[25] M. Liu, J. M. J. Frechet, *Pharm. Sci. Technol Today* **1999**, *2*, 393–401.

[26] L. J. Twyman, A. E. Beezer, R. Esfand, M. J. Hardy, J. C. Mitchell, *Tetrahedron Lett.* **1999**, *40*, 1743–1746.

[27] A. K. Patri, I. J. Majoros, J. R. Baker, *Curr. Opin. Chem. Biol.* **2002**, *6*, 466–471.

[28] G. Decher, *Science* **1997**, *277*, 1232–1237.

[29] E. Donath, G. B. Sukhorukov, F. Caruso, S. A. Davis, H. Mohwald, *Angew. Chem. Int. Ed.* **1998**, *37*, 2202–2205.

[30] F. Caruso, *Chemistry – A European Journal*, **2000**, *6*, 413–419.

[31] P. Bertrand, A. Jonas, A. Laschewsky, R. Legras, *Macromol. Rapid Commun.* **2000**, *21*, 319–348.

[32] F. Caruso, *Adv. Mater.* **2001**, *13*, 11–22.

[33] F. Caruso, R. A. Caruso, H. Moehwald, *Chem. Mater.* **1999**, *11*, 3309–3314.

[34] D. I. Gittins, F. Caruso, *Adv. Mater.* **2000**, *12*, 1947–1949.

[35] N. G. Balabushevitch, G. B. Sukhorukov, N. A. Moroz, D. V. Volodkin, N. I. Larionova, E. Donath, H. Mohwald, *Biotechnol. Bioeng.* **2001**, *76*, 207–213.

[36] P. Rilling, T. Walter, R. Pommersheim, W. Vogt, *J. Membr. Sci.* **1997**, *129*, 283–287.

[37] C. Schueler, F. Caruso, *Biomacromolecules* **2001**, *2*, 921–926.

[38] F. Caruso, W. Yang, D. Trau, R. Renneberg, *Langmuir* **2000**, *16*, 8932–8936.

[39] D. Wang, A. L. Rogach, F. Caruso, *Nano Lett.* **2002**, *2*, 857–861.

[40] B. Lehr, M. Seufert, G. Wenz, G. Decher, *Supramol. Sci.* **1996**, *2*, 199–207.

[41] M.-K. Park, C. Xia, R. C. Advincula, P. Schuetz, F. Caruso, *Langmuir* **2001**, *17*, 7670–7674.

[42] O. P. Tiourina, I. Radtchenko, G. Sukhor-ukov, H. Moehwald, *J. Membr. Biol.* **2002**, *190*, 9–16.

[43] I. Pastoriza-Santos, B. Scholer, F. Caruso, *Adv. Function. Mater.* **2001**, *11*, 122–128.

[44] S. A. Jenekhe, X. L. Chen, *Science* **1998**, *279*, 1903–1907.

[45] L. Zhang, A. Eisenberg, *Science* **1995**, *268*, 1728–1731.

[46] D. D. Lasic, D. Needham, *Chem. Rev.* **1995**, *95*, 2601–2628.

[47] K. B. Thurmond, II, T. Kowalewski, K. L. Wooley, *J. Am. Chem. Soc.* **1996**, *118*, 7239–7240.

[48] S. Liu, J. V. M. Weaver, M. Save, S. P. Armes, *Langmuir* **2002**, *18*, 8350–8357.

[49] J.-F. Gohy, N. Willet, S. Varshney, J.-X. Zhang, R. Jerome, *Angew. Chem. Int. Ed.* **2001**, *40*, 3214–3216.

[50] Q. Zhang, E. E. Remsen, K. L. Wooley, *J. Am. Chem. Soc.* **2000**, *122*, 3642–3651.

[51] X. Liu, M. Jiang, S. Yang, M. Chen, D. Chen, C. Yang, K. Wu, *Angew. Chem. Int. Ed.* **2002**, *41*, 2950–2953.

[52] C. Allen, D. Maysinger, A. Eisenberg, *Colloids Surfaces, B: Biointerfaces* **1999**, *16*, 3–27.

[53] A. V. Kabanov, E. V. Batrakova, N. S. Melik-Nubarov, N. A. Fedoseev, T. Y. Dorodnich, V. Y. Alakhov, V. P. Chekhonin, I. R. Nazar-ova, V. A. Kabanov, *J. Controll. Rel.* **1992**, *22*, 141–157.

[54] A. V. Kabanov, V. P. Chekhonin, V. Y. Ala-khov, E. V. Batrakova, A. S. Lebedev, N. S. Melik-Nubarov, S. A. Arzhakov, A. V. Leva-shov, G. V. Morozov, et al., *FEBS Lett.* **1989**, *258*, 343–345.

[55] G. Kwon, S. Suwa, M. Yokoyama, T. Okano, Y. Sakurai, K. Kataoka, *J. Controll. Rel.* **1994**, *29*, 17–23.

[56] G. Kwon, M. Naito, M. Yokoyama, T. Okano, Y. Sakurai, K. Kataoka, *J. Controll. Rel.* **1997**, *48*, 195–201.

[57] M. Yokoyama, G. S. Kwon, T. Okano, Y. Sakurai, H. Ekimoto, K. Kataoka, *J. Controll. Rel.* **1994**, *28*, 336–337.

[58] M. Yokoyama, S. Fukushima, R. Uehara, K. Okamoto, K. Kataoka, Y. Sakurai, T. Okano, *J. Controll. Rel.* **1998**, *50*, 79–92.

[59] N. S. Cameron, A. Eisenberg, R. G. Brown, *Biomacromolecules* **2002**, *3*, 124–132.

[60] P. L. Soo, L. Luo, D. Maysinger, A. Eisen-berg, *Langmuir* **2002**, *18*, 9996–10004.

[61] C. Nardin, T. Hirt, J. Leukel, W. Meier, *Langmuir* **2000**, *16*, 1035–1041.

[62] R. Stoenescu, W. Meier, *Chem. Commun.* **2002**, *24*, 3016–3017.

[63] W. Meier, C. Nardin, M. Winterhalter, *Angew. Chem. Int. Ed.* **2000**, *39*, 4599–4602.

[64] W. Meier, *Chimia* **2002**, *56*, 20.

[65] K. Velonia, A. E. Rowan, R. J. M. Nolte, *Polymer Preprints (American Chemical Society, Division of Polymer Chemistry)*, **2002**, *43*, 686.

[66] K. Velonia, A. E. Rowan, R. J. M. Nolte, *J. Am. Chem. Soc.* **2002**, *124*, 4224–4225.

[67] H. Kukula, H. Schlaad, M. Antonietti, S. Foerster, *J. Am. Chem. Soc.* **2002**, *124*, 1658–1663.

[68] F. Checot, S. Lecommandoux, Y. Gnanou, H.-A. Klock, *Angew. Chem. Int. Ed.* **2002**, *41*, 1339–1343.

[69] C. Nardin, S. Thoeni, J. Widmer, M. Win-terhalter, W. Meier, *Chem. Commun.* **2000**, 1433–1434.

[70] C. Nardin, W. Jorg, W. Mathias, M. Wolf-gang, *Eur. Physical J. E* **2001**, *4*, 403–410.

[71] A. Graff, M. Sauer, P. Van Gelder, W. Meier, *Proc. Natl. Acad. Sci. USA* **2002**, *99*, 5064–5068.

[72] D. A. Hammer, D. E. Disher, *Annu. Rev. Mater. Res.* **2001**, *31*, 387–404.

[73] A. Pohorille, D. Deamer, *Trends Biotechnol.* **2002**, *20*, 123–128.

[74] M. Sauer, D. Streich, W. Meier, *Adv. Mater.* **2001**, *13*, 1649–1651.

[75] G. Pistolis, A. Malliaris, D. Tsiourvas, C. M. Paleos, *Chemistry – A European Journal* **1999**, *5*, 1440–1444.

[76] H. Stephan, H. Spies, B. Johannsen, C. Kauffmann, F. Voegtle, *Org. Lett.* **2000**, *2*, 2343–2346.

[77] Z. Sideratou, D. Tsiourvas, C. M. Paleos, *Langmuir* **2000**, *16*, 1766–1769.

[78] M. Kramer, J.-F. Stumbe, H. Turk, S. Krause, A. Komp, L. Delineau, S. Prok-horova, H. Kautz, R. Haag, *Angew. Chem. Int. Ed.* **2002**, *41*, 4252–4256.

[79] G. B. Sukhorukov, A. A. Antipoc, A. Voigt, E. Donath, H. Moehwald, *Macromol. Rapid Commun.* **2001**, *22*, 44–46.

[80] G. B. Sukhorukov, M. Brumen, E. Donath, H. Moehwald, *J. Phys. Chem. B* **1999**, *103*, 6434–6440.

[81] G. B. Sukhorukov, E. Donath, S. Moya, A. S. Susha, A. Voigt, J. Hartmann,

H. Mohwald, *J. Microencapsulation* **2000**, *17*, 177–185.

[82] L. Daehne, S. Leporatti, E. Donath, H. Moehwald, *J. Am. Chem. Soc.* **2001**, *123*, 5431–5436.

[83] A. M. Lowman, M. Morishita, M. Kajita, T. Nagai, N. A. Peppas, *J. Pharm. Sci.* **1999**, *88*, 933–937.

[84] J. E. Chung, M. Yokoyama, T. Okano, *J. Controll. Rel.* **2000**, *65*, 93–103.

[85] M. Sastry, M. Rao, K. N. Ganesh, *Acc. Chem. Res.* **2002**, *35*, 847–855.

[86] J. Liu, Q. Zhang, E. E. Remsen, K. L. Wooley, *Biomacromolecules* **2001**, *2*, 362–368.

[87] T. Azzam, A. Raskin, A. Makovitzki, H. Brem, P. Vierling, M. Lineal, A. J. Dom, *Macromolecules* **2002**, *35*, 9947–9953.

[88] T. Segura, L. D. Shea, *Annu. Rev. Mater. Res.* **2001**, *31*, 2–46.

[89] J. Dennig, E. Duncan, *Rev. Molec. Biotechnol.* **2002**, *90*, 339–347.

[90] W. Meier, unpublished results.

13
Biomolecular Motors Operating in Engineered Environments

Stefan Diez, Jonne H. Helenius, and *Jonathon Howard*

13.1
Overview

Recent advances in understanding how biomolecular motors work has raised the possibility that they might find applications as nanomachines. For example, they could be used as molecule-sized robots that:

- work in molecular factories where small, but intricate structures are made on tiny assembly lines;
- construct networks of molecular conductors and transistors for use as electrical circuits;
- or that continually patrol inside "adaptive" materials and repair them when necessary.

Thus, biomolecular motors could form the basis of bottom-up approaches for constructing, active structuring and maintenance at the nanometer scale. We will review the current status of the operation of biomolecular motors in engineered environments, and discuss possible strategies aimed at implementing them in nanotechnological applications. We cite reviews whenever possible for the biochemical and biophysical literature, and include primary references to the nanotechnological literature.

Biomolecular motors are the active workhorses of cells [1]. They are complexes of two or more proteins that convert chemical energy – usually in the form of the high-energy phosphate bond of ATP – into directed motion. The most familiar motor is the protein myosin which moves along filaments, formed from the protein actin, to drive the contraction of muscle. In fact, all cells – not just specialized muscle cells – contain motors that move cellular components such as proteins, mitochondria, and chromosomes from one part of the cell to another. These motors include relatives of muscle myosin (that also move along actin filaments), as well as members of the kinesin and dynein families of proteins. The latter motors move along another type of filament called the microtubule. The reason that motors are necessary in cells is that diffusion is too slow to transport molecules efficiently from where they are made (which typically is near the nucleus) to where they are used (which is often at the periphery of the cell). For example, the passive diffusion of a small protein to the end of a 1 meter-long neuron would take approximately 1000 years, yet kinesin moves it in a week. This corresponds to a speed of 1–2 μm s^{-1}, which is typical

Nanobiotechnology. Edited by Christof Niemeyer, Chad Mirkin
Copyright © 2004 WILEY-VCH Verlag GmbH & Co. KgaA, Weinheim
ISBN 3-527-30658-7

for biomolecular motors [2]. Actin filaments and microtubules form a network of highways within cells, and localized cues are used to target specific cargoes to specific sites in the cell [3]. By using filaments and motors, cells build highly complex and active structures on the molecular (nanometer) scale. Little imagination is needed to envisage employing biomolecular motors to build molecular robots [4].

Biomolecular motors are unusual machines that do what no man-made machines do: they convert chemical energy to mechanical energy directly rather than via an intermediate such as heat or electrical energy. This is essential because the confinement of heat, for example, on the nanometer scale is not possible because of its high diffusivity in aqueous solutions [2]. As energy converters, biomolecular machines are highly efficient. The chemical energy available from the hydrolysis of ATP is 100×10^{-21} J = 100 pN nm^{-1} (under physiological conditions, where the ATP concentration is 1 mM and the concentrations of the products ADP and phosphate are 0.01 mM and 1 mM, respectively). With this energy, a kinesin molecule is able to perform an 8-nm step against a load of 6 pN [2]. The energy efficiency is therefore almost 50%. For the rotary motor F_1F_0-ATPase synthase which uses the electrochemical gradient across mitochondrial and bacterial membranes to generate ATP, the efficiency is reported to be between 80 and 100% [5, 6]. This high efficiency demonstrates that, like other biological systems, the operation of biological motors has been optimized through evolution.

High efficiency is but one feature that makes biomolecular motors attractive for nanotechnological applications. Other features are:

1. They are small and can therefore operate in a highly parallel manner.
2. They are easy to produce and can be modified through genetic engineering.
3. They are extremely cheap. For example, 20×10^9 kinesin motors can be acquired for 1 US cent from commercial suppliers (1 mg = 3.3×10^{15} motors cost \$1500; Cytoskeleton, Inc., Colorado, USA) and the price could be significantly decreased if production were scaled up.
4. A wide array of biochemical tools have been developed to manipulate these proteins outside the cell.

This review focuses on two broad categories of molecular motors:

- *Linear motors* generate force as they move along intracellular filaments. In addition to myosin and kinesin mentioned above, linear motors also include enzymes that move along DNA and RNA.
- *Rotary motors* generate torque via the rotation of a central core within a larger protein complex. They include ATP synthase, mentioned above, as well as the motor that drives bacterial motility.

Representatives of both categories have been used to manipulate molecules and nanoparticles. Mechanical and structural properties of relevant filaments are contained in Table 13.1, and those of several associated motors in Table 13.2.

The general set-ups for studying motor proteins outside cells – the so-called motility assays – are depicted in Figure 13.1. In the gliding assay, the motors are immobilized on a surface and the filaments glide over the assembly (Figure 13.1A). In the stepping assay, the filaments are laid out on the surface where they form tracks for the motors

Table 13.1 Physical attributes of actin filaments, microtubules, DNA, and RNA. The persistence length (*Lp*) is related to the flexural rigidity (*EI*) by: *Lp* = *EI* / *kT*, where *k* is the Boltzmann constant and *T* is absolute temperature. Young's modulus (*E*) is calculated assuming that the filament is homogenous and isotropic. The repeat length describes the periodicity along a strand of the filament.

Filament	Diameter	Strands per filament	Repeat length	Persis- tence length	Young's modulus	Maximum length	Motors	Reference
Actin filament	6 nm	2	5.5 nm	10 µm	2 GPa	100 µm	Myosin	77
Micro- tubule	25 nm	13	8 nm	5 mm	2 GPa	10 cm	Kinesin, Dynein	78
DNA	2 nm	2	0.34 nm	50 nm	1 GPa	100 mm	RNA polymerase, DNA helicase, topoisomerase	79
RNA	2 nm	2	0.34 nm	75 nm	1.5 GPa	30 µm	Ribosome	80

Table 13.2 Values characterizing the operation of several important biomolecular motors. The filaments along which the linear motors operate are indicated in Table 13.1. The sizes refer to the motor domains. Dynamic parameters were determined by in-vitro experiments at high ATP concentration.

Motor	Filament	Size* [nm]	Step size [nm]	Maximum speed [nm s^{-1}]	Maximum force [pN]	Effi- ciency [%]	Refe- rence(s)
Myosin II	Actin	16	5	30000	10 pN	50	2, 81
Myosin V	Actin	24	36	300	1.5 pN	50	82
Conventional kinesin	Microtubule	6	8	800	6 pN	50	2, 83
Dynein	Microtubule	24		6400	6 pN		84, 85
T7 DNA poly- merase (exonu- clease activity)	DNA		0.34	>100 bps	34 pN	NA	86
RNA poly- merase	DNA	15	0.34	5	25 pN	NA	87, 88
Topoisomerase	DNA		up to 43 nm/turn	–		NA	89, 90
Bacteriophage portal motor	DNA		0.34	100 bps	57 pN		91
Type IV pilus retraction motor	pilus			1000	110 pN		92, 93
F$_1$-ATPase	NA	8 × 14	120°	8 rps	100 pN nm	80	6
Flagellar motor	NA	45		300 rps	550 pN nm		94

NA, not applicable.
* The sizes refer to the motor domains.

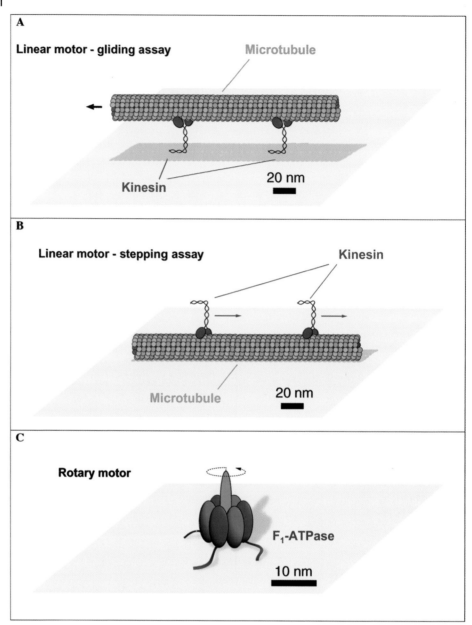

Figure 13.1 Biomolecular motor systems currently applicable for nanotechnological developments. (A) Linear transport of filaments by surface bound motor molecules (gliding assay). (B) Linear movement of motor proteins along filaments (stepping assay). (C) Rotation generated by a rotary motor.

to move along (Figure 13.1B). In both assays, movement is observed under the light microscope using fluorescence markers or high-contrast techniques. Variations on these assays have been used to reconstitute linear motility on the four types of filaments – actin filaments, microtubules, DNA, and RNA.

The gliding motility assay has provided detailed data on the directionality, speed, and force generation of purified molecular motors [2, 7]. However, for use in nanotechnological applications, the movement of gliding filaments must be controllable in space and time. For example, a simple application would be to employ a moving filament to pick up cargo at point A, move it along a user-defined path to point B, and then release it.

A number of methods for the spatial and temporal control of filament movement have been developed. Spatial control has been achieved using topographical features [8–11], chemical surface modifications [10, 12–14], and a combination of both [15–18]. Electrical fields [19–21] and hydrodynamic flow [22, 23] have also been used to direct the motion of gliding filaments. An example from our laboratory of gliding microtubules that are guided by channels is shown in Figure 13.2. Temporal control has been achieved by manipulating the ATP concentration [9, 24].

In addition to these basic techniques for controlling motion, some simple applications of the gliding assay have been demonstrated. These include the transport of streptavidin-

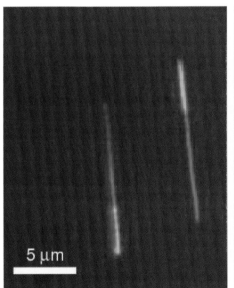
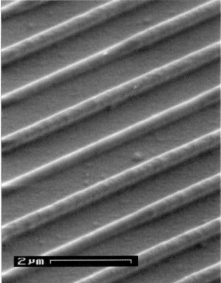

Figure 13.2 (A) Directed movement of gliding microtubules along microstructured polyurethane channels on the surface of a coverslip. The initial positions of the microtubules are shown in orange, while the paths they traveled over the subsequent 12 s are shown in green. (B) Scanning electron microscopy image of the polyurethane channels. The channels are a replica mold of a Si-master (channel width 500 nm, periodicity 1000 nm, depth 300 nm) produced using a poly(dimethylsiloxane) (PDMS) stamp as an intermediate. Note, that the ridges have been "undercut". This probably aids the guiding of the microtubules in the channels. (Silicon master provided by T. Pompe, Institute of Polymer Research, Dresden, Germany.)

coated beads [9], the transport and stretching of individual DNA molecules [25], the measurement of forces in the pN range [26], and the imaging of surfaces [27].

The stepping assay opens up additional possibilities. Initially, micrometer-sized beads were coated with motor proteins and visualized as they moved along filaments. The movement of beads can be tracked with nanometer precision to determine the speed and step size [2], and the use of optical tweezers allows forces to be measured [28]. In addition to beads, 10 μm-diameter glass particles [29] and Si-microchips [30] have been transported and membrane tubes have been pulled [32a] along filaments. In another variation, high-sensitivity fluorescence microscopy is used to visualize individual motor molecules as they step along filaments [31, 32]. An example from our laboratory of a single kinesin motor fused to the green fluorescent protein moving along a microtubule is shown in Figure 13.3, see p. 192. Despite the power of single-molecule techniques, they have yet to be exploited for nanotechnological applications.

Rotary motors can be studied in vitro by fixing the stator to a surface and following the movement of the rotor (see Figure 13.1C). Rotation can be visualized under the light microscope by attaching a fluorescent label or a microscopic marker to the rotor. Both techniques have been used to investigate the stepwise rotation generated by F_1-ATPase, which is a component of the F_1F_0-ATP synthesis machinery [5, 33]. Individual motors have been integrated into nanoengineered environments by arraying them on a nanostructured surface and using them to rotate fluorescent microspheres [34] or to drive Ni-nanopropellers [6].

13.2
Methods

There are many challenges in applying biomolecular motors to nanotechnology. Motility must be robust, it must be controlled both spatially and temporally, and the motors must be hitched to and unhitched from their cargoes. This section summarizes key techniques towards these ends.

13.2.1
General Conditions for Motility Assays

Motility assays are performed in aqueous solutions that must fulfill a number of requirements. We will illustrate these requirements with the kinesin/microtubule system. Kinesin uses ATP as its fuel; the maximum speed is reached at ~0.5 mM, approximately equal to the cellular concentration. Other nucleotides such as GTP, TTP, and CTP can substitute for ATP, but the speed is lower [35]. Motility also requires divalent cations, with magnesium preferred over calcium, and strontium and barium unable to substitute [36]. Optimal motility, assessed by gliding speed, occurs over a range of pH, between 6 and 9 [35, 37], and over a range of ionic strengths, between 50 mM and 300 mM [37]. The speed increases with temperature, doubling for each 10 °C between 5 °C and 50 °C [24, 38]; motility fails at higher temperatures. The force is independent of temperature between 15 °C and 35 °C [39]. When assays are performed in the middle of these ranges, motility is robust and only a small drop in the mean velocities is seen after 3 hours [24, 37]. If fluorescent markers are used, then an oxygen-scavenging enzyme system must be present in order to prevent photodamage. Many experimental details, including a discussion of the densities of the motors, can be found in Ref. [7].

13.2.2
Temporal Control

Motors can be reversibly switched off and on by regulating the concentration of fuel, or by adding and removing inhibitors. The ATP concentration can be rapidly altered by flowing in a new solution. In such a set-up, the kinesin-dependent movement of microtubules can be stopped within 1 s and restarted within 10 s (unpublished data from our laboratory). Similarly, inhibitors such as AMP-PNP (a non-hydrolyzable analogue of ATP [40]), adociasulfate-2 (a small molecule isolated from sponge [41]) and monastrol [42] can be perfused to stop motility.

An alternative method to control energy supply is to use photoactivatable ATP. In this method, a flash of UV light is used to release ATP from a derivatized, nonfunctional precursor; an ATP-consuming enzyme is also present to return the ATP concentration to low levels following release. Using such a system, microtubule movement has been repeatedly started and stopped [9], though the start-up and slow-down times were slow, on the order of minutes. The advantage of this method is that the solution in the flow cell does not have to be exchanged.

Fortuitously, many proteins possess natural regulatory mechanisms and, once understood, these might offer additional means to regulate the motors in vitro. Examples include the regulation of myosins by phosphorylation and calcium/calmodulin [43] and the inhibition of kinesin by its cargo-binding "tail" domain [44]. Because such natural controls might not always be applicable in a synthetic environment, there is strong interest in the development of artificial control mechanisms for motor proteins. Towards this end, metal-ion binding sites have been genetically engineered into the F_1-ATPase motor. The binding of ions at the engineered site immobilizes the moving parts of the motor, thus inhibiting its rotation [45]. ATP-driven rotation can be restored by the addition of metal ion chelators. Clever genetic engineering of motors could provide temporal control mechanisms that may be switched by temperature, light, electrical fields, or buffer composition.

13.2.3
Spatial Control

In order to control the path along which filaments glide – a process that we call "guiding" – it is necessary to restrict the location of active motors to specific regions of a surface. This can be done by coating a glass or silicon surface with resist polymers such as PMMA, SU-8, or SAL601 and using UV, electron beam or soft lithography to remove resist from defined regions [12–19]. The motor-containing solution is then perfused across the surface. By choosing appropriate properties of this solution [e. g., the concentration of motors, salts, other blocking proteins such as casein and bovine serum albumin (BSA),

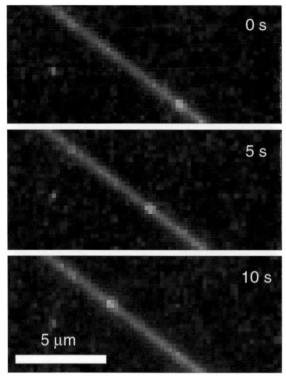

Figure 13.3 Movement of a single kinesin molecule (labeled with the green fluorescent protein) along a microtubule (red). Micrographs were acquired at the indicated times using total-internal-reflection fluorescence microscopy.

and detergents such as Triton X-100], motility can be restricted to either the unexposed, resist surface or to the exposed, underlying substrate. For example, it has been found that myosin motility is primarily restricted to the more hydrophobic resist surfaces while kinesin motility is primarily restricted to the more hydrophilic non-resist surfaces. However, the detailed interactions of the motors with these surfaces are not well understood. One limitation of this approach to binding proteins to surfaces is that the motors tend to bind everywhere, so it is difficult to attain good contrast. A proven method to prevent motor binding is to coat a surface with polyethylene oxide (PEO) [10, 46]. Techniques to bind motors and filaments via affinity tags to surfaces are summarized in section 13.2.4.

While chemical patterning can restrict movement of filaments to areas with a high density of active motors, walking off the trails is not prevented. This was demonstrated by Hess et al. [10], who showed that microtubules move straight across a boundary between high motor density (non-PEO) and low motor density (PEO), where they dissociate from the surface. The problem with a purely chemical pattern is that if a rigid filament is propelled by several motors along its length, there is nothing to stop the motors at the rear from pushing the filament across a boundary into an area of low motor density.

The behavior of microtubules colliding with the walls of channels imprinted in polyurethane has been studied by Clemmens et al. [11]. They found that the probability of a filament being guided by the walls decreased as the approach angle increased. At high inci-

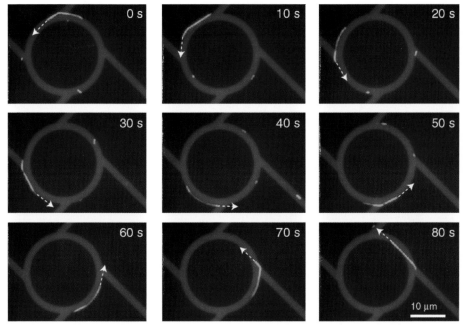

Figure 13.4 Sequence of fluorescent images showing the kinesin-driven, unidirectional movement of a rhodamine-labeled microtubule (red) along a chemically and topographically structured Si-chip. The bottom of the channels (green), the depths of which are 300 nm, is coated with kinesin. The surrounding regions are blocked by polyethylene glycol. (Research in collaboration with R. M. M. Smeets, M. G. L. van den Heuvel, and C. Dekker, Delft University of Technology, The Netherlands.)

dent angles, guiding was not observed and instead the microtubules climbed the walls. Combining chemical and topographic features – as occurs in the lithographic studies described above – leads to more efficient guiding. For example, in the study of Moorjani et al. [18], filaments remained at the bottom of the channels formed in the SU-8 even when they collided with the walls at angles above 80°. When the leading end of the microtubule hits the wall, the motors at the rear force the microtubule to bend into the region of high motor density, and in this way the motion is guided by the boundary (see Figure 13.4, unpublished results from our laboratory).

While it is possible to use chemical and topographical patterning to guide filaments – that is, to restrict their movement to particular paths – it is more difficult to control the direction of movement along the path. The difficulty arises because the orientation in which motors bind to a uniform surface is not controlled. Some motors will be oriented so that they propel filaments in one direction along the path, whereas others will propel filaments in the opposite direction. The reason that motors do not counteract each other is that filaments are polar structures: the orientation of the proteins that form up the filaments is maintained all along the length of the filament (see Figure 13.1). Because the motors bind stereospecifically to the filament, they will exert force in only one direction. Thus, the orientation of the filament determines its direction of motion; one end always leads.

The direction of filament gliding can be controlled by the application of external forces. Actin filaments and microtubules both possess negative net charges and, consequently, in the presence of a uniform electric field, will experience a force directed towards the positive electrode. It is possible to apply high enough electric fields to steer motor-driven filaments in a specified direction [19, 21]. Because the refractive index of protein differs from that of water, filaments become electrically polarized in the presence of an electric field, and consequently in a nonuniform field they move in the direction of highest field strength. This so-called dielectrophoretic force has been used to direct the gliding of actin filaments on a myosin-coated substrate [20]. It is even possible to manipulate a microtubule using optical gradients produced by focusing a laser beam (i.e., an optical tweezers) [47]. Directional control of microtubule gliding has also been achieved using hydrodynamic flow fields [23, 29].

An alternative approach to directionality relies on more sophisticated guiding concepts. For example, unidirectional movement of filaments can be achieved if guiding geometries based on arrow and ratchet structures are employed [10, 15]. An example of the unidirectional movement of a microtubule on a topographically and chemically structured silicon chip is depicted in Figure 13.4.

To control the direction of motion in stepping assays, the orientation of the filaments on the surface must be controlled. Towards this end, the generation of isopolar filament arrays has been achieved by binding specific filament ends to a surface, and using hydrodynamic flow to align the filaments along the surface to which they are subsequently adhered to [30, 48–50]. Alternatively, moving filaments can be aligned in a particular orientation by a flow field prior to fixation by glutaraldehyde [23, 29], which has been shown not to interfere with kinesin motility [51]. Fluid flow has also been used to align microtubules binding to patterned silane surfaces, though the orientation of the microtubules was not controlled [52].

13.2.4
Connecting to Cargoes and Surfaces

Cargoes can be attached to filaments using several different approaches. The prospective cargo can be coated with an antibody to the filament [53] or to a filament-binding protein such as gelsolin [54]. A clever refinement of this technique is genetically to fuse gelsolin with a cargo protein, thereby generating a dual-functional protein [55]. Alternatively, the cargo can be coated with streptavidin which binds to filaments that have been derivatized with biotin [56]. There are many other possibilities which have not yet been realized.

Analogous methods can be used to couple motors to surfaces. For example, the motor can be fused with the bacterial biotin-binding protein [57] and in this way bound to streptavidin-coated cargoes or surfaces. There any many peptide tags that can be fused to proteins to aid their purification [58, 59]. These tags can be used to couple these proteins to surfaces coated with the complementary ligand. A popular tag is the hexahistidine tag which binds Ni^{2+} and other metals that are chelated to nitriloamines (NTA). A nice approach is to couple the NTA to the terminal ethyleneoxides of triblock copolymers containing PEO. In principle, this provides specific binding of a his-tagged motor (or another protein) to a surface while the PEO groups block nonspecific binding [46, 60].

Controlled unloading of cargo has not been demonstrated, but ought to be feasible. For example, there are biotins that can be irreversibly cleaved by light and reversibly cleaved by reducing agents, and the histidine-Ni^{2+}–NTA connection can be broken by sequestering the Ni^{2+} with EDTA.

13.3
Outlook

Although the first steps have been made towards the operation of biomolecular motors in engineered environments, many advances are necessary before these motors can be used in nanotechnological applications such as working in molecular factories and building circuits.

An immediate task is to improve the spatial and temporal control over the motors. By combining improved surface techniques with the application of external electric, magnetic, and/or optical fields it should be possible, in the near future, to stretch and collide single molecules, to control cargo loading and unloading, and to sort and pool molecules.

Another goal is to control the position and orientation of motors with molecular precision. This means placing motors with an accuracy of ~ 10 nm on a surface and controlling their orientation within a few degrees. In this way both the location and the direction of motion of filaments can be controlled. One approach to molecular patterning is to "decorate" filaments with stereospecifically bound motors. Once aligned along the filament matrix, the motors can be transferred to another surface. This approach was taken by Spudich et al. [48, 61] and should be followed up. A further development of this idea is to directly produce (perhaps by stamping a mold made with a filament) surfaces that have structures functionally similar to motor-binding sites. An alternative approach is to use dip-pen lithography or other AFM techniques [62] to directly pattern motors on surfaces.

The robustness of motors must be increased. Motors operate only in aqueous solutions and under a restricted range of solute concentrations and temperatures. While it is inconceivable that protein-based motors could operate in a nonaqueous environment, two approaches to increasing their robustness can be envisaged. First, motors could be purified from thermophilic or halophilic bacteria, some of which grow at temperatures up to 112 °C and salinities above 5 M. There are also extreme eukaryotes that grow at up to 62 °C or 5 M NaCl. This approach has already been taken for ATP synthase [63], but not with linear motors because no obvious homologues of myosins or kinesins have been found in bacteria. Second, a genetic screening approach might reveal mutations that allow motors to operate in less restrictive or different conditions. A longer-term goal is to use the design principles learnt from the study of biomolecular motors to build purely artificial nanomotors that can operate in air or vacuum. This is a daunting prospect however, and it is not even clear what fuel(s) might be used. A potential way forward is to use chemical energy from a surface: for example, it was demonstrated that tin particles slide across copper surfaces driven by the formation of bronze alloy [64], this being analogous to paraffin-driven toy boats.

Besides the motor systems discussed so far, other biomechanical assemblies are good candidates for nanotechnological applications. In addition to providing paths along which motors move, active biological filaments on their own might find use in nanotechnological applications. The pushing and pulling forces generated by the polymerization and depolymerization of actin filaments and microtubules provide an alternative method of moving molecules [2, 65]. This ability is of particular interest because bacteria possess actin- [66] and microtubule-like [67] filaments and, as mentioned above, the proteins of extremophilic bacteria function in extreme environmental conditions. Filaments and motors can also self-organize under certain conditions [68–71]. On a side note, the flagellar filament in conjunction with the flagellar motors allow the bacteria to move in three-dimensional liquid space [72].

In addition to the motors that we have described so far, cells contain numerous biomolecular machines that can also be thought of as motors (for example, see Ref. [3]). These machines use chemical energy to replicate DNA (DNA polymerases) and process it (recombinases, topoisomerases and endonucleases), to produce RNA (RNA polymerases) and splice it (spliceosomes), to make proteins (ribosomes) and fold them (chaperones) and move them across membranes (translocases), and finally destroy them (proteasomes). The energy is provided by another group of machines that generate the electrochemical gradients (electron transport system, bacteriorhodopsin) used by the F_1F_0-ATP synthase to make ATP or by flagellar motors to propel bacteria. All these machines are candidates for nanotechnological applications, and a recent report of the use of chaperones to maintain nanoparticles in solution [73] is a step in this general direction.

We finish up by pointing out that the high order and nanometer-scale periodicity of DNA, actin filaments and microtubules make them ideal scaffolds on which to erect three-dimensional nanostructures. While these features have been exploited to make DNA-based structures [74] (see chapter 20), the use of DNA motors to address specific sites (based on nucleotide sequence) has not, to our knowledge, been realized. Some years ago it was proposed that the regular lattice of microtubules might serve as substrates for molecular computing and information storage [75, 76]. While these ideas seem crazy in the context of the living organism, they may be realizable for biomolecular motors operating in engineered environments. At the moment, anything is possible!

Acknowledgements

The authors thank U. Queitsch for help with experiments on guiding microtubules, T. Pompe, R.M.M. Smeets, M.G.L. van den Heuvel, and C. Dekker for fabricating microstructured channels, and F. Friedrich for assistance with the illustrations.

References

[1] B. Alberts, *Cell* **1998**, *92*, 291.

[2] J. Howard, *Mechanics of Motor Proteins and the Cytoskeleton*, Sinauer Associates, Sunderland, MA, **2001**.

[3] B. Alberts, *Molecular biology of the cell*, 4th edition, Garland Science, New York, **2002**.

[4] M. Crichton, *Prey*, HarperCollins, Toronto, **2002**.

[5] K. Kinosita, Jr., R. Yasuda, H. Noji, S. Ishiwata, M. Yoshida, *Cell* **1998**, *93*, 21.

[6] R. K. Soong, G. D. Bachand, H. P. Neves, A. G. Olkhovets, H. G. Craighead, C. D. Montemagno, *Science* **2000**, *290*, 1555.

[7] J. M. Scholey, *Motility assays for motor proteins*, Academic Press, San Diego, **1993**.

[8] J. R. Dennis, J. Howard, V. Vogel, *Nanotechnology* **1999**, *10*, 232.

[9] H. Hess, J. Clemmens, D. Qin, J. Howard, V. Vogel, *Nano Lett.* **2001**, *1*, 235.

[10] H. Hess, J. Clemmens, C. M. Matzke, G. D. Bachand, B. C. Bunker, V. Vogel, *Appl. Physics A (Materials Science Processing)* **2002**, *A75*, 309.

[11] J. Clemmens, H. Hess, J. Howard, V. Vogel, *Langmuir* **2003**, *19*, 1738.

[12] H. Suzuki, A. Yamada, K. Oiwa, H. Nakayama, S. Mashiko, *Biophys. J.* **1997**, *72*, 1997.

[13] D. V. Nicolau, H. Suzuki, S. Mashiko, T. Taguchi, S. Yoshikawa, *Biophys. J.* **1999**, *77*, 1126.

[14] J. Wright, D. Pham, C. Mahanivong, D. V. Nicolau, M. Kekic, C. G. dos Remedios, *Biomed. Microdevices* **2002**, *4*, 205.

[15] Y. Hiratsuka, T. Tada, K. Oiwa, T. Kanayama, T. Q. P. Uyeda, *Biophys. J.* **2001**, *81*, 1555.

[16] C. Mahanivong, J. P. Wright, M. Kekic, D. K. Pham, C. dos Remedios, D. V. Nicolau, *Biomed. Microdevices* **2002**, *4*, 111.

[17] R. Bunk, J. Klinth, L. Montelius, I. A. Nicholls, P. Omling, S. Tagerud, A. Mansson, *Biochem. Biophys. Res. Commun.* **2003**, *301*, 783.

[18] S. G. Moorjani, L. Jia, T. N. Jackson, W. O. Hancock, *Nano Lett.* **2003**, *3*, 633.

[19] D. Riveline, A. Ott, F. Julicher, D. A. Winkelmann, O. Cardoso, J. J. Lacapere, S. Magnusdottir, J. L. Viovy, L. Gorre-Talini, J. Prost, *Eur. Biophys. J. Biophys. Lett.* **1998**, *27*, 403.

[20] S. B. Asokan, L. Jawerth, R. L. Carroll, R. E. Cheney, S. Washburn, R. R. Superfine, *Nano Lett.* **2003**, *3*, 431.

[21] R. Stracke, K. J. Bohm, L. Wollweber, J. A. Tuszynski, E. Unger, *Biochem. Biophys. Res. Commun.* **2002**, *293*, 602.

[22] P. Stracke, K. J. Bohm, J. Burgold, H. J. Schacht, E. Unger, IOP Publishing. *Nanotechnology* **2000**, *11*, 52.

[23] I. Prots, R. Stracke, E. Unger, K. J. Bohm, *Cell Biol. Int.* **2003**, *27*, 251.

[24] K. J. Bohm, R. Stracke, M. Baum, M. Zieren, E. Unger, *FEBS Lett.* **2000**, *466*, 59.

[25] S. Diez, C. Reuther, C. Dinn, R. Seidel, M. Mertig, W. Pompe, J. Howard, *Nano Lett.* **2003**, *3*, 1251–1254.

[26] H. Hess, J. Howard, V. Vogel, *Nano Lett.* **2002**, *2*, 1113.

[27] H. Hess, J. Clemmens, J. Howard, V. Vogel, *Nano Lett.* **2002**, *2*, 113.

[28] A. D. Mehta, M. Rief, J. A. Spudich, D. A. Smith, R. M. Simmons, *Science* **1999**, *283*, 1689.

[29] K. J. Bohm, R. Stracke, P. Muhlig, E. Unger, IOP Publishing. *Nanotechnology* **2001**, *12*, 238.

[30] L. Limberis, R. J. Stewart, Proceedings of Spie: the International Society for Optical Engineering, *SPIE Int. Soc. Opt. Eng.* **1998**, *3515*, 66.

[31] R. D. Vale, T. Funatsu, D. W. Pierce, L. Romberg, Y. Harada, T. Yanagida, *Nature* **1996**, *380*, 451.

[32] A. Yildiz, J. N. Forkey, S. A. McKinney, T. Ha, Y. E. Goldman, P. R. Selvin, *Science* **2003**, *300*, 2061.

[32a] A. Roux, G. Cuppello, J. Cartaud, J. Prost, B. Goud, P. Bassereau, *Proc. Natl. Acad. Sci. USA*, **2002**, *99*, 5394–5399.

[33] H. Noji, R. Yasuda, M. Yoshida, K. Kinosita, Jr., *Nature* **1997**, *386*, 299.

[34] C. Montemagno, G. Bachand, *Nanotechnology* **1999**, *10*, 225.

[35] S. A. Cohn, A. L. Ingold, J. M. Scholey, *J. Biol. Chem.* **1989**, *264*, 4290.

[36] K. J. Bohm, P. Steinmetzer, A. Daniel, M. Baum, W. Vater, E. Unger, *Cell. Motil. Cytoskeleton* **1997**, *37*, 226.

[37] K. J. Bohm, R. Stracke, E. Unger, *Cell Biol. Int.* **2000**, *24*, 335.

[38] K. Kawaguchi, S. Ishiwata, *Cell. Motil. Cytoskeleton* **2001**, *49*, 41.

[39] K. Kawaguchi, S. Ishiwata, *Biochem. Biophys. Res. Commun.* **2000**, *272*, 895.

[40] B. J. Schnapp, B. Crise, M. P. Sheetz, T. S. Reese, S. Khan, *Proc. Natl. Acad. Sci. USA* **1990**, *87*, 10053.

[41] R. Sakowicz, M. S. Berdelis, K. Ray, C. L. Blackburn, C. Hopmann, D. J. Faulkner, L. S. Goldstein, *Science* **1998**, *280*, 292.

[42] T. U. Mayer, T. M. Kapoor, S. J. Haggarty, R. W. King, S. L. Schreiber, T. J. Mitchison, *Science* **1999**, *286*, 971.

[43] J. R. Sellers, H. V. Goodson, *Protein Profile* **1995**, *2*, 1323.

[44] D. L. Coy, W. O. Hancock, M. Wagenbach, J. Howard, *Nature Cell Biol.* **1999**, *1*, 288.

[45] H. Liu, J. J. Schmidt, G. D. Bachand, S. S. Rizk, L. L. Looger, H. W. Hellinga, C. D. Montemagno, *Nature Mater* **2002**, *1*, 173.

[46] M. J. deCastro, C. H. Ho, R. J. Stewart, *Biochemistry* **1999**, *38*, 5076.

[47] H. Felgner, R. Frank, J. Biernat, E. M. Mandelkow, E. Mandelkow, B. Ludin, A. Matus, M. Schliwa, *J. Cell Biol.* **1997**, *138*, 1067.

[48] J. A. Spudich, S. J. Kron, M. P. Sheetz, *Nature* **1985**, *315*, 584.

[49] L. Limberis, J. J. Magda, R. J. Stewart, *Nano Lett.* **2001**, *1*, 277.

[50] L. Limberis, R. J. Stewart, IOP Publishing. *Nanotechnology* **2000**, *11*, 47.

[51] T. B. Brown, W. O. Hancock, *Nano Lett.* **2002**, *2*, 1131.

[52] D. Turner, C. Y. Chang, K. Fang, P. Cuomo, D. Murphy, *Anal. Biochem.* **1996**, *242*, 20.

[53] D. C. Turner, C. Y. Chang, K. Fang, S. L. Brandow, D. B. Murphy, *Biophys. J.* **1995**, *69*, 2782.

[54] Y. Wada, T. Hamasaki, P. Satir, *Mol. Biol. Cell* **2000**, *11*, 161.

[55] C. Veigel, L. M. Coluccio, J. D. Jontes, J. C. Sparrow, R. A. Milligan, J. E. Molloy, *Nature* **1999**, *398*, 530.

[56] J. Yajima, M. C. Alonso, R. A. Cross, Y. Y. Toyoshima, *Curr. Biol.* **2002**, *12*, 301.

[57] F. Gittes, E. Meyhofer, S. Baek, J. Howard, *Biophys. J.* **1996**, *70*, 418.

[58] E. Berliner, H. K. Mahtani, S. Karki, L. F. Chu, J. E. Cronan, Jr., J. Gelles, *J. Biol. Chem.* **1994**, *269*, 8610.

[59] J. W. Jarvik, C. A. Telmer, *Annu. Rev. Genet.* **1998**, *32*, 601.

[60] K. Terpe, *Appl. Microbiol. Biotechnol.* **2003**, *60*, 523.

[61] J. P. Bearinger, S. Terrettaz, R. Michel, N. Tirelli, H. Vogel, M. Textor, J. A. Hubbell, *Nature Mater.* **2003**, *2*, 259.

[62] F. Jiang, K. Khairy, K. Poole, J. Howard, D. Muller, submitted **2003**.

[63] A. Hazard, C. Montemagno, *Arch. Biochem. Biophys.* **2002**, *407*, 117.

[64] A. K. Schmid, N. C. Bartelt, R. Q. Hwang, *Science* **2000**, *290*, 1561.

[65] M. Dogterom, B. Yurke, *Science* **1997**, *278*, 856.

[66] F. van den Ent, J. Moller-Jensen, L. A. Amos, K. Gerdes, J. Lowe, *EMBO J.* **2002**, *21*, 6935.

[67] J. Lowe, L. A. Amos, *Nature* **1998**, *391*, 203.

[68] F. J. Nedelec, T. Surrey, A. C. Maggs, S. Leibler, *Nature* **1997**, *389*, 305.

[69] T. Surrey, M. B. Elowitz, P. E. Wolf, F. Yang, F. Nedelec, K. Shokat, S. Leibler, *Proc. Natl. Acad. Sci. USA* **1998**, *95*, 4293.

[70] K. Kruse, F. Julicher, *Phys. Rev. Lett.* **2000**, *85*, 1778.

[71] D. Humphrey, C. Duggan, D. Saha, D. Smith, J. Kas, *Nature* **2002**, *416*, 413.

[72] W. S. Ryu, R. M. Berry, H. C. Berg, *Nature* **2000**, *403*, 444.

[73] D. Ishii, K. Kinbara, Y. Ishida, N. Ishii, M. Okochi, M. Yohda, T. Aida, *Nature* **2003**, *423*, 628.

[74] N. C. Seeman, *Nature* **2003**, *421*, 427.

[75] S. R. Hameroff, R. C. Watt, *J. Theoret. Biol.* **1982**, *98*, 549.

[76] R. Penrose, *Ann. N. Y. Acad. Sci.* **2001**, *929*, 105.

[77] P. Sheterline, J. Clayton, J. Sparrow, *Protein Profile* **1995**, *2*, 1.

[78] E. Nogales, *Annu. Rev. Biochem.* **2000**, *69*, 277.

[79] C. Bustamante, Z. Bryant, S. B. Smith, *Nature* **2003**, *421*, 423.

[80] P. J. Hagerman, *Annu. Rev. Biophys. Biomol. Struct.* **1997**, *26*, 139.

[81] C. Ruegg, C. Veigel, J. E. Molloy, S. Schmitz, J. C. Sparrow, R. H. Fink, *News Physiol. Sci.* **2002**, *17*, 213.

[82] A. Mehta, *J. Cell Sci.* **2001**, *114*, 1981.

[83] F. J. Kull, *Essays Biochem.* **2000**, *35*, 61.

[84] C. Shingyoji, H. Higuchi, M. Yoshimura, E. Katayama, T. Yanagida, *Nature* **1998**, *393*, 711.

[85] S. A. Burgess, M. L. Walker, H. Sakakibara, P. J. Knight, K. Oiwa, *Nature* **2003**, *421*, 715.

[86] G. J. Wuite, S. B. Smith, M. Young, D. Keller, C. Bustamante, *Nature* **2000**, *404*, 103.

[87] M. D. Wang, M. J. Schnitzer, H. Yin, R. Landick, J. Gelles, S. M. Block, *Science* **1998**, *282*, 902.

[88] N. R. Forde, D. Izhaky, G. R. Woodcock, G. J. Wuite, C. Bustamante, *Proc. Natl. Acad. Sci. USA* **2002**, *99*, 11682.

[89] J. J. Champoux, *Annu. Rev. Biochem.* **2001**, *70*, 369.

[90] T. R. Strick, G. Charvin, N. H. Dekker, J. F. Allemand, D. Bensimon, V. Croquette, *Comptes Rendus Physique* **2002**, *3*, 595.

[91] D. E. Smith, S. J. Tans, S. B. Smith, S. Grimes, D. L. Anderson, C. Bustamante, *Nature* **2001**, *413*, 748.

[92] A. J. Merz, M. So, M. P. Sheetz, *Nature* **2000**, *407*, 98.

[93] B. Maier, L. Potter, M. So, H. S. Seifert, M. P. Sheetz, *Proc. Natl. Acad. Sci. USA* **2002**, *99*, 16012.

[94] D. J. DeRosier, *Cell* **1998**, *93*, 17.

14

Nanoparticle–Biomaterial Hybrid Systems for Bioelectronic Devices and Circuitry

Eugenii Katz and *Itamar Willner*

14.1
Introduction

The unique electronic [1, 2], optical [3–6], and catalytic [7–9] properties of metal and semi-conductor nanoparticles (1–200 nm) pave the way to new generations of devices [10–14] and materials [15–17] that exhibit novel properties and functions. A variety of synthetic methods for the preparation of metal or semiconductor nanoparticles and their stabilization by functional monolayers [18–23], thin films, or polymers [24–26] are available. The functionalized metal or semiconductor nanoparticles provide exciting building blocks for the emerging and rapidly progressing field of nanotechnology.

The chemical functionalities associated with nanoparticles enable the assembly of two- and three-dimensional nanoparticle architectures on surfaces [27–30]. Composite layered or aggregated structures of molecular or macromolecular crosslinked nanoparticles on surfaces have been prepared, and the nanostructures assembled on surfaces were applied for the specific sensing of substrates [31–34], the generation of tunable electrochemilumi-nescence [35] and enhanced laser systems [36], and the tailoring of tunable enhanced photoelectrochemical systems [37]. The assembly of nanoparticle architectures on surfaces has also led to the fabrication of nanoscale devices such as single electron transistors [38–40], nanoparticle-based molecular switches [41], computing devices [42], metal-insu-lator-nanoparticle-insulator-metal (MINIM) capacitors [43], receptor–nanoparticle ISFET devices [44], and others. Several reviews have addressed the synthesis, properties, and functions of nanoparticles [44–48] and the progress in the integration of composite nanoparticle systems on surfaces and their use as functional devices [48–50].

The conjugation of nanoparticles with biomaterials is a tempting research project as it may provide new dimensions into the area of nanobiotechnology [51]. Bioelectronics is a rapidly progressing research field in modern science [52, 53]. It involves the integration of biomaterials such as enzymes [54–57], antigen–antibodies [58–60], DNA [61–63], neurons [64] or cells [65] with electronic elements such as electrodes, field-effect-transistors, or piezoelectric crystals with the aim to transduce biological events occurring on these elements by electronic signals or, alternatively, activate the biomaterial function by electronic stimuli. Different applications of such bioelectronic systems have been suggested includ-

Nanobiotechnology. Edited by Christof Niemeyer, Chad Mirkin
Copyright © 2004 WILEY-VCH Verlag GmbH & Co. KgaA, Weinheim
ISBN 3-527-30658-7

ing the development of biosensors [66, 67], biofuel cell elements [68–71] or artificial organs [72]. The discovery of nanoparticles or nano-rods and the elucidation of their unique optical, photonic, and electronic properties suggests that the coupling of biomaterials and the nanoparticles (or nano-rods) into hybrid systems could yield new materials and add new perspectives to the field of bioelectronics. Evolution has optimized fascinating macromolecular structures exhibiting unique recognition, transport, catalytic, and replication properties. Thus, the generation of hybrid systems between man-made nanoparticles and biomaterials may lead to a new generation of materials where electronic and photonic signals read-out biological phenomena or electronic, and photonic stimuli activate biological functionalities.

Enzymes, antigens and antibodies, nucleic acids and receptors have dimensions in the range of 2 to 100 nm. These dimensions are comparable to those of nanoparticles, and thus the synthetic nanostructures and the biomaterial units exhibit structural compatibility. Several features of biomaterials seem to be attractive for their future applications as building blocks for nanoparticle architectures:

1. Biomaterials reveal specific and strong complementary recognition interactions, for example, antigen–antibody, nucleic acid–DNA or hormone–receptor interactions. The functionalization of similar nanoparticles or different nanoparticles with complementary biomaterials could thus lead to nanoparticle aggregation or self-assembly. The unique interparticle electronic coupling, such as interparticle plasmon coupling of Au-nanoparticles may then provide photonic transduction of biological recognition processes [73, 74].
2. Various biomaterials include several binding sites, for example, the two Fab-chains of antibodies, the four binding domains of avidin or concanavalin A. This allows the multidirectional growth of nanoparticle architectures in predesigned geometries [75, 76].
3. Proteins may be genetically engineered and modified with specific anchoring groups [77]. This facilitates the specific binding of proteins to nanoparticles and the formation of defined and aligned structures of the nanoparticle–protein conjugates. As a result, directional growth of nanoparticle structures can be dictated. Furthermore, other biomaterials such as double-stranded DNA may be synthesized in the form of complex rigidified structures, and these may act as templates for the formation of nanoparticle patterns. For example, the association of nanoparticles to double-stranded DNA by intercalation, electrostatic binding to phosphate units or by the covalent binding of the nanoparticle to chemical functionalities tethered to the DNA, may lead to the formation of predesigned nanoparticle structures [78, 79].
4. Enzymes provide catalytic tools for the manipulation of biomaterials [80]. For example, ligase stimulates the ligation of nucleic acids, endonucleases lead to the scission of nucleic acids, and telomerase effects the elongation of DNA with constant telomer repeat units. Such enzymes provide effective tools for controlling the shape and structure of biomaterial hybrid systems.

It is the aim of this chapter to summarize recent advances in the organization of functional nanoparticle–biomaterial hybrid systems. Specifically, we discuss the use of biomaterial–nanoparticle hybrid systems in bioelectronics. The assembly of electronic, electrochemical, and photoelectrochemical sensor devices based on nanoparticle–biomaterial systems are presented, and the possibility of using nanoparticle–biomaterial hybrids for electronic circuitry are discussed.

14.2
Biomaterial–Nanoparticle Systems for Bioelectronic and Biosensing Applications

Metal nanoparticles such as gold or silver nanoparticles exhibit plasmon absorbance bands in the visible spectral region that are controlled by the size of the respective particles. Numerous studies reported on the labeling of bioassays and the staining of biological tissues by metal particles as a means to image and visualize biological processes [81]. The spectral shifts originating from adjacent or aggregated metal nanoparticles (e. g., Au-nanoparticles [82]) find increasing interest in the development of optical biosensors based on biomaterial–nanoparticle hybrid systems. For example, by applying two kinds of nucleic acid-functionalized nanoparticles that are complementary to two segments of an analyzed DNA, the hybridization of the nanoparticles with the analyzed DNA leads to aggregation of the nanoparticle and to the detection of a red-shifted interparticle plasmon absorbance of the nanoparticle aggregate [73]. An alternative approach reported for the optical detection of biorecognition processes involved the use of metallic nanoparticles as local quenchers of the fluorescence of dyes [83]. For example, in a beacon DNA terminated at its ends with a Au-nanoparticle and a dye, respectively, intramolecular quenching of the dye fluorescence persists. Opening of the DNA molecular beacon by hybridization with an analyte DNA regenerates the dye fluorescence because of the spatial separation of the nanoparticle and dye units [84]. Similarly, semiconductor nanoparticles exhibit size-dependent tunable fluorescence. The high fluorescence quantum yields of semiconductor nanoparticles, their photostability and their tunable fluorescence bands evoked substantial research effort for using the semiconductor nanoparticles as fluorescence labels for biorecognition processes. The extensive use of metal and semiconductor nanoparticles in biosensing suggests that the unique catalytic or photoelectrochemical properties of the nanoparticles could be used to develop electronic biosensors. For example, the catalytic electroless deposition of metals on nanoparticle-hybrid labels could be used to generate conductive domains and surfaces, and the conductivity properties of the systems may then transduce the biosensing processes [84]. The following sections address recent advances in the application of nanoparticle–biomaterial conjugates as active components in bioelectronic and biosensing systems.

14.2.1
Bioelectronic Systems Based on Nanopaticle–Enzyme Hybrids

Electrical contacting of redox-enzymes with electrodes is a key process in the tailoring of enzyme-electrodes for bioelectronic applications such as biosensors [53, 66, 67, 85–88] or biofuel cell elements [68–71]. While redox-enzymes usually lack direct electrical communication with electrodes, the application of diffusional electron mediators [89], the tethering of redox-relay groups to the protein [55, 90–93], or the immobilization of the enzymes in redox-active polymers [94, 95] were applied to establish electrical communication between the redox-proteins and the electrodes. Nonetheless, relatively inefficient electrical contacting of the enzymes with the electrode is achieved due to the nonoptimal modification of the enzymes by the redox units [96], or the lack of appropriate alignment of the enzymes in respect to the electrode. Very efficient electrical communication between redox-proteins and electrodes was achieved by the reconstitution of apo-enzymes on relay-cofactor monolayers associated with electrodes [56, 97–101]. For example, apo-glucose oxidase was reconstituted on a relay-FAD monolayer [56, 97, 98], and apo-glucose dehydrogenase was reconstituted on a pyrroloquinoline quinone (PQQ)-modified polyaniline film associated with an electrode [101]. Effective electrical communication between the redox-centers of the biocatalysts and the different electrodes was observed and reflected by high turnover electron transfer rates from the redox-sites to the electrode. The effective electrical contacting of these redox enzymes was attributed to the alignment of the proteins on the electrodes and to optimal positioning of the intermediary electron-relay units between the enzyme redox centers and the electrode.

A few biocatalytic electrodes have been prepared for biosensor applications by co-deposition of redox-enzymes and Au-nanoparticles on electrode supports [102–104]. The biocatalytic electrodes were reported to operate without electron transfer mediators, but the random and nonoptimized positioning of the redox proteins on the conductive nanoparticles did not allow the efficient electron transfer between the enzyme active sites and the electrode support. Highly efficient electrical contacting of the redox-enzyme glucose oxidase through a single Au-nanoparticle was accomplished by the reconstitution of the apo-flavoenzyme, apo-glucose oxidase, (apo-GOx) with a 1.4 nm Au_{55}-nanoparticle functionalized with N^6-(2-aminoethyl)-flavin adenine dinucleotide (FAD cofactor amino-derivative) **1**. The conjugate produced was assembled on a thiolated monolayer using different dithiols **2–4** as linkers (Figure 14.1A) [105]. Alternatively, the FAD-functionalized Au-nanoparticle could be assembled on a thiolated monolayer associated with an electrode, with apo-GOx subsequently reconstituted on the functional nanoparticles (Figure 14.1B). The enzyme-electrodes prepared by these two routes reveal similar protein surface coverages of ca. 1 $\times$ 10^{-12} mol cm^{-2}. The nanoparticle-reconstituted glucose oxidase layer was found to be electrically contacted with the electrode without any additional mediators, and the enzyme assembly stimulates the bioelectrocatalyzed oxidation of glucose (Figure 14.1C). The resulting nanoparticle-reconstituted enzyme electrodes revealed unprecedented efficient electrical communication with the electrode (electron transfer turnover rate ca. 5000 s^{-1}). This electrical contacting makes the enzyme-electrode insensitive to oxygen or to common oxidizable interferants such as ascorbic acid. The electron transfer from the enzyme active center through the Au-nanoparticle is rate-limited by the structure of the dithiol molecular linker that bridges the particle to the electrode. The conjugated benzene dithiol **4** was found as the most efficient electron transporter unit among the linkers **2–4**. The future application of effective molecular wires such as oligophenylacetylene units could further improve the electrical contacting efficiency.

While the previous system employed the metal nanoparticle as a nanoelectrode that communicates electronically the enzyme redox-site with the macroscopic electrode, one may use enzyme–nanoparticle hybrid systems where the product generated by the biocatalytic process activates the functions of the nanoparticle. This has recently been demonstrated by tailoring an acetylcholine esterase (AChE)-CdS nanoparticle hybrid monolayer

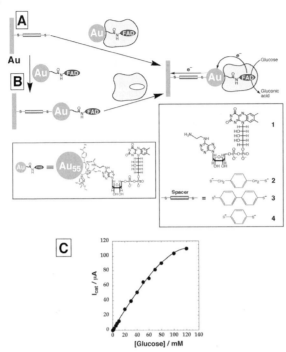

Figure 14.1 Electrical "wiring" of glucose oxidase (GOx) by the apo-enzyme reconstitution with a Au-nanoparticle-functionalized with a single FAD cofactor unit. (A) Reconstitution process performed in a solution followed by the assembly adsorption onto a dithiol-modified Au electrode. (B) Au-FAD conjugate adsorption onto a dithiol-modified Au electrode followed by the reconstitution of the apo-GOx at the interface. (C) Calibration plot of the electrocatalytic current developed by the reconstituted GOx electrode in the presence of different concentrations of glucose.

on a Au-electrode, and the activation of the photoelectrochemical functions of the nano-particles by the biocatalytic process [106]. The CdS–AChE hybrid interface was assembled on the Au-electrode by the stepwise coupling of cystamine-functionalized CdS to the electrode, and the secondary covalent linkage of the enzyme AChE to the particles (Figure 14.2A). In the presence of acetylthiocholine **5** as a substrate, the enzyme catalyzes the hydrolysis of **5** to thiocholine **6** and acetate. Photoexcitation of the CdS semiconductor yields the electron-hole pair in the conduction-band and the valence-band, respectively. The enzyme-generated thiocholine **6** acts as an electron donor for valence-band holes. The scavenging of the valence-band holes results in the accumulation of the electrons in the conduction-band and their transfer to the electrode with the generation of a photo-current (Figure 14.2B). The addition of enzyme inhibitors such as 1,5-bis(4-allyldimethyl-ammoniumphenyl)pentane-3-one dibromide **7** blocks the biocatalytic functions of the enzyme and, as a result, inhibits the photocurrent formation in the system (Figure 14.2C). Thus, the hybrid CdS–AChE system provides a functional interface for sensing of the AChE inhibitors (e.g., chemical warfare) by means of photocurrent measurements.

A similar system composed of photoactivated CdS nanoparticles and co-immobilized formaldehyde dehydrogenase that utilizes formaldehyde as an electron donor has been reported [107]. In this hybrid system the direct electron transfer from the enzyme active center to the CdS photogenerated holes was achieved and the steady-state photocurrent signal in the system was reported to be directly related to the substrate concentration.

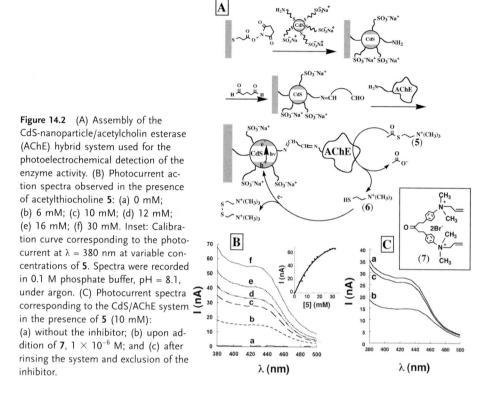

Figure 14.2 (A) Assembly of the CdS-nanoparticle/acetylcholin esterase (AChE) hybrid system used for the photoelectrochemical detection of the enzyme activity. (B) Photocurrent action spectra observed in the presence of acetylthiocholine **5**: (a) 0 mM; (b) 6 mM; (c) 10 mM; (d) 12 mM; (e) 16 mM; (f) 30 mM. Inset: Calibration curve corresponding to the photocurrent at $\lambda = 380$ nm at variable concentrations of **5**. Spectra were recorded in 0.1 M phosphate buffer, pH = 8.1, under argon. (C) Photocurrent spectra corresponding to the CdS/AChE system in the presence of **5** (10 mM): (a) without the inhibitor; (b) upon addition of **7**, 1×10^{-6} M; and (c) after rinsing the system and exclusion of the inhibitor.

14.2.2

Bioelectronic Systems for Sensing of Biorecognition Events Based on Nanoparticles

The unique optical [3–6], photophysical [48], electronic [1, 2], and catalytic [7–9] properties of metal and semiconductor nanoparticles turn them into ideal labels for biorecognition and biosensing processes. For example, the unique plasmon absorbance features of Au-nanoparticles, and specifically the interparticle-coupled plasmon absorbance of conjugated particles, have been widely used for DNA [108] and antibody–antigen [109–111] analyses. Similarly, the tunable fluorescence properties of semiconductor nanoparticles were used for the photonic detection of biorecognition processes [112].

Metal and semiconductor nanoparticles coupled to biomaterials generate solubilized entities. Nonetheless, even nanoscale particulate clustered systems include many atoms/molecules in the clusters. The solubility of the nanoparticle–biomaterial structures allows the application of washing procedures on surfaces that include a sensing interface, and thus nonspecific adsorption processes are eliminated. On the other hand, the specific capturing of biomaterial–nanoparticles on the respective sensing interfaces allows the secondary dissolution of the captured nanoparticles, and thus enables the amplified detection of the respective analyte by the release of many ions/molecules as a result of a single recognition event.

For example, an electrochemical method was employed for the Au-nanoparticle-based quantitative detection of the 406-base human cytomegalovirus DNA sequence (HCMV DNA) [113]. The HCMV DNA was immobilized on a microwell surface and hybridized with the complementary oligonucleotide-modified Au-nanoparticle. The resulting surface-immobilized Au-nanoparticle double-stranded assembly was treated with HBr/Br_2, resulting in the oxidative dissolution of the gold particles. The solubilized Au^{3+}-ions were then electrochemically reduced and accumulated on the electrode and subsequently determined by anodic stripping voltammetry using a sandwich-type screen-printed microband electrode (SPMBE). The combination of the sensitive detection of Au^{3+}-ions at the SPMBE due to nonlinear mass transport of the ions, and the release of a large number of Au^{3+}-ions upon the dissolution of the particle associated with a single recognition event provides an amplification path that enables the detection of the HCMV DNA at a concentration of 5×10^{-12} M.

Biomaterial-functionalized magnetic particles (e.g., Fe_3O_4) have been extensively applied in a broad variety of bioelectronic applications [114]. For example, the electronic detection of DNA utilized magnetic particles for the separation and concentration of the target DNA [115]. Avidin-modified magnetic particles were functionalized with a biotinylated DNA probe, and the hybridized target DNA was separated from the analyzed sample by means of an external magnet. Release of the hybrid DNA under basic conditions, followed by the chronopotentiometric stripping of the released guanine residue, enabled quantitative analysis of the target DNA. The sequence specific to the breast cancer gene *BRCA1* was analyzed using this method, with the reversible magnetically controlled oxidation of DNA being accomplished in the presence of nucleic acid-modified magnetic particles [116]. Avidin-modified magnetic particles were functionalized with the biotinylated probe nucleic acid, and subsequently hybridized with the complementary DNA. Two carbon-paste electrodes were patterned on a surface and applied as working electrodes. Spatial deposition of the functionalized magnetic particles on the right (R) or left (L) electrode enabled the magneto-controlled oxidation of the DNA by chronopotentiometric experiments (potential pulse from 0.6 V to 1.2 V) (Figure 14.3). Changing the position of the magnet (below planar printed electrodes) was thus used for "ON" and "OFF" switching of the DNA oxidation (through attraction and removal of DNA-functionalized magnetic particles). The process was reversed and repeated upon switching the position of the magnet, with and without oxidation signals in the presence and absence of the magnetic field, respectively. Such magnetic triggering of the DNA oxidation holds great promise for DNA arrays.

An interesting approach for the magneto-controlled amplified detection of DNA was introduced by Wang and colleagues using nucleic acid-modified metal [117–119] or semiconductor nanoparticles [120] as tags for the amplified electrochemical detection of DNA. Using this approach, magnetic particles functionalized with a biotinylated nucleic acid by an avidin bridge act as the capturing particles. Hybridization of the analyzed DNA with the capturing nucleic acid is followed by the secondary association of metal or semiconductor nanoparticles functionalized with a nucleic acid that is complementary to a free segment of the analyzed DNA. The binding of the nanoparticle labels to the biorecognition assay then provides amplifying clustered tags that by dissolution enable the release of many ion/molecule units. Also, the metal nanoparticles associated with the sen-

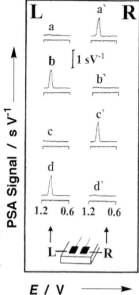

Figure 14.3 Chronopotentiometric signals for the DNA oligomer-functionalized magnetic particles (100 µg) using the dual-carbon paste electrode assembly: (a), (b), (c), and (d) are potentiograms at the "left" (L) electrode, while (a′), (b′), (c′), and (d′) are potentiograms obtained at the "right" (R) electrode. (a), (b′), (c), and (d′) are potentiograms obtained in the absence of the magnet, while (a′), (b), (c′), and (d) are potentiograms recorded in the presence of the magnet. (Reproduced from Ref. [115], Figure 2, with permission.)

sing interface may act as catalytic sites for the electroless deposition of other metals, thus leading to the amplified detection of DNA by the intermediary accumulation of metals that are stripped off, or by generating an enhanced amount of dissolved product that can be electrochemically analyzed [121, 122]. Figure 14.4A depicts the amplified detection of DNA by the application of nucleic acid-functionalized magnetic beads and Au-nanoparticles as catalytic seeds for the deposition of silver [117]. A biotin-labeled nucleic acid **8** was immobilized on the avidin-functionalized magnetic particles and hybridized with the complementary biotinylated nucleic acid **9**. The hybridized assembly was then reacted with the Au-nanoparticle–avidin conjugate. Treatment of the magnetic particles–DNA–Au-nanoparticle conjugate with silver ions (Ag$^+$) in the presence of hydroquinone results in the electroless catalytic deposition of silver on the Au-nanoparticles acting as catalyst. This process provides an amplification route since the catalytic accumulation of silver on the Au-nanoparticle originates from a single DNA recognition event. The magnetic separation of the particles by an external magnet concentrated the hybridized assembly from the analyzed sample. The current originating from the potential stripping off of the accumulated silver provided then the electronic signal that transduced the analysis of the target DNA.

In a related system (Figure 14.4B) [118], the electrochemical detection of the DNA was accomplished by the use of Au-nanoparticles as electroactive and catalytic tags. The primer biotinylated nucleic acid **10** was linked to magnetic beads through an avidin bridge. The hybridization of the nucleic acid **11** functionalized with Au-nanoparticle was then detected by the dissolution of the Au-nanoparticles with a HBr/Br$_2$ solution, followed by the electrochemical reduction of the generated Au^{3+} ions at the electrode, and the subsequent electrochemical stripping off of the surface-generated gold (Figure 14.4B; route a). This

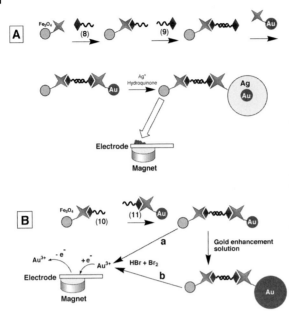

Figure 14.4 Electrochemical analysis of DNA upon the assembly of DNA molecules at magnetic particles followed by their association with Au-nanoparticles. (A) The Au-nanoparticles are used for silver deposition, and the DNA analysis is performed by electrochemical Ag stripping. (B) The Au-nanoparticles are chemically dissolved, the resulting Au^{3+} ions electrochemically reduced, and the deposited gold is electrochemically stripped, route (a). The intermediate enlargement of the Au-nanoparticles, route (b), results in the further amplification of the signal.

analytical procedure was further amplified by the intermediary deposition of gold on the Au-nanoparticles (Figure 14.4B; route b). The higher gold content after the catalytic deposition leads to a higher chronopotentiometric signal. Figure 14.5 shows the potentiograms corresponding to the stripping of the gold generated on the electrode upon the analysis of the nucleic acid associated with 5 nm Au-nanoparticles (Figure 14.5; route a), and after the deposition of gold on the Au-nanoparticles for 10 minutes (route b).

Similarly, CdS-semiconductor nanoparticles modified with a nucleic acid, were employed as tags for the detection of hybridization events of DNA [120]. Dissolution of the CdS (in the presence of 1 M HNO_3) followed by the electrochemical reduction of the Cd^{2+} to Cd^{0} that accumulates on the electrode, and the stripping off of the generated

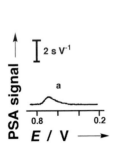

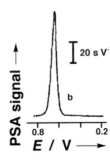

Figure 14.5 Effect of the gold enhancement upon the stripping response for the analyte DNA 11 (10 µg mL^{-1}): (a) Stripping signal of the Au-nanoparticle label. (b) Stripping signal after 10 min in the gold enhancement solution. Hybridization time, 25 min; amount of magnetic beads, 90 µg; amount of 5-nm avidin-coated gold particles, 7.6 $\times 10^{10}$; gold oxidation time, 5 min. (Reproduced from Ref. [118], Figure 9, with permission.)

Cd^0 (to Cd^{2+}) provided the electrical signal for the DNA analysis. Figure 14.6 shows the chronopotentiograms resulting in the analysis of different concentrations of the complementary target DNA using the CdS-nanoparticles as tags. The methods outlined in Figure 14.4 used the magnetic particles as carriers for the metal or semiconductor nanoparticles and as a vehicle to concentrate the analyzed DNA. An interesting aspect of these systems is, however, the future possibility of using a combination of different metal or semiconductor tags linked to different nucleic acids for the simultaneous analysis of different DNA targets (a library) linked to different magnetic beads. By using this approach [123], different nucleic acid probes complementary to different DNA targets are linked to magnetic particles. Similarly, different semiconductor or metallic nanoparticle tags complementary to segments of the different target DNAs are used as amplifying detection units for the primary hybridization process. The hybridization of the nucleic acid-functionalized semiconductor or metal particle to the specific DNA targets, followed by the dissolution of the nanoparticles and the electrochemical accumulation and stripping off of the metal, enables the determination of the specific DNA targets present in the sample. That is, the characteristic potentials needed to strip off the metal provide electrochemical indicators for the nature of the analyzed DNA. Indeed, a model system that follows this principle was developed [123] where three different kinds of magnetic particles modified by three different nucleic acids were hybridized with three different kinds of semiconductor nanoparticles, ZnS, CdS, PbS, that were functionalized with nucleic acids complementary to the nucleic acids associated with the magnetic particles. The magnetic particles allow easy transportation and purification of the analyte sample; whereas the semiconductor particles provide nonoverlapping electrochemical signals that transduce the specific kind of hybridized DNA. Stripping voltammetry of the semiconductive nanoparticles yields well-defined and resolved stripping peaks, for example, at −1.12 V (Zn), −0.68 V (Cd), and −0.53 V (Pb) (versus Ag/AgCl reference), thus allowing simultaneous electrochemical analysis of several DNA analytes tagged with the labeling semiconductive nanoparticles. For example, Figure 14.7 depicts stripping voltammograms for a solution containing three DNA samples labeled with the ZnS, CdS, and PbS nanoparticle tracers.

The catalytic features of metal nanoparticles that enable the electroless deposition of metals on the nanoparticle clusters allow the enlargement of the particles to conductive interparticle-connected entities. The formation of conductive domains as a result of biorecognition events then provides an alternative path for the electrical transduction of biorecognition events. This was exemplified by the design of a miniaturized immunosensor based on Au-nanoparticles and their catalytic properties [84] (Figure 14.8A). Latex particles stabilized by an anionic protective layer were attracted to a gap between micron-sized Au-electrodes by the application of a nonuniform alternating electric field between the electrodes (dielectrophoresis). Removal of the protective layer from the latex particles by an oppositely charged polyelectrolyte resulted in the aggregation of the latex particles and their fixation in the gap domain. Adsorption of protein A on the latex surface yielded a sensing interface for the specific association of the human immunoglobulin (IgG) antigen. The association of the human immunoglobulin on the surface was probed by the binding of the secondary Au-labeled anti-human IgG antibodies to the surface, followed by the catalytic deposition of a silver layer on the Au-particles. The silver layer bridged the gap between the two microelectrodes, resulting in a conductive "wire". Typical resis-

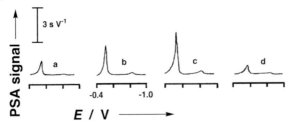

Figure 14.6 Stripping potentiograms measured upon sensing of different concentrations of DNA which is bound to magnetic particles and labeled with CdS-nanoparticles: (a) 0.2 mg L⁻¹; (b) 0.4 mg L⁻¹; (c) 0.6 mg L⁻¹; (d) Control experiment with noncomplementary DNA, 0.6 mg L⁻¹. Amount of magnetic particles, 20 μg; concentration of the DNA-functionalized CdS-nanoparticles, 0.01 mg mL⁻¹; hybridization time, 10 min; accumulation potential, −0.9 V; accumulation time, 2 min; stripping current, 1 μA. (Reproduced from Ref. [120], Figure 2, with permission.)

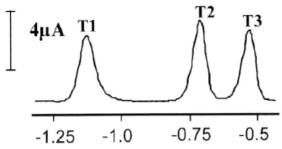

Figure 14.7 Stripping voltammogram recorded upon analysis of three DNA samples (54 nM) labeled: (T1) with ZnS nanoparticles; (T2) with CdS nanoparticles; and (T3) with PbS nanoparticles. The measurements were performed on a mercury-coated glassy-carbon electrode, with 1-min pre- treatment at 0.6 V; 2-min accumulation at −1.4 V; 15 s rest period (without stirring); square-wave voltammetric scan with a step potential of 50 mV; amplitude, 20 mV; frequency, 25 Hz. (Reproduced from Ref. [123], Figure 2(E), with permission.)

tances between the microelectrodes were 50–70 Ω, whereas control experiments that lack the specific catalytic enlargement of the domain by the Au-nanoparticle-antibody conjugate generated resistances > 10³ Ω. The method enabled the analysis of human IgG with a detection limit of ca. 2 × 10⁻¹³ M.

A similar DNA detection has also been performed using microelectrodes fabricated on a silicon chip [124] (Figure 14.8B). A probe nucleic acid **12** was immobilized on the SiO₂ interface in the gap separating the microelectrodes. The target 27-mer-nucleotide **13** was then hybridized with the probe interface, and subsequently a nucleic acid-functionalized Au-nanoparticle **14** was hybridized with the free 3′-end of the target DNA. The Au-nanoparticle catalyzed hydroquinone-mediated reduction of Ag⁺-ions, resulting in the deposition of silver on the particles, lowering the resistance between the electrodes. Single-base mutants of the analyte oligonucleotide **13** were washed off from the capture-nucleic acid **12** by the use of a buffer with the appropriate ionic strength. A difference of 10⁶ in the gap resistance was found between the analyte and the mutants. The low resistances be-

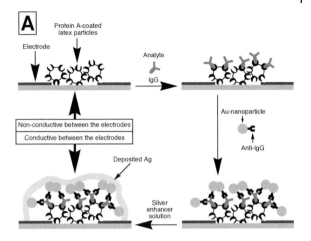

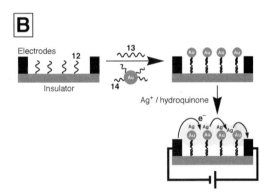

Figure 14.8 (A) Immunosensing at micro-sized Au-electrodes based on the change of conductivity between the Au strips upon binding of Au nanoparticles followed by silver deposition. (B) The use of a DNA-nanoparticle conjugate and subsequent silver deposition to connect two microelectrodes, as a means to sense a DNA analyte.

tween the microelectrodes were found to be controlled by the concentration of the target DNA, and the detection limit for the analysis was estimated to be ~5 × 10^{-13} M. This sensitivity translates to ~1 µg µL^{-1} of human genomic DNA or ~0.3 ng µL^{-1} of DNA from a small bacterium. These concentrations suggest that the DNA may be analyzed with no pre-PCR amplification. The simultaneous analysis of a collection of DNA targets was accomplished with a chip socket that included 42 electrode gaps, and appropriate different nucleic acid sensing probes between the electrode gaps [125].

Photoelectrochemical transduction of DNA recognition processes has been demonstrated by using semiconductor (CdS) nanoparticles modified with nucleic acids [126]. Semiconductor CdS nanoparticles (2.6 ± 0.4 nm) were functionalized with one of the two thiolated nucleic acids **15** or **16** that are complementary to the 5′ and 3′ ends of a target DNA **17**. An array of CdS nanoparticle layers was then constructed on a Au-electrode by a layer-by-layer hybridization process (Figure 14.9A). A primary thiolated DNA monolayer of **18** was assembled on a Au-electrode and the target DNA **17** acted as a cross-linking unit for the association of the **15**-CdS nanoparticles to the electrode by the hybridization of the ends of **17** to the **18**-modified surface and the **15**-functionalized CdS particles, respectively. The subsequent association of the second type of **16**-modified CdS particles hybridized to the first generation of the CdS particles resulted in the second generation of CdS particles. By the stepwise application of the two different kinds of nucleic acid-functionalized CdS nanoparticles hybridized with **17**, an array with a controlled number of nanoparticle generations could be assembled on the electrode. This array was characterized by spectroscopic means (absorption, fluorescence) upon the assembly of the array on glass supports, and by microgravimetric quartz crystal microbalance analyses on Au-quartz piezoelectric crystals. Illumination of the array resulted in the generation of a photocurrent. The photocurrents increased with the number of CdS nanoparticle generations associated with the electrode, and the photocurrent action spectra followed the absorbance features of the CdS nanoparticles, implying that the photocurrents originated from the photoexcitation of the CdS nanoparticles. That is, photoexcitation of the semiconductor induced the transfer of electrons to the conduction-band and the formation of an electron-hole pair. Transfer of the conduction band electrons to the bulk electrode, and the concomitant transfer of electrons from a sacrificial electron donor to the valence-band holes, yielded the steady-state photocurrent in the system. The ejection of the conduction-band electrons into the electrode occurred from nanoparticles in intimate contact with the electrode support. This was supported by the fact that $Ru(NH_3)_6^{3+}$ units ($E° = -0.16$ V versus SCE) that were electrostatically bound to the DNA enhanced the photocurrent from the DNA–CdS array. That is, the $Ru(NH_3)_6^{3+}$ units acted as electron wiring elements that facilitated electron hopping of conduction-band electrons of CdS particles that lacked contact with the electrode through the DNA tether. The system is important not only because it demonstrates the use of photoelectrochemistry as a transduction method

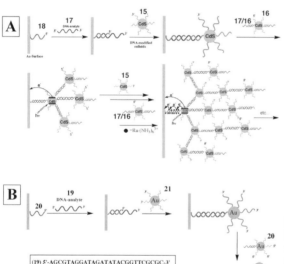

Figure 14.9 (A) The construction of CdS-nanoparticle/DNA superstructures, and their use for generation of photocurrents. (B) Dendritic amplified DNA-sensing using oligonucleotide-functionalized Au-nanoparticles assembled on a quartz crystal microbalance (QCM) electrode.

(19) 5'-AGCGTAGGATAGATATACGGTTCGCGC-3'
(20) 5'-TCTATCCTACGCT-(CH₂)₆-SH-3'
(21) 5'-HS-(CH₂)₆-GCGCGAACCGTATA-3'

for DNA sensing, but also because the system reveals the nanoengineering of organized DNA-tethered semiconductor nanoparticles on conductive supports. These latter nanoengineered structures are the first step towards electronic nanocircuitry (see discussion in section 14.3.2).

Nanoparticles as components of metal–nanoparticle–nucleic acid hybrids represent high molecular-weight units that make these conjugates ideal labels for microgravimetric quartz-crystal analyses of biorecognition processes on the surfaces of piezoelectric crystals. Furthermore, as nanoparticles act as catalysts for the deposition of metals, even higher mass changes may be stimulated and thus amplified microgravimetric detection of biorecognition processes may be accomplished. For a quartz piezoelectric crystal (AT-cut), the crystal resonance frequency changes by Δf when a mass change Δm occurs on the crystal according to Equation 1 (the Sauerbrey equation), where f_o is the fundamental frequency of the quartz crystal, Δm is the mass change, A is the piezoelectrically active area, ρ_q is the density of quartz (2.648 g cm^{-3}), and μ_q is the shear modulus (2.947 $\times$ 10^{11} dyne cm^{-2} for AT-cut quartz).

$$\Delta f = -2 \cdot f_o^2 \frac{\Delta m}{A \cdot (\mu_q \cdot \rho_q)^{1/2}} \tag{1}$$

Microgravimetric (QCM) DNA detection using nucleic acid-functionalized Au-nanoparticles as "nano-weights" was accomplished by the hybridization of a target DNA 19 to an Au-quartz crystal modified with a probe oligonucleotide 20, followed by the hybridization with Au-nanoparticles functionalized with DNA 21 that is complementary to the free 3'-segment of the target DNA 19 [127, 128] (Figure 14.9B). Further amplification of the response was reported by the use of a secondary Au-nanoparticle that is functionalized with the nucleic acid 20 that is complementary to the 5'-segment of the target DNA 19 and enables a layer-by-layer deposition of the Au nanoparticles. The hybridization of the 20-modified Au-nanoparticles with the analyzed DNA 19 followed by hybridization of the complex to the primary nanoparticle layer, yielded a "second generation" of Au-nanoparticles reminiscent of the growth of dendrimers [128, 129]. Concentrations as low as 1 $\times$ 10^{-10} M of DNA could be sensed by the amplification of the target DNA by the nucleic acid-functionalized Au-nanoparticle labels. It has been shown that the increase of the size of the Au nanoparticles labels from 10 nm up to ca. 40–50 nm results in an enhanced Δf signal, thus increasing the amplification factor in the DNA analysis [130]. Further increase of the Au nanoparticle size, however, resulted in smaller changes in the microgravimetric signal because of incomplete hybridization of the DNA analyte due to too large a size of the labeling particles.

A different approach for the amplified quartz-crystal-microbalance analysis of DNA utilizes the catalytic metal deposition on the nanoparticle labels [131]. Figure 14.10A depicts the amplified detection of the 7229-base M13ϕ DNA using the catalytic deposition of gold on a Au-nanoparticle conjugate [132]. The DNA primer 22 was assembled on a Au/quartz crystal. After hybridization with M13ϕ DNA 23, the double-stranded assembly was replicated in the presence of the mixture of nucleotides (dNTP-mix) dATP, dGTP, dUTP, biotinylated-dCTP (B-dCTP) and polymerase (Klenow fragment). The resulting biotin-labeled replica was then reacted with a streptavidin–Au-nanoparticle conjugate 24, and the resulting Au-labeled replica was subjected to the Au-nanoparticle-catalyzed deposition of gold by the NH_2OH-stimulated reduction of $AuCl_4^-$. The replication process represents the primary amplification since it increases the mass associated with the crystal, and simultaneously generates a high number of biotin labels for the association of the streptavidin–Au-nanoparticle conjugate. The binding of the nanoparticle conjugate represents the secondary amplification step for the analysis of the M13φ DNA. The third step, involving the catalyzed precipitation of the metal, led to the highest amplification in the sensing process. The M13φ DNA could be sensed by this method with a lower detection limit of ca. 1×10^{-15} M. This amplification route was also applied for the analysis of a single base mismatch in DNA as exemplified in Figure 14.10B). This is exemplified by the analysis of the DNA mutant **25a** that includes the single base substitution of the A-base in the normal gene **25** with a G-base. The analysis of the mutant was performed by the immobilization of the probe DNA **26** that is complementary to the normal gene **25** as well as to the mutant **25a** up to one base prior to the mutation site, on the Au-quartz crystal. Hybridization of the normal gene or the mutant with this probe interface, followed by the reaction of the hybridized surfaces with biotinylated-dCTP (B-dCTP) in the presence of polymerase (Klenow fragment) incorporated the biotin-labeled base only into the assembly that included the mutant **25a**. The subsequent association of the streptavidin–Au conjugate **24** followed by the catalyzed deposition of gold on the Au-nanoparticles, amplified the analysis of the single base mismatch in **25a**. Figure 14.11 (curve a) shows the micro-

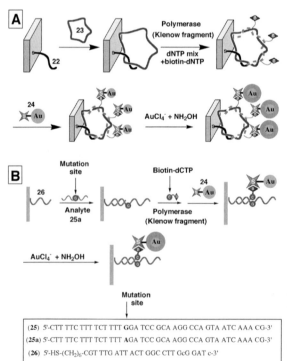

(25) 5'-CTT TTC TTT TCT TTT GGA TCC GCA AGG CCA GTA ATC AAA CG-3'
(25a) 5'-CTT TTC TTT TCT TTT AGA TCC GCA AGG CCA GTA ATC AAA CG-3'
(26) 5'-HS-(CH₂)₆-CGT TTG ATT ACT GGC CTT GcG GAT c-3'

Figure 14.10 (A) Amplified detection of the 7229-base M13Φ DNA using the catalytic deposition of gold on a Au-nanoparticle conjugate. (B) Analysis of a single base mismatch in DNA using the catalytic deposition of gold on a Au-nanoparticle conjugate.

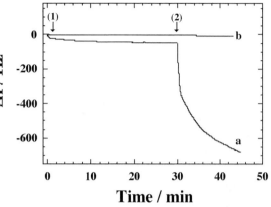

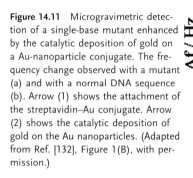

Figure 14.11 Microgravimetric detection of a single-base mutant enhanced by the catalytic deposition of gold on a Au-nanoparticle conjugate. The frequency change observed with a mutant (a) and with a normal DNA sequence (b). Arrow (1) shows the attachment of the streptavidin–Au conjugate. Arrow (2) shows the catalytic deposition of gold on the Au nanoparticles. (Adapted from Ref. [132], Figure 1(B), with permission.)

gravimetric detection of the mutant **25a** by this method, while the normal gene **25** does not alter the frequency of the crystal (Figure 14.11; curve b). Using this method, the mutant could be detected with a detection limit of 5×10^{-13} M.

The immobilization of nanoparticles on surfaces may also be used to yield high surface area electrodes [133]. Enhanced electrochemical detection of nucleic acids was reported by the roughening of flat gold electrodes with a Au-nanoparticle monolayer [134]. The roughening of a Au-quartz crystal with a monolayer consisting of Au-nanoparticles was also employed for the enhanced microgravimetric analysis of DNA [135].

14.3
Biomaterial-based Nanocircuitry

The miniaturization of objects by lithographic methods reaches its theoretical limits. It is generally accepted that different miniaturization methods need to be developed in order to overcome this barrier. While the lithographic methods use "top" to "down" miniaturization of patterns, the alternative approach of the "bottom-up" construction of objects has been suggested as a means to overcome the lithographic limitations. That is, the construction of objects on molecular or supramolecular templates could generate nanometer-sized features.

Biomaterials such as proteins or nucleic acids may act as attractive templates for the generation of nanostructures and nano-objects that may act as building blocks of nanocircuitry. Among the different biomaterials, proteins and nucleic acids may act as important building blocks for functional nanocircuitry, and eventually may provide nano-elements for the construction of nano-devices. In the present section the potential use of proteins for the generation of nanocircuitry will be discussed, while the subsequent section will address the use of DNA as a template for nanocircuitry.

14.3.1
Protein-based Nanocircuitry

Many proteins include well-defined structural channels (e. g., ferritin [136]) or assemblies in well-organized pore-containing layers (e. g., S-layers [137–141]). These channel- or pore-containing materials may act as templates for the generation of nanostructures, nanorods and even circuitry. For example, ferritin consists of a hollow polypeptide shell with 8 nm internal diameter and 12 nm external diameter, and a 5 nm diameter ferric oxide $(5Fe_2O_3 \cdot 9H_2O)$ core [136]. Reduction of the ferric oxide core causes it to be washed from the protein, and the apo-ferritin channel can be re-mineralized with different inorganic oxides, sulfides, or selenides that form nanorods (e. g., MnO, FeS, CdS, CdSe) [142–145]. Crystalline bacterial cell-surface-layers (S-layers) [137–141] reveal a broad application potential:

The pores passing through S-layers show identical size and morphology and are in the range of ultrafiltration membranes, and thus provide identical structural containers for generating the nanostructures. Furthermore the functional groups on the surface and in the pores are aligned in well-defined positions and orientations and are accessible for chemical modifications and binding functional molecules in very precise fashion. Also, isolated S-layers subunits from a variety of organisms are capable of recrystallizing as closed monolayers onto solid supports (e. g., metals, polymers, silicon wafers), thus allowing the assembly of the nanostructures on surfaces. In addition, functional domains can be incorporated in S-layers proteins by genetic engineering, thus allowing the control of the pore dimensions

Thus, S-layer technologies particularly provide new approaches for biotechnology, biomimetics, molecular nanotechnology, nanopatterning of surfaces and formation of ordered arrays of metal clusters or nanoparticles as required for nanoelectronics. S-layers can be patterned by deep ultraviolet radiation or by the application of a well-known soft lithography technique, micromolding in capillaries (MIMIC) [146] (Figure 14.12A). The patterned S-layers could be used as immobilization matrices for biologically functional molecules or templates in the formation of ordered arrays of nanoparticles, which are required for nanoelectronics and nonlinear optics. Monodisperse gold particles of 4–5 nm diameter were formed in the pore region of the S-layer, and the interparticle spacing of the gold superlattice resembled the S-layer lattice (Figure 14.12B). Patterning the S-layers using MIMIC technology was applied to produce nanocircuits of high complexity. Following S-layer recrystallization and mold removal, human IgG was covalently attached to active carboxylate groups on the S-layer track surface. Subsequent binding of fluoroscein isothiocyanate-labeled anti-human IgG enabled fluorescence imaging of the pattern produced by the modified S-layer (Figure 14.12C).

The specific binding interactions of proteins may provide specific crosslinking or bridging elements for generating nanoparticle circuitries. For example, streptavidin (Sav) is a homotetrameric protein that is characterized by four high-affinity ($k_a > 10^{14}$ M^{-1}) binding sites for biotin. By the appropriate functionalization of Au-nanoparticles [76, 147] (or of ferritin particles [148]) with biotin units, three-dimensional nanoparticle aggregates were generated (Figure 14.13A). Nanoparticle aggregates that exhibit three-dimensional ordering were also generated by the use of complementary antigen–antibody binding in-

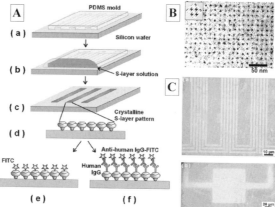

Figure 14.12 (A) Schematic representation of S-layer protein patterning and assembly by MIMIC. (a) Channels are formed when a poly(dimethylsiloxane) (PDMS) mold contacts a silicon wafer support. (b) Channels are filled with a protein solution by capillary forces. (c–d) Following mold removal, crystalline protein patterns are observed on the support surface. (e) S-layer patterns are labeled with a fluorescence marker or (f) used as substrates for an antibody–antigen immunoassay. (B) Electron micrograph of a nanometer-scale gold superlattice on a S-layer with square lattice symmetry. The inset shows a scanning-force image of the native S-layer at the same scale. (C) Fluorescence images of a S-layer patterned using a "circuit-like" PDMS mold. (Parts A and C reproduced from Ref. [139], Figures 1 and 5, and Part B reproduced from Ref. [141], Figure 5 with permission.)

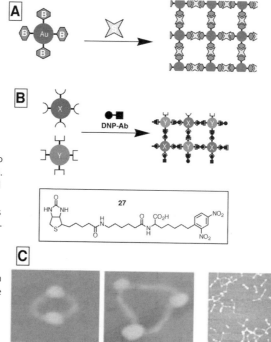

Figure 14.13 The use of specific protein interactions to build nanoparticle networks. (A) Using streptavidin (S) to link biotin-functionalized nanoparticles. (B) Formation of a network composed of two kinds of nanoparticles functionalized with different recognition units and crosslinked with a bifunctional linker. (C) AFM images of nanoparticle networks prepared by using double-stranded DNA (ds-DNA) as spacer groups. The ds-DNA fragments contain two biotin-binding sites attached to the two 5'-ends of the ds-DNA that allow crosslinking of the biotin-binding protein streptavidin as a model nano-object. (Reproduced from Ref. [112], Figure 10, with permission).

teractions [75, 149]. The simplest antigen–antibody–metal nanoparticle ordered aggregate involves the crosslinking of Au-nanoparticles functionalized with the anti-dinitrophenyl antibody (DNP-Ab) with a bifunctional bisdinitrophenyl antigen (bis-N-2, 4-dinitrophenyl-octamethylene diamine) that crosslinks the antibody-functionalized gold particles. Circuitries of further complexity may be generated by the application of different kinds of protein receptors and different kinds of metals for the ordered organization of the nanoparticle systems [75, 149]. For example, Figure 14.13B shows the formation of a nanoparticle aggregate consisting of two different nanoparticles that are functionalized with Sav and the DNP-Ab, respectively. The molecular hetero-bifunctional crosslinker **27** that includes the biotin and dinitrophenyl-antigen units then bridges the two kinds of nanoparticles. The oligomeric aggregates generated from bioorganic Sav particles and bis-biotinylated double-stranded DNA spacers were used as model systems to study the properties of complex particle networks. The Sav functions as a 5-nm model particle that can undergo a limited number of interconnections to other particles within the network. Either one, two, three, or four biotinylated DNA fragments were conjugated with Sav by means of the high-affinity Sav–biotin interaction resulting in dimers, trimers, and oligomers interconnected in a network observed by scanning force microscopy (SFM) [150] (Figure 14.13C).

14.3.2
DNA as Functional Template for Nanocircuitry

Among the different biomaterials, DNA is of specific interest as a template for the construction of nanocircuitries. Several arguments support the use of DNA as a future building block of nanostructures:

1. Nucleic acids of predesigned lengths, base-orderings and shapes can be synthesized, and complex structures were generated by self-assembly methods [151, 152].
2. Nature provides us with an arsenal of biocatalysts that can manipulate DNA. These enzymes may be considered as tools for shaping the desired DNA and eventually for the generation of nanocircuitry. For example, ligase ligates nucleic acids, endonucleases affect the specific scission of nucleic acids, telomerase elongates single-stranded nucleic acids by telomer units, and polymerase replicates DNA. These biocatalysts represent "cut" and "paste" tools for the formation of DNA templates, and by the application of the replication biocatalyst, the design of future "factories" of nanowires may be envisaged.
3. The intercalation of molecular components into DNA and the binding of cationic species, for example, metal ions to the phosphate units of nucleic acids allow the assembly of chemically active functional complexes on the DNA template.
4. Different proteins bind specifically to certain nucleic acid sequences. This allows the addressable assembly of complex DNA–protein structures. Such protein–DNA complexes may either act as addressable domains other than the bare DNA for the selective deposition of metals, or alternatively, may act as temporary shielding domains that protect the DNA from metal deposition. Such insulated domains may then be used for the deposition of other metals or semiconductors, thus enabling the fabrication of patterned complex structures.

With the vision that DNA may act as a template for the generation of nanocircuitry, attempts were made to explore the possibility to organize DNA-crosslinked semiconductor nanoparticles and DNA-based metal nanoparticle nanowires on surfaces. Since nanoparticles are loaded on the DNA template with gaps between them, the issue of electrical conductivity of DNA matrix itself is important. Electron transport through DNA has been one of the most intensively debated subjects in chemistry over the past ten years [153–155], and is still under extensive theoretical and experimental investigation. Despite some optimistic observations showing highly conductive properties of the entire DNA backbone [156, 157], most of the studies report on poorly conductive [158–160] or insulating [161–163] DNA properties.

In order to examine the electrical conductivity of DNA, the conductivity of double-stranded DNA that connects Au-nanoparticles was analyzed. Towards this goal, thiol-derivatized oligonucleotides were linked to Au-nanoparticles, and then the DNA-functionalized Au-nanoparticles were bridged with DNA chains composed of double-stranded helices of various lengths (24, 48, or 72 bases). These helices were terminated on both sides with single-stranded domains complementary to the oligonucleotides bound to the Au-nanoparticles [164]. The resulting Au-nanoparticle aggregates linked with double-stranded DNA spacers were deposited on an electrically nonconductive solid support, and their conductivity was measured by the four-probe method. Surprisingly, the conductivities of the aggregates generated by all three linkers ranged from 10^{-5} to 10^{-4} S cm^{-1} at room temperature, and they showed similar temperature-dependent behavior. The similarity of the electrical properties of the aggregates originates from the fact that the DNA spacers are compressed, thus providing small and similar distances between the Au-nanoparticles. Accordingly, the measured conductivity parameters reflect the electrical properties of the metallic nanoparticles separated with short gaps.

The conductivity of metallic nanoparticle aggregates on a DNA template can be enhanced upon the chemical deposition of another metal (e. g., Ag deposition on Au aggregates) filling the gaps and forming a continuous conductive nanowire. The general concept of DNA metallization is based on the association of metal complexes or metallic nanoparticles onto the DNA template and the further growth of the metal seeds with the same or different metal upon chemical reduction of the respective metal ions in the presence of strong reductants, such as dimethylaminoborane, hydroquinone, or sodium borohydride [155]. When metal ions or metal complexes (e. g., silver or copper ions [165], platinum or palladium complexes [166]) electrostatically associated with a DNA template are used as the catalytic centers for the metallization of the DNA, the process of the continuous nanowire formation proceeds from several hours up to one day [167, 168]. Application of pre-prepared metal nanoparticles associated with the DNA template makes this process significantly shorter [169].

The binding of the primary metallic clusters to the template DNA for the subsequent catalytic deposition of wires on the DNA frame may be accomplished by several means: These include the reduction of metal ions linked to the phosphate groups to metallic seeds linked to the DNA. Alternatively, metal or semiconductor nanoparticles functionalized with intercalator units may be used, where the intercalation of the molecular components into double-stranded DNA leads to the association of the nanoparticle to the DNA template. Also, the synthesis of DNA with functional tethers that enable the covalent attachment of the metal or semiconductor nanoparticles to the DNA provides a means to introduce the the the catalytic sites into the DNA. Furthermore, the synthesis of single-stranded DNA that includes constant repeat units (e. g., telomers) and the hybridization of metal and semiconductor nanoparticles functionalized with short nucleic acids that are complementary to the single-stranded DNA repeat units may provide a versatile route for the generation of nanowires.

Figure 14.14A exemplifies the method for assembling the Au-nanoparticles on the DNA template using Au-nanoparticles functionalized with an intercalator [170]. The amino-functionalized psoralen **28** is reacted with the Au$_{55}$-nanocluster (diameter 1.4 nm) that includes a single *N*-hydroxysuccinamide active ester functionality, to yield the psoralen-functionalized Au$_{55}$-nanoparticle. As psoralen acts as a specific intercalator for A–T base pairs, the functionalized Au$_{55}$-nanoparticles were reacted with the pA/pT-double-stranded DNA. Subsequently, the assembly was irradiated with UV light to induce the $2\pi + 2\pi$ cycloaddition reaction between the psoralen units and the thymine base sites of DNA. This latter process fixes covalently the Au$_{55}$-nanoparticles to the DNA matrix. Figure 14.14B depicts the AFM image of the resulting nanoparticle wire. A ca. 600–700 nm-long nanoparticle wire is formed, the width of which corresponds to ca. 3.5–8 nm and is controlled by the width of the DNA template. The height of the wire is ca. 3–4 nm; this is consistent with the fact that the Au-nanoparticles intercalate into the DNA on opposite sides of the double-stranded DNA template. The continuous appearance of the Au-nanoparticle wire is due to the dimensions of the scanning AFM-tip and, in reality, most of the particles are not in intimate contact one with another. The possibility of arranging the Au-nanoparti-

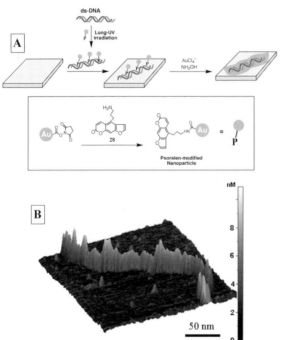

Figure 14.14 (A) Assembly of an Au-nanoparticle wire in the polyA/poly T template. (B) AFM image of an Au-nanoparticle wire in the polyA/polyT template.

cles on the DNA template allows further catalytic enlargement of the particles by an electroless deposition process (e. g., reaction of $AuCl_4^{2-}$ with NH_2OH) to yield continuous conductive nanowires.

Practical applications of the nanowires require their electrical contacting with macro- or micro-electrodes. Towards this goal, a single silver nanowire on a DNA template was generated, and the metallic wire bridged two micro-size electrodes [171, 172]. A similar approach was used to generate highly conductive nanowires of other metals that bridged macroscopic Au electrodes [173]. The binding of proteins (e. g., RecA) to DNA has been used as a means for the patterning of nanoscale DNA-based metal wires with nonconductive or semiconductive gaps [174]. Such metal/semiconductor/metal nanowire represents a nanometric transistor configuration.

14.4
Conclusions and Perspectives

This chapter has summarized recent advances in the preparation of biomaterial–nanoparticle hybrid systems and their application as active components in bioelectronic and electronic nanocircuitry systems. Several fundamental features of biomaterial–nanoparticle hybrid systems make these assemblies attractive for future bioelectronic applications:

- Biomaterial-functionalized nanoparticles are water-soluble, and exhibits high surface-area. These properties allow the effective labeling of the respective complementary analyte and the efficient capturing of the labeled biomaterial complexes on surfaces.
- The catalytic properties of nanoparticles associated with biomaterials enable enlargement of the particles by catalytic deposition of the same metal, or eventually another metal. This enables the generation of metal cluster aggregates, and ultimately leads to the formation of a continuous conductive film. This allows the generation of metallic clusters in the form of the biomaterial, and the conductivity of the resulting film may be used as an electrical readout signal for the primary biorecognition event.
- The unique photophysical and electronic properties of nanoparticles, and the possibilities of tuning and controlling these properties by the dimensions of the nanoparticle, suggest different electrical, electrochemical, photoelectrochemical, and piezoelectric transduction of biosensing processes that involve biomaterial–nanoparticle hybrids.

Several applications of biomaterial–nanoparticle hybrid systems were discussed. Some of these applications are viable technologies, while others are still in an embryonic phase that requires further fundamental research. The analytical applications of nanoparticle–biomaterial systems have advanced tremendously during the past decade, particularly in the labeling and electronic detection of DNA and antigen–antibody interactions. Arrays consisting of DNA spotted interdigitated electrodes for the electrical detection of DNA are already marketed. On the other hand, the use of nanoengineered metal or semiconductor nanoparticle-enzyme systems is still at the level of fundamental research. The progress in the characterization of these systems suggests, however, that new electroactive and photoelectrochemically active electrodes will soon be applied for tailoring new biosensor systems and biofuel cell devices.

The use of biomaterials as templates for the generation of nanostructures and nanocircuitries is still at the early phases of development. The viability of the concept of using biomaterials as molds for metal or semiconductor nanostructures has been proven, and nanowires of controlled shapes and functions have been generated. The major challenges in the area of biomaterial–nanoparticle-based nanocircuitry are, however, ahead of us and several key goals can be identified:

1. The development of synthetic routes for the preparation of semiconductor/metallic nanoparticle wires in high quantities is essential. Methods for the replication of nanoparticle nanowires or eventually the in-situ telomerase-induced preparation of nanoparticle-functionalized telomers using cancer cells may provide new strategies for the production of nanoparticle structures.
2. The identification of the electronic properties of the metal/semiconductor nanowires is a vital problem. The connection of the nanowires to the external macroscopic world is a key step for the characterization of the electronic properties of the systems. Strange electronic characteristics were observed for some metal nanowires, and several studies attributed the resulting properties to connect-points barriers with the macroscopic electrodes.
3. The fabrication of operating nanoscale devices based on nanoparticle–biomaterial hybrid systems is vital to support the promise of these materials in future nanoelectronics. Towards this goal it will be important to fabricate in a reproducible fashion biomaterial–nanoparticle-based logic devices. For example, a nanowire consisting of an ordered metal-semiconductor-metal structure could act as a nanotransistor that may be further applied for nanoscale sensing events.

The unique properties of nanoparticles and biomaterials provide a unique platform for physicists, chemists, biologists, and material scientists to mold this new area of nanobioelectronics. The world-wide interdisciplinary activities in the field, coupled with recent advances in the area, suggest that exciting new science and systems will emerge from these efforts.

Acknowledgments

The authors' research program on biomaterial–nanoparticle hybrid systems is supported by the German-Israeli Program (DIP).

References

[1] R. F. Khairutdinov, *Colloid J.* **1997**, *59*, 535–548.

[2] "Single Charge Tunnelling and Coulomb Blockade Phenomena in Nanostructures": H. Grabert, M. H. Devoret (Eds.) NATO ASI Ser. B **1992**, Plenum Press, New York.

[3] P. Mulvaney, *Langmuir* **1996**, *12*, 788–800.

[4] M. M. Alvarez, J. T. Khoury, T. G. Schaaff, M. N. Shafigullin, I. Vezmar, R. L. Whetten, *J. Phys. Chem. B* **1997**, *101*, 3706–3712.

[5] A. P. Alivisatos, *J. Phys. Chem.* **1996**, *100*, 13226–13329.

[6] L. E. Brus, *Appl. Phys. A* **1991**, *53*, 465–474.

[7] L. N. Lewis, *Chem. Rev.* **1993**, *93*, 2693–2730.

[8] V. Kesavan, P. S. Sivanand, S. Chandrasekaran, Y. Koltypin, A. Gedankin, *Angew. Chem. Int. Ed.* **1999**, *38*, 3521–3523.

[9] R. Ahuja, P.-L. Caruso, D. Möbius, W. Paulus, H. Ringsdorf, G. Wildburg, *Angew. Chem. Int. Ed.* **1993**, *32*, 1033–1036.

[10] D. L. Klein, R. Roth, A. K. L. Kim, A. P. Alivisatos, P. L. McEuen, *Nature* **1997**, *389*, 699–701.

[11] T. Sato, H. Ahmed, D. Brown, B. F. G. Johnson, *J. Appl. Phys.* **1997**, *82*, 696–701.

[12] R. S. Ingram, M. J. Hostetler, R. W. Murray, T. G. Schaaff, J. T. Khoury, R. L. Whetten, T. P. Bigioni, D. K. Guthrie, P. N. First, *J. Am. Chem. Soc.* **1997**, *119*, 9279–9280.

[13] T. Sato, H. Ahmed, *Appl. Phys. Lett.* **1997**, *70*, 2759–2761.

[14] H. Weller, *Angew. Chem. Int. Ed.* **1998**, *37*, 1658–1659.

[15] J. H. Fendler, *Chem. Mater.* **1996**, *8*, 1616–1624.

[16] A. K. Boal, F. Ilhan, J. E. DeRouchey, T. Thurn-Albrecht, T. P. Russell, V. M. Rotello, *Nature* **2000**, *404*, 746–748.

[17] V. I. Chegel, O. A. Raitman, O. Lioubashevski, Y. Shirshov, E. Katz, I. Willner, *Adv. Mater.* **2002**, *14*, 1549–1553.

[18] A. C. Templeton, D. E. Cliffel, R. W. Murray, *J. Am. Chem. Soc.* **1999**, *121*, 7081–7089.

[19] M. J. Hostetler, S. J. Green, J. J. Stokes, R. W. Murray, *J. Am. Chem. Soc.* **1996**, *118*, 4212–4213.

[20] R. S. Ingram, R. W. Murray, *Langmuir* **1998**, *14*, 4115–4121.

[21] J. J. Pietron, R. W. Murray, *J. Phys. Chem. B* **1999**, *103*, 4440–4446.

[22] H. Imahori, S. Fukuzumi, *Adv. Mater.* **2001**, *13*, 1197–1199.

[23] H. Imahori, M. Arimura, T. Hanada, Y. Nishimura, I. Yamazaki, Y. Sakata, S. Fukuzumi, *J. Am. Chem. Soc.* **2001**, *123*, 335–336.

[24] N. Herron, D. L. Thorn, *Adv. Mater.* **1998**, *10*, 1173–1184.

[25] J. F. Ciebien, R. T. Clay, B. H. Sohn, R. E. Cohen, *New J. Chem.* **1998**, *22*, 685–691.

[26] R. Gangopadhyay, A. De, *Chem. Mater.* **2000**, *12*, 608–622.

[27] K. V. Sarathy, P. J. Thomas, G. U. Kulkarni, C. N. R. Rao, *J. Phys. Chem. B* **1999**, *103*, 399–401.

[28] R. Blonder, L. Sheeney, I. Willner, *Chem. Commun.* **1998**, 1393–1394.

[29] T. Zhu, X. Zhang, J. Wang, X. Fu, Z. Liu, *Thin Solid Films* **1998**, *327–329*, 595–598.

[30] K. Bandyopadhyay, V. Patil, K. Vijayamohanan, M. Sastry, *Langmuir* **1997**, *13*, 5244–5248.

[31] M. Lahav, A. N. Shipway, I. Willner, *J. Chem. Soc. Perkin Trans. 2* **1999**, 1925–1931.

[32] A. N. Shipway, M. Lahav, R. Blonder, I. Willner, *Chem. Mater.* **1999**, *11*, 13–15.

[33] J. Liu, J. Alvarez, W. Ong, A. E. Kaifer, *Nano Lett* **2001**, *1*, 57–60.

[34] A. K. Boal, V. M. Rotello, *J. Am. Chem. Soc.* **2002**, *124*, 5019–5024.

[35] W. Chen, D. Grouquist, J. Roak, *J. Nanosci. Nanotechnol.* **2002**, *2*, 47–53.

[36] E. Corcoran, *Sci. Am.* **1990**, *263*, 74–83.

[37] A. N. Shipway, I. Willner, *Chem. Comm.* **2001**, 2035–2045.

[38] T. W. Kim, D. C. Choo, J. H. Shim, S. O. Kang, *Appl. Phys. Lett.* **2002**, *80*, 2168–2170.

[39] D. L. Klein, P. L McEuen, J. E. Bowen, R. Roth, A. P. Alivisatos, *Appl. Phys. Lett.* **1996**, *68*, 2574–2576.

[40] R. P. Andres, T. Bein, M. Dorogi, S. Feng, J. I. Henderson, C. P. Kubiak, W. Mahoney, R. G. Osifchin, R. Reifenberger, *Science* **1996**, *272*, 1323–1325.

[41] J. Liu, M. Gomez-Kaifer, A. E. Kaifer, In: J.-P. Sauvage (ed.), *Structure and Bonding*,

Springer, Berlin, **2001**, Vol. 99, pp. 141–162.

[42] A. O. Orlov, I. Amlani, G. H. Berstein, C. S. Lent, G. L. Snider, *Science* **1997**, *277*, 928–930.

[43] D. L. Feldheim, K. C. Grabar, M. J. Natan, T. E. Mallouk, *J. Am. Chem. Soc.* **1996**, *118*, 7640–7641.

[44] A. B. Kharitonov, A. N. Shipway, I. Willner, *Anal. Chem.* **1999**, *71*, 5441–5443.

[45] T. Trindade, P. O'Brien, N. L. Pickett, *Chem. Mater.* **2001**, *13*, 3843–3858.

[46] J. T. Lue, *J. Phys. Chem. Solids* **2001**, *62*, 1599–1612.

[47] K. Grieve, P. Mulvaney, F. Grieser, *Cur. Opin. Colloid Interface Sci.* **2000**, *5*, 168–172.

[48] A. N. Shipway, E. Katz, I. Willner, *ChemPhysChem.* **2000**, *1*, 18–52.

[49] M. Brust, C. J. Kiely, Colloid Surf. A **2002**, *202*, 175–186.

[50] W. P. McConnell, J. P. Novak, L. C. Brousseau, R. R. Fuierer, R. C. Tenent, D. L. Feldheim, *J. Phys. Chem. B* **2000**, *104*, 8925–8930.

[51] E. Katz, A. N. Shipway, I. Willner, In: *Nanoparticles – From Theory to Applications*, G. Schmid (Ed.), Wiley-VCH, Weinheim, **2003**, in press.

[52] I. Willner, *Science* **2002**, *298*, 2407–2408.

[53] I. Willner, E. Katz, *Angew. Chem. Int. Ed.* **2000**, *39*, 1180–1218.

[54] A. Heller, *J. Phys. Chem.* **1992**, *96*, 3579–3587.

[55] I. Willner, A. Riklin, B. Shoham, D. Rivenzon, E. Katz, *Adv. Mater.* **1993**, *5*, 912–915

[56] I. Willner, V. Heleg-Shabtai, R. Blonder, E. Katz, G. Tao, A. F. Bückmann, A. Heller, *J. Am. Chem. Soc.* **1996**, *118*, 10321–10322.

[57] A. Bardea, E. Katz, A. F. Bückmann, I. Willner, *J. Am. Chem. Soc.* **1997**, *119*, 9114–9119.

[58] A. Warsinke, A. Benkert, F. W. Scheller, *Fresen. J. Anal. Chem.* **2000**, *366*, 622–634

[59] R.-I. Stefan, J. F. van Staden, H. Y. Aboul-Enein, *Fresen. J. Anal. Chem.* **2000**, *366*, 659–668.

[60] A. Bardea, E. Katz, I. Willner, *Electroanalysis* **2000**, *12*, 1097–1106.

[61] F. Patolsky, E. Katz, I. Willner, *Angew. Chem. Int. Ed.* **2002**, *41*, 3398–3402.

[62] F. Patolsky, A. Lichtenstein, I. Willner, *Nature Biotechnol.* **2001**, *19*, 253–257.

[63] J. Wang, *Chem. Eur. J.* **1999**, *5*, 1681–1685.

[64] P. Fromherz, *Physica E* **2003**, *16*, 24–34.

[65] T. H. Park, M. L. Shuler, *Biotechnol. Prog.* **2003**, *19*, 243–253.

[66] I. Willner, E. Katz, B. Willner, In: *Biosensors and Their Applications*, V. C. Yang, T. T. Ngo (Eds.), Kluwer Academic Publishers, New York, **2000**, Chapter 4, pp. 47–98.

[67] I. Willner, B. Willner, E. Katz, *Rev. Molec. Biotechnol.* **2002**, *82*, 325–355.

[68] E. Katz, I. Willner, A. B. Kotlyar, *J. Electroanal. Chem.* **1999**, *479*, 64–68.

[69] E. Katz, A. N. Shipway, I. Willner, in: *Handbook of Fuel Cells – Fundamentals, Technology, Applications*, W. Vielstich, H. Gasteiger, A. Lamm (eds), Wiley, **2003**, Vol. 1, Part 4, Chapter 21, pp. 355–381.

[70] E. Katz, I. Willner, *J. Am. Chem. Soc.* **2003**, *125*, 6803–6813.

[71] T. Chen, S. C. Barton, G. Binyamin, Z. Gao, Y. Zhang, H.-H. Kim, A. Heller, *J. Am. Chem. Soc.* **2001**, *123*, 8630–8631.

[72] A. Prokop, *Annals NY Acad. Sci.* **2001**, *944*, 472–490.

[73] C. A. Mirkin, R. L. Letsinger, R. C. Mucic, J. J. Storhoff, *Nature* **1996**, *382*, 607–609.

[74] R. C. Mucic, J. J. Storhoff, C. A. Mirkin, R. L. Letsinger, *J. Am. Chem. Soc.* **1998**, *120*, 12674–12675.

[75] S. Mann, W. Shenton, M. Li, S. Connoly, D. Fitzmaurice, *Adv. Mater.* **2000**, *12*, 147–150.

[76] S. Connoly, D. Fitzmaurice, *Adv. Mater.* **1999**, *11*, 1202–1205.

[77] V. Pardo-Yissar, E. Katz, I. Willner, A. B. Kotlyar, C. Sanders, H. Lill, *Faraday Discuss.* **2000**, *116*, 119–134.

[78] A. P. Alivisatos, K. P. Johnsson, X. Peng, T. E. Wilson, C. J. Loweth, M. P. Bruchez, Jr., P. G. Schultz, *Nature* **1996**, *382*, 609–611.

[79] C. J. Loweth, W. B. Caldwell, X. Peng, A. P. Alivisatos, P. G. Schultz, *Angew. Chem. Int. Ed.* **1999**, *38*, 1808–1812.

[80] L. Alfonta, I. Willner, *Chem. Comm.* **2001**, 1492–1493.

[81] R. E. Palmer, Q. Guo, *Phys. Chem. Chem. Phys.* **2002**, *4*, 4275–4284.

[82] J. C. Riboh, A. J. Haes, A. D. McFarland, C. R. Yonzon, R. P. Van Duyne, *J. Phys. Chem. B* **2003**, *107*, 1772–1780.

[83] B. Dubertret, M. Calame, A. J. Libchaber, *Nature Biotechnol.* **2001**, *19*, 365–370.

[84] O. D. Velev, E. W. Kaler, *Langmuir* **1999**, *15*, 3693–3698.

[85] F. A. Armstrong, G. S. Wilson, *Electrochim. Acta* **2000**, *45*, 2623–2645.

[86] L. Habermuller, M. Mosbach, W. Schuhmann, W. *Fresenius' J. Anal. Chem.* **2000**, *366*, 560–568.

[87] I. Willner, B. Willner, *Trends Biotechnol.* **2001**, *19*, 222–230.

[88] F. A. Armstrong, H. A. Heering, J. Hirst, *Chem. Soc. Rev.* **1997**, *26*, 169–179.

[89] P. N. Bartlett, P. Tebbutt, R. G. Whitaker, *Prog. React. Kinet.* **1991**, *16*, 55–155.

[90] Y. Degani, A. Heller, *J. Phys. Chem.* **1987**, *91*, 1285–1289.

[91] Y. Degani, A. Heller, *J. Am. Chem. Soc.* **1988**, *110*, 2615–2620.

[92] W. Schuhmann, T. J. Ohara, H.-L. Schmidt, A. Heller, *J. Am. Chem. Soc.* **1991**, *113*, 1394–1397.

[93] I. Willner, E. Katz, A. Riklin, R. Kasher, *J. Am. Chem. Soc.* **1992**, *114*, 10965–10966.

[94] S. A. Emr, A. M. Yacynych, *Electroanalysis* **1995**, *7*, 913–923.

[95] A. Heller, *Acc. Chem. Res.* **1990**, *23*, 128–134.

[96] A. Badia, R. Carlini, A. Fernandez, F. Battaglini, S. R. Mikkelsen, A. M. English, *J. Am. Chem. Soc.* **1993**, *115*, 7053–7060.

[97] E. Katz, A. Riklin, V. Heleg-Shabtai, I. Willner, A. F. Bückmann, *Anal. Chim. Acta* **1999**, *385*, 45–58.

[98] O. A. Raitman, E. Katz, A. F. Bückmann, I. Willner, *J. Am. Chem. Soc.* **2002**, *124*, 6487–6496.

[99] H. Zimmermann, A. Lindgren, W. Schuhmann, L. Gorton, *Chem.- Eur. J.* **2000**, *6*, 592–599.

[100] L.-H. Guo, G. McLendon, H. Razafitrimo, Y. Gao, *J. Mater. Chem.* **1996**, *6*, 369–374.

[101] O. A. Raitman, F. Patolsky, E. Katz, I. Willner, *Chem. Commun.* **2002**, 1936–1937.

[102] J. Zhao, R. W. Henkens, J. Stonehurner, J. P. O'Daly, A. L. Crumbliss, *J. Electroanal. Chem.* **1992**, *327*, 109–119.

[103] A. L. Crumbliss, S. C. Perine, J. Stonehurner, K. R. Tubergen, J. Zhao, R. W. Henkens, J. P. O'Daly, *Biotech. Bioeng.* **1992**, *40*, 483–490.

[104] J. Zhao, J. P. O'Daly, R. W. Henkens, J. Stonehurner, A. L. Crumbliss, *Biosens. Bioelectron.* **1996**, *11*, 493–502.

[105] Y. Xiao, F. Patolsky, E. Katz, J. F. Hainfeld, I. Willner, *Science* **2003**, *299*, 1877–1881.

[106] V. Pardo-Yissar, E. Katz, J. Wasserman, I. Willner, *J. Am. Chem. Soc.* **2003**, *125*, 623–624.

[107] M. L. Curri, A. Agostiano, G. Leo, A. Mallardi, P. Cosma, M. D. Monica, *Mater. Sci. Eng. C* **2002**, *22*, 449–452.

[108] L. He, M. D. Musick, S. R. Nicewarner, F. G. Salinas, S. J. Benkovic, M. J. Natan, C. D. Keating, *J. Am. Chem. Soc.* **2000**, *122*, 9071–9077.

[109] S. Kubitschko, J. Spinke, T. Brückner, S. Pohl, N. Oranth, *Anal. Biochem.* **1997**, *253*, 112–122.

[110] L. A. Lyon, M. D. Musick, M. J. Natan, *Anal. Chem.* **1998**, *70*, 5177–5183.

[111] P. Englebienne, A. V. Hoonacker, M. Verhas, *Analyst* **2001**, *126*, 1645–1651.

[112] C. M. Niemeyer, *Angew. Chem. Int. Ed.* **2001**, *40*, 4128–4158.

[113] L. Authier, C. Grossirod, P. Brossier, B. Limoges, *Anal. Chem.* **2001**, *73*, 4450–4456.

[114] I. Willner, E. Katz, *Angew. Chem. Int. Ed.* **2003**, *42*, 4576–4588.

[115] J. Wang, A.-N. Kawde, A. Erdem, M. Salazar, *Analyst* **2001**, *126*, 2020–2024.

[116] J. Wang, A.-N. Kawde, *Electrochem. Commun.* **2002**, *4*, 349–352.

[117] J. Wang, D. Xu, R. Polsky, *J. Am. Chem. Soc.* **2002**, *124*, 4208–4209.

[118] J. Wang, D. Xu, A.-N. Kawde, R. Polsky, *Anal. Chem.* **2001**, *73*, 5576–5581.

[119] J. Wang, R. Polsky, D. Xu, *Langmuir* **2001**, *17*, 5739–5741.

[120] J. Wang, G. Liu, R. Polsky, A. Merkoci, *Electrochem. Commun.* **2002**, *4*, 722–726.

[121] J. Wang, O. Rincón, R. Polsky, E. Dominguez, *Electrochem. Commun.* **2003**, *5*, 83–86.

[122] H. Cai, Y. Wang, P. He, Y. Fang, *Anal. Chim. Acta* **2002**, *469*, 165–172.

[123] J. Wang, G. Liu, A. Merkoçi, *J. Am. Chem. Soc.* **2003**, *125*, 3214–3215.

[124] S.-J. Park, T. A. Taton, C. A. Mirkin, *Science* **2002**, *295*, 1503–1506.

[125] M. Urban, R. Müller, W. Fritzsche, *Rev. Sci. Instr.* **2003**, *74*, 1077–1081.

[126] I. Willner, F. Patolsky, J. Wasserman, *Angew. Chem. Int. Ed.* **2001**, *40*, 1861–1864.

[127] X. C. Zhou, S. J. O'Shea, S. F. Y. Li, *Chem. Commun.* **2000**, 953–954.

[128] F. Patolsky, K. T. Ranjit, A. Lichtenstein, I. Willner, *Chem. Commun.* **2000**, 1025–1026.

[129] S. Han, J. Lin, M. Satjapipat, A. J. Beca, F. Zhou, *Chem. Commun.* **2001**, 609–610.

[130] T. Liu, J. Tang, H. Zhao, Y. Deng, L. Jiang, *Langmuir* **2002**, *18*, 5624–5626.

[131] I. Willner, F. Patolsky, Y. Weizmann, B. Willner, *Talanta* **2002**, *56*, 847–856.

[132] Y. Weizmann, F. Patolsky, I. Willner, *Analyst* **2001**, *126*, 1502–1504.

[133] A. Doron, E. Katz, I. Willner, *Langmuir* **1995**, *11*, 1313–1317.

[134] H. Cai, C. Xu, P. He, Y. Fang, *J. Electroanal. Chem.* **2001**, *510*, 78–85.

[135] H. Lin, H. Zhao, J. Li, J. Tang, M. Duan, L. Jiang, *Biochem. Biophys. Res. Comm.* **2000**, *274*, 817–820.

[136] P. M. Harrison, P. Arosio, *Biochim. Biophys. Acta* **1996**, *1275*, 161–203.

[137] M. Sára, U. B. Sleytr, *J. Bacteriol.* **2000**, *182*, 859–868.

[138] U. B. Sleytr, P. Messner, D. Pum, M. Sára, *Angew. Chem. Int. Ed.* **1999**, *38*, 1034–1054.

[139] E. S. Györvary, A. O'Riordan, A. J. Quinn, G. Redmond, D. Pum, U. B. Sleytr, *Nano Lett.* **2003**, *3*, 315–319.

[140] U. B. Sleytr, M. Sára, D. Pum, B. Schuster, *Prog. Surf. Sci.* **2001**, *68*, 231–278.

[141] D. Pum, U. B. Sleytr, *Nanotechnology* **1999**, *17*, 8–12.

[142] F. C. Meldrum, B. R. Heywood, S. Mann, *Science* **1992**, *257*, 522–523.

[143] T. Douglas, D. P. E. Dickson, S. Betteridge, J. Charnock, C. D. Garner, S. Mann, *Science* **1995**, *269*, 54–57.

[144] F. C. Meldrum, T. Douglas, S. Levi, P. Arosio, S. Mann, *J. Inorg. Biochem.* **1995**, *8*, 59–68.

[145] K. K. W. Wong, S. Mann, *Adv. Mater.* **1996**, *8*, 928–931.

[146] Y. Xia, G. M. Whitesides, *Angew. Chem., Int. Ed. Engl.* **1998**, *37*, 550–575.

[147] S. Connoly, S. Cobbe, D. Fitzmaurice, *J. Phys. Chem. B* **2001**, *105*, 2222–2226.

[148] M. Li, K. W. Wong, S. Mann, *Chem. Mater.* **1999**, *11*, 23–26.

[149] W. Shenton, S. A. Davis, S. Mann, *Adv. Mater.* **1999**, *11*, 449–452.

[150] C. M. Niemeyer, M. Adler, S. Gao, L. Chi, *Angew. Chem. Int. Ed.* **2000**, *39*, 3056–3059.

[151] N. C. Seeman, *Acc. Chem. Res.* **1997**, *30*, 357–363.

[152] N. C. Seeman, *Angew. Chem. Int. Ed.* **1998**, *37*, 3220–3238.

[153] S. O. Kelley, J. K. Barton, *Science* **1999**, *283*, 375–381.

[154] M. Ratner, *Nature* **1999**, *397*, 480–481.

[155] J. Richter, *Physica E* **2003**, *16*, 157–173.

[156] D. D. Eley, R. B. Leslie, *Nature* **1963**, *197*, 898–899.

[157] H. W. Fink, C. Schönenberger, *Nature* **1999**, *398*, 407–410.

[158] P. Tran, B. Alavi, G. Gruner, *Phys. Rev. Lett.* **2000**, *85*, 1564–1567.

[159] L. T. Cai, H. Tabata, T. Kawai, *Appl. Phys. Lett.* **2000**, *77*, 3105–3106.

[160] Y. Okahata, T. Kobayashi, K. Tanaka, M. Shimomura, *J. Am. Chem. Soc.* **1998**, *120*, 6165–6166.

[161] P. J. de Pablo, F. Moreno-Herrero, J. Colchero, J. Gomez-Herrero, P. Herrero, A. M. Baro, P. Ordejon, J. M. Soler, E. Artacho, *Phys. Rev. Lett.* **2000**, *85*, 4992–4995.

[162] M. Bockrath, N. Markovic, A. Shepard, M. Tinkham, L. Gurevich, L. P. Kouwenhoven, M. S. W. Wu, L. L. Sohn, *Nano Lett.* **2002**, *2*, 187–190.

[163] A. J. Storm, J. van Noort, S. de Vries, C. Dekker, *Appl. Phys. Lett.* **2001**, *79*, 3881–3883.

[164] S.-J. Park, A. A. Lazarides, C. A. Mirkin, P. W. Brazis, C. R. Kannewurf, R. L. Letsinger, *Angew. Chem. Int. Ed.* **2000**, *39*, 3845–3848.

[165] C. F. Monson, A. T. Woolley, *Nano Lett.* **2003**, *3*, 359–363.

[166] M. Mertig, L. C. Ciacchi, R. Seidel, W. Pompe, A. De Vita, *Nano Lett.* **2002**, *2*, 841–844.

[167] J. Richter, R. Seidel, R. Kirsch, M. Mertig, W. Pompe, J. Plaschke, H. K. Schackert, *Adv. Mater.* **2000**, *12*, 507–509.

[168] W. E. Ford, O. Harnack, A. Yasuda, J. M. Wessels, *Adv. Mater.* **2001**, *13*, 1793–1797.

[169] O. Harnack, W. E. Ford, A. Yasuda, J. M. Wessels, *Nano Lett.* **2002**, *2*, 919–923.

[170] F. Patolsky, Y. Weizmann, O. Lioubashevski, I. Willner, *Angew. Chem. Int. Ed.* **2002**, *41*, 2323–2327.

[171] E. Braun, Y. Eichen, U. Sivan, G. Ben-Yoseph, *Nature* **1998**, *391*, 775–778.

[172] Y. Eichen, E. Braun, U. Sivan, G. Ben-Yoseph, *Acta Polymerica* **1998**, *49*, 663–670.

[173] J. Richter, M. Mertig, W. Pompe, I. Münch, H. K. Schackert, *Appl. Phys. Lett.* **2001**, *78*, 536–538.

[174] K. Keren, M. Krueger, R. Gilad, G. Ben-Yoseph, U. Sivan, E. Braun, *Science* **2002**, *297*, 72–75.

15
DNA–Protein Nanostructures

Christof M. Niemeyer

15.1
Overview

15.1.1
Introduction

Nature has evolved incredibly functional assemblages of proteins, nucleic acids, and other (macro)molecules to perform complicated tasks that are still daunting for us to try to emulate. Biologically programmed molecular recognition provides the basis of all natural systems, and the spontaneous self-assembly of the ribosome from its more than 50 individual building blocks is one of the most fascinating examples of such processes. The ribosome is a cellular nanomachine, capable of synthesizing the polypeptide chains using an RNA molecule as the informational template. The ribosome spontaneously self-assembles from its more than 50 individual building blocks, driven by an assortment of low-specificity, noncovalent contacts between discrete amino acids of the protein components interacting with distinct nucleotide bases and the phosphate backbone of the ribosomal RNAs. The structure of the ribosomal subunits have recently been resolved at atomic resolution, and the atomic structures of this subunit and its complexes with two substrate analogs revealed that the ribosome is in fact a ribozyme [1]. Knowledge of the atomic structure of this complex biological nanomachine not only satisfies our demand to fundamentally understand the molecular basis of life, but it also further motivates research to emulate natural systems in order to produce artificial devices of entirely novel functionality and performance.

Biological self-assembly has stimulated biomimetic "bottom-up" approaches for the development of artificial nanometer-scaled elements which are required commercially to produce microelectronics and micromechanical devices of increasingly small dimensions in the range of ~5 to 100 nm. In this regard, researchers had suggested some time ago that synthetic nanometer-sized elements might be fabricated from biomolecular building blocks [2, 3], and today DNA is being used extensively as a construction material for the fabrication of nanoscaled systems (e. g., see Chapters 19 and 20). The simple A–T and G–C hydrogen-bonding interaction allows the convenient programming of DNA receptor

Nanobiotechnology. Edited by Christof Niemeyer, Chad Mirkin
Copyright © 2004 WILEY-VCH Verlag GmbH & Co. KgaA, Weinheim
ISBN 3-527-30658-7

moieties, highly specific for the complementary nucleic acid. Another very attractive feature of DNA is the great mechanical rigidity of short double helices and its comparatively high physico-chemical stability. Moreover, Nature provides a comprehensive toolbox of highly specific ligases, nucleases, and other DNA-modifying enzymes, all of which can be used to process and manipulate the DNA with atomic precision, and thus create molecular constructions on the nanometer-length scale [4, 5].

The generation of semisynthetic DNA–protein conjugates allows one to combine the unique properties of DNA with the almost unlimited variety of protein components, which have been tailored by billions of years of evolution to perform highly specific functions such as catalytic turnover, energy conversion, or translocation of other components [6]. In this chapter, the current state of the art of both the preparation and application of such hybrid DNA–protein conjugates in life sciences and nanobiotechnology will be summarized. In particular, DNA–protein conjugates are applied to the self-assembly of high-affinity reagents for immunoassays, nanoscale biosensor elements, the fabrication of laterally microstructured biochips, and the biomimetic "bottom-up" synthesis of nanostructured supramolecular devices.

15.1.2
Oligonucleotide–Enzyme Conjugates

Pioneering studies in the preparation of semisynthetic DNA–protein conjugates were carried out by Corey and Schultz [7], who reported the synthesis of an oligonucleotide–*Staphylococcus* nuclease (SN) conjugate that can be used as a synthetic nuclease. The single-stranded DNA (ss-DNA) moiety of the DNA–SN conjugate was designed to form specific triple helices at complementary target regions of a plasmid DNA, thus enabling its site-specific cleavage. Later studies established that DNA–SN conjugates reveal enhanced kinetic rate constants for the hybridization with double-stranded DNA (ds-DNA) targets [8, 9]. This accelerated hybridization is probably due to coulomb attraction between the basic SN moiety and the target DNA, leading to an increased effective concentration of the conjugate near its target site. Similar effects have been observed for oligonucleotide–peptide conjugates [8, 10].

Earlier studies reported on the preparation of oligonucleotide–enzyme conjugates using 5′-thiolated oligonucleotides and calf intestine alkaline phosphatase (AP), horseradish peroxidase, or beta-galactosidase. These conjugates were used as hybridization probes in the detection of nucleic acids [11]. For example, the DNA–AP conjugate allowed the detection of attomol-amounts of target DNA, while covalent adducts of AP and streptavidin were labeled with short 10mer biotinylated oligonucleotides, and the resulting DNA–protein conjugates were used as probes in nucleic acid hybridization detection [12]. An oligonucleotide conjugate of fungal lipase was also synthesized to generate a thermostable probe for hybridization assays and biosensors [13].

An interesting way to organize several proteins along a one-dimensional ds-DNA fragment is based on the specific binding of DNA (cytosine-5)-methyltransferases to distinct recognition sequences of ds-DNA [14]. Covalent adducts are formed if the synthetic DNA base analog 5-fluorocytosine (^{F}C) is present in the recognition site. The sequence-specific covalent attachment of two representative methyltransferases, M. *Hha*I and M. *Msp*I, at their target sites, G^FCGC and ^FCCGG, respectively, was demonstrated. Since the methyl-transferases can be modified with additional binding domains by recombinant techniques, this concept might be useful for generating a variety of DNA–protein conjugates, applicable as chromatin models or other functional biomolecular devices [14]. Interestingly, methyltransferases have been used to modulate the charge transport that can occur along the DNA double helix [15, 16]. Because the methyltransferase extrudes the target base cytosine completely from the double helix, this local distortion effectively reduces the photoinduced charge transfer through the DNA duplex. These studies give rise to the development of electrochemical sensors for the analysis of protein–DNA interactions, and to the elaboration of novel bioelectronic devices.

15.1.3
DNA Conjugates of Binding Proteins

Antibodies have been coupled with DNA oligomers for immunoassay applications. Covalent conjugates of ss-DNA and ds-DNA fragments and immunoglobulin (IgG) molecules were used as probes in immuno-PCR (IPCR) [17, 18], which is a method for the ultrasensitive detection of proteins and other antigens. IPCR was originally developed by Sano et al. [19], and is a combination of the conventional enzyme-linked immunosorbent assay (ELISA) with the amplification power of the PCR (Figure 15.1a). In general, IPCR allows one for an ~1000-fold enhancement of the detection limit of the analogous ELISA system [20–23]. Coupling of the DNA marker with the antibody was achieved using bispecific fusion proteins [19], chemical crosslinking [17, 18], streptavidin–biotin interaction [20], and

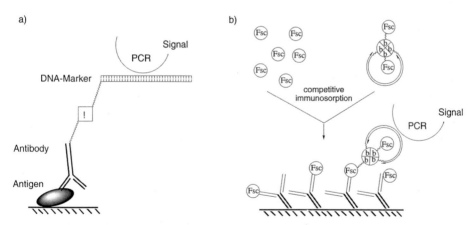

Figure 15.1 (a) Application of DNA–protein conjugates in immuno-PCR (IPCR). The surface-immobilized antigen is coupled with a specific primary antibody, which is conjugated to a DNA-marker fragment. Amplification of the DNA marker by means of PCR and subsequent analysis of the PCR amplicons enables highly sensitive detection of the antigen. The square between DNA and the antibody points out the necessity for an efficient chemical conjugation strategy. (b) Detection of low molecular-weight analytes by competitive immuno-PCR (cIPCR), illustrated with the model analyte fluorescein (Fsc). The free analyte competes the immunosorptive binding of the signal generating hapten–streptavidin conjugate to an antibody-modified surface. Subsequent to competitive binding and washing, the surface-immobilized hapten–DNA conjugate is detected using PCR.

self-assembled oligomeric DNA–streptavidin conjugates [21–24]. Other applications of antibody–DNA conjugates concern the use of antibody–polyadenylic acid conjugates in magnetic bead-based cell sorting [25], the preparation of microstructured biochips by the DNA-directed immobilization technique [26, 27], and chip-based protein detection by means of immuno-rolling circle amplification [28, 29].

Conjugates of DNA and receptor proteins have been developed for both therapeutic and bioanalytical purposes. The synthesis of an oligonucleotide–asialoglycoprotein conjugate was reported for applications in antisense technology [30]. The site-specific covalent coupling of a controlled number of signal peptides to plasmid DNA was attained using psoralen–oligonucleotide–peptide conjugates, which bind to the plasmid DNA by triple helix formation and are covalently coupled by photoactivation. The peptide–plasmid conjugates were then used for transfection of NIH-3T3 cells [31]. Other applications of oligonucleotide–(poly)peptide conjugates are involved in the cellular delivery of antisense reagents, stabilization of nucleic acid hybrids, and the recognition of biomolecular structures, and have recently been summarized by Tung and Stein [10].

Two groups have reported an unconventional approach to covalently link nucleic acids with proteins, such as single-chain antibody fragments [32, 33]. The principle of this scheme is based on the in-vitro translation of mRNA, covalently modified with a puromycin group at its 3′-end. The peptidyl-acceptor antibiotic puromycin covalently couples the mRNA with the polypeptide chain grown at the ribosome particle, leading to the specific conjugation of the informative (mRNA) with the functional (polypeptide) moiety. This approach has implications on the high-throughput screening of peptide and protein libraries, as well as on the generation of diverse protein microarrays [34].

The site-specific coupling of thiolated oligonucleotides to proteins can be achieved by using recombinant proteins, which are genetically engineered with additional cysteine residues for the coupling by disulfide bond formation. This approach has, for instance, been used to synthesize well-defined DNA–protein conjugates ("DNA-nanopores") which consist of an individual DNA oligonucleotide covalently attached within the lumen of the alpha-hemolysin pore [35]. As detailed in Chapter 7, the resulting bioconjugates are capable of identifying individual DNA strands with single-base resolution. Synthetic channels, comprised of rigid rod β-barrel peptides can also act as host structures for the noncovalent inclusion of ds-DNA fragments, thereby forming static transmembrane B–DNA complexes [36]. Site-specific disulfide formation was also used in the synthesis of DNA conjugates from recombinant streptavidin [37], thus complementing statistical covalent conjugation of streptavidin using heterobispecific crosslinkers (see Section 15.1.5) [26].

Recently, we have established a novel approach for synthesizing well-defined conjugates of nucleic acids and proteins [38]. The method is based on the native chemical ligation of recombinant proteins containing a C-terminal thioester with cysteine conjugates of nucleic acids. Expressed protein ligation had previously been used for the synthesis of a variety of proteins [39, 40]. To this end, the target protein fused to the construct of an intein and a chitin-binding domain (CBD) is expressed in *Escherichia coli*. This latter domain allows the affinity purification of the intein-fusion protein using a chitin matrix. Liberation from the column is achieved by cleaving the intein with mercaptoethansulfonic acid, thereby producing a C-terminal thioester of the target protein. The thioester containing protein can be ligated to peptides containing an N-terminal cysteine, or to cysteine-nucleic acid conjugates [38]. Due to the convenient synthesis of peptide nucleic acid (PNA)–Cys conjugates, as opposed to DNA–Cys conjugates, PNA was chosen as a model system. The mild and highly efficient chemical ligation led to nucleic acid–protein conjugates which are well-defined with respect to stoichiometric composition and regiospecific linkage. This method has several advantages over conventional chemical coupling techniques, and thus, constitutes a major improvement for further developments of DNA-directed immobilization as well as of artificial multi-protein arrangements and nanostructured hybrid assemblies (see Section 15.1.5). Since DNA–peptide conjugates are also available [10], it is now possible to produce – rapidly and automatically – PNA–protein as well as DNA–protein conjugates from libraries of recombinant proteins. Thus, the chemical ligation will be useful for a wide variety of applications, ranging from proteome research and clinical diagnostics to the arising field of nanobiotechnology.

15.1.4
Noncovalent DNA–Streptavidin Conjugates

In addition to covalent coupling chemistry, a convenient approach for the ready production of semisynthetic DNA–protein conjugates is based on the remarkable biomolecular recognition of biotin by the homotetrameric protein streptavidin. Due to the outstanding affinity constant of the streptavidin–biotin interaction of $\sim 10^{14}$ dm^3 mol^{-1}, the extreme chemical and thermal stability of the streptavidin, and the availability of numerous biotin-derivatives and mild biotinylation procedures, biotin–streptavidin conjugates form the basis of many diagnostic and analytical tests. Although biotinylated oligonucleotides are routinely prepared by automated DNA synthesis and are broadly applied in molecular biology and nucleic acid analyses, the formation and structure of conjugates of streptavidin and biotinylated nucleic acids are not yet fully exploited.

We have studied the self-assembly of streptavidin 1 and 5′,5′-bisbiotinylated ds-DNA fragments 2 (Figure 15.2) [24]. The bivalent ds-DNA molecules interconnect the tetravalent streptavidin, thereby generating three-dimensionally linked networks. Gel electrophoresis and scanning force microscopy (SFM) indicated that the oligomers 3 predominantly contain bivalent streptavidin molecules bridging adjacent DNA fragments. Despite the tetravalency of the streptavidin protein, trivalent streptavidin molecules occur as branch points with a low frequency, and the presence of tetravalent streptavidin in the supramolecular networks is scarce (see Figure 15.2b). As a consequence of the streptavidin's low valency, the oligomeric conjugates have a large residual biotin-binding capacity. This can be utilized for further functionalization of the complexes. For example, biotinylated antibodies have been coupled with 3, leading to functional conjugates applicable as powerful reagents in IPCR assays [21–24].

The oligomeric DNA–streptavidin conjugates 3 might also serve as a molecular framework for the generation of DNA-based nanomaterials. For instance, biotinylated macromolecules, such as enzymes, antibodies, peptides, fluorophores [41, 42], and even inorganic metal nanoclusters [43] or polymers [37] can be arranged at the nanometer length scale. Due to their size, connectivity and topography, the oligomers 3 can also be used as model systems for DNA-linked nanoparticle networks, for instance, serving to establish basic immobilization and characterization techniques for such assemblages [44]. As the

DNA fragments within the networks are susceptible to external stimuli, nanomechanical devices (e. g., ion-switchable nanoparticle aggregates) can be fabricated (Figure 15.2d) [44]. Potential applications include the fabrication of functional supramolecular nanomaterials which are useful, for example, in controlling the optical and electronic properties of nanoparticles, in regulating accessibility of the DNA to enzymes, or in the manufacture of addressable supports for sensors and bioelectronic devices.

We have established that the oligomers **3** can be effectively transformed into well-defined supramolecular DNA–streptavidin nanocircles **4** by thermal treatment (Figure 15.2) [45]. Due to their readily availability and well-defined stoichiometry and structure, the nanocircles **4** form the basis of a supramolecular construction kit for generating hapten–DNA conjugates. For example, functionalization of **4** with biotinylated haptens allows

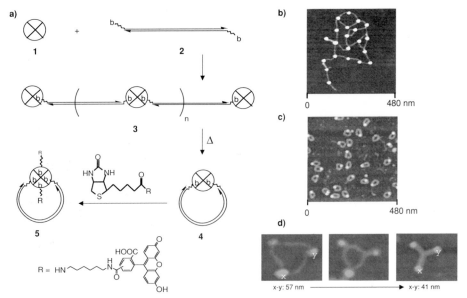

Figure 15.2 (a) Schematic representation of the self-assembly of oligomeric DNA–streptavidin (STV) conjugates **3** from 5′,5′-bis-biotinylated DNA **2** and STV **1**. Note that the schematic structure of **3** is simplified since a fraction of the STV molecules function as tri- and tetravalent linker molecules between adjacent DNA fragments (see SFM image (b)). The supramolecular networks of **3** can be disrupted by thermal treatment, leading to the efficient formation of DNA–STV nanocircles **4** (see SFM image (c)). The nanocircles **4** can be functionalized by the coupling of biotinylated hapten groups, such as fluorescein (Fsc) to yield nanocircle **5**. For simplification, complementary DNA strands are drawn as parallel lines. The 3′-ends are indicated by the arrow heads. (b) SFM images of the oligomeric DNA–STV conjugates **3**. (c) SFM images of the DNA–STV nanocircles **4**. (Adapted from Ref. [45], with kind permission.) (d) Ionic-switching of the oligomers **3**. The relative distance of the STV particles is altered by increased supercoiling of the interconnecting DNA linkers. The SFM images indicate structural changes observed for representative DNA$_3$–STV$_3$ elements, occurring within the random oligomeric networks **3**. Note that the structure at the left represents the extended species, occurring under low-salt conditions, while the structure at the right contains fully supercoiled DNA fragments, present under high-salt conditions. The structure in the middle represents an intermediate formed by partial supercoiling of the DNA linkers. (Adapted from Ref. [44], with kind permission.)

the generation of hapten conjugates **5**, which can be used as reagents in a novel competitive IPCR (cIPCR) assay for the ultra-sensitive detection of low-molecular weight analytes (see Figure 15.1b) [46]. Results obtained from model studies have suggested that cIPCR allows for between ~10- and 1000-fold improvements of the detection limit of conventional antibody-based assays, such as the competitive ELISA [46].

With respect to nanobiotechnology, **3** and **4** can also be used as soft materials topography standardization reagents for SFM analyses [47, 48]. As the two different biopolymers of DNA and proteins occur in a highly characteristic, well-defined composition and supramolecular structure, this allows direct comparisons to be made, for example, of the deformation properties of the two biopolymers, depending on the SFM measurement modes applied [48].

A second nanobiotechnological application of the DNA–streptavidin conjugates **3** and **4** involves the area of "biomolecular templating" (see also Chapters 16 and 17). In this approach, the electrostatic and topographic properties of biological macromolecules and supramolecular complexes comprised thereof are used for the synthesis and assembly of organic and inorganic components. In this context, the necessity for generating complex well-defined biomolecular architecture is clearly evident, which might be used for the templated growth of inorganic components. DNA–streptavidin conjugates **3** and **4** have re-

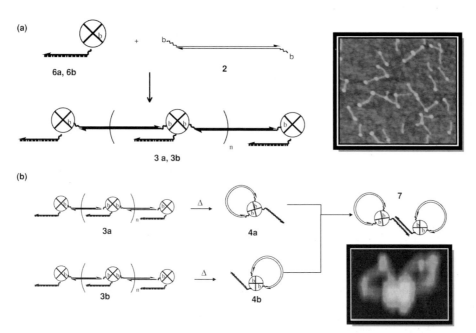

Figure 15.3 Construction of supramolecular DNA and streptavidin (STV) conjugates. (a) Networks **3a** and **3b** were generated from *bis*-biotinylated ds-DNA **2** and covalent oligonucleotide–STV conjugates **6a** and **6b**, respectively, containing complementary oligonucleotides covalently coupled to the STV. The inset shows a typical SFM image of **3a**. (b) Oligomers **3a** and **3b** are transformed into the corresponding nanocircles **4a** and **4b**, which hybridize with each other to form the nanocircle dimer **7**. The inset shows a typical SFM image of dimer **7**.

cently been used as building blocks for generating complex biomolecular nanostructures [49, 50]. As indicated in Figure 15.3, networks **3a** and **3b** were generated from *bis*-biotinylated ds-DNA **2** and the covalent oligonucleotide–streptavidin conjugates **6a** and **6b** [26] (for details on the covalent conjugates **6**, see below). Interestingly, the comparison with the analogous oligomers **3** (prepared from native streptavidin) revealed that the covalent streptavidin–oligonucleotide hybrid conjugates **3a** and **3b** assemble with the *bis*-biotinylated DNA to generate oligomeric aggregates of significant smaller size, containing no branch points and, on average, only about 2.5 times less ds-DNA fragments per aggregate (see SFM images in Figures 15.2 and 15.3) [49]. This phenomenon was attributed to electrostatic and steric repulsion between the ds-DNA and the covalently attached single-stranded oligomer moiety of **6**. Nevertheless, the single-stranded oligonucleotide moieties could be used for further functionalization of **3a** and **3b** by hybridization with complementary oligonucleotide-tagged macromolecules. For instance, oligomers **3a** were transformed into the corresponding nanocircles **4a**, which can hybridize with analogous circles **4b**, containing a complementary oligonucleotide sequence, to form the nanocircle dimers **7**. The SFM image of the dimeric conjugate **7** indicates the predicted structure [50] (Figure 15.3b).

15.1.5
Multifunctional Protein Assemblies

The concept of using DNA as a framework for the precise spatial arrangement of molecular components, as initially suggested by Seeman [3], was demonstrated experimentally by positioning several of the covalent DNA–streptavidin conjugates **6** along a single-stranded nucleic acid carrier molecule containing a set of complementary sequences (Figure 15.4) [26]. Covalent conjugates **6** were synthesized from thiolated oligonucleotides **8** and streptavidin **1** using the heterobispecific crosslinker sulfosuccinimidyl 4-[*p*-maleimidophenyl]butyrate (sSMPB, **9**). Conjugates **6** can be used as versatile molecular adaptors because the covalent attachment of an oligonucleotide moiety to the streptavidin provides a specific recognition domain for a complementary nucleic acid sequence in addition to the four native biotin-binding sites. For instance, supramolecular DNA nanostructures (e. g., **10** in Figure 15.4) have been assembled as model systems to investigate the basic principles of the DNA-directed assembly of proteins [26, 43]. These studies showed that, in particular, the formation of intramolecular secondary structures of the nucleic acid components often interferes with an effective intermolecular formation of the supramolecular DNA–protein assemblies [5].

The DNA-directed assembly of proteins can be applied to fabricate artificial spatially well-defined multienzyme constructs, which are not accessible by conventional chemical crosslinking. In biological systems, multienzyme complexes reveal mechanistic advantages during the multistep catalytic transformation of a substrate because reactions limited by the rate of diffusional transport are accelerated by the immediate proximity of the catalytic centers. Furthermore, the "substrate-channeling" of intermediate products avoids side reactions. Recently, the conjugates **6** were used to assemble surface-bound bienzymic complexes **13** from biotinylated luciferase and oxidoreductase [51]. The total enzymatic activities of the oxidoreductase/luciferase bienzymic complexes, which catalyze

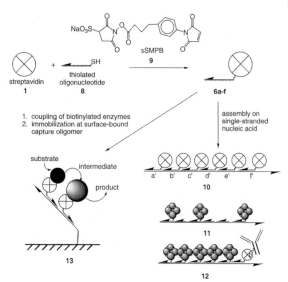

Figure 15.4 Synthesis of covalent DNA–streptavidin (STV) conjugates **6** by coupling of 5′-thiol-modified oligonucleotides **8** and STV **1** using the heterobispecific crosslinker sSMPB **9**. A number of conjugates **6** with individual oligonucleotide sequences (e. g., **6a–f**) self-assemble in the presence of a single-stranded carrier nucleic acid, containing complementary sequence stretches, to form supramolecular conjugates, such as **10** [26]. Following this strategy, biometallic aggregate **11** has been fabricated from **6** loaded with biotinylated 1.4 nm gold clusters [43]. The functional antibody-containing biometallic construct **12** was obtained from gold-labeled **6** and a conjugate obtained from **6** and a biotinylated immunoglobulin. The gold cluster- and IgG-conjugates were previously coupled in separate reactions [43]. In a related approach, two STV conjugates **6** were coupled with biotinylated enzymes to allow for the spatially controlled DNA-directed immobilization of the functional bienzymic complex **13** [51].

the consecutive reactions of flavin mononucleotide reduction and aldehyde oxidation, depended on the absolute and relative spatial orientation of the two enzymes. Such studies are useful to explore proximity effects in biochemical pathways, as well as to investigate the artificial multienzymes that will allow the development of novel catalysts for enzyme process technology, the regeneration of cofactors, and/or the performance of multistep chemical transformations to convert cheap precursors into drugs and fine chemicals.

With respect to synthetic nanosystems and materials science, developments of the DNA-directed organization of semiconductor and metal nanoclusters [52] (see also Chapter 19) have stimulated the use of DNA–streptavidin conjugates **6** to organize biotinylated gold nanoclusters to generate novel biometallic nanostructures, such as **11** in Figure 15.4 [43]. Given that the conjugates **6** can be used like components of a molecular construction kit, functional proteins (e. g., immunoglobulins) can be conveniently incorporated into the biometallic nanostructures. The proof of feasibility was achieved by the assembly of IgG-containing construct **12** (Figure 15.4), which is capable of binding specifically to surface-immobilized complementary antigens [43].

These experiments clearly demonstrate the applicability of the DNA-directed assembly to construct inorganic/bioorganic hybrid nanomaterials. Similar approaches should even allow the fabrication of highly complex supramolecular structures. Seeman and coworkers have impressively demonstrated the power of DNA in the rational construction of complex molecular framework [4] (see also Chapter 20). For example, Seeman's group synthesized the "truncated octahedron", a DNA polyhedron containing 24 individual oligonucleotide arms at its vertices which can, in principle, be used as a framework for the selective spatial positioning of 24 different proteins, inorganic nanoclusters, and/or other functional molecular devices.

Future applications of the DNA-directed assembly will focus, for example, on the generation of oligospecific antibody constructs. The presence of multiple binding domains leads to an enhanced affinity for the target structure owing to polyvalent binding interactions. Thus, supramolecular constructs, containing a DNA structural backbone to control the spatial arrangement of the binding sites, should allow for specific recognition of the target's topography, even when the individual epitopes are either not in close proximity, or reveal only weak antibody–antigen interactions. An additional major advantage of nanoscaled DNA–receptor constructs is that the backbone can be modified and detected by enzymatic means; hence, the supramolecular constructs are traceable at extremely low levels and even in rather complex environments due to the enormous detection potential of PCR techniques [24].

15.1.6
DNA–Protein Conjugates in Microarray Technologies

In addition to being investigated as tools in proteome research, protein biochips have the added attraction of serving as miniaturized multianalyte immunosensors in clinical diagnostics [53–55]. The miniaturization of ligand-binding assays not only reduces costs by decreasing reagent consumption but also leads to enhanced sensitivity in comparison with macroscopic techniques. Recent applications of protein microarrays include high-throughput gene expression and antibody screening [56], analysis of antibody–antigen interactions [57], or identification of the protein targets of small molecules [58]. Whilst DNA microarrays can be easily fabricated by automated deposition techniques [59], the step-wise, robotic immobilization of multiple proteins at chemically activated surfaces is often obstructed by the instability of most biomolecules, which usually reveal a significant tendency for denaturation. DNA-directed immobilization (DDI) provides a chemically mild process for the highly parallel binding of multiple delicate proteins to a solid support, using DNA microarrays as immobilization matrices (Figure 15.5) [26, 27, 38]. Because the lateral surface structuring is carried out at the level of stable nucleic acid oligomers, the DNA microarrays can be stored almost indefinitely, functionalized with proteins of interest via DDI immediately prior to use and, subsequent to hybridization, they can be regenerated by alkaline denaturation of the double helical DNA linkers. As an additional advantage of DDI in immunoassay applications, the intermolecular binding of the target antigens by antibodies can be carried out in homogeneous solution, rather than in a less efficient heterogeneous solid-phase immunosorption. Subsequently, the immunocomplexes formed are site-specifically captured at the microarray by nucleic acid hybridization [23].

The reversibility and site-selectivity of DDI provides the system with a variety of applications, including the recovery and reconfiguration of biosensor surfaces, the fabrication

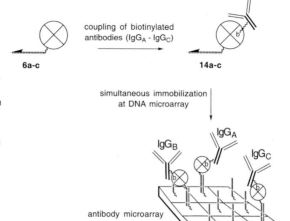

Figure 15.5 Schematic drawing of "DNA-directed immobilization" (DDI). A set of covalent DNA–streptavidin (STV) conjugates **6a–c** is coupled with biotinylated antibodies to generate IgG conjugates **14a–c**, respectively. A microarray of capture oligonucleotides is used as the immobilization matrix. Note that due to the specificity of Watson–Crick base pairing, many different compounds can be site-specifically immobilized simultaneously in a single step. (Adapted from Ref. [41], with kind permission.)

of mixed arrays containing both nucleic acids and proteins for genome and proteome research, and the generation of miniaturized biochip elements [41]. Recent adaptations of DDI have included the use of synthetic DNA analogs, pyranosyl-RNA oligomers, as recognition elements for the addressable immobilization of antibodies and peptides [60], and the DNA-directed immobilization of hapten groups for the immunosensing of pesticides [61]. Recently, the DDI method was applied in functional genomics to identify the members of a small molecule split-pool library which bind to protein targets [62] and to trace functional members of G-protein-coupled cellular signal transduction [38].

DDI has been applied to inorganic gold nanoparticles, thereby enabling the highly sensitive scanometric detection of nucleic acids in DNA-microarray analyses through the gold particle-promoted silver development [63, 64]. Further uses of the DDI of DNA-functionalized gold nanoparticles include the signal enhancement in the DNA hybridization detection with the quartz crystal microbalance [65] and by surface plasmon resonance (SPR) [66]. In the latter case, an ~1000-fold improvement in sensitivity was obtained, suggesting that the detection limit of SPR might soon reach that of traditional fluorescence-based DNA hybridization methods. The DDI technique has also been used to functionalize DNA-coated gold nanoparticles with DNA–antibody conjugates (Figure 15.6) [67]. These hybrid components were used as reagents in sandwich immunoassays, and the read-out by gold particle-promoted silver development allowed the spatially addressable detection of fmol quantities of chip-immobilized antigens [67]. Recently, Mirkin et al. have adapted the DDI approach to surface nanostructuring by "dip-pen" nanolithography (DPN), which employs an SFM tip to "write" thiolated compounds with less than 30 nm linewidth resolution on gold substrates [68] (see also Chapter 19). In addition, the direct writing of thiol- and acrylamide-modified oligonucleotides [69] enables the production of nanostructured matrices for use in DDI.

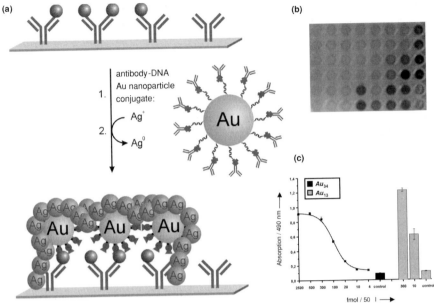

Figure 15.6 Array-based scanometric detection of proteins using DNA- and protein-modified nanoparticle probes. (a) Schematic drawing of the utilization of antibody/DNA-functionalized gold nanoparticles as reagents in a sandwich immunoassay. Surface-attached capture antibodies are used to bind the antigen (gray circles) through specific immunosorption. Subsequently, the antigen is labeled with anti-mouse IgG-modified gold nanoparticles. The IgG–DNA particle conjugate was previously assembled from Au-nanoparticles modified with ss-DNA and IgG–oligonucleotide conjugates by means of the DDI method. For signal generation, a silver development is carried out. (b) The generation of silver precipitate is monitored spectrophotometrically, or by imaging with a CCD camera or a flatbed scanner. (c) Typical dose–response curve obtained from the sandwich assay, depicted in (a). The absorbance at 490 nm of the silver film depends on the amount of antigen present in various samples analyzed. The black sets of signals were obtained using 34-nm gold particles (Au_{34}), while the gray histograms represent signals obtained with 13-nm gold particles (Au_{13}). (Adapted from Ref. [67].)

15.2
Methods

15.2.1
Conjugation of Nucleic Acids and Proteins

Various methods for the chemical coupling of DNA with protein molecules are described in the references cited in Section 15.1. Amongst these, covalent coupling most often relies on the use of heterobispecific crosslinkers, such as sulfoSMPB (Figure 15.4), which is first reacted with the protein to provide thiol-reactive maleimido groups. The maleimido-activated protein is usually purified and subsequently reacted with thiol-modified DNA oligonucleotides. If desired, longer DNA fragments can be obtained by PCR using a thiol-modified primer. This method of covalent conjugation requires extensive purification of the conjugates to remove excessive protein and oligonucleotides after each chemical coupling step. For examples, see Refs. [18, 26].

Since the conjugation of biotinylated nucleic acids with streptavidin, and conjugates thereof, is usually carried out in situ by mixing stoichiometric amounts of the nucleic acid and streptavidin, a purification is not required in the case of many applications (see Refs. [12, 24, 70]).

In both of the above approaches, the coupling of the protein and the nucleic acid is statistical, and thus basically no control of the stoichiometry and regioselectivity of the attachment site can be achieved. Directed conjugation – and in particular the regioselective and stoichiometrically controlled conjugation of nucleic acids to proteins – is much more difficult to achieve, and it is necessarily associated with the cloning and expression of recombinant proteins that contain reactive groups which can be selectively addressed by chemical means.

Protein engineering has been carried out to incorporate cysteine groups which have been subsequently used for the coupling with thiol-modified oligonucleotides by disulfide linkage (for examples, see Refs. [7, 35, 37]). A novel approach for the easily controlled site-selective linkage of nucleic acids with proteins is based on the native chemical ligation of recombinant proteins containing a C-terminal thioester which are selectively coupled to cysteine conjugates of nucleic acids under very mild conditions [38].

Additional approaches for the selective coupling of nucleic acids and proteins rely on chemically modified nucleic acid conjugates which contain, for example, the synthetic DNA base analog 5-fluorocytosine [14] or the peptidyl-acceptor antibiotic puromycin [32, 33].

15.2.2
Immuno-PCR

The quantification of proteins and other antigens by means of the IPCR method is very powerful, as IPCR not only leads to a ~1000- to 10000-fold gain in sensitivity (compared with conventional ELISA), but also reveals a very broad linear dynamic range of up to six orders of magnitude. However, none of the various IPCR applications described so far has crossed the border between a research method and a highly standardized, GLP (good laboratory practice) -compatible application, suitable for the routine analysis of large numbers of samples. The major reason for the current lack of standardization and robustness of IPCR is intimately associated with the enormous amplification power of PCR. Thus, IPCR is very sensitive to contamination and false-positive signals which result from the nonspecific binding of reagents. To overcome these problems, particular emphasis must be given to the minimization of nonspecific binding of DNA-containing marker conjugates. This can be achieved by using IPCR reagents comprising the oligomeric DNA–streptavidin conjugates (see Section 15.1.4 [21–24]). Other technical problems may arise during the read-out of the IPCR amplicons, which can be most effectively achieved by using either microplate-based ELISA [20] or solid-phase hybridization assay [21], or the real-time PCR platform based on the TaqMan principle [22]. Further improvements with respect to minimization of experimental errors and optimization of the limit of detection can be achieved by extensive standardization using internal competitor fragments during the PCR amplification step of the IPCR [21].

15.2.3
Supramolecular Assembly

One of the major technical problems in the supramolecular assembly of several DNA-tagged proteins relates to characterization of the assembly products. The application of suitable methods is essential not only to prove successful synthesis but also to understand the fundamental principles of reactivity and assembly properties of the biomolecular building blocks. A particularly suitable method which is often used is based on gel electrophoresis [24, 26, 42, 43]. This technique allows the covalent coupling to be monitored and the DNA:protein coupling ratio to be estimated [26], and can also be used to demonstrate directed supramolecular assembly (e. g., see Figure 14.4) [26, 43]. Although gel electrophoresis allows the circular structures of compounds (e. g., conjugates **4**) to be clearly determined [50], it fails in the case of multimeric DNA–protein conjugates (e. g., conjugates **3** in Figure 15.2). In such cases, high-resolution SFM has been proven to be the method of choice [24, 42]. However, the current resolution capabilities of SFM are by far insufficient to resolve supramolecular assemblies with a tight coupling of the individual components (see e. g., conjugates **10–13** in Figure 15.4). In these cases, characterization might be achieved by labeling and subsequent analysis with transmission electron microscopy [43] in addition to electrophoresis. If functional components are integrated within the supramolecular DNA–protein conjugates, such as antibodies or enzymes (conjugates **12** and **13** in Figure 15.4, respectively), then functional assays can often be designed which allow one to prove the intactness of the conjugate [43, 51].

15.2.4
DNA-directed Immobilization

The chemical and structural features of the nucleic acid constituents employed in both DDI and DNA-directed supramolecular assembly play an essential role, as they determine the hybridization efficiency of the individual components [26, 27, 43, 70]. Solid-phase hybridization studies have shown that the attachment of voluminous proteins to an oligonucleotide decreases the kinetics of intermolecular hybridization [70]. However, the individual oligonucleotide sequences induce even larger variations in the hybridization efficiency, depending on the formation of secondary structures, such as intramolecular hairpin loops [70]. Intramolecular folding also affects the supramolecular assembly of several proteins along a single-stranded nucleic acid carrier backbone. At moderate temperatures, carrier strands may form stable secondary structures, and thus, an equilibrium is formed between the uncomplexed carrier and also the protein–carrier conjugate [43]. A completion of the previously incomplete supramolecular aggregation of several compounds can be achieved by means of helper-oligonucleotides, capable of binding to uncomplexed sequence stretches of the carrier, thereby disrupting its secondary structure [43, 71]. It has been shown that DNA molecules are superior templates for the supramolecular assembly than RNA carriers, probably due to the lower stability of the intramolecular folding [71]. To overcome eventually the problem of nucleic acid secondary structure, future developments will focus on the use of synthetic DNA analogs in nucleic acid-directed supramolecular synthesis. Thus, despite their current high costs, peptide nucleic acids [72] or pyranosyl-RNA oligomers [60] for example may be employed due to their high specificity of binding, major (bio)chemical and physical robustness, and low tendency to form secondary structures.

An additional important parameter for successful application of the DDI method concerns the use of suitable DNA microarrays as a capture matrix. High densities of capture oligonucleotides are required for optimal performance of DDI, and to this end we have developed a novel method for the surface immobilization of DNA, using prefabricated PAMAM starburst dendrimers as mediator moieties. Dendrimers containing 64 primary amino groups in their outer sphere were covalently attached to silylated glass supports, and subsequently the dendritic macromolecules were chemically activated with glutaric anhydride and N-hydroxysuccinamide. Due to the dendritic PAMAM linker system, the surfaces reveal both a very high immobilization efficiency for amino-modified DNA-oligomers and also a remarkably high stability during repeated regeneration and reusing cycles [73, 74]. The latter feature, in particular, is very important for DDI-based applications.

15.3
Outlook

Commercial applications of semisynthetic DNA–protein conjugates are currently focused on the bioanalytics sector. In addition, these chimeric components show promise for the fabrication of nanostructured molecular arrangements, and thus their development will contribute to the rapid establishment of the novel discipline, now termed "nanobiotechnology" [5, 75]. Future perspectives include their use as high-affinity diagnostic reagents, artificial multienzymes, light-harvesting devices, or even autonomous drug-delivery systems. In addition, the self-assembling nanoscale fabrication of technical elements, such as dense arrays of molecular switches, transistors and logical parts, as well as inorganic/bioorganic hybrid devices for biomedical diagnostics and interface structures between electronic and living systems might be foreseen. To realize these fascinating biotechnological perspectives, however, a variety of serious technical obstacles remain to be solved. In particular, powerful analytical techniques as well as the refinement of bioconjugation and biomolecular evolution strategies are crucial to eventually attain comprehensive understanding and capabilities for tailoring the structure and reactivity of semisynthetic nucleic acid–protein conjugates. Given that the initial steps summarized here have already clearly demonstrated feasibility, the future developments in this new field of research promise much excitement.

Acknowledgments

The author thanks his coworkers for their motivated contributions, and Deutsche Forschungsgemeinschaft and Fonds der Chemischen Industrie are gratefully acknowledged for financially supporting these studies.

References

[1] D. M. J. Lilley, *ChemBioChem* **2001**, *2*, 31–35.

[2] K. E. Drexler, *Proc. Natl. Acad. Sci. USA* **1981**, *78*, 5275–5278.

[3] N. C. Seeman, *J. Theoret. Biol.* **1982**, *99*, 237–247.

[4] N. C. Seeman, *Trends Biotechnol.* **1999**, *17*, 437–443.

[5] C. M. Niemeyer, *Curr. Opin. Chem. Biol.* **2000**, *4*, 609–618.

[6] C. M. Niemeyer, *Trends Biotechnol.* **2002**, *20*, 395–401.

[7] D. R. Corey, P. G. Schultz, *Science* **1987**, *238*, 1401–1403.

[8] D. R. Corey, *J. Chem. Am. Soc.* **1995**, *117*, 9373–9374.

[9] S. V. Smulevitch, C. G. Simmons, J. C. Norton, T. W. Wise, D. R. Corey, *Nature Biotechnol.* **1996**, *14*, 1700–1704.

[10] C. H. Tung, S. Stein, *Bioconjug. Chem.* **2000**, *11*, 605–618.

[11] S. S. Ghosh, P. M. Kao, A. W. McCue, H. L. Chappelle, *Bioconjug. Chem.* **1990**, *1*, 71–76.

[12] A. Guerasimova, I. Ivanov, H. Lehrach, *Nucleic Acids Res.* **1999**, *27*, 703–705.

[13] E. Kynclova, A. Hartig, T. Schalkhammer, *J. Mol. Recognit.* **1995**, *8*, 139–145.

[14] S. S. Smith, L. M. Niu, D. J. Baker, J. A. Wendel, S. E. Kane, D. S. Joy, *Proc. Natl. Acad. Sci. USA* **1997**, *94*, 2162–2167.

[15] S. R. Rajski, S. Kumar, R. J. Roberts, J. K. Barton, *J. Am. Chem. Soc.* **1999**, *121*, 5615–5615.

[16] H.-A. Wagenknecht, S. R. Rajski, M. Pascaly, E. D. A. Stemp, J. K. Barton, *J. Am. Chem. Soc.* **2001**, *123*, 4400–4407.

[17] E. R. Hendrickson, T. M. Hatfield, T. M. Truby, R. D. Joerger, W. R. Majarian, R. C. Ebersole, *Nucleic Acids Res.* **1995**, *23*, 522–529.

[18] R. D. Joerger, T. M. Truby, E. R. Hendrickson, R. M. Young, R. C. Ebersole, *Clin. Chem.* **1995**, *41*, 1371–1377.

[19] T. Sano, C. L. Smith, C. R. Cantor, *Science* **1992**, *258*, 120–122.

[20] C. M. Niemeyer, M. Adler, D. Blohm, *Anal. Biochem.* **1997**, *246*, 140–145.

[21] M. Adler, M. Langer, K. Witthohn, J. Eck, D. Blohm, C. M. Niemeyer, *Biochem. Biophys. Res. Commun.* **2003**, *300*, 757–763.

[22] M. Adler, R. Wacker, C. M. Niemeyer, *Biochem. Biophys. Res. Commun.* **2003**, *308*, 240–250.

[23] C. M. Niemeyer, R. Wacker, M. Adler, *Nucleic Acids Res.* **2003**, *31*, e90.

[24] C. M. Niemeyer, M. Adler, B. Pignataro, S. Lenhert, S. Gao, L. F. Chi, H. Fuchs, D. Blohm, *Nucleic Acids Res.* **1999**, *27*, 4553–4561.

[25] W. H. Scouten, P. Konecny, *Anal. Biochem.* **1992**, *205*, 313–318.

[26] C. M. Niemeyer, T. Sano, C. L. Smith, C. R. Cantor, *Nucleic Acids Res.* **1994**, *22*, 5530–5539.

[27] C. M. Niemeyer, L. Boldt, B. Ceyhan, D. Blohm, *Anal. Biochem.* **1999**, *268*, 54–63.

[28] B. Schweitzer, S. Wiltshire, J. Lambert, S. O'Malley, K. Kukanskis, Z. Zhu, S. F. Kingsmore, P. M. Lizardi, D. C. Ward, *Proc. Natl. Acad. Sci. USA* **2000**, *97*, 10113–10119.

[29] B. Schweitzer, S. Roberts, B. Grimwade, W. Shao, M. Wang, Q. Fu, Q. Shu, I. Laroche, Z. Zhou, V. T. Tchernev, J. Christiansen, M. Velleca, S. F. Kingsmore, *Nature Biotechnol.* **2002**, *20*, 359–365.

[30] S. B. Rajur, C. M. Roth, J. R. Morgan, M. L. Yarmush, *Bioconjug. Chem.* **1997**, *8*, 935–940.

[31] C. Neves, G. Byk, D. Scherman, P. Wils, *FEBS Lett.* **1999**, *453*, 41–45.

[32] N. Nemoto, E. Miyamoto-Sato, Y. Husimi, H. Yanagawa, *FEBS Lett.* **1997**, *414*, 405–408.

[33] R. W. Roberts, J. W. Szostak, *Proc. Natl. Acad. Sci. USA* **1997**, *94*, 12297–12302.

[34] M. Kurz, K. Gu, A. Al-Gawari, P. A. Lohse, *ChemBioChem* **2001**, *2*, 666–672.

[35] S. Howorka, S. Cheley, H. Bayley, *Nature Biotechnol.* **2001**, *19*, 636–639.

[36] N. Sakai, B. Baumeister, S. Matile, *ChemBioChem* **2000**, *1*, 123–125.

[37] R. B. Fong, Z. Ding, C. J. Long, A. S. Hoffman, P. S. Stayton, *Bioconjug. Chem.* **1999**, *10*, 720–725.

[38] M. Lovrinovic, R. Seidel, R. Wacker, H. Schroeder, O. Seitz, M. Engelhard, R. Goody, C. M. Niemeyer, *Chem. Commun.* **2003**, 822–823.

[39] R. M. Hofmann, T. W. Muir, *Curr. Opin. Biotechnol.* **2002**, *13*, 297–303.

[40] R. S. Goody, K. Alexandrov, M. Engelhard, *ChemBioChem* **2002**, *3*, 399–403.

[41] C. M. Niemeyer, *Chem. Eur. J.* **2001**, *7*, 3188–3195.

[42] J. M. Tomkins, B. K. Nabbs, K. Barnes, M. Legido, A. J. Blacker, R. A. McKendry, C. Abell, *ChemBioChem* **2001**, *2*, 375–378.

[43] C. M. Niemeyer, W. Bürger, J. Peplies, *Angew. Chem. Int. Ed. Engl.* **1998**, *37*, 2265–2268, *Angew. Chemie* **1998**, *110*, 2391–2395.

[44] C. M. Niemeyer, M. Adler, S. Lenhert, S. Gao, H. Fuchs, L. F. Chi, *ChemBioChem* **2001**, *2*, 260–265.

[45] C. M. Niemeyer, M. Adler, S. Gao, L. F. Chi, *Angew. Chem. Int. Ed.* **2000**, *39*, 3055–3059, *Angew. Chem.* **2000**, *112*, 3183–3187.

[46] C. M. Niemeyer, R. Wacker, M. Adler, *Angew. Chem. Int. Ed.* **2001**, *40*, 3169–3172; *Angew. Chem.* **2001**, *113*, 3262–3265.

[47] S. Gao, L. F. Chi, S. Lenhert, B. Anczy-kowsky, C. M. Niemeyer, M. Adler, H. Fuchs, *ChemPhysChem* **2001**, *2*, 384–388.

[48] B. Pignataro, L. F. Chi, S. Gao, B. Anczy-kowsky, C. M. Niemeyer, M. Adler, H. Fuchs, *Appl. Phys. A* **2002**, *74*, 447–452.

[49] C. M. Niemeyer, M. Adler, S. Gao, L. F. Chi, *Bioconjug. Chem* **2001**, *12*, 364–371.

[50] C. M. Niemeyer, M. Adler, S. Gao, L. F. Chi, *J. Biomol. Struct. Dyn.* **2002**, 223–230.

[51] C. M. Niemeyer, J. Koehler, C. Wuerde-mann, *ChemBioChem* **2002**, *3*, 242–245.

[52] J. J. Storhoff, C. A. Mirkin, *Chem. Rev.* **1999**, *99*, 1849–1862.

[53] H. Zhu, M. Snyder, *Curr. Opin. Chem. Biol.* **2001**, *5*, 40–45.

[54] C. A. Borrebaeck, *Immunol. Today* **2000**, *21*, 379–382.

[55] M. F. Templin, D. Stoll, M. Schrenk, P. C. Traub, C. F. Vohringer, T. O. Joos, *Trends Biotechnol.* **2002**, *20*, 160–166.

[56] A. Lueking, M. Horn, H. Eickhoff, K. Bussow, H. Lehrach, G. Walter, *Anal. Biochem.* **1999**, *270*, 103–111.

[57] R. M. de Wildt, C. R. Mundy, B. D. Gorick, I. M. Tomlinson, *Nature Biotechnol.* **2000**, *18*, 989–994.

[58] G. MacBeath, S. L. Schreiber, *Science* **2000**, *289*, 1760–1763.

[59] C. M. Niemeyer, D. Blohm, *Angew. Chem. Int. Ed.* **1999**, *38*, 2865–2869; *Angew. Chem.* **1999**, *111*, 3039–3043.

[60] N. Windhab, C. Miculka, H.-U. Hoppe, *Chemical Abstracts* **1999**, *130*, 249124x.

[61] F. F. Bier, F. Kleinjung, E. Ehrentreich-Forster, F. W. Scheller, *Biotechniques* **1999**, *27*, 752–756.

[62] N. Winssinger, J. L. Harris, B. J. Backes, P. G. Schultz, *Angew. Chem. Int. Ed.* **2001**, *40*, 3152–3155.

[63] T. A. Taton, C. A. Mirkin, R. L. Letsinger, *Science* **2000**, *289*, 1757–1760.

[64] S. J. Park, T. A. Taton, C. A. Mirkin, *Science* **2002**, *295*, 1503–1506.

[65] F. Patolsky, K. T. Ranjit, A. Lichtenstein, I. Willner, *Chem. Commun.* **2000**, 1025–1026.

[66] L. He, M. D. Musick, S. R. Nicewarner, F. G. Salinas, S. J. Benkovic, M. J. Natan, C. D. Keating, *J. Am. Chem. Soc.* **2000**, *122*, 9071–9077.

[67] C. M. Niemeyer, B. Ceyhan, *Angew. Chem. Int. Ed.* **2001**, *40*, 3685–3688, *Angew. Chem.* **2001**, *113*, 3798–3801.

[68] L. M. Demers, S.-J. Park, T. A. Taton, Z. Li, C. A. Mirkin, *Angew. Chem. Int. Ed.* **2001**, *40*, 3071–3073, *Angew. Chem.* **2001**, *113*, 3161–3163.

[69] L. M. Demers, D. S. Ginger, S. J. Park, Z. Li, S. W. Chung, C. A. Mirkin, *Science* **2002**, *296*, 1836–1838.

[70] C. M. Niemeyer, W. Bürger, R. M. J. Hoedemakers, *Bioconjug. Chem.* **1998**, *9*, 168–175.

[71] C. M. Niemeyer, L. Boldt, B. Ceyhan, D. Blohm, *J. Biomol. Struct. Dyn.* **1999**, *17*, 527–538.

[72] P. E. Nielsen, *Annu. Rev. Biophys. Biomol. Struct.* **1995**, *24*, 167–183.

[73] R. Benters, C. M. Niemeyer, D. Wöhrle, *ChemBioChem* **2001**, *2*, 686–694.

[74] R. Benters, C. M. Niemeyer, D. Drutsch-mann, D. Blohm, D. Wöhrle, *Nucleic Acids Res.* **2002**, *30*, E10.

[75] C. M. Niemeyer, *Angew. Chem. Int. Ed.* **2001**, *40*, 4128–4158, *Angew. Chem.* **2001**, *113*, 4254–4287.

16
DNA-templated Electronics

Erez Braun and *Uri Sivan*

16.1
Introduction and Background

Since the publication of the celebrated work by Aviram and Ratner [1] in 1974, dozens of other molecular wires, rectifier, switches, storage elements, etc., have been discovered [2]. Yet, almost 30 years later, there is not a single proven scheme for assembling even just two molecules to form a functional electronic circuit. By comparison, it took less than 25 years from the discovery of the transistor (1947) and less than 15 years from the inception of the integrated circuit (1958) to produce a commercial processor (e. g., Intel 4004) with 2500 interconnected transistors.

In order to appreciate the competition faced by molecular electronics, it should be noted that present-day silicon-based processors contain over 40 million transistors, and the road map to a billion-transistor chip in 2007 is already laid. Moreover, those transistors will be limited by fundamental, quantum mechanical barriers that apply equally [3] to molecular devices. Full exploitation of molecular devices will require even higher levels of integration.

At the heart of the microelectronics revolution sits the integrated circuit and photolithography – the technology developed to embed the vast complexity of microelectronic circuits in a virgin silicon wafer. At present, there is no equivalent concept for the assembly of complex structures from molecular building blocks. It is clear that the invention of such strategy is critical for turning molecular electronics – and more generally nanotechnology – into reality. Failure to invent such a concept will render molecular engineering a curiosity limited to very simple structures.

Various engineering strategies for assembling molecular electronics are being pursued, from conventional approaches based on molecular monolayers deposited between electrodes [4], through grid-like structures defined by microelectronic techniques [5] and semiconducting nanowire grids [6], to sophisticated self-assembly schemes based on molecular recognition. The attempts to harness self-assembly are inspired by molecular biology – the only paradigm we have for the assembly of complex structures from molecular building blocks. A parallel effort has been put into the development of self-assembly capabilities with nonbiological molecules [7], but the attainable complexity there is at present signifi-

Nanobiotechnology. Edited by Christof Niemeyer, Chad Mirkin
Copyright © 2004 WILEY-VCH Verlag GmbH & Co. KgaA, Weinheim
ISBN 3-527-30658-7

cantly inferior when compared to that achieved with biological molecules. It is therefore only natural to try and harness the remarkable molecular recognition capabilities of the latter – in particular nucleic acids and proteins – for the self-assembly of nonbiological constructs. Indeed, a number of groups have utilized DNA molecules for the organization of nanoparticles [8, 9], the construction of DNA substrates [10] (see Chapter 20) and their subsequent decoration with nanoparticles [11], and templated growth of semiconductor nanoparticles [12]. Other attempts have included the use of proteins [13, 14], and some of these initial exploratory investigations are reviewed in Refs. [15, 16].

Any attempt to harness biological molecules for the direct realization of molecular scale electronics faces a fundamental difficulty in that most biological molecules do not conduct electronic currents. DNA conductivity, for instance, was investigated extensively in recent years. Notwithstanding several reports on reasonable electronic conduction of DNA molecules [17, 18], and even proximity-induced superconductivity [19], there is now a growing body of evidence indicating that DNA molecules are, for all practical purposes, insulating in character [20–28]. In fact, for fundamental reasons, molecular recognition and electronic conduction in the same location in space seem to be mutually exclusive [29], and it is therefore impossible to realize molecular electronics by biological molecules alone. However, any attempt to use electronic materials for the same purpose reveals another barrier. The superior electronic properties of these materials are accompanied by very simple chemistry, which does not allow encoding of the large amounts of information needed for the self-assembly of complex constructs. One is therefore faced with a fundamental dichotomy – electronic conduction, or self-assembly.

In an attempt to overcome this barrier, we proposed some years ago [27, 28] to hybridize between the two material systems to form a broader system where electronic conduction is provided by metals, semiconductors, and conducting polymers while self-assembly is provided by DNA molecules and their related proteins. As electronic materials and biological molecules are typically alien to each other, the schemes described below consist of two steps. First, the biological components are utilized to assemble a template or a scaffold with well-defined molecular addresses, which serves in a second step to localize the electronic devices and interconnect them electrically. In the latter step, electronic functionality is instilled to the DNA substrate. Unlike the conventional semiconductor substrates used in microelectronics, the DNA scaffold contains a high density of addresses, thereby enabling the localization of electronic devices with nanometer-scale resolution. Our bottom-to-top approach takes self-assembly a substantial step forward compared with previous state-of-the-art strategies, and also highlights the tremendous barrier that must be surmounted on the way to a meaningful self-assembly of complex systems.

An outline of the DNA-templated electronics concept is presented in the next section. We then introduce a new concept, sequence-specific molecular lithography. The protein RecA, which is normally responsible for homologous recombination in *Escherichia coli* bacteria, is utilized as a sequence-specific resist, analogous to photoresist in conventional photolithography. Here, however, the patterning information is encoded in the underlying DNA substrate rather than in glass masks. The same process facilitates precise localization of molecular devices on the DNA substrate and formation of molecularly accurate junctions. Our perspective on DNA-templated electronics and more generally biologically directed assembly of nonbiological constructs is briefly discussed in the last section.

16.2
DNA-templated Electronics

A possible scheme for the DNA templated assembly of molecular-scale electronics is depicted in Figure 16.1. This heuristic scheme addresses three major challenges on the way to molecular electronics. First, it allows the precise localization of a large number of devices at molecularly accurate addresses on the substrate. Second, it provides, by construction, inter-device wiring. Finally, it wires the molecular network to the macroscopic world, thus bridging between the nanometer and macroscopic scales. The same scheme also highlights some fundamental difficulties, which we discuss below.

There are four major obstacles to the realization of this concept. First, biological processes need to be adopted and modified to enable the in-vitro construction of stable DNA junctions and networks with well-defined connectivity. Second, the hybridization of electronic materials with biological molecules needs to be advanced to the point where precise localization of electronic devices on the network is made possible. Third, appropriate nanometer-scale electronic devices need to be developed. These devices should be compatible with the assembly and functionalization chemistry. Finally, in our scheme the DNA molecules serve to wire devices and interconnect between the circuit and the external world. Since DNA molecules are insulating, they need to be converted into conductive wires.

As demonstrated below, homologous recombination addresses the first two problems very effectively; it can generate molecularly accurate DNA junctions and networks, and also localize molecular objects at precise addresses on the DNA scaffold. There are various candidates for molecular-scale switching devices. One of the promising possibilities is the single-electron transistor, which is based on charging effects.

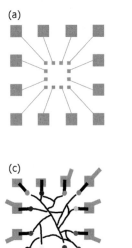

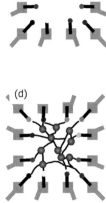

Figure 16.1 Heuristic scheme of a DNA-templated electronic circuit.
(a) Gold pads are defined on an inert substrate. (b–d) correspond to the circle of (a) at different stages of circuit construction. (b) Oligonucleotides of different sequences are attached to the different pads. (c) DNA network is constructed and bound to the oligonucleotides on the gold electrodes.
(d) Metal clusters or molecular electronic devices are localized on the DNA network. The DNA molecules are finally converted into metallic wires, rendering the construct into a functional electronic circuit. Note that the figures are not to scale; the metallic clusters are nanometer-sized, while the electrode pads are micrometer-sized.

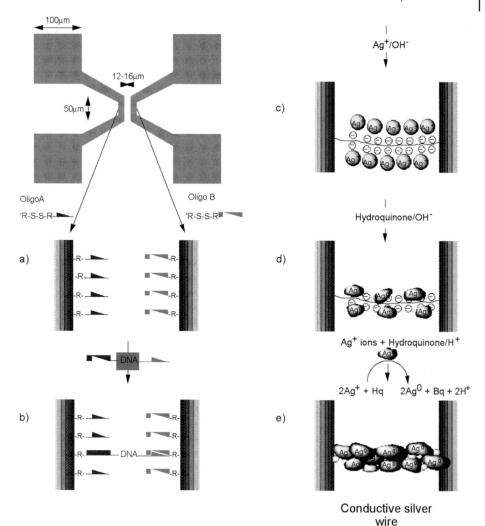

Figure 16.2 Gold pattern, 0.5 × 0.5 mm in size, was defined on a passivated glass using microelectronics techniques. The pattern comprised four bonding pads, 100 μm in size, connected to two 50 μm-long parallel gold electrodes, 12–16 μm apart. (a) The electrodes were each wetted with a 10⁻⁴ μL droplet of disulfide-derivatized oligonucleotide solution of a given sequence (Oligos A and B). (b) After rinsing, the structure was covered with 100 μL of a solution of λ-DNA having two sticky ends that are complementary to Oligos A and B. A flow was applied to stretch the λ-DNA molecule between the two electrodes, allowing its hybridization. (c) The DNA bridge was loaded with silver ions by Na⁺/Ag⁺ ion exchange. (d) The silver ion–DNA complex was reduced using a basic hydroquinone solution to form metallic silver aggregates bound to the DNA skeleton. (e) The DNA-templated wire was "developed" using an acidic solution of hydroquinone and silver ions. (Reprinted with permission from Ref. [27]; © 1998, *Nature*.)

The insulating nature of DNA molecules is circumvented by their selective coating with metal [27, 28]. The experimental procedure used to demonstrate DNA-templated assembly and electrode attachment of a conductive silver wire is depicted in Figure 16.2. First, 12-base oligonucleotides, derivatized with a disulfide group at their 3'-end, were attached to the electrodes through thiol–gold interaction. Each of the two electrodes was marked with a different oligonucleotide sequence. The electrodes were then bridged by hybridization of a 16 μm-long λ-DNA molecule containing two 12-base sticky ends, each complementary to one of the two sequences attached to the gold electrodes. Figure 16.3 presents a single DNA molecule bridge observed by fluorescence microscopy. To instill electrical functionality, the DNA bridge was coated with silver, the three-step chemical deposition process being based on selective localization of silver ions along the DNA molecule through Ag^+/Na^+ ion-exchange [30] and formation of complexes between the silver and the DNA bases [31–33]. The silver ion-exchanged DNA was then reduced to form nanometer-size silver aggregates bound to the DNA skeleton. These aggregates were further 'developed', much as in standard photographic procedure, using an acidic solution of hydroquinone and silver ions under low-light conditions [34, 35]. Although this solution was metastable, and spontaneous metal deposition was normally very slow, the silver aggregates on the DNA catalyzed the process. Under the experimental conditions, metal deposition therefore occurred only along the DNA skeleton, leaving the passivated substrate practically clean of silver. Figure 16.3 shows an AFM image of a segment of a 100 nm-wide, 12 μm-long wire connecting the two gold electrodes. Two-terminal electrical measurements of the DNA-templated silver wire produced the I–V curves depicted in Figure 16.3. These curves were nonlinear and asymmetric with respect to zero bias. Other deposition conditions led to ohmic behavior, albeit with high resistance. The insets to the right panel of Figure 16.3 present two I–V curves of control samples. Neither the bare DNA

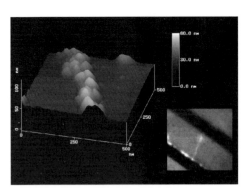
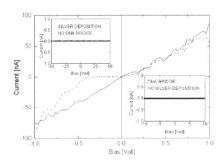

Figure 16.3 Left: AFM image of a silver wire connecting two gold electrodes 12 μm apart. Field size, 0.5 μm. Inset: Fluorescently labeled λ-DNA molecule stretched between two gold electrodes (dark strips), 16 μm apart. The electrodes are connected to large bonding pads 0.25 mm away (see Figure 16.2). Right: Two-terminal I–V curves of the silver wire shown on the left. Note the current plateau (dashed dotted line), on the order of 0.5 V. By ap-

plying 50 V to the wire, the plateau has been permanently eliminated to give an ohmic behavior (solid line) over the whole measurement range. I–V curves of a DNA bridge with no silver deposition, and silver deposition without a DNA bridge, are depicted at the bottom and top insets, respectively. Clearly, the sample is insulating in both cases. (Reprinted with permission from Ref. [27]; © 1998, *Nature*.)

molecule bridge nor the sample without DNA conducted current. Other DNA metalization schemes have been developed by other groups (see Chapter 17), and a better approach to DNA metalization with gold is presented below.

DNA metalization suffers from a serious drawback. The coating of DNA molecules with metal destroys their molecular-recognition capabilities and hence, further utilization of molecular biology after metalization. It therefore prevents, among other things, the construction of intrinsic feedback loops from the electronic properties back to biology, namely, a genuine interface between biology and electronics. This is probably the major obstacle along the road to bioelectronic molecular devices and circuits, and this point is elaborated on later in the chapter.

16.3
Sequence-specific Molecular Lithography

DNA-templated electronics requires elaborate manipulations on scales ranging from nanometers to many micrometers, including the formation of complex DNA geometries, wire patterning at molecular resolution, and molecularly accurate device localization. In conventional microelectronics, the circuit structure is dictated by lithography, but the use of an elaborate molecular assembly requires an alternative approach which is applicable at considerably smaller scales. To that end, we have recently developed the process of sequence-specific molecular lithography [36] that utilizes homologous recombination processes by RecA protein, operating on the double-stranded DNA (ds-DNA) substrate molecules [37]. The information guiding the lithography is encoded in the DNA substrate molecules and in short auxiliary probe DNA molecules. The RecA protein provides the resist function as well as the assembly capabilities. The same process can also be used to generate the DNA junctions that characterize elaborated DNA scaffolds and for the localization of molecular-scale objects at arbitrary positions along the DNA substrate.

Homologous recombination is a protein-mediated reaction by which two DNA molecules, possessing some sequence homology, crossover at equivalent sites. RecA is the major protein responsible for this process in *Escherichia coli* [37]. In our procedure, RecA proteins are polymerized on a probe DNA molecule to form a nucleoprotein filament, which is then mixed with the substrate molecules. The nucleoprotein filament binds to the DNA substrate at homologous probe–substrate locations as shown schematically in Figure 16.4i and ii. Note that RecA polymerization on the probe DNA is not sensitive to sequence. The binding specificity of the nucleoprotein filament to the substrate DNA is dictated by the probe's sequence and its homology to the substrate molecule.

Homologous recombination can be harnessed for sequence-specific patterning of DNA metal coating (Figure 16.4). The preferred metalization process differs from the scheme described above, and yields excellent wires. The key is localization of the reducing agent on the DNA substrate. In the demonstration presented in Ref. [36], DNA molecules were first aldehyde-derivatized by reacting them with glutaraldehyde, which left the DNA intact and biologically active. Next, the sample was incubated in an $AgNO_3$ solution. The reduction of silver ions by the DNA-bound aldehyde in the unprotected segments of the substrate molecule resulted in tiny silver aggregates along the DNA skeleton. The aggregates catalyzed subsequent electroless gold deposition, which produced a continuous,

(i) Polymerization

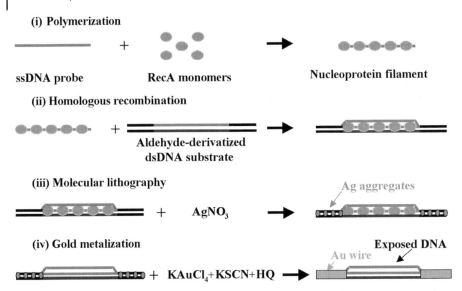

(ii) Homologous recombination

Aldehyde-derivatized
dsDNA substrate

(iii) Molecular lithography

Ag aggregates

+ AgNO₃ →

(iv) Gold metalization

Exposed DNA

Au wire

+ KAuCl₄+KSCN+HQ →

Figure 16.4 Schematics of the homologous recombination reaction and molecular lithography. (i) RecA monomers polymerize on a ss-DNA probe molecule to form a nucleoprotein filament. (ii) The nucleoprotein filament binds to an aldehydederivatized ds-DNA substrate molecule at an homologous sequence. (iii) Incubation in AgNO₃ solution results in the formation of silver aggregates along the substrate molecule at regions unprotected by RecA. (iv) The silver aggregates catalyze specific gold deposition on the unprotected regions. A highly conductive gold wire is formed, with a gap in the protected segment. (Reprinted with permission from Ref. [36]; © 2002, The American Association for the Advancement of Science.)

highly conductive gold wire. A scanning electron microscopy (SEM) image of a DNA-templated wire produced this way is shown in Figure 16.5 (inset). Electrode deposition and direct electrical measurements revealed a ~25 Ω resistance and ohmic characteristics up to currents on the order of 200 nA. The wire conductivity was only seven times lower than that of polycrystalline gold, and four orders of magnitude higher compared with the DNA-templated silver wires discussed above [27]. The reducing agent (aldehyde) localization on the DNA scaffold resulted in a very low background metalization.

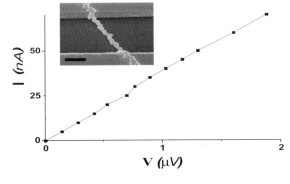

Figure 16.5 Two-terminal I–V curve of a DNA-templated gold wire. The wire's resistivity (1.5×10^{-7} Ωm) is only seven times higher than that of polycrystalline gold (2.2×10^{-8} Ωm). Inset: SEM image of a typical DNA-templated gold wire stretched between two electrodes deposited by electron-beam lithography. Scale bar, 1 μm. (Reprinted with permission from Ref. [36]; © 2002, The American Association for the Advancement of Science.)

This metalization process was used in conjunction with the molecular lithography process to produce sequence-specific, patterned metallic wires [36]. RecA was first used to localize a 2027-base single-stranded (ss) probe molecule on the homologous section in the middle of a 48502-bp aldehyde-derivatized λ-DNA substrate molecule. Sequence-specific nucleoprotein binding to the substrate molecule was confirmed by protection against restriction-enzyme digestion [38]. The efficiency and specificity of the homologous recombination reaction were not affected by aldehyde derivatization of the substrate DNA. Following the recombination reaction, the molecules were stretched on a passivated silicon wafer. The AFM image in Figure 16.6a shows a RecA nucleoprotein filament bound to the homologous location on the DNA substrate. Next, the sample was incubated in an AgNO$_3$ solution. The localized RecA proteins, serving as a resist, prevented Ag deposition on the protected aldehyde-derivatized DNA segment and led to a gap of exposed sequence between the Ag-loaded segments of the substrate molecule (Figure 16.6b). A subsequent electroless gold deposition produced two continuous gold wires separated by the predesigned gap (Figure 16.6c and d). Extensive AFM and SEM confirmed that the metalization gap was located where expected. The position and size of the insulating gap can be tailored at will by choosing the probe's sequence and length.

In our molecular lithography concept, exposed sequences of ds-DNA provide addresses for the localization of molecular objects. A RecA-driven recombination reaction can be utilized to localize arbitrary labeled molecular objects at specific locations along the ds-DNA

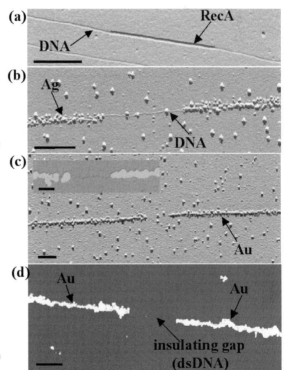

Figure 16.6 Sequence-specific molecular lithography on a single DNA molecule. (a) AFM image of a 2027-base RecA nucleoprotein filament bound to an aldehydederivatized λ-DNA substrate molecule. (b) AFM image of the sample after silver deposition. Note the exposed DNA at the gap between the silver-loaded sections. (c) AFM image of the sample after gold metalization. Inset: zoom on the gap. The height of the metalized sections is ~50 nm. (d) SEM image of the wire after gold metalization. All scale bars, 0.5 μm; inset to (c), 0.25 μm. The variation in the gap length is mainly due to variability in DNA stretching on the solid support. The very low background metalization in the SEM image compared with the AFM images indicates that most of the background is insulating. (Reprinted with permission from Ref. [36]; © 2002, The American Association for the Advancement of Science.)

substrate. The specificity [39] in localizing a biotin-labeled, 500 base-long ss-DNA probe on a fragmented λ-DNA substrate was demonstrated [36], and is described briefly here. The RecA-probe nucleoprotein filament bound specifically to an homologous address on a long ds-DNA molecule. The deproteinized reaction products were incubated with strepta-vidin-conjugated 1.4 nm gold particles (nanogold, Nanoprobes), whereupon the streptavi-din-conjugated nanogold bound specifically to the biotinylated probe. An AFM image (Figure 16.7a) shows the nanogold-bound ss-DNA probe localized on the substrate ds-DNA, while Figure 16.7b proves that electroless gold deposition [36] resulted in gold growth only around the catalyzing nanogold particles.

RecA-mediated recombination can be harnessed to generate the molecularly accurate DNA junctions required for the realization of elaborate DNA scaffolds. Artificial DNA junctions produced by hybridization have been demonstrated previously, but this required precise design and ss-DNA (oligonucleotides) synthesis [40] (see also Chapter 20). RecA, on the other hand, generates junctions between any pair of ds-DNA molecules having homologous regions [41]. Junction generation by RecA protein was demonstrated in Ref. [36]. Briefly, two types of DNA molecules which were 15 kbp and 4.3 kbp long respec-tively, were prepared. The short molecule was homologous to a 4.3 kbp segment at one end of the long molecule. The RecA was first polymerized on the short molecules and then reacted with the long molecules. The recombination reaction led to the formation of a stable, three-armed junction with two 4.3 kbp-long arms and an 11 kbp-long third arm. Junction formation was confirmed by biotin labeling of the 4.3-kbp molecules and

(a　　　　　　　　**(b)**

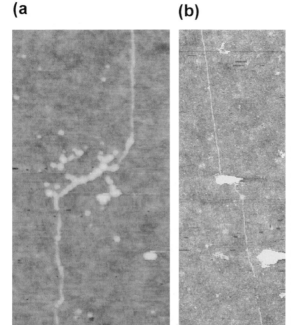

Figure 16.7 Sequence-specific localiza-tion of labeled nanometer-scale objects on a ds-DNA substrate. (a) AFM image of a nanogold-labeled 500 base-long probe bound to a λ-DNA substrate molecule. The height of the central features is 5 nm. Scale bar, 0.2 μm. (b) Sample after electroless gold me-talization with nanogold particles serving as nucleation centers. The metalized object at the center of the DNA molecule is over 60 nm high. Scale bar, 1 μm. (Reprinted with permission from Ref. [36]; © 2002, The American Association for the Advancement of Science.)

Figure 16.8 Stable three-armed junction. AFM images of a three-armed junction, which can serve as a scaffold for a three-terminal device. Scale bars: upper panel, 0.25 µm; lower panel, 50 nm. The lengths of the arms are consistent with the expected values considering the variations due to interaction with the substrate in the combing process. (Reprinted with permission from Ref. [36]; © 2002, The American Association for the Advancement of Science.)

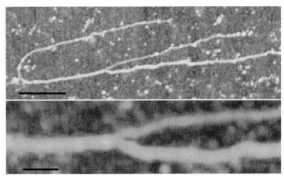

gel electrophoresis. AFM images of the resulting junction are shown in Figure 16.8; these junctions can serve to template a three-terminal device.

16.4
Summary and Perspectives

Sequence-specific molecular lithography constitutes an important step toward integrated DNA-templated electronics. Homologous recombination by RecA operates on scales varying between a few bases (nanometers) to thousands of bases (micrometers) with essentially single-base accuracy (~0.3 nm). It should be emphasized that the various molecular lithography processes demonstrated above can be carried out sequentially. For example, junction definition followed by specific metalization of the unprotected junction's arms and colloid localization at the junction represent three levels of lithography. Molecular lithography can obviously be applied to other DNA programmed constructs, perhaps for mechanical applications. The resist function provided by the RecA protein can most likely be extended to operations other than metalization as the protein apparently blocks the access of even small molecules (silver ions in the present case) to the DNA substrate.

More generally, the substantial research conducted thus far on the utilization of DNA molecules and their related proteins to the self-assembly of nonbiological nanometer scale objects demonstrates the strength of this approach. Yet, all realizations of molecular-scale electronics were confined to simple constructs, namely, periodic arrays or elementary circuits. The generalization of self-assembly to complex structures requires new elements including error correction mechanisms, check points and feed-back loops. Nature utilizes these elements from the molecular level to systems. As self-assembly is prone to errors, there must be a context-specific scale beyond which the assembled structure must be checked and fixed or discarded. Such decisions require structural or functional tests and feedback to the assembly process. Molecular biology provides such tools for certain operations, but as soon as the assembly departs from the biological pathway many of these tools prove useless. Feed-back mechanisms based on the desired electronic functionality are even more difficult, especially in the two-step assembly process adapted here. Once the DNA-templated substrate is instilled with electronic functionality, it loses its biological features and is no longer available for further biological manipulations. At present, we are unable to feed back from the desired electronic functionality to the biological realm, though a possible strategy might exploit gene expression and possibly its control by electronic signals.

References

[1] A. Aviram, M. A. Ratner, *Chem. Phys. Lett.* **1974**, *29*, 277.

[2] C. Joachim, J. K. Gimzewski, A. Aviram, *Nature* **2000**, *408*, 541.

[3] The field effect transistor size is dictated by the minimal thickness of the dielectric layer between the gate and the channel. In the 1 billion transistor chip that layer will measure three atomic layers or about 0.8 nm. The unavoidable electron tunneling from the gate to the channel limits further thinning of that layer. The gate length, typically 30–50 times the dielectric layer thickness, is then determined by the gain required for the operation of the transistor in a circuit. The subnanometer thickness of the dielectric layer dictates this way a much larger transistor size. The same constraint applies to any field effect devices, including for instance, molecular Coulomb blockade devices.

[4] M. A. Reed, et al., *Appl. Phys. Lett.* **2001**, *78*, 3735–3737.

[5] C. P. Collier, et al., *Science* **1999**, *285*, 391–394.

[6] Yu Huang, et al., *Science* **2001**, *294*, 1313–1317.

[7] J.-M. Lehn, *Proc. Natl. Acad. Sci. USA* **2002**, *99*, 4763–4768.

[8] A. P. Alivisatos, et al., *Nature* **1996**, *382*, 609.

[9] C. A. Mirkin, R. L. Letsinger, R. C. Mucic, J. J. Storhoff, *Nature* **1996**, *382*, 607.

[10] E. Winfree, F. Liu, L. A. Wenzler, N. C. Seeman, *Nature* **1998**, *394*, 539–544.

[11] S. Xiao, et al., *J. Nanoparticle Res.* **2002**, *4*, 313–317.

[12] J. L. Coffer, et al., *Appl. Phys. Lett.* **1996**, *69*, 3851.

[13] C. M. Niemeyer, W. Burger, J. Peplies, *Angew. Chem. Int. Ed. Eng.* **1998**, *37*, 2265.

[14] R. Ross, et al., *Adv. Mater.* **2002**, *14*, 1453–1457.

[15] J. J. Storhoff, C. A. Mirkin, *Chem. Rev.* **1999**, *99*, 1849–1862.

[16] C. M. Niemeyer, *Curr. Opin. Chem. Biol.* **2000**, *4*, 609–618.

[17] H. W. Fink, C. Schonenberger, *Nature* **1999**, *398*, 407–410.

[18] L. Cai, H. Tabata, T. Kawai, *Appl. Phys. Lett.* **2000**, *77*, 3105–3106.

[19] A. Yu Kasumov, et al., *Science* **2001**, *291*, 280–282.

[20] B. Gaehwiler, I. Zschokke-Graenacher, E. Baldinger, H. Luethy, *Biophysik* **1970**, *6*, 331–344.

[21] C. Gomez-Navarro, et al., *Proc. Natl. Acad. Sci. USA* **2002**, *99*, 8484–8487.

[22] P. J. de Pablo, et al., *Phys. Rev. Lett.* **2000**, *85*, 4992–4995.

[23] M. Bockrath, et al., *Nano Lett.* **2002**, *2*, 187–190.

[24] D. Porath, A. Bezryadin, S. De Vries, C. Dekker, *Nature* **2000**, *403*, 635–638.

[25] A. J. Storm, J. van Noort, S. de Vries, C. Dekker, *Appl. Phys. Lett.* **2001**, *79*, 3881–3883.

[26] Y. Zhang, R. H. Austin, J. Kraeft, E. C. Cox, N. Pong, *Phys. Rev. Lett.* **2002**, *89*, 198102.

[27] E. Braun, Y. Eichen, U. Sivan, G. Ben-Yoseph, *Nature* **1998**, *391*, 775–778.

[28] Y. Eichen, E. Braun, U. Sivan, G. Ben-Yoseph, *Acta Polym.* **1998**, *49*, 663–670.

[29] Molecular recognition requires binding energies on the order of a few $k_B T$. Higher binding energies will result in nonselective sticking, while lower binding energies do not facilitate binding due to excessive thermal fluctuations. Nature provides the right energy scale by screening binding energies on the order of one Rydberg by the high dielectric constant of water, $\epsilon \approx 80$. When free electrons are added to the same space, the interaction is further screened to a point where the binding energy drops well below $k_B \times 300 K$. Binding is then impossible.

[30] J. K. Barton, in: Bertini I., et al. (eds), *Bioinorganic Chemistry*, University Science Books, Mill Valley, Chapter 8 and references therein, **1994**.

[31] T. G. Spiro (ed.), *Nucleic Acid-Metal Ion Interaction*, Wiley Interscience, New York, **1980**.

[32] L. G. Marzilli, T. J. Kistenmacher, M. Rossi, *J. Am. Chem. Soc.* **1977**, *99*, 2797–2798, and references therein.

[33] G. L. Eichorn (ed.), *Inorganic Biochemistry*, Vol. 2, Elsevier Press, Amsterdam, Chapters 33–34, **1973**.

[34] C. S. Holgate, et al., *J. Histochem. Cytochem.* **1983**, *31*, 938–944.

[35] G. B. Birrell, et al., *J. Histochem. Cytochem.* **1986**, *34*, 339–345.

[36] K. Keren, et al., *Science* **2002**, *297*, 72.

[37] M. M. Cox, *Prog. Nucleic Acid Res. Mol. Biol.* **2000**, *63*, 311–366.

[38] W. Szybalski, *Curr. Opin. Biotechnol.* **1997**, *8*, 75–81.

[39] S. M. Honigberg, R.B Jagadeeshwar, C. M. Radding, *Proc. Natl. Acad. Sci. USA* **1986**, *83*, 9586–9590.

[40] C. Mao, S. Weiqiong, N. C. Seeman, *J. Am. Chem. Soc.* **1999**, *121*, 5437–5443.

[41] B. Muller, I. Burdett, S. C. West, *EMBO J.* **1992**, *11*, 2685–2693.

17

Biomimetic Fabrication of DNA-based Metallic Nanowires and Networks

Michael Mertig and *Wolfgang Pompe*

17.1
Introduction

In his textbook on supramolecular chemistry, the French Noble prize laureate Jean-Marie Lene emphasizes the existence of an obvious gap between chemistry and biology with respect to two basic parameters, complexity and breadth [1]. Structures, developed during biological evolution, are of both extremely rich variety and highest level of complexity, yet the building blocks on which these structures are based belong only to limited, well-defined classes. For example, the complete "blueprint" of the human body is based entirely on the rich chemistry of only four DNA bases, guanine (G), cytosine (C), adenine (A), and thymine (T) [2]. In comparison to biology, today's chemistry is still of relatively low complexity, but the diversity of its structural building blocks is unlimited. Therefore, Lene concludes that the white area in the complexity–diversity diagram has to be filled by supramolecular chemistry aiming on the synthesis of more complex chemical architectures.

However, a second option to fill the so-far white diagram area arises from the tremendous progress in molecular biology. By virtue of the fundamental explorations and methodical progress in genomics and proteomics during the last 20 years, new tools are available now which allow a novel, biomolecular-based approach to manufacture abiotic materials of highly complex architecture. Pioneered by Stephen Mann, a new field – biomimetic materials chemistry – has been developed [3]. It is understood as part of organized-matter chemistry which is established on the concept that molecular-based interactions can be integrated into higher levels of organization across a wide range of length scales.

Today, the main areas of biomimetic material synthesis are: the implementation of biological concepts, such as supramolecular pre-organization or interfacial molecular recognition, into the synthesis of novel materials; the investigation of the role of biomolecules and matrixes for the nucleation and architecture of inorganic material on various length scales; and the use of both cellular processes and molecular machines for technical purposes. Among others, one of the key issues of this development is the template-directed materials synthesis, which involves specific nucleation and growth of inorganic phases on the surface of biomolecular structures under strict biological control. At this point, the

Nanobiotechnology. Edited by Christof Niemeyer, Chad Mirkin
Copyright © 2004 WILEY-VCH Verlag GmbH & Co. KgaA, Weinheim
ISBN 3-527-30658-7

so-called "bottom-up" approach of nanostructure fabrication comes into play, for two major reasons:

1. The controllability of nucleation and growth processes is also the main prerequisite for any build-up of nanostructures based on the assembly of atomic or molecular building blocks.
2. Template-directed materials synthesis allows directly physical patterning when individual biomolecules or biomolecular structures with well-defined architectures are used as the templates.

Following this concept, the use of biological macromolecules for the directed formation of regular patterns of inorganic nanoparticles has become an inspiring field of modern materials science [4–12]. The ultimate goal of this particular approach to nanostructure fabrication is to combine the unique self-assembly capabilities of biopolymers with the optical, electronic, and quantum properties of small particles to develop a new generation of devices at the nanometer scale. With this aim, biological macromolecules can serve as almost ideal templates because of their well-determined properties.

Here, we address three fundamental issues of a controlled bottom-up fabrication of artificial nanostructures by biomolecular templating which shall demonstrate both the prevailing advantage and the enormous potential of using biomolecules in an engineering context for future technological applications:

1. The site-specific integration of biomolecular structures into microelectronic or micro-reaction systems being realized through molecular recognition between the molecules and locally functionalized binding pads on technical substrates.
2. The build-up of artificially designed biomolecular structures, which do not exist in this particular form in nature, with the aim to tailor the complexity of desired biomolecular templates by making use of their specific self-assembly capabilities.
3. The growth of metallic clusters at biomolecular structures, promoted and controlled by the template itself, with the goal of transforming the template structures into stable, artificial nanostructures with a functionality other than that of the biomolecule used as the template.

We will discuss the multifunctional use of selective properties, intrinsic to biomolecules and important for a controlled bottom-up processing, along one particular example, the engineering of nano-scaled electronic circuits when DNA is used as the biomolecular template.

We will show how to interconnect microscopic gold electrodes with single DNA molecules in a site-specific manner, and how to fabricate artificial DNA building blocks with desired complexity for the build-up of DNA networks taking full advantage of the unique recognition, association, and binding properties of DNA. After this, we will focus on a novel method which allows the fabrication of ultra-thin, uniform, and continuous metal cluster chains [8] or nanowires with metallic conductivity [7, 13]. This method permits the achievement of a template-promoted growth of small metal particles along the biopolymer. Simultaneously, it allows the suppression, kinetically, of any spurious homogeneous metal nucleation in bulk solution, and results therefore in a perfectly clean, in-place metallization of biomolecules.

17.2
Template Design

17.2.1
DNA as a Biomolecular Template

The peculiarity of DNA among other biological macromolecules relates to the specificity of the Watson–Crick base pairing. The ability to synthesize DNA with arbitrary base sequence [14] permits programming of intra- and intermolecular associations, and thus to build-up artificially engineered supramolecular structures and networks. Different strategies have been scrutinized to fabricate periodic two-dimensional (2D) or sophisticated three-dimensional (3D) DNA structures exploiting both hybridization of base-complementary, single-stranded DNA and ligation of "sticky" DNA ends [15]. The polymerase chain reaction (PCR) [16] allows amplification of the basic DNA building blocks to amounts which are necessary for the successful engineering of nanoscaled nucleic acid assemblies.

A linear double-stranded DNA has a diameter of about 2 nm and a length up to tens of micrometers [17]. Because of its large geometrical aspect ratio, its remarkable mechanical properties [18], and its large variety of binding sites for different ions [19], it constitutes an ideal template for nanowires fabricated by direct growth of metals [6, 8, 20, 21] or semiconductors [22] on DNA. Moreover, DNA can be used to organize preformed nanoparticles at the nanometer scale. In this case, the nanoparticles can either be bound to the phosphate backbone of DNA via electrostatic interactions [23], where DNA acts as mere linear support, or the nanoparticles can be assembled into 3D arrays when the particles are functionalized with specific oligonucleotides which allow stable next-neighbor binding via sequence-specific hybridization [4, 5, 24, 25].

In addition, various methods have been developed to manipulate DNA at macroscopic level by molecular combing [26] or transfer-printing [27], and to manipulate DNA at single molecular level by AFM [28], optical trapping [18, 29], or magnetic tweezers [30].

Both, the specific properties of the molecule and the methods developed for DNA handling, make DNA the most promising candidate to serve as template for the fabrication of nanowires and more complex networks.

17.2.2
Integration of DNA into Microelectronic Contact Arrays

First of all, the construction of functional nanoscale electronic devices by biomimetic bottom-up approaches requires a stable electrical interfacing of biomolecules to macroscopic electrodes. To this aim, self-assembly as well as in-situ control of the interconnections are desired, and these are the major challenges in the development of this approach. In this context, site-specific molecule attachment and the possibility to address particular contact pads are the main issues. Both have been investigated for DNA as a model system making use of the unique interfacial molecular recognition capabilities of this molecule [6, 31]. In the simplest conceivable case, electrical interfacing of a single DNA molecule is achieved by attaching its both ends at two different microfabricated contacts pads, thus "bridging" adjacent electrodes.

For a specific single-end anchoring of DNA at patterned gold contacts, different methods have been developed based on the recognition and binding to chemically or biologically functionalized electrode surfaces (Figure 17.1).

1. One method is the hybridization of the "sticky ends" of DNA, which constitute single stranded overhangs of defined lengths and sequences on both sides of the molecule, with oligonucleotides of complementary sequences which have been previously immobilized on the Au contacts via a thiol group (Figure 17.1a) [6, 32]. This method also allows the two DNA ends to be addressed separately, allowing, for example, the interconnection of two microfabricated contacts with a DNA molecule in a predefined orientation when the two electrodes are functionalized differently. In this case, each contact should be specific to one particular DNA end.

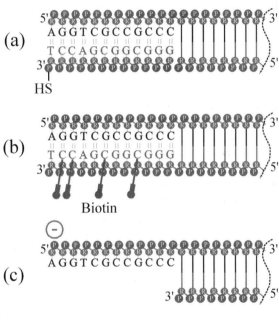

Figure 17.1 Schematic viewgraph of three different methods of specific single-end anchoring of DNA at patterned gold contacts. (a–c) Different functionalizations of the DNA ends for their attachment via (a) hybridization of the sticky ends of DNA to an oligonucleotide-terminated Au surface; (b) antibody binding by formation of a streptavidin bridge between biotinylated DNA ends and a biotinylated Au surface; and (c) electrostatic bonding between amino-thiol groups on Au and negatively charged DNA ends. (d) Corresponding functionalizations of the contact pads (a = left; b = middle; c = right).

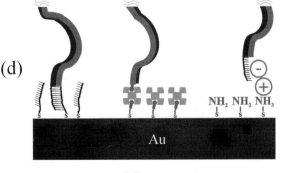

2. In a second method, specific antibody binding is used for the anchoring of DNA (Figure 17.1b). For this, biotinylated DNA ends are anchored to biotinylated Au surfaces by the formation of a streptavidin bridge [33].

3. A third, relatively robust method to realize end-specific DNA attachment is the electrostatic bonding between positively charged amino-thiol groups immobilized on Au contacts and the negatively charged ends of DNA molecules at a pH where the molecule body is not attracted to the surface (Figure 17.1c) [31, 34].

Similarly to the first-described method, an addressable, oriented integration of DNA molecules can be obtained by any combination of the three anchoring methods.

A second important consideration of DNA wiring is stretching of the molecule. DNA forms a random coil in solution and, for this reason, a DNA molecule will never interconnect two adjacent electrodes which are several microns apart, just by itself. This will require stretching of the molecule to form a linear, wire-like geometry. Therefore, in a typical anchoring experiment, a flow of a diluted DNA solution is directed over the substrate. Once one end of a DNA molecule attaches on a functionalized Au pad, it is stretched by the hydrodynamic flow and its conformation changes from a random coil to a linear 'wire'. When the free end of the stretched molecule anchors at an adjacent pad, a single-molecule bridge is formed between the two contacts (Figure 17.2). This procedure can be applied equally well to all three binding protocols described above. In all cases, the single-end anchoring of DNA to a gold contact is strong enough to withstand flow velocities up to ~100 μm s^{-1} and repeated changes of the flow direction of 180° [31]. Stable binding of the DNA molecule on both ends can be easily proven by applying a hydrodynamic flow crosswise to the attached molecule (Figure 17.2b). In this case, the molecule bends slightly as a consequence of the applied hydrodynamic force, but only when both ends are anchored. When the flow is switched off, the elastic DNA molecules recovers into the shape of a straight wire (Figure 17.2a).

This relatively simple example clearly shows that interfacial molecular recognition, in this case by the formation of specific biotin–streptavidin bridges, can be used to control

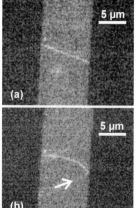

Figure 17.2 A single λ-DNA molecule stretched between two gold contact pads. The functionalized electrodes appear dark, the fluorescently labeled DNA appears bright. The molecule was anchored to the contacts via biotin–streptavidin–biotin bridges by applying a hydrodynamic flow in the direction of the attached molecule (a). Changing the flow direction, as indicated by the arrow, causes bending of the elastic molecule into the form of a bow (b), which gives clear experimental evidence that the molecule is connected on its both ends to the adjacent contacts.

the site-specific positioning of biopolymers in a preformed, microfabricated electrode structure. With the techniques discussed here, it is in general possible to integrate more complex DNA networks into microfabricated electrode arrays, though this particular aspect has not yet been demonstrated.

17.2.3
DNA Branching for Network Formation

An important aspect of the biomimetic approach to nanostructure fabrication is the possibility to tailor the biomolecular templates in a controlled manner to increase gradually the complexity of the desired nanostructures. An example would be to pass over from linear DNA wires to more complex DNA networks with several nodal points up to sophisticated 3D architectures for future nanoelectronic applications. Towards this end, it is also necessary to engineer the biomolecular templates – that is, to construct modified, more advanced templates, which do not exist in this particular form in nature. The advantage of this technique is two-fold. First, template tailoring can be achieved by applying the same biological principles as used for the normal template handling, such as molecular recognition and self- assembly. Second, template engineering can be accomplished while preserving those basic template properties which are important for the subsequent metallization process (see section 17.3). In the following section, these advantages will be demonstrated along one particular example, namely the fabrication of three-armed DNA junctions which can serve as nodal points necessary to construct DNA networks.

Recently, Seeman and co-workers have developed different strategies to fabricate elaborated DNA structures [15]. These are mainly based on the synthesis of multibranched DNA-linker elements which are used as basic elements for the formation of 2D or 3D DNA arrays. Sticky-ended cohesion is exploited to direct the associations of the branched molecules. The linker elements are formed by hybridization of single-stranded synthetic oligonucleotides. However, as the size of synthetic oligonucleotides is currently restricted to about 100 bases, the length of the single branches is limited to a few nanometers. This size is not large enough for the applications discussed here. Such tripods are too small for the integration into currently available contact arrays (see section 17.2.2), and their size is of the same order as the smallest structure obtained by metallization of DNA so far [8] (see section 17.3). Therefore, they would be completely embedded into a few clusters in a subsequent metallization step. For those reasons, we have constructed larger three-armed junctions in a two-step process [35]. First, small three-armed DNA tripods are assembled by using ~40 bp-long DNA oligonucleotides (Figure 17.3a). Thereafter, the tiny arms of these tripods – which will serve as central elements of the junctions – are elongated by ligation with ~500 bp-long fragments possessing sticky ends which are complementary to those of the tripod molecule. The expected morphology of the junctions is confirmed by scanning force microscopy (SFM) imaging (Figure 17.3b). As expected, the lengths of the three arms of the junction are the same. In good agreement with values calculated from the number of base pairs of the arms, each arm is about 200 nm long. Interestingly, the statistics of the measurement of the angles between adjacent arms showed a standard deviation of less than 15° from 120°. These geometrical values derived from the SFM investigations clearly indicate that synthesized junctions are stable against manipulation in

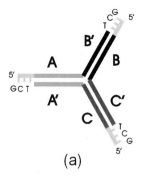

(a)

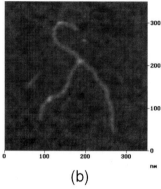

(b)

Figure 17.3 (a) Schematic viewgraph of the DNA tripod used as central element for the construction of a DNA junction. Lines with same gray values represent complementary DNA duplex strands. The end of each arm consists of a GCT 5'-overhang used for the elongation of the arms by ligation of DNA fragments with sticky ends complementary to the tripod overhangs. (b) Scanning force microscopy of a three-armed DNA junction. The mean apparent height of the DNA strands is 0.6 nm.

liquid and deposition onto substrate surfaces, and that they are also relatively stiff. Both properties are necessary to employ the fabricated junctions for the future build-up of integrated, conducting DNA networks. A generic feature of the basic design principle discussed here is that higher-branched nodes can be easily constructed [35].

17.3
Metallization

In general, if DNA were to be electrically conducting, then DNA networks integrated into microfabricated contact arrays could be readily used for nanoelectronic applications. However, the electric long-range performance of native DNA is poor [36], and so a direct use of DNA in electrical circuits is not possible. One way to overcome this situation is chemical metal deposition on DNA in order to instill electrical conductivity. Simultaneously, by metallization the template structure can be transferred into an artificial nanostructure with greater stability, which could for example be operated at temperatures higher than the melting temperature of DNA.

In the following we will present a template-directed method of DNA metallization generic for the preparation of precious-metal nanostructures on biomolecular templates.

17.3.1
Controlled Cluster Growth on DNA Templates

The main challenge of the bottom-up fabrication of nanostructures by biomolecular templating is to achieve conditions for a controlled, selective growth of the inorganic phase at the surface of biomolecules. On the one hand, this requires a "guided" heterogeneous nucleation of metal on the biomolecule, directed by the template itself. On the other hand, homogeneous nucleation of metal clusters in the bulk liquid should be suppressed simultaneously to prevent spurious metal deposition on the template. In other words, to obtain a pure, in-place metallization of biomolecules, the balance between heterogeneous and homogeneous nucleation must be shifted to conditions, where a preferential heterogeneous nucleation takes place.

In a series of metallization experiments, we have recently shown that DNA is capable of promoting selectively the heterogeneous nucleation of metallic nanoparticles approved by chemical reduction of dissolved, hydrolyzed tetrachloroplatinate (K_2PtCl_4) in the presence of DNA molecules [8, 21, 31]. Preferred heterogeneous nucleation at the template, and simultaneous kinetic suppression of homogeneous nucleated cluster growth are obtained, when the formation of specific nucleation sites along the DNA molecule is initiated prior to the chemical reduction of the metal salt complexes. The formation of nucleation centers at the DNA is accomplished in an activation step by binding the Pt(II) complexes covalently to specific nucleotides (Figure 17.4). The in-vitro Pt(II) complexation with DNA has been intensively studied for the anticancer drug cisplatin because of its unique antimitotic properties [37, 38]. In the initial binding, the complexes coordinate primarily to the N7 positions of G bases to form monofunctional adducts. Subsequently, those are converted to bifunctional adducts by reacting with a second nucleophile. The characteristic binding time is in the order of minutes when working with aquated Pt complexes where one or both chlorines are hydrolyzed via association of solvent water [38]. The initial reaction of K_2PtCl_4 with DNA takes also selectively place at the G–C planes, but the sequence specificity is less due to a higher reactivity of tetrachloroplatinate compared to cisplatin. For long interaction times and large drug-to-nucleotide (D/N) ratios, a saturation of

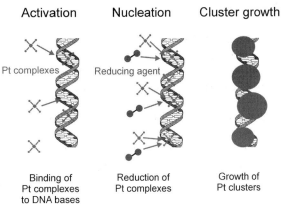

Figure 17.4 Schematic viewgraph of the DNA metallization procedure. In a first step of activation, DNA is incubated with a Pt salt solution. During incubation, part of the metal complexes bind covalently to the DNA bases. The Pt–DNA adducts act as active nucleation centers in the following step of chemical reduction; this leads to a preferred heterogeneous nucleation, and thus growth of metal clusters along the biomolecular template.

Activation   Nucleation   Cluster growth

Pt complexes   Reducing agent

Binding of Pt complexes to DNA bases   Reduction of Pt complexes   Growth of Pt clusters six Pt complexes per (A–T, G–C) unit is reported [39]. In this case, the saturation time is ≥ 10 hours [40].

In a typical metallization experiment, DNA is first activated by incubating the biomolecules with a hydrolyzed K_2PtCl_4 solution, keeping a D/N ratio at 65:1. Subsequent chemical reduction of the DNA–metal complex solution with dimethylamine borane (DMAB) leads to the formation of regular, continuous chains of Pt clusters with a relatively narrow diameter distribution of 4 ± 1 nm, as shown in the transmission electron microscopy (TEM) image presented in Figure 17.5a. All nanoparticles formed are completely aligned along the biomolecular template, thus reflecting the linear morphology of the biopolymer. This also demonstrates that the template stabilizes the nanoclusters grown on DNA, which is usually only achieved when capping agents are used to prevent agglomeration of clusters. In addition, high-resolution TEM (Figure 17.5b) confirms the metallic character of the clusters, covalently bound to the DNA molecule, by identifying the (111)-lattice-planes distance with d = 2.27 Å corresponding to that of bulk platinum. A remarkable feature of Figure 17.5a is the completely clean background of the image, as could be repeatedly found in the metallization experiments. This provides clear experimental evidence that the presence of activated DNA stabilizes the formation of small clusters on the DNA, and prevents simultaneously both homogeneous nucleation and their subsequent agglomeration into larger particles, as is usually observed in the absence of ligands. This observation leads to the conclusion that DNA indeed plays an active role in the nucleation and growth process of Pt nanoparticles. However, the fact that in the described experiment less than 3 % of the solved complexes bind to the DNA during activation, raises the following question: Why does the presence of activated DNA gives rise to the observed complete suppression of homogenous nucleation although more than 97 % of the Pt(II) complexes are still in bulk solution when the chemical reduction is started? This seems only to be possible if the Pt-DNA adducts formed constitute highly efficient heterogeneous nucleation centers.

In order to obtain more evidence on this unique behavior, the reaction kinetics of platinum metallization was thoroughly investigated in combination with the study of the morphology of the reduction products by SFM which allows to image metal particles and DNA simultaneously, even when the cluster density is relatively low [21]. The reaction kinetics of the subsequent reduction step accelerates when the activation time is increased (Figure 17.6a). This corresponds to the expected behavior taking into account that the number of nucleation centers at the template, facilitated by the binding of Pt(II) complexes to the bases of DNA, increases with increasing activation time. At the same time, however, the morphology of the reaction products changes dramatically. Whereas large aggregates of homogeneously nucleated particles in coexistence with a few isolated clusters at the template are observed to form after short activation times (Figure 17.6b), exclusively heterogeneous nucleation takes place after long activation (Figure 17.6c). Again, continuous Pt cluster chains form, which can be considered as quasi one-dimensional nanocluster aggregates tailored by the DNA template, and homogeneous nucleation is suppressed. These observations suggest that the Pt-DNA adducts formed during the activation promote the heterogeneous nucleation and growth of Pt on the DNA template by locally enhancing the reaction rate, so that the homogeneous reaction channel becomes kinetically suppressed.

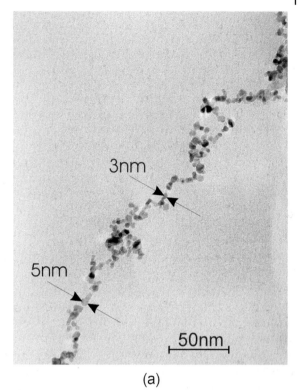

(a)

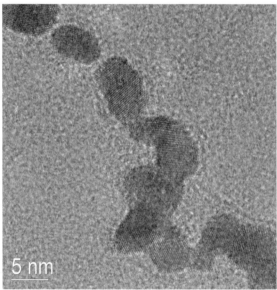

(b)

Figure 17.5 (a) A continuous chain of platinum clusters grown selectively on a single λ-DNA molecule and imaged by TEM [8]. (b) High-resolution TEM image of the platinum clusters grown on the template. The DNA is partly covered with clusters, which are grown together to form a short metal wire with a diameter of about 5 nm.

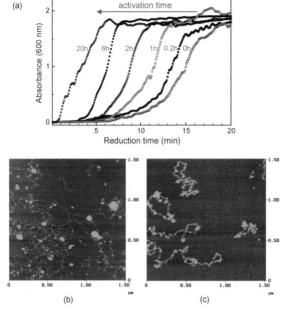

Figure 17.6 (a) Time evolution of the 600-nm absorbance of a hydrolyzed K_2PtCl_4 solution in the presence of DNA at different activation times. The development of absorbance indicates the formation of colloidal platinum and the aggregation of small particles into bigger structures during the reduction process [45]. (b) AFM image of the reaction products without activation [21]. (c) Reaction products after 20 hours of activation [21].

For a complete suppression of heterogeneous nucleation in solution, relatively long activation times are necessary. We assume this to have two reasons. First, the density of nucleation centers along DNA needs to be sufficiently high to be effective enough for a suppression of homogeneous nucleation. In connection with this, one must take into account that, for example, the time constant to achieve the saturation density is larger than 10 hours [40]. Second, bifunctional Pt(II) adducts to DNA must undergo further hydrolysis before they can be reduced by the addition of electrons. Hydrolysis is likewise known to have a time constant of several hours [38].

The hypothesis that the Pt clusters nucleate and grow directly at the activated DNA is further supported by the observation that the dependence of the reaction kinetics on the activation time is directly correlated with the content of G–C base pairs of DNA. Studying the reaction kinetics of activated DNA with different content of G–C base pairs, one would expect the following behavior. For zero activation time, the characteristic reaction time should be equal for all samples because binding of complexes did not yet occur. The same behavior is expected for very long activation times when complex binding to DNA becomes saturated accompanied by a loss in binding specificity [39]. However, for intermediate activation times, in the regime where the Pt complexes coordinate primarily to G bases [38], one expects a larger reaction rate for DNA with a higher G–C base plane content. This expected behavior was verified in the experiment (Figure 17.7).

These two experiments show that adjusting the functionality of the activated DNA allows entire control over the cluster nucleation process. First, changing the activation time permits control of heterogeneous nucleation at the DNA template versus homogeneous nucleation in solution until the point where the latter process is completely sup-

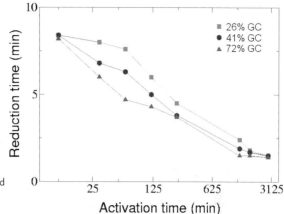

Figure 17.7 Influence of the GC content of the DNA on the metallization kinetics [31]. The characteristic reduction time, defined as the reaction time where the kinetics curve reaches half of its saturation value [8], is plotted via the DNA activation time in a semi-logarithmic plot.

pressed. This condition is particularly necessary for any severe fabrication procedure of thoroughly predefined inorganic nanostructures based on in-place metallization of biomolecules or assembled biomolecular structures. Second, a correlation between the DNA composition and the reaction kinetics of DNA metallization is revealed by varying the content of G–C versus A–T base pairs of the used DNA. This result suggests that a defined DNA template design may lead to a structure formation, where only predefined portions of the DNA strands become metalized, while the other parts of the template remain in their native form. Thus, it is conceivable that the metallization technique presented here will allow in future to design the electrical performance of DNA-based electronic devices already via the sequence of the DNA strands used to build up the employed template structure, which would demonstrate the full potential of biomimetic materials synthesis for bottom-up nanostructure fabrication.

Here, it should be noted that the described method of DNA metallization can – with slight modifications – also be successfully employed for other biomolecular templates, such as microtubules [9, 10] or regular bacterial surface layers [12].

17.3.2
First-Principle Molecular Dynamics Calculations of DNA Metallization

To gain a deeper insight into the molecular mechanism of cluster formation, and in particular to elucidate why DNA is capable of promoting the heterogeneous cluster nucleation, first-principle molecular dynamics (FPMD) calculations of the nucleation and growth of Pt clusters in solution [41, 42] and on biopolymers [8, 43, 44] have been carried out. Since the formation of transition metal clusters is an autocatalytic process, which is nucleation limited [45, 46], the very first step of cluster nucleation is the decisive event of cluster formation. Therefore, the FPMD simulations have been focused on the formation of Pt dimers. The subsequent growth of metal clusters in solution proceeds primarily through a surface-growth mechanism where metal complexes are adsorbed at the cluster surface and reduced in situ [42, 46]. This process is increasingly catalyzed with growing cluster size, leading to an autoaccelerating growth kinetics.

In the following, we will compare the formation of Pt dimers in bulk solution, which mainly models the homogeneous reaction channel, with the dimer formation, where one of the interacting Pt complexes is covalently bound either to one or to two G bases before the chemical reduction is started. The latter process models the heterogeneous cluster nucleation at the DNA template. Interestingly, in both cases we find that the Pt dimer formation in solution does not obey the classical nucleation picture, in which stable dimers are assumed to be formed by an aggregation of two Pt atoms in zerovalent oxidation state – a process which would require four electrons to form one dimer. In contrast, we will find that under mild reducing conditions stable dimers already form after a single reduction step, meaning that only one electron is needed instead of four. Thus, Pt dimer formation does not necessarily involve atoms in the zerovalent oxidation state. This result is for example in agreement with the observation of stable $[Pt_2Cl_4(CO)_2]^{2-}$ ions, which are monovalent Pt dimers with a Pt–Pt distance of 2.6 Å [47].

Dimer formation in bulk solution is investigated by simulating the reduction of two $PtCl_2(H_2O)_2$ complexes surrounded by randomly placed water molecules [41]. These complexes are the hydrolysis products which take an active part in the process of metal cluster formation [48]. The first unpaired electron, when added to the system, localizes at one of the complexes, and this leads to the formation of a linear $PtCl_2^-$ complex with a central Pt atom (see, e. g., Figure 17.8(a2)) caused by the detachment of both water ligands. Immediately after this, a stable Pt–Pt bond with a bonding distance of ~2.9 Å and a bonding energy of 1.5 eV is formed between the linear Pt(I) complex and the unreduced Pt(II) complex. Adding a second electron causes the loss of a chlorine ligand from the Pt(II) complex and a shortening of the Pt–Pt distance to 2.6 Å, which is a typical value for Pt(I) dimers [47, 49]. The strengthening of the Pt–Pt bond is accompanied with an increase of the bonding energy to a value of 1.8 eV.

The heterogeneous reaction channel is scrutinized by calculating the dimer formation of one $PtCl_2(H_2O)_2$ complex with a preformed Pt(II) · DNA adduct [8]. The adducts are modeled by hydrolyzed Pt complexes covalently bound either to one guanine or to two stacked guanines, taking the N7 positions of the G bases as coordination sites [37]. Now, in principle two different pathways are possible in the calculations depending on where the added electron goes: (pathway A) the bound Pt complex will be reduced, and then react with the unreduced free complex, or – vice versa – (pathway B) the unbound complex will be reduced and react with the unreduced Pt(II) · DNA adduct. The calculations show that the reduction of the bound complex is preferred in comparison to the reduction of the free complex in solution due to the presence of delocalized orbital states on the heterocyclic ligands which allow to accommodate the additional electron in a more favorable way. Nevertheless, both pathways have to be considered in the calculations. This is because the number of free complexes in solution is more than 30 times larger than the number of bound complexes at any experimental conditions (compare section 17.3.1). However, the observed pattern of events is similar for both pathways (see, e. g., Figure 17.8). Immediately after one of the complexes is reduced, a Pt–Pt bond is formed, which leads to the detachment of a water ligand from the Pt(II) complex accompanied by a considerable strengthening of the Pt–Pt bond. The obtained equilibrium distances between the Pt atoms are ~2.6 Å in all calculated cases, and the bond energies are – depending on the hydrolysis state of the formed dimers – between 1.8 and 2.6 eV. Higher bond energies were found for doubly hydrolyzed Pt(II) · DNA adduct as for single hydrolyzed adducts [43], indicating that further hydrolysis after binding the complexes to the DNA bases is important for the strength and the stability of the initially formed Pt dimers, and it might also explain why the activation time necessary to achieve pure heterogeneous nucleation of Pt cluster along the DNA template is so long [8] (compare section 17.3.1).

The results of the FPMD calculations clearly indicate that heterogeneously formed dimers are more stable than homogeneously formed ones. Dimers formed at DNA have considerably larger bonding energies. In addition, already after applying *one* electron, they do exhibit the same Pt–Pt-bond distance as homogeneously formed dimers take only on after adding *two* electrons to the system. The higher stability gained in the heterogeneous dimer formation process is caused by the observed detachment of a water ligand. The latter is only possible in the presence of heterocyclic ligands with strong electron donor character. The charge density accumulation on the Pt atoms induced by the presence of heterocyclic ligands leads to a highly repulsive antibonding interaction between the Pt(II) atom and one of its water ligands, which in turn causes a strengthening of

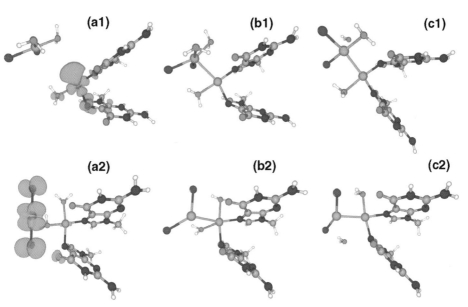

Figure 17.8 Snapshots of two FPMD simulations of the formation of platinum dimers bound to two stacked DNA bases [8], following the reaction paths A (top) and B (bottom) as described in the text. Pt = yellow; Cl = green; O = red; N = blue; C = gray; H = white. The orange iso-dense surface at 0.002 a. u. is associated with the unpaired orbital state of the reactant species reduced before dimer formation. The steric constraint exerted by the DNA molecule is modeled by binding the N9 atom of each guanine to a methyl group, which is kept fixed. Neither the complementary strand nor the backbone are expected to influence qualitatively the mechanism of dimer formation. This is justified by the observed selective affinity of Pt complexes for the bases of DNA [37]. The simulated reaction occurring in the major groove of DNA is expected not to be sterically hindered by the DNA structure. (a1) and (a2) are initial states after accommodation of the additional electron at one of the complexes; (b1) and (b2) are early states of the bond formation between the two Pt atoms; (c1) and (c2) are final states of dimer formation after the detachment of a water molecule.

the Pt–Pt bond when the water molecule detaches. Thus, the first-formed Pt–Pt bonds are mainly stabilized by a ligand-to-metal electron donation mechanism at DNA. Moreover, the calculations show that complexes bound to DNA are easier to reduce than complexes in solution, and that dimers formed at DNA exhibit a higher electron affinity [8, 43] than homogeneously formed ones, which favors further steps of reduction and complex addition to the Pt nucleus growing at DNA. Consequently, the initial Pt cluster nucleation at DNA should occur more easily than the corresponding process in solution, when Pt complexes are bound to DNA before chemical reduction is started.

This result is in full agreement with the experimental observations, where we obtained selective growth of Pt particles at activated DNA (see Figure 17.5). Preferred heterogeneous nucleation of Pt clusters at DNA combined with the autocatalytic growth behavior indeed allows conditions to be defined in which the first-formed heterogeneous nuclei can quickly develop into bigger clusters, thus "consuming" the feedstock of metal complexes from solution. As a result, the density of metal complexes in solution is substantially reduced. This leads in turn to a kinetic suppression of the homogeneous particle formation, as was experimentally observed. Therefore, the results of the combined experimental and theoretical investigations presented here strongly suggest that a novel property of the studied template molecule has been discovered: that DNA can act as very efficient molecular promoter of cluster nucleation. This specific property of DNA is one basic prerequisite for the achievement of a selective metallization of DNA. Recent calculations suggest that proteins containing heterocyclic amino acids (e. g., histidine) may exhibit similar properties as DNA [43, 44]. This emphasizes that organometallic complexes, formed before the reduction is started, play an important role for cluster nucleation, and can be used to control the nucleation and growth conditions of metals on biomolecular templates.

17.4
Conductivity Measurements on Metalized DNA Wires

As shown previously in Figure 17.5b, the thinnest metal wires yet fabricated by direct Pt growth on DNA have a diameter of ~5 nm, which is only three times the diameter of the template itself. Continuous wires are formed at the DNA template, where Pt clusters are located close enough to grow together. However, until now the wires with diameters below 10 nm are relatively short, and therefore not suited for two-terminal electron transport measurements where the contact pads are several microns apart. Mainly for this reason, all conductance measurements reported for metalized DNA are carried out on wires with diameters > 50 nm [6, 7, 13]. It is interesting to note that the Pd wires reported in Refs. [13a, 13b] exhibit a relatively coarse and irregular structure, although they are fabricated following basically the Pt metallization scheme described in section 17.3.1. The main reason for this behavior is the much faster reaction rate of the Pd cluster formation by chemical reduction compared to Pt. In distinction to the characteristic reduction time of Pt, which is in the order of several minutes (see Figure 17.6a), the time constant of Pd cluster formation is in the order of seconds. Therefore, it becomes difficult – even in the presence of DNA – to suppress kinetically the homogeneous cluster nucleation in a Pd solution [50]. As a result, homogeneously nucleated clusters are deposited on the template in an uncontrolled manner, giving rise to the observed irregular structures. Nevertheless, those Pd depositions led to several microns-long, continuous wires suited for two-terminal conductivity measurements.

The examined Pd nanowires showed pure ohmic electron transport behavior. Linear I–V curves were detected at all investigated temperatures, from 300 K down to 4.2 K. Neither nonconducting barriers nor Coulomb blockade behavior [51] were observed, giving clear evidence that the fabricated nanowires are indeed continuous. The specific electrical conductivity of the nanowires was found to be only one order of magnitude less than that of bulk palladium, indicating metallic transport behavior. The obtained value of 2×10^4 S cm^{-1} is in good agreement with a simple Drude-model-based estimation assuming grain boundary scattering with a mean free electron path of 2 nm, corresponding to the typical grain size of the metal deposition, as the dominant electron scattering mechanism. The metallic transport behavior is further confirmed by the observed linear resistance decrease with decreasing temperature (Figure 17.9a). As a characteristic feature, a resistance minimum at about 30 K was found; below that, the resistance increases again following a logarithmic temperature dependence (Figure 17.9b). This indicates the occurrence of quantum behavior in the low temperature conductivity caused by weak electron localization and/or electron–electron interactions [52]. Weak electron localization takes place in disordered systems at temperatures where the mean free path for the inelastic electron scattering be-

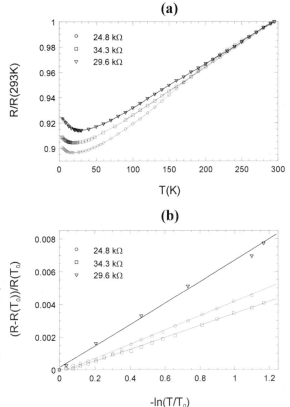

Figure 17.9 (a) Temperature dependence of the resistance of three palladium wires, normalized to their room temperature resistances of 24.8, 29.6, and 34.3 kΩ [13b]. Above 50 K the resistances decrease linearly with decreasing temperature. Below 30 K the resistance increases with decreasing temperature. (b) Low temperature resistance plotted as a function of $-\ln(T/T_0)$, with $T_0 = 13.5$ K.

comes larger than the mean free path of the elastic scattering. Under this condition, electrons can undergo impurity scattering without loosing their phase coherence. This results in an enhanced back-scattering, leading to an increase of resistance with decreasing temperature. Interestingly, the observed logarithmic temperature dependence does not correspond to the behavior expected for one-dimensional (1D) wires, but to that predicted and measured for 2D disordered films [53]. A similar behavior has been observed in wires with diameters below 100 nm, patterned from disordered Pd films [54]. 2D rather than 1D transport behavior is observed in these wires because their diameter is still too large. To perceive 1D transport behavior, the wire diameter should be considerably smaller than the characteristic localization length, which has been estimated to be less than 50 nm [55]. Annealing of the samples for 3 hours at 200 °C led to a decrease in the room temperature resistance, to an increase of the temperature coefficient, and made the low temperature quantum behavior disappear.

The overall electron transport behavior makes metalized DNA wires well suited for specific applications in DNA-based nanoelectronic networks.

17.5
Conclusions and Outlook

Today, knowledge concerning the structural and functional properties of individual biomolecules or biomolecular structures is expanding rapidly due to the enormous progress being made in the modern biological methods, which in turn increasingly allows the study of biological processes at the single-molecule level. Simultaneously, biological molecules and cellular machines are becoming increasingly attractive for materials and nanosciences. In particular, a strong tendency is currently observed to implement biological concepts – including interfacial molecular recognition, self-assembly, supramolecular preorganization, or structural evolution – into the synthesis of advanced materials. Thus, the application of biomolecules in an engineering context is increasingly studied. Here, we have examined generic aspects for the development of nanoelectronic circuitry when DNA is used as the biomolecular template. On the one hand, this example confirms that the discussed biomimetic materials-synthesis approach clearly has the potential of a controlled bottom-up fabrication of nanostructures with characteristic dimensions below 10 nm. On the other hand, this relatively new field of research has only just commenced, and will undoubtedly offer much greater possibilities in future than might be anticipated today.

The central issue of biomolecular templating is the use of well-determined, highly specific molecule properties, which have been developed and optimized during the long process of biological evolution, to control strictly the build-up of thoroughly predefined inorganic nanostructures. One of the main advantages of this technique is that biomolecules can simultaneously accomplish several functions, all of which are essential for the bottom-up assembly of nanostructures. For example, biomolecules provide affinity sites that are spatially determined, and also possess predefined chemical and physical properties which can be used for the selective nucleation and growth of inorganic phases at the molecule. In addition, the morphologies of the fabricated structures are determined by the geometric structure of the templates, which simultaneously act as scaffolds to stabilize the desired structures. Thus, the template molecule fulfills assignments within this novel approach to the formation of nanostructures, and which are very similar to the specific functions of proteins involved in biomineralization processes. Moreover, the finding that a single particular property of the biotemplate could be used multifunctionally within this approach seems to offer another clear advantage. As shown here, for example, sequence-specific binding of DNA has been exploited to obtain: (i) site-specific integration of DNA into microelectronic contact arrays; (ii) controlled tailoring of branched templates; and (iii) tuning of the properties of DNA as a metallization template. The latter has been demonstrated by the observed strong influence of the GC content of the DNA sequence on the kinetics of DNA metallization.

The fabrication of cluster chains and nanowires by chemical metal deposition at DNA is a generic example of a template-directed formation of engineered nanostructures. Selective metallization has been obtained because heterogeneous cluster nucleation at activated DNA is preferred over homogeneous cluster nucleation in bulk solution. When combined with the autocatalytic growth behavior of transition metal clusters, this leads – under appropriate conditions – to complete suppression of the unwanted homogeneous reaction channel. As a result, clean in-place metallization of the DNA molecules is obtained. Preferred heterogeneous nucleation at activated DNA is a consequence of the presence of heterocyclic ligands with strong electron donor characteristics, and this both enhances the electron affinity of the metal nuclei and induces the formation of metal–metal bonds which are stronger than those obtained in solution. Because the amino acid histidine exhibits the same behavior, the metallization method described is likewise expected to be successfully employable for protein templates.

Regardless of the recent major progress that has been made in the development of DNA-based nanoelectronics, many fundamental problems remain unsolved today. Indeed, some such issues will have to be addressed in near future, and include:

- How can a template design be developed which will allow a degree of complexity and reliability similar to that of current microelectronics?
- Is it possible to design stable three-dimensional DNA structures for nanoelectronic applications?
- Can the manifold of template-design tools be enlarged by the use of proteins which specifically bind to DNA proteins?
- Can chemical linkers [56] be used to build branched DNA elements?
- Is it possible to exchange DNA by chemically synthesized structures with similar properties in future?
- Can methods be developed which allow the integration of highly complex DNA templates into semiconducting microelectronic structures?
- Can highly integrated DNA templates be positioned in parallel? Can molecular motors [57] be used for such template build-up?
- Can single DNA or DNA networks be used to position other molecular building blocks such as carbon nanotubes [58] or colloid particles [24, 25] into microelectronic circuits?
- How can a low-resistance interfacing of metalized DNA to metallic or semiconducting contact materials be achieved?
- Do sequence-specific molecular lithography methods, as very recently developed [7], allow the fabrication of functional electrical devices such as transistors?

17.6
Methods

17.6.1
Site-Specific DNA Attachment

In all binding experiments, λ-phage DNA (New England Biolabs) was used which has a length of ~16 μm. The DNA integration into microscopic contact arrays was accomplished in a home-made flow cell mounted on the table of an inverted optical microscope. Glass samples with patterned Au contact pads were used as the substrates. A DNA solution was injected into the cell via glass capillaries, which were connected to a micromanipulator to control the flow direction. The DNA molecules were visualized in a fluorescence microscope after labeling with YOYO1 (Molecular Probes). For the anchoring of the DNA ends at patterned gold contacts, three different methods have been applied: (i) hybridization of the sticky ends of DNA at Au terminated with complementary thiol oligomers; (ii) specific binding via the formation of a Au–biotin–streptavidin–biotin–DNA bridge; and (iii) end-specific electrostatic binding between DNA and amino-thiol groups immobilized on the Au contacts. The basic contact functionalization and binding protocols are described in Refs. [6], [33], and [34], respectively.

17.6.2
DNA Junctions

Junctions with three arms were fabricated in a two-step procedure. First, a central DNA tripod serving as nodal point of the junction was assembled by hybridization of three, 37 bases-long, partially complementary, single-stranded DNA oligomers (MWG Biotech AG, Germany). In the second step, the three arms of the central tripod were elongated by ligation of 540 bp-long DNA fragments having sticky ends complementary to the three GCT overhangs at the 5′-ends of the tripod. The formation of basic tripods is accomplished by heating the oligomer solution for 5 min to 90 °C and cooling it subsequently to 20 °C at a rate of –0.1 K min^{-1}. The tripods were then phosphorylated using T4 polynucleotide kinase, and the arms ligated to the central tripods for 16 hours at 16 °C using T4 ligase (both from New England Biolabs). The reaction products were verified and separated according to their mass by gel electrophoresis. For SFM imaging of the junctions, the corresponding bands were eluted from the gel and subsequently adsorbed onto freshly cleaved mica applying HEPES buffer + 5 mM MgCl$_2$ at pH 7.6. SFM imaging was performed using a NanoScope IIIa (Digital Instruments, Santa Barbara, CA).

17.6.3
DNA Metallization

In the metallization experiments, the DNA template was first 'activated' by incubation with dissolved [PtCl$_4$]$^{2-}$ ions. The deposition of metallic platinum was then induced by chemical reduction of the whole solution. The reduction kinetics was monitored by ultra-violet–visible (UV-VIS) spectroscopy, and the products were imaged by SFM and/or TEM. For the TEM investigations, a 5 µg mL^{-1} solution of λ-DNA was incubated for ~20 hours with a 1 mM aged solution of K$_2$PtCl$_4$ (Fluka, Switzerland). The D/N ratio was maintained at 65:1. Metallization of the DNA molecules was accomplished at temperatures between 27 °C and 37 °C by addition of a 10 mM solution of DMAB (Fluka) in stoichiometric excess with respect to the amount of Pt(II) complexes. For the combined UV-VIS and SFM studies, a 20 µg mL^{-1} solution of DNA from salmon testes (Sigma-Aldrich, Germany) was used instead of λ-DNA. In the experiments at variable DNA composition, DNA was used from *Clostridium perfringens* (containing 26.5 % of GC base pairs), salmon testes (41.2 % GC), and *Micrococcus luteus* (72 % GC) (all purchased from Sigma-Aldrich).

Acknowledgments

The authors thank Lucio Colombi Ciacchi, Alexander Huhle, Ralf Seidel, Alessandro De Vita, Reiner Wahl, and Michael Weigel for fruitful collaboration and intensive co-working. Helpful discussions with all members of the BioNanotechnology and Structure Formation Group of the Max Bergmann Center of Biomaterials, Dresden, are kindly acknowledged. They also thank Remo Kirsch, Jan Richter, and Hans K. Schackert for their participation in the early metallization experiments, and Peter Seibel for many discussions about junction formation. Ingolf Mönch and Hartmut Vinzelberg are thanked for their help in the conductance measurements, and Uri Sivan and Erez Braun for stimulating discussions about the experiments of DNA anchoring. Financial support from the DFG, the BMBF and the SMWK is also acknowledged. All computational resources were provided by the Center for High-Performance Computing of the Technical University Dresden.

References

[1] J.-M. Lene, *Supramolecular Chemistry: Concepts and Perspectives*, VHC, Weinheim, 1995, p. 204.

[2] E. Schrödinger, *What is life?*, Cambridge University Press, Cambridge, 1967.

[3] S. Mann, *Biomineralization: principles and concepts in bioinorganic materials chemistry*, Oxford University Press, Oxford, 2001.

[4] R. Elghanian, J. J. Storhoff, R. C. Mucic, R. L. Letsinger, C. A. Mirkin, *Science* 1997, 277, 1078–1081.

[5] A. P. Alivisatos, K. P. Johnsson, X. G. Peng, T. E. Wilson, C. J. Loweth, M. P. Bruchez, P. G. Schultz, *Nature* 1996, 382, 609–611.

[6] E. Braun, Y. Eichen, U. Sivan, G. Ben-Yoseph, *Nature* 1998, 391, 775–778.

[7] K. Keren, M. Krueger, R. Gilad, G. Ben-Yoseph, U. Sivan, E. Braun, *Science* 2002, 297, 72–75.

[8] M. Mertig, L. Colombi Ciacchi, R. Seidel, W. Pompe, A. De Vita, *Nano Lett.* 2002, 2, 841–844.

[9] a) R. Kirsch, M. Mertig, W. Pompe, R. Wahl, G. Sadowski, K. J. Boehm, E. Unger, *Thin Solid Films* 1997, 305, 48–253; b) M. Mertig, R. Kirsch, W. Pompe, *Appl. Phys. A* 1998, 66, S723–S727; c) W. Fritzsche, J. M. Köhler, K. J. Boehm, E. Unger, T. Wagner, R. Kirsch, M. Mertig, W. Pompe, *Nanotechnology* 1999, 10, 331–335.

[10] S. Behrens, K. Rahn, W. Habicht, K.-J. Boehm, H. Rösner, E. Dinjus, E. Unger, *Adv. Mater.* 2002, 14, 1621–1625.

[11] a) K. Douglas, N. O. Clark, K. J. Rothschild, *Appl. Phys. Lett.* 1990, 56, 692–694; W. Shenton, D. Pum, U. B. Sleytr, S. Mann, *Nature* 1997, 389, 585–587; S. R. Hall,

W. Shenton, H. Engelhardt, and S. Mann, *ChemPhysChem.* **2001**, *3*, 184–186.

[12] a) M. Mertig, R. Kirsch, W. Pompe, H. Engelhardt, *Eur. Phys. J. D* **1999**, *9*, 45–48; b) M. Mertig, R. Wahl, M. Lehmann, P. Simon, W. Pompe, *Eur. Phys. J. D* **2001**, *16*, 317–320.

[13] a) J. Richter, M. Mertig, W. Pompe, I. Mönch, H. K. Schackert, *Appl. Phys. Lett.* **2001**, *78*, 536–538; b) J. Richter, M. Mertig, W. Pompe, H. Vinzelberg, *Appl. Phys. A* **2002**, *74*, 725–728.

[14] M. H. Caruthers, *Science* **1985**, *230*, 281–285.

[15] a) Y. Wang, J. E. Mueller, B. Kemper, N. C. Seeman, *Biochemistry* **1991**, *30*, 5667–5674; b) N. C. Seeman, *Annu. Rev. Biophys. Biomol. Struct.* **1998**, *27*, 225–248; c) N. C. Seeman, *Trends Biotechnol.* **1999**, *17*, 437–433; d) C. Mao, W. Sun, N. C. Seeman, *J. Am. Chem. Soc.* **1999**, *121*, 5437–5444; e) N. C. Seeman, *Nano Lett.* **2001**, *1*, 22–26.

[16] R. K. Saiki, S. Scharf, F. Faloona, K. B. Mullis, G. T. Horn, H. A. Erlich, N. Arnheim, *Science* **1985**, *230*, 1350–1354.

[17] H. G. Hansma, R. L. Sinsheimer, M. Q. Li, and P. K. Hansma, *Nucleic Acids Res.* **1992**, *20*, 3585–3590.

[18] S. B. Smith, Y. Cui, C. Bustamante, *Science* **1996**, *271*, 795–799.

[19] T. G. Spyro (ed.), *Nucleic Acid – Metal Ion Interactions*, Wiley, New York, **1980**.

[20] J. Richter, R. Seidel, R. Kirsch, M. Mertig, W. Pompe, J. Plaschke, H. K. Schackert, *Adv. Mater.* **2000**, *12*, 507–510.

[21] R. Seidel, M. Mertig, W. Pompe, *Surface and Interface Analysis* **2002**, *33*, 151–154.

[22] J. L. Coffer, S. R. Bogham, X. Li, R. F. Pinizzotto, Y. G. Rho, R. M. Pirtle, I. L. Pirtle, *Appl. Phys. Lett.* **1996**, *69*, 3851–3583.

[23] T. Torimoto, M. Yamashita, S. Kuwabata, T. Sakata, H. Mori, H. J. Yoneyama, *Phys. Chem. B* **1999**, *103*, 8799–8803.

[24] a) C. A. Mirkin, R. L. Letsinger, R. C. Mucic, J. J. Storhoff, *Nature* **1996**, *382*, 607–609; b) R. Jin, G. Wu, Z. Li, C. A. Mirkin, G. C. Schatz, *J. Am. Chem. Soc.* **2003**, *125*, 1643–54.

[25] D. Zanchet, C. M. Micheel, W. J. Parak, D. Gerion, A. P. Alivisatos, *Nano Lett.* **2001**, *1*, 32–35.

[26] D. Bensimon, A. J. Simon, V. Croquette, A. Bensimon, *Phys. Rev. Lett.* **1995**, *74*, 4754–4757.

[27] H. Nakao, H. Gad, S. Sugiyama, K. Otobe, T. Ohtani, *J. Am. Chem. Soc.* **2003**, *125*, 7162–7163.

[28] a) M. Rief, H. Clausen-Schaumann, H. E. Gaub, *Nature Struct. Biol.* **1999**, *6*, 346–349; b) J. Hu, Y. Zhang, H. Gao, M. Li, U. Hartmann, *Nano Lett.* **2002**, *2*, 55–57.

[29] Y. Arai, R. Yasuda, K. Akashi, Y. Harada, H. Miyata, K. Kinosita Jr., H. Itoh, *Nature* **1999**, *399*, 446–448.

[30] a) S. B. Smith, L. Finzi, C. Bustamante, *Science* **1992**, *258*, 1122–1126: b) C. Gosse, V. Croquette, *Biophys. J.* **2002**, *82*, 3314–3329.

[31] M. Mertig, R. Seidel, L. Colombi Ciacchi, W. Pompe, *AIP Conference Proceedings* **2002**, *633*, 449–453.

[32] M. Chee, R. Yang, E. Hubbell, A. Berno, X. C. Huang, D. Stern, J. Winkler, D. Lockhart, M. S. Moris, S. P. A. Fodor, *Science* **1996**, *274*, 610–614.

[33] R. Zimmermann, E. C. Cox, *Nucleic Acids Res.* **1994**, *22*, 492–497.

[34] J.-F. Allemand, D. Bensimon, L. Jullien, A. Bensimon, V. Croquette, *Biophys. J.* **1997**, *73*, 2064–2070.

[35] A. Huhle, R. Seidel, W. Pompe, M. Mertig, unpublished results.

[36] A. J. Storm, S. J. T. van Noort, S. de Vries, C. Dekker, *Appl. Phys. Lett.* **2001**, *79*, 3881–3883.

[37] B. Lippert, *Cisplatin: chemistry and biochemistry of a leading anticancer drug*, VCH, Weinheim, **1999**.

[38] D. P. Bancroft, C. A. Lepre, S. J. Lippard, *J. Am. Chem. Soc.* **1990**, *112*, 6860–6868.

[39] a) J.-P. Macquet, T. Theophanides, *Biopolymers* **1975**, *14*, 781–799; b) J.-P. Macquet, T. Theophanides, *Inorg. Chim. Acta* **1976**, *18*, 189–194; c) J.-P. Macquet, J.-L. Butour, *Eur. J. Biochem.* **1978**, *83*, 375–387.

[40] E. T. Sacharenko, Yu. S. Moschkowski, *Biophysica* **1972**, *7*, 373–378 (in Russian).

[41] L. Colombi Ciacchi, W. Pompe A. De Vita, *J. Am. Chem. Soc.* **2001**, *123*, 7371–7380.

[42] L. Colombi Ciacchi, W. Pompe, A. De Vita, *J. Phys. Chem. B* **2003**, *107*, 1755–1764.

[43] L. Colombi Ciacchi, M. Mertig, R. Seidel, W. Pompe, A. De Vita, *Nanotechnology* **2003**, *14*, 840–848.

[44] L. Colombi Ciacchi, M. Mertig, W. Pompe, S. Meriani, A. De Vita, *Platinum Metals Review* **2003**, *47*, 98–107.

[45] A. Henglein, B. G. Ershov, M. Malow, *J. Phys. Chem.* **1995**, *99*, 14129–14136.

[46] M. A. Watzky, R. G. Finke, *J. Am. Chem. Soc.* **1997**, *119*, 10382–10400.

[47] P. L. Gogging, R. J. Goodfellow, *J. Chem. Soc. Dalton Trans.* **1973**, 2355.

[48] A. Henglein, M. Giersing, *J. Phys. Chem. B* **2000**, *104*, 6767–6772.

[49] T. E. Müller, F. Ingold, S. Menzer, D. M. P. Mingos, D. J. Wiiliams, *J. Organomet. Chem.* **1997**, *528*, 163–178.

[50] R. Seidel, L. Colombi-Ciacchi, M. Weigel, W. Pompe, M. Mertig, unpublished results.

[51] H. Grabert, M. H. Devoret, *Single Charge Tunneling: Coulomb Blockade Phenomena in Nanostructures*, Kluwer Academic Publishers, New York, **1992**.

[52] G. Bergmann, *Phys. Rep.* **1984**, *107*, 1–58.

[53] a) M. J. Burns, W. C. McGinnis, R. W. Simon, G. Deutscher, P. M. Chaikin, *Phys. Rev. Lett.* **1981**, *47*, 1620–1624; b) D. J. Bishop, D. C. Tsui, R. C. Dynes, *Phys. Rev. Lett.* **1980**, *44*, 1153–1156.

[54] G. J. Dolan, D. D. Osheroff, *Phys. Rev. Lett.* **1979**, *43*, 721–724.

[55] W. C. McGinnis, P. M. Chaikin, *Phys. Rev. B* **1985**, *32*, 6319–6330.

[56] M. Scheffler, A. Dorenbeck, S. Jordan, M. Wüstefeld, G. von Kiedrowski, *Angew. Chem. Int. Ed.* **1999**, *38*, 3311–3315.

[57] S. Diez, C. Reuther, R. Seidel, M. Mertig, W. Pompe, J. Howard, *Nano Letters* **2003** *3*, 1251–1254.

[58] K. A. Williams, P. T. M. Veenhuizen, B. G. de la Torre, R. Eritja, C. Dekker, *Nature* **2002**, *420*, 761.

18
Mineralization in Nanostructured Biocompartments: Biomimetic Ferritins For High-Density Data Storage

Eric L. Mayes and *Stephen Mann*

18.1
Overview

Transcription of the supramolecular structure of self-assembled biomolecules into new types of organized inorganic matter has resulted in a proliferation of novel materials, often with biomimetic form and complexity [1, 2]. While a variety of methods for the production of nanostructured materials exist (such as vapor-phase synthesis), biomolecular-mediated synthesis can offer precise control over composition and morphology. Biomolecules offer a variety of morphologies and many are robust enough to support reactions to produce metals, metal oxides, metal sulfides, and silicates. Further, unit uniformity enables small size distributions and often leads to self-organization of superlattice structures. For example, ordered two-dimensional (2D) arrays of multisubunit proteins, such as S-layer proteins [3, 4] and chaperonins [5], have been exploited for nanoparticle superlattice assembly.

Organized arrays of nanoparticles have also been prepared on filamentous biotemplates such as tubulin [6] or self-assembled chiral phospholipids [7]. At the mesoscale level, liquid crystalline phases consisting of oriented arrays of rod-shaped virus particles [8], or multicellular bacterial filaments [9, 10] have been used in the template-directed synthesis of meso- and macro-porous silica monoliths, respectively.

One potential template for the production of meso-porous silica is the rod-shaped tobacco mosaic virus (TMV), which is 300 nm long, 18 nm wide with a 4 nm-diameter inner channel. TMV liquid crystals have been used to template amorphous silica by hydrolyzing and condensing mixtures of tetraethoxysilane (TEOS) and aminopropyltriethoxysilane (APTES) [8]. Whereas mesoporous replicas of the biological liquid crystal were produced under standard conditions, nanoparticulate meso-porous silica spheres with porous channels radiating from their core were formed in the presence of low levels of the silica precursor. The ends of the TMV–silica complexes were bound together during nucleation, followed by radial silica growth such that bioinorganic nanoparticles were produced as the viral rods sheared during the processing. The nanoparticles remained intact upon calcination, which removed the remaining viral proteins to leave a radially oriented mesoporous interior.

Nanobiotechnology. Edited by Christof Niemeyer, Chad Mirkin
Copyright © 2004 WILEY-VCH Verlag GmbH & Co. KgaA, Weinheim
ISBN 3-527-30658-7

Other studies have focused on the use of elongated biomolecular structures, such as bacterial threads [11], DNA filaments [12, 13], rhapidosomes [14], cylindrical viroids [15], and lipid microtubules [16] to prepare bioinorganic materials with high shape anisotropy. Recently, helicoid ribbons and tubes of the synthetic phospholipid 1, 2-bis(10, 12-tricosadiyonyl)-*sn*-glycero-3-phosphatidylcholine ($DC_{8,9}PC$) have received much attention, as their high aspect ratio coupled with the smaller size enabled their utility in the pharmaceutical and electronics industries [17, 18]. In addition, the transcription of the helicoid form into inorganic structures consisting of silica or metal microtubules [19, 20] or chiral silica–lipid lamellar mesophases [21] has been undertaken.

Nanoscopic cages of ferritin and ferritin-like proteins [22–28], enzymes [29], and capsids [30–32] have been used for the synthesis of bioencapsulated quantum-sized inorganic particles. Researchers have used spherical cowpea chlorotic mottle viruses (CCMV) as nanoscale reactors. CCMV virions are 28 nm in diameter, with an 18 nm inner cavity. In their native state, the inner cavity is used to carry the RNA that biochemically describes the structural components of the virus. However, the virion can undergo reversible swelling at high pH [30] such that at pH values >6.5 the native RNA can be removed. Following the addition of aqueous molecular tungstate (WO_4) species, nanoparticles of paratungstate ($H_2W_{12}O_4$) can be precipitated within the virion by lowering the pH, which also seals the protein cage [30, 31]. This technique may prove useful for aqueous nanoparticle reactions that occur above pH 6.5, especially as a wide variety of virus capsids exist.

In this chapter, we focus specifically on the iron storage protein, ferritin, and how recent fundamental research on biomimetic ferritins has led to an innovative approach to high-density nanoparticle-based magnetic storage devices with promising commercial viability as developed by NanoMagnetics Ltd., UK.

18.2
Biomimetic Ferritins

Ferritin is a self-assembled, 12 nm-diameter multi-subunit protein involved in biological functions such as iron storage and heme production. The protein is constructed from 24 nearly identical subunits, which self-assemble to form a spherical shell that encloses an 8.0 nm-diameter internal cage [33] (Figure 18.1). The cavity contains up to 4500 Fe atoms in the form of a poorly crystalline iron (III) oxy-hydroxide mineral, ferrihydrite. The size of the ferrihydrite core is constrained by the 8 nm inner cavity of the protein, suggesting that the protein cage could be exploited for the production of alternative mineral nanoparticles. In this regard, the iron oxide core can be readily removed by reductive dissolution to produce an intact demetalized protein referred to as apoferritin [34]. Apoferritin is robust, being able to withstand relatively high temperatures systems (65 °C) and wide pH variations (approximately 4.0–9.0) for limited periods of time without significant disruption to its quaternary structure.

Together, these features provide a generic reaction vessel that can be used for the synthesis and confinement of a variety of protein-encapsulated inorganic nanoparticles. For example, biomimetic ferritins have been prepared by reductive dissolution of the iron oxide cores followed by reconstitution of the empty cage of apoferritin with manganese oxide [22, 23], uranyl oxide nanoparticles [22], cadmium sulfide quantum dots [26], as well as

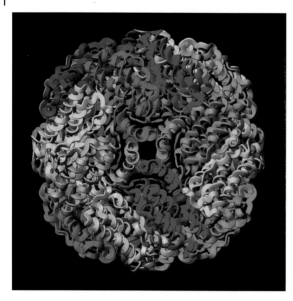

Figure 18.1 Structural representation of the 12 nm-diameter ferritin macromolecule from high-resolution studies [36, 37], prepared using MolMol [38] and viewed along the 4-fold axis. The 24 individual protein subunits are colored for differentiation.

the mixed-valence iron oxide, magnetite (Fe_3O_4), which significantly increases the associated magnetic moment compared with the native protein [24, 27]. Alternatively, the native iron oxide cores of ferritin can be transformed in situ to protein-encapsulated FeS nanoparticles [25]. Here, we focus on a recent breakthrough involving the synthesis of the magnetic alloy cobalt platinum (CoPt) within apoferritin and demonstrate that this advanced functionality can support specific applications such as high-density data storage [35].

18.3
High-Density Magnetic Data Storage

In hard disk drives, data are stored as rectangular magnetized regions or 'bits' on a thin metal film supported by a glass or aluminum disk substrate. Commercially available disk drives currently store data at a maximum areal density of around 11 Gbits cm^{-2} (i. e., billions of bits per cm^2). With improvements in the recording film and electronics, densities of 23 Gbits cm^{-2} have been demonstrated in the laboratory [39]. At 11 Gbits cm^{-2}, bits are roughly 30 nm wide by 300 nm long, and are composed from hundreds of nano-scaled magnetic alloy grains oriented longitudinally in the plane of a disk. These grains are produced through a process of sputtering that results in a polydisperse grain size distribution. As bit size reduces with increasing density, it becomes critical to have a narrower grain size distribution, particularly when bit dimensions approach the average size of the grains (currently around 9 nm in diameter). Reducing the average grain size is initially beneficial, but encounters a restraint called the "superparamagnetic limit" – below a certain volume, a grain cannot maintain a preferred magnetization at room temperature. The maximum density possible for this longitudinally-oriented recording convention is anticipated to be in the region of 30 to 80 Gbits cm^{-2}.

While sputtered thin films may have an average grain size larger than the superparamagnetic limit, their inherent size distribution leaves a percentage which is thermally unstable. Recent multilayer thin films that exploit antiferromagnetic coupling help to stabilize smaller grains, and may extend longitudinal recording to 80 Gbits cm^{-2}. However, new materials with higher magnetocrystalline anisotropy are required for continually smaller, thermally stable grains. Alternative recording conventions such as the perpendicular magnetic orientation of grains have been proposed to extend areal densities up to 155 Gbit cm^{-2} [40], but perpendicular recording is also limited by disordered, polydisperse grains.

The most likely candidates to extend areal densities significantly beyond 30 Gbit cm^{-2} are thin films comprising monodisperse high anisotropy nanoparticles. Chemically synthesized magnetic nanoparticles exhibit extremely narrow size distributions that also encourage self-organized patterning, and could potentially support bit-per-particle densities from 1550 to 7750 Gbit cm^{-2} [41].

Equiatomically alloyed $L1_0$ phase CoPt and FePt are being considered for ultrahigh-density recording films because of their high magnetocrystalline anisotropy that provides thermal stability for grains 3–4 nm in diameter [42, 43]. Researchers have used organic stabilizers in nonaqueous solutions to synthesize arrays of monodisperse, 4 nm diameter nanoparticles that form $L1_0$ FePt upon annealing [44, 45]. The resulting nanoparticle array films have supported areal densities corresponding to 0.2 Gbit cm^{-2} [46], but are highly vulnerable to sintering during the annealing process [47]. Sintering of the nanoparticles not only increases the average grain size and widens the size distribution, but it also destroys any self-organized patterning.

Recently, Mayes and co-workers have attempted to address the above considerations through a novel biomimetic approach to the synthesis and fabrication of magnetic nanoparticle-based ultrahigh-density films in which monodisperse precursors of the $L1_0$ phase of CoPt were prepared within the supramolecular cage of ferritin [35]. Significantly, ferritin strictly regulates the diameter of the grains synthesized and moreover as a consequence of its external uniformity, it decouples the formation of highly regular self-organized patterned films from perturbations in the particle size distribution associated with the encapsulated nanoparticles. Most importantly, the 2 nm-thick protein coating discourages sintering of nanoparticle grains at the temperatures required for the transformation to the $L1_0$ phase [48]. The initial results are extremely promising, with thin films showing a moderate increase in areal density at 0.34 Gbit cm^{-2} [49]. On sufficiently smooth substrates, ferritin is known to produce self-organized, large-area hexagonally close-packed (HCP) arrays [50], indicating that an array of biomimetic ferritin CoPt nanoparticles could ultimately support bit-per-particle recording up to 1550 Gbit cm^{-2} – a goal currently being pursued by NanoMagnetics Ltd., UK.

In the next section, we briefly describe the experimental protocol for the ferritin-controlled synthesis of CoPt nanoparticles, followed by a summary of recent results.

18.4
Methods

Dispersions of protein-encapsulated CoPt nanoparticles can be prepared by sequestration of metal ions and complexes within the cavity of apoferritin, followed by chemical reduction. Apoferritin was prepared by demineralizing native ferritin (lyophilized horse spleen ferritin; New Zealand Pharmaceuticals) using standard procedures of reductive dissolution (Figure 18.2a) [34]. Typically, 250 mL apoferritin (0.5 mg mL^{-1}) were buffered to pH 8.0 in deaerated HEPES (N-2-hydroxyethylpiperazine-N'-2-ethanesulfonic acid, 0.025 M; 99%, Sigma), and the solution was stirred and maintained at a constant temperature of 45 °C using a water bath. Ammonium tetrachloroplatinate solutions (0.1 M; 99.9%, AlfaAesar) and cobalt acetate tetrahydrate (98+%, Aldrich) were added slowly to the buffered and stirred apoferritin in 1.8:1 Co:Pt ratios (0.18 mL Co; 0.10 mL Pt). After 30 minutes, a stoichiometric amount of sodium borohydride (0.01 M; Aldrich) was added over a period of 1 minute to reduce the metal ion precursors. This process was repeated multiple times to increase the size of the encapsulated metallic nanoparticles until a high yield of fully loaded protein molecules was obtained (Figure 18.2b).

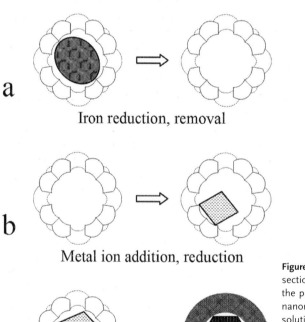

a

Iron reduction, removal

b

Metal ion addition, reduction

c

Annealing

Figure 18.2 Schematic showing cross-section of a ferritin molecule and use in the production of biomimetic ferritin nanomagnets. (a) Reduction and dissolution of native ferrihydrite core to form apoferritin; (b) reconstitution of apoferritin with Co(II) and [PtCl$_4$]$^{2-}$, and chemical reduction to form a metal alloy core; (c) annealing of CoPt core to produce the $L1_0$ CoPt phase encased in a carbonized matrix.

The resulting dispersion was washed with water using a filtration unit to remove buffer and salts, and then immediately used for thin film production.

Films of CoPt nanoparticles were prepared on washed glass disk substrates (65 mm diameter; O'Hara). While these substrates were not smooth enough to support ultrahigh-quality self-organized arrays typically achieved using thermally oxidized Si substrates [44, 45], an ink-jetted dispersion adhered strongly to the glass as a smooth and dense layer. The films were then annealed at 500–650 °C for 60 minutes with a 19 kPa partial pressure of H_2 to form the ferromagnetic $L1_0$ phase (Figure 18.2c).

18.5
Results

Transmission electron microscopy (TEM) studies indicated that borohydride reduction of Co(II) and $[PtCl_4]^{2-}$ resulted in discrete electron dense CoPt nanoparticles that were specifically located within the 8 nm-diameter cavity of apoferritin. Prior to annealing, both electron and X-ray diffraction confirmed that these precursor nanoparticles were crystalline with a face-centered-cubic (fcc) CoPt structure. This was consistent with the corresponding superparamagnetic behavior shown in room temperature magnetometer analysis. Significantly, annealing of nanoparticle-containing films above 500 °C for 60 minutes in the presence of H_2 resulted in a phase change. Diffraction studies of the resulting films showed a transition to the ferromagnetically ordered $L1_0$ CoPt structure. The films exhibited a magnetic coercivity (H_c) of 9500 Oe consistent with the $L1_0$ phase, but the remnant to saturation magnetization ratio (M_r/M_s) was 0.72 both perpendicular and parallel to the plane of the film, This value is unexpected for a putative isotropic distribution of crystalline axes, and suggests that exchange interactions occur within the film. Indeed, exchange interactions have recently been observed in organic stabilizer-derived FePt nanoparticle films by monitoring δM curves at increasing annealing temperatures [47]. As the protein coating in the biomimetic CoPt-containing ferritin is both thick (~2 nm) and dense, even a moderate reduction in thickness due to the formation of a carbonized shell during annealing should not lead to coupled exchange interactions. However, TEM images of dispersions annealed at 550 °C showed that although the majority of the CoPt nanoparticles remained spatially separated (Figure 18.3, region 'a'), some sintering occurred, probably associated with indistinct material formed external to the protein (Figure 18.3, region 'b').

Resolving the issue of sintering is essential if self-organized, patterned thin films are to be viable at the densities proposed. Current studies indicate that sintering can be addressed by more thoroughly cleaning of the dispersions to remove material not encapsulated within protein. With recent refinements, films consisting of ferritin-derived CoPt nanoparticles have been produced that can support stable magnetization reversal transitions at densities up to 1.24 Gbit cm^{-2} (unpublished data). Using densely-packed, but disordered films which were ~100 nm thick, a contact drag tester with a 0.384 μm-wide magneto-resistive head recorded transitions corresponding to linear densities of 1190, 2381, 3571, and 4760 flux changes per millimeter (fc mm^{-1}) (Figure 18.4, left to right). These linear densities correspond to 0.31, 0.62, 0.93, and 1.24 Gbit cm^{-2}, respectively.

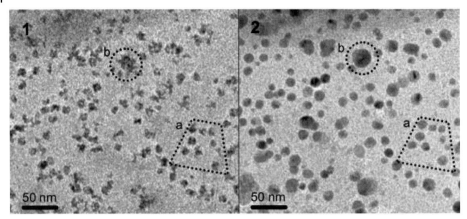

Figure 18.3 TEM images of a dispersion of biomimetic ferritin CoPt nanoparticles on a Si₃N₄ window prior to (1) and after (2) annealing at 550 °C for 60 minutes in H₂. Particles that are obviously separated by a protein coating prior to annealing remain discrete afterwards (a), but indistinct material exhibits sintering (b).

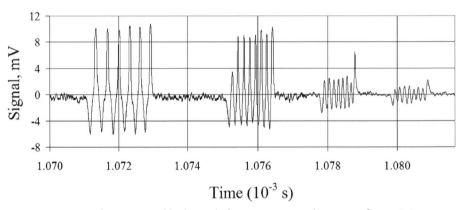

Figure 18.4 Contact drag tester read-back signals from a continuous, biomimetic ferritin CoPt nanoparticle film. Dc-erased regions separate magnetization reversal transitions at linear densities corresponding to 0.31, 0.62, 0.93, and 1.24 Gbit cm⁻², respectively (left to right).

The magnetic ferritin films currently prepared do not exhibit any preferential magnetization direction, and could therefore gain signal amplitude through magnetic orientation. Further, the films were not mechanically polished like conventional thin films, such that asperities occasionally degraded the recorded transitions due to spacing loss. Many refinements are therefore required to extend recording densities from 1.24 Gbit cm⁻² to the commercially relevant 30 Gbit cm⁻² and beyond, but these are clearly within the scope of current technological methodologies.

**18.6
Outlook**

In this chapter, we have described a novel biomimetic route to CoPt nanoparticles and films with promising potential in ultrahigh-density data storage. Several milestones have already been surpassed, and with more stringent regulation over the composition, cleanliness and uniformity of the ferritin-derived nanomagnets, as well as control over their magnetic orientation, competitive materials are likely to reach the market in the near future.

The use of biomimetic ferritins for nanoparticle synthesis, assembly, and application offers additional significant advantages compared with more conventional technologies. Self-assembled ferritin cages are readily available and naturally monodisperse, providing a reproducible system for exacting control over the synthesis of a wide range of nanoparticles with uniform morphology and size. These features also facilitate self-organized patterning and the assembly of higher-order structures over large areas. Moreover, for materials applications that require thermal processing of nanoparticle films or powders, the protein coating helps to prevent interparticle sintering by carbon-shell formation. In addition, the inherent biocompatibility and conjugation properties of the external surface of the protein cage make these bioinorganic magnetic ferritins ideal for pharmaceutical and medical applications such as contrast agents, targeted drug delivery, and immunomagnetic labeling [51–54]. Biologically targeted semiconducting nanoparticles can also be used for rapid protein or DNA sequencing.

Although initial applications of ferritin-derived nanoparticles are likely to exploit the inherent material characteristics of the inorganic component in conjunction with a subset of potential benefits offered by the biomolecular properties, the wide range and availability of bionanostructures indicate that the biomimetic strategy described in this chapter could have generic application. The biological expression and replication of unique supramolecular morphologies with inherent self-organization, patterning and reactivity, in combination with the increasing demand for advanced small-scale materials and devices, offers an exciting future for bio-inspired strategies in the emerging fields of nanoscience and nanotechnology.

References

[1] S. Mann, *Nature* 1993, 365, 499.

[2] S. Mann, S. L. Burkett, S. A. Davis, C. E. Fowler, N. H. Mendelson, S. D. Sims, D. Walsh, N. T. Whilton, *Chem. Mater.* 1997, 9, 2300.

[3] W. Shenton, D. Pum, U. B. Sleytr, S. Mann, *Nature* 1997, 389, 585.

[4] S. R. Hall, W. Shenton, H. Engelhardt, S. Mann, *ChemPhysChem.* 2001, 3, 184–186.

[5] R. A. McMillan, C. D. Paavola, J. Howard, S. L. Chan, J. Zaluzec, J. D. Trent, *Nature Mater.* 2002, 1, 247–252.

[6] S. Behrens, K. Rahn, W. Habicht, K. J. Bohm, H. Rosner, E. Dinjus, E. Unger, *Adv. Mater.* 2002, 14, 1621–1625.

[7] S. L. Burkett, S. Mann, *Chem. Commun.* 1996, 321–322.

[8] C. E. Fowler, W. Shenton, G. Stubbs, S. Mann, *Adv. Mater.* 2001, 13, 1266–1269.

[9] S. A. Davis, S. L. Burkett, N. H. Mendelson, S. Mann, *Nature* 1997, 385, 440.

[10] B. J. Zhang, S. A. Davis, N. H. Mendelson, S. Mann, *Chem. Commun.* 2000, 781, 9.

[11] N. H. Mendelson, *Science* 1992, 258, 1633.

[12] J. Coffer, S. Bigham, X. Li, R. Pinnizzotto, Y. Rho, R. Pirtle, I. Pirtle, *Appl. Phys. Lett.* 1996, 69, 3851.

[13] E. Braun, Y. Eichen, U. Sivan, G. Ben-Yoseph, *Nature* 1998, 391, 775.

[14] M. Pazirandeh, S. Baral, J. Campbell, *Biomimetics* 1992, 1, 41.

[15] W. Shenton, T. Douglas, M. Young, G. Stubbs, S. Mann, *Adv. Mater.* 1999, 11, 253.

[16] D. D. Archibald, S. Mann, *Nature* 1993, 364, 430.

[17] J. M. Schnur, *Science* 1993, 262, 1669–1676.

[18] R. R. Price, M. Patchan, A. Clare, D. Rittschof, J. Bonaventura, *Biofouling* 1992, 6, 207.

[19] J. M. Schnur, R. Price, P. Scheon, P. Yager, J. M. Calvert, J. George, A. Singh, *Thin Solid Films* 1987, 152, 181–206.

[20] S. Baral, P. Schoen, *Chem. Mater.* 1993, 5, 145–147.

[21] M. Seddon, H. M. Patel, S. L. Burkett, S. Mann, *Angew. Chem. Int. Ed.* 2002, 41, 2988–2991.

[22] F. C. Meldrum, T. Douglas, S. Levi, P. Arosio, S. Mann, *J. Inorg. Biochem.* 1995, 58, 59–68.

[23] F. C. Meldrum, V. J. Wade, D. L. Nimmo, B. R. Heywood, S. Mann, *Nature* 1991, 349, 684–687.

[24] F. C. Meldrum, B. R. Heywood, S. Mann, *Science* 1992, 257, 522–523.

[25] T. Douglas, D. P. E. Dickson, S. Betteridge, J. Charnock, C. D. Garner, S. Mann, *Science* 1995, 269, 54–57.

[26] K. K. W. Wong, S. Mann, *Adv. Mater.* 1996, 8, 928–933.

[27] K. K. W. Wong, T. Douglas, S. Gider, D. D. Awschalom, S. Mann, *Chem. Mater.* 1998, 10, 279–285.

[28] M. Allen, D. Willits, J. Mosolf, M. Young, T. Douglas, *Adv Mater.* 2002, 14, 1562–1565.

[29] W. Shenton, S. Mann, H. Coelfen, A. Bacher, M. Fischer, *Angew. Chemie Int. Ed.* 2001, 40, 442–445.

[30] T. Douglas, M. Young, *Nature* 1998, 393, 152–155.

[31] T. Douglas, M. Young, *Adv. Mater.* 1999, 11, 679–681.

[32] E. Dujardin, C. Peet, G. Stubbs, J. N. Culver, S. Mann., *Nano Lett.* 2003, 3, 413–417.

[33] P. Harrison, P. Artymiuk, G. Ford, D. Lawson, J. Smith, A. Treffry, J. White, in: S. Mann, J. Webb, R. J. P. Williams (eds), *Biomineralization: chemical and biochemical perspectives.* VCH, Weinheim, 1989, pp. 257–294.

[34] F. Funk, J. Lenders, R. Crichton, W. Scheinder, *Eur. J. Biochem.* 1985, 152, 167–172.

[35] B. Warne, O. Kasyutich, E. Mayes, J. Wiggins, K. K. W. Wong, *IEEE Trans. Mag.* 2000, 36, 3009–3011.

[36] PDB ID: 1AEW. P. Hempstead, S. Yewdall, A. Fernie, D. Lawson, P. Artymiuk, D. Rice, G. Ford, P. Harrison, *J. Mol. Biol.* 1997, 268, 424.

[37] F. Bernstein, T. Koetzle, G. Williams, E. Meyer, Jr., M. Brice, J. Rogers, O. Kennard, T. Shimanouchi, M. Tasumi, *J. Mol. Biol.* 1977, 112, 535.

[38] R. Koradi, M. Billeter, K. Wüthrich, *J. Mol. Graphics* 1996, 14, 51.

[39] G. Choe, J. Zhou, B. Demczyk, M. Yu, M. Zheng, R. Weng, A. Chekanov,

K. Johnson, F. Liu, K. Stoev, *TMRC*, **2002**, Paper B1.

[40] R. Wood, *IEEE Trans. Mag.* **2000**, 36, 36–42.

[41] S. Sun, D. Weller, C. Murray, in: The physics of ultra-high-density magnetic recording, M. Plumer, J. van Ek, D. Weller (eds), Springer, New York, **2001**, pp. 249–276.

[42] D. Sellmyer, C. Luo, M. Yan, Y. Liu, *IEEE Trans. Magn.* **2001**, 37, 1286–1291.

[43] D. Weller, A. Moser, L. Folks, M. Best, W. Lee, M. Toney, M. Schwickert, J. Thiele, M. Doerner, *IEEE Trans. Magn.* **2000**, 36, 10–15.

[44] S. Sun, D. Weller, *J. Mag. Soc. Japan* **2001**, 25, 1434–1440.

[45] S. Sun, C. Murray, D. Weller, L. Folks, A. Moser, *Science* **2002**, 287, 1989–1992.

[46] D. Weller, Seagate Technology, Pittsburgh, PA, personal communication, 2001.

[47] H. Zeng, S. Sun, T. Vedantam, J. Liu, Z. Dai, Z. Wang, *Appl. Phys. Lett.* **2002**, 80, 2583–2585.

[48] S. Tsang, J. Qiu, P. Harris, Q. Fu, N. Zhang, *Chem. Phys. Lett.* **2000**, 322, 553–560.

[49] E. Mayes, *J. Mag. Soc. Japan* **2002**, 26, 932–935.

[50] K. Nagayama, S. Takeda, S. Endo, H. Yoshimura, *Jpn. J. Appl. Phys.* **1995**, 34, 3947–3954.

[51] J. W. M. Bulte, T. Douglas, S. Mann, R. B. Frankel, B. M. Moskovitz, R. A. Brooks, C. D. Baumgarner, J. Vymazel, M.-P. Strub, J. A. Frank, *J. Magnet. Reson. Imag.* **1994**, 4, 497–505.

[52] J. W. M. Bulte, T. Douglas, S. Mann, R. B. Frankel, B. M. Moskovitz, R. A. Brooks, C. D. Baumgarner, J. Vymazel, M.-P. Strub, J. A. Frank, *Invest. Radiol.* **1994**, 29, S214–216.

[53] J. W. M. Bulte, T. Douglas, S. Mann, J. Vymazal, B. S. Laughlin, J. A. Frank, *Acad. Radiol.* **1995**, 2, 871–878.

[54] M. Zborowski, C. Bor Fur, R. Green, N. J. Baldwin, S. Reddy, T. Douglas, S. Mann, J. J. Chalmers, *Cytometry* **1996**, 24, 251–259.

19
DNA–Gold-Nanoparticle Conjugates

C. Shad Thaxton and *Chad A. Mirkin*

19.1
Overview

19.1.1
Introduction

Genomics and proteomics are two fields of research that hold great promise for the unraveling and subsequent understanding of complex biological processes. Although the number of human genes is still debated, it is generally agreed that there are on the order of tens-of-thousands. With the human genome sequence complete [1–3], relatively easy access to sequence information and expressed sequence tags (ESTs), commercially available cloned complementary DNA (cDNA) libraries, and a growing number of proteins available in purified form, researchers are left with the great task of deciphering the information contained within the human genome and its translated protein products on a global scale. Further, in an age of bioterrorist threats and growing microbial antibiotic resistance, the detection of microbial pathogens, with their own unique DNA and protein fingerprints, is of utmost importance for the accurate identification and diagnosis of infectious diseases and the subsequent delivery of timely treatment. Ultimately, the in-depth study of complex biological processes for its own sake – as well as for the prevention, diagnosis, and treatment of human disease – depends on one's ability to detect sensitively and selectively multiple DNA and protein targets of interest when present in complex media. Therein, directed point-of-care testing and treatment, and therapy based on genomic and proteomic information will ultimately offer more solutions and better answers to medical problems.

Historically, the detection of DNA has been carried out using either radioactive or organic fluorophore labels. Indeed, most commercial DNA detection assays still rely on the use of organic fluorophores, and radioactive labels are still used for some DNA detection applications. In addition, protein detection is traditionally performed using an antibody specific for the target of interest in enzyme-linked immunosorbent assays (ELISAs) or in a blotting format (e.g., Western blot). Such labeling techniques have allowed researchers to compile a tremendous amount of DNA- and protein-specific information,

Nanobiotechnology. Edited by Christof Niemeyer, Chad Mirkin
Copyright © 2004 WILEY-VCH Verlag GmbH & Co. KgaA, Weinheim
ISBN 3-527-30658-7

and this has had a major impact on the field of medicine, forensics, and molecular biology. However, there are inherent drawbacks to the use of both organic fluorophore and radioactive labeling techniques, as will be discussed below. Therefore, the technology for the detection of DNA and protein targets is also moving forward. Namely, advances in nanotechnology and DNA-nanoparticle conjugate systems (including quantum dots and metal nanoparticles) have emerged as novel and extremely powerful tools for DNA and protein detection. They have numerous advantages over traditional methods and could represent the next generation of biomarkers.

In 1996, it was shown that the distance-dependent optical properties of oligonucleotide functionalized gold nanoparticles could be used in colorimetric assays for DNA detection [4]. Further investigation of the properties of these novel nanostructured probes showed that they exhibited unusually sharp melting transitions when hybridized to complementary DNA, and also displayed catalytic properties that could be used in a variety of high-sensitivity and -selectivity assays for DNA [5–8]. Subsequently, numerous research groups began taking advantage of DNA-nanoparticle conjugates for the labeling of DNA and proteins such that the specific nanoparticle tag allowed for the detection of target molecules based on the novel properties of the material or materials that made up the nanoparticle [9–13]. Specifically, metal nanoparticles with well-defined sizes, shapes, and compositions are currently being exploited by research groups for DNA and protein detection [13–17]. The following is a review of the progress made in this area, the methods which we employ in our laboratories, and future prospects with regard to this growing and exciting field.

19.1.2
Nanoparticles

The first step in synthesizing DNA-nanoparticle conjugates is the identification of a suitable nanoparticle. In this regard, there are numerous nanoparticles from which to choose, and one can do so from metal, semiconductor, magnetic, and polymer particle representatives (Table 19.1).

The size of such particles can be tailored from 1 nm to 1 μm in diameter with moderate to excellent control over size dispersity, depending upon chosen composition. Interestingly, nanoparticle size and size dispersity often dictates the physical properties of colloids made from such materials. For example, by tuning the size, composition and shape of a nanoparticle, one can tailor the wavelength of scattered light from such particles (Figure 19.1).

Colloidal gold nanoparticles (Au-NPs) ranging in size from 3 to 100 nm, have been the focus of intense research. These particles are stable, environmentally benign, and their chemical properties can be easily tailored by chemically modifying their surfaces. Gold nanoparticles in the 1 to 100 nm range can be made in relatively monodisperse forms via a variety of synthetic methods [18, 19]. Typically, they are charged particles that are very sensitive to changes in solution dielectric. Indeed, for typical citrate stabilized particles, the addition of NaCl shields the surface charge and leads to a concomitant decrease in interparticle distance and eventual particle aggregation. This property becomes important in certain nanoparticle biomolecule detection schemes as discussed below.

Table 19.1 Representative nanoparticle compositions and sizes

Particle composition	Available particle size [nm]	
Metals		
	Au	2–150
	Ag	1–180
	Pt	1–20
	Cu	1–150
Semiconductors		
	CdX (X = S, Se, Te)	1–20
	ZnX (X = S, Se, Te)	1–20
	PbS	2–18
	TiO_2	3–50
	ZnO	1–30
	GaAs, InP	1–15
	Ge	6–30
Magnetic		
	Fe_3O_4	6–40
Polymer		
	Many compositions	50–1000

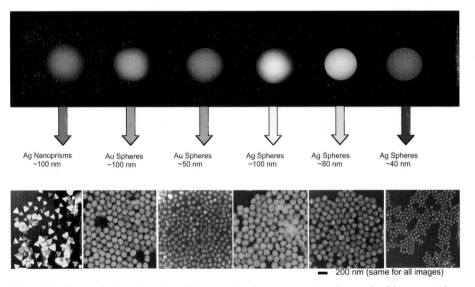

Figure 19.1 Size- and shape-dependent light scattering by representative silver and gold nanoparticles with corresponding transmission electron microscopic (TEM) images of the particles.

19.1.3
DNA-functionalized Gold Nanoparticles

The novel properties of Au-NPs can be taken advantage of for the detection of DNA and proteins. Gold nanoparticles have long been conjugated with antibodies and other proteins (e. g., streptavidin) for use in protein and DNA detection assays, and such chemistry has resulted in the commercial development of some systems, primarily for proteins (see reference and references therein) [14]. However, as demonstrated in all of the detection assays described below, the direct functionalization of the Au-NP surface with DNA "probe" strands (DNA–Au-NP) that recognize cDNA targets of interest can be used in highly sensitive and selective DNA and protein detection assays with direct DNA and protein target labeling. This is accomplished by functionalizing citrate stabilized Au-NPs with mercaptoalkyloligonucleotides, as discussed in the methods section. The probe strand is designed to be complementary to a target of interest and is attached to the Au-NP through chemisorption of the thiol group onto the surface of the gold nanoparticle.

The surface plasmon resonance (SPR) of Au-NPs is responsible for their intense colors and this can be taken advantage of in biodetection assays. In solution, monodisperse Au-NPs appear red and exhibit a relatively narrow surface plasmon absorption band centered at 520 nm in the UV-Visible spectrum (Figure 19.2). In contrast, a solution containing aggregated Au-NPs appears purple in color, corresponding to a characteristic red shift in the surface plasmon resonance of the particles from 520 to 574 nm (Figure 19.2). Using a DNA or protein target as a linking molecule to aggregate Au-NPs allows one to take advantage of the novel optical properties of disperse versus aggregated gold particles for use in DNA and protein detection assays. The SPR of protein- and DNA-functionalized Au-NPs has also been taken advantage of in surface-based DNA and protein detection assays due to the Au-NPs' ability to amplify changes in the SPR of a noble metal surface-film when the two are brought in close proximity after the binding of a targeted analyte [20–24]. Using Au-NPs in such a way enabled researchers to increase the sensitivity of SPR-based biomolecule sensing techniques used extensively for studying biomolecular interactions [25–27].

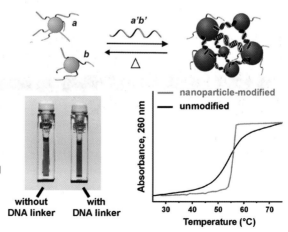

Figure 19.2 Optical properties of monodisperse versus aggregated gold nanoparticles. Two probe species (a and b) are linked by a complementary DNA target strand (a′b′). Thermal denaturation profiles of DNA-nanoparticles exhibit substantially sharper melting transitions when compared to unmodified DNA.

Further, Au-NPs promote the catalytic reduction of silver ions to silver metal in the presence of a reducing agent such as hydroquinone. The process of silver staining Au-NP conjugates has been taken advantage for antibody–, protein–, and DNA–nanoparticle conjugates [17, 28–30]. This property is used to increase the sensitivity of chip-based DNA and protein detection assays where silver development acts to amplify the signal provided by the gold nanoparticle labels bound to complementary target molecules.

Since metal nanoparticles also conduct electricity, they can be used in chip-based assays with electrical readout schemes. Such schemes are attractive since they lend themselves well to miniaturization and assay multiplexing. In addition, they are ideal prospects for point-of-care detection schemes [6, 31, 32].

Finally, size and shape-dependent light-scattering properties of nanoparticles allow one to prepare multi-color labeling schemes based upon them [5, 13].

It should be pointed out here that throughout this chapter, "probe" strand will refer to the DNA-Au-NP "probe," the target will be the DNA or protein for detection, and DNA oligomers immobilized on a chip surface will be referred to as "capture" strands. This is in contrast to some texts which reserve the term "probe" for the DNA immobilized to the chip surface.

19.1.4
Nanoparticle Based DNA and RNA Detection Assays

19.1.4.1 **Homogeneous DNA Detection**
In 1996, Mirkin and co-workers reported the use of mercaptoalkyloligonucleotide-modified gold nanoparticle probes (DNA–Au-NP probes) for the colorimetric detection of cDNA target sequences [4, 10]. Assays were conducted in which single-stranded oligonucleotide targets were detected using DNA–Au-NP probes. Two species of probes were present such that each was functionalized with a DNA–oligonucleotide complementary to one half of a given target oligonucleotide (Figure 19.2). Mixing the two DNA–Au-NP probes with the target resulted in the formation of a polymeric network of DNA–Au-NPs with a concomitant red-to-purple color change. This hybridization signal is governed by the optical properties of the nanoparticles as discussed previously. Importantly, melting analyses of the polymeric structures revealed extraordinarily sharp melting transitions [full width at half maximum (FWHM) as low as 1 °C] such that imperfect targets could be readily differentiated from complementary targets on the basis of color and temperature (Figure 19.3, see p. 294). The basis for the sharp melting transition is believed to be due to a cooperative mechanism that results from the presence of multiple DNA target strands between each pair of DNA–Au-NPs and a decrease in the aggregate melting temperature as DNA strands melt due to the concomitant reduction in local salt concentration [33]. Transfer of the hybridization mixture to a reverse-phase silica plate resulted in a permanent and easily readable record of the hybridization state at any given temperature (Figure 19.3, see p. 294). Subsequent research demonstrated that one could selectively detect target DNA sequences with single base imperfections, regardless of position, using DNA–Au-NP probes oriented in a tail-to-tail arrangement [7]. These initial studies were important because they identified the chemistry required to stabilize nanoparticles to the point where they could be used under assay-relevant conditions. Moreover, these studies iden-

tified novel and unanticipated properties of the nanoparticle conjugates that point to major advantages in detection.

19.1.4.2 Chip-based (Heterogeneous) DNA Detection Assays

Scanometric DNA detection

Chip-based quantitative DNA assays, otherwise known as "gene-chips" or "DNA-micro-arrays," are having a major impact in genomic studies. Developed by Brown and co-work-ers in 1995, this technology has been commercialized and now is widely used in many research laboratories [34]. With such arrays, DNA labeling is typically carried out with or-ganic fluorophore labels. As will be pointed out later in this chapter, fluorophore labeling carries with it several drawbacks, many of which can be avoided by using DNA–nanopar-ticle conjugate labels.

Numerous research groups have taken advantage of Au-NP probes for chip-based detec-tion schemes. As mentioned previously, Keating and co-workers used Au-NP probes to amplify changes in the SPR of gold films during DNA hybridization to capture strands [21]. In addition, Genicon offers a commercially available DNA chip-based assay in which Au-NPs functionalized with anti-biotin IgG are used to label biotinylated DNA tar-gets of interest in chip-based format whereupon the light-scattering properties of Au-NPs with different sizes and shapes are used as the readout [13, 35]. Further, Mirkin and co-workers demonstrated the direct labeling of DNA target strands using DNA–Au-NP probes. In their work, a sandwich assay was devised such that DNA "capture" strands, replacing the probe strands from their homogeneous solution-based experiments, were covalently attached to the surface of a glass slide in microarray fashion. Subsequently, a mixture of DNA target strands and cDNA–Au-NP probes were added to the chip surface for hybridization (Figure 19.4). Since the Au-NPs promote the reduction of silver ions to silver metal, a silver-developing solution was added to the chip surface in order to amplify probe signal and visualize the surface bound capture–target–probe complex. Results visi-ble to the naked eye could be recorded with a conventional flatbed scanner. As discussed below, light scattering can also be used as a readout mechanism, in some cases obviating the need for silver enhancement of surface immobilized DNA–Au-NP probes [5, 13]. Using stringency washes of increasing temperature, it was demonstrated that single-base imperfections could be detected with increased selectivity (factor of 4) and sensitivity (10 000 times), respectively, when compared to fluorophore labeling methods [6, 8, 28].

Two-color labeling of oligonucleotide arrays using Au-NP size-dependent light scattering

Combinatorial microarrays are constructed in order to both detect the presence of a DNA/RNA target of interest and also to compare the relative abundance of a target in two or more samples. In chip-based analyses in which a single chip is used to determine the relative abundance of a DNA or RNA target, multiple "colors," – typically organic fluoro-phore labels – are needed to distinguish between or among target samples and their selective hybridization to complementary strands immobilized on a chip surface.

As pointed out previously, Au-NP probes of different size and shape scatter light at dif-ferent wavelengths. Therefore, instead of using two different organic fluorophore labels, one can use Au-NPs with different size dimensions to achieve multi-color capabilities.

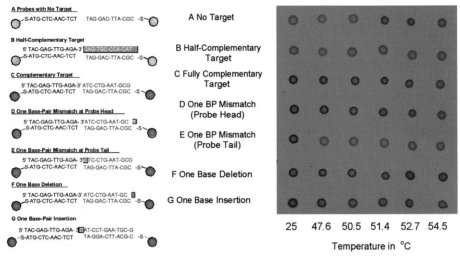

Figure 19.3 Homogeneous single-nucleotide poly-
morphism (SNP) discrimination using DNA–nanoparticle probes. The sharp thermal denaturation
profiles for gold nanoparticle probes allows for SNP

detection. The Northwestern spot test is shown in
which samples of the aggregate mixture are spotted
to a reverse-phase silica plate at a given temperature permanently recording the hybridization status.

A commercial system based upon this concept utilizes gold probes functionalized with
anti-biotin IgG which subsequently label biotinylated DNA targets of interest [13].
DNA–Au-NP probes of differing dimensions allow for the direct labeling of DNA targets
of interest in multi-color fashion.

We have demonstrated that two-color labeling can be accomplished using the size-
dependent light-scattering properties of two different sizes of gold nanoparticles. Impor-

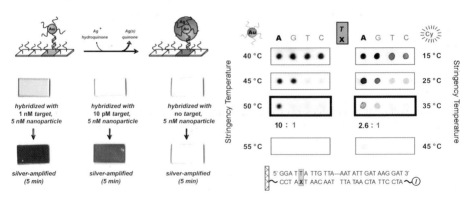

Figure 19.4 Scanometric chip-based DNA detection sandwich assay using silver amplification of the
detection signal. The scanometric DNA detection
scheme is depicted on the left. A direct SNP de-
tection/discrimination comparison of DNA-nano-

particles versus organic fluorophore labeling probes
is shown on the right. The DNA-nanoparticle system is shown here to have a selective advantage of
approximately 4:1.

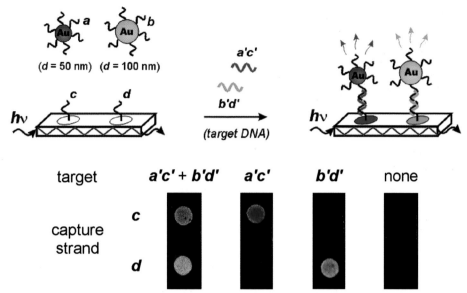

Figure 19.5 Two-color DNA sandwich assay using light scattering from different diameter gold nanoparticle probes (a and b) specific for a given target (a'c' and b'd', respectively). When the target sequence is present and hybridizes with the appropriate chip-immobilized capture strand (c and d), the nanoparticle probes are bound to the surface. Evanescent illumination of the chip surface and dark-field visualization allows for the detection of specific hybridization events.

tantly, as these assays are based upon the same types of nanoparticles that exhibit the sharp melting transitions when hybridized to cDNA, they show the high selectivity of the other assays based upon them [7, 28]. The light-scattering method is more sensitive than the analogous fluorescence-based assay by a couple of orders of magnitude, but it is difficult to calibrate due to the background and complexity of the scattering response. Indeed, the signal from nanoparticles and assemblies of nanoparticles on the detection surface are often overlapping and must be deconvoluted in any assay based upon such structures (Figure 19.5) [5].

Electrical detection of DNA with Au-NP probes in a chip-based assay
The detection of DNA targets in a chip-based assay with an electrical readout scheme allows for many potential advantages over conventional readout and detection methods. First, massive multiplexing is possible where on-chip electronic circuitry can be fabricated such that an individual circuit corresponds to an individual DNA target of interest. Changes in the electrical behavior of such a circuit, monitored as a change in the resistance or conductance, can be used as an extremely convenient and straightforward detection readout method. In addition, the sensitivity of such a system can be extraordinarily high depending on the design of the assay (e. g., circuit dimensions and number of hybridization events necessary for signal generation). One, in principle, also can obtain a sense of the relative number of target molecules present by the strength of the readout signal. In light of such advantages, much progress has been made on the development of such an assay.

Mirkin and co-workers developed the following detection assay based on the ability of surface-bound DNA–Au-NPs to conduct electricity, especially when exposed to silver-developing solution. Chips with microelectrode gaps were fabricated such that DNA capture strands were situated in the gap. In sandwich array format, capture–target–probe complexes were hybridized in the gap and subsequently exposed to silver-developing solution. When a target complementary to the capture and probe strand was present, silver developing resulted in a decrease in the resistance of the gap, an increased conductivity, and subsequent electrical readout (Figure 19.6). As with the previous assays, it was shown that the DNA–Au-NP probes had the ability to distinguish perfectly complementary targets from ones with base mismatches. Importantly, it was also shown that thermal stringency washing was not the only method available to remove mismatched targets. Stringency washes at

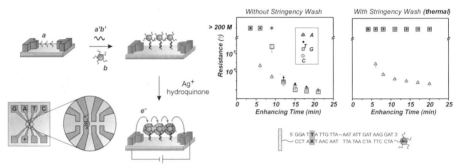

Figure 19.6 Electrical detection of DNA targets using DNA-nanoparticle probes. Immobilized capture strands (a) hybridize to complementary targets (a′b′). cDNA-nanoparticle probes (b) then hybridize and, upon silver enhancement, alter the conductance through the electrode gaps for target detection (shown on the left). A 50 °C thermal stringency wash (0.3 M PBS) allows for the specific detection of the complementary target strand and SNP discrimination (shown on the right).

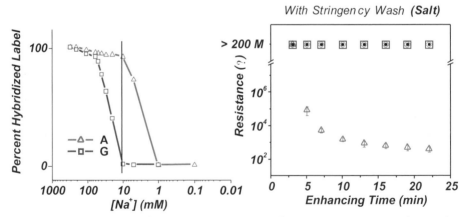

Figure 19.7 Salt stringency wash to discriminate SNPs. In order to increase stringency and SNP selectivity, a salt-based stringency wash (0.01 M PBS at room temperature) was demonstrated to increase the selectivity for the perfectly matched target sequence (same as in Figure 19.6).

room temperature, but with different buffered salt concentrations, were used such that the ability to remove mismatched targets was enhanced when compared to thermal stringency washes with the same buffered salt concentration (Figure 19.7). Further, the electrical detection assay was shown to have an unoptimized sensitivity of approximately 500 fM. In principle, that number can be dramatically decreased through miniaturization of the electrode gap and the number of nanoparticles required to close it [6].

Willner and co-workers developed a surface-based DNA assay where Faradaic impedance spectroscopy or microgravimetric measurements using Au-electrodes and Au-quartz crystals were used for target hybridization and detection. Importantly, DNA-functionalized liposomes and biotinylated liposomes were shown to dendritically amplify the detection signal such that target concentrations as low as 1×10^{-12} M were detected and single-base mismatches were identified [31, 36].

Nanoparticles with Raman spectroscopic fingerprints for DNA and RNA detection in chip-based format

As discussed in the section regarding multi-color labeling of DNA targets using size-dependent light scattering by DNA–Au-NP probes, multiplexing and ratioing are two key aspects with respect to the successful use of microarray technology for several applications (e. g., SNP detection, gene expression). Indeed, the reliable use of such powerful technology depends on the use of sensitive and specific labels for DNA targets, as well as the ability to use multiple types of labels with addressable and individual labeling information. Quantum dot biolabels display decreased photobleaching rates, and more narrow emission spectra when compared to organic fluorophores [9, 37, 38]. In addition, size and composition control allow for the production of multi-colored labels. Therefore, some of the problems inherent to the use of organic fluorophore labels will be remedied by using quantum dots (see Chapter 22). However, the increased stability, sensitivity and selectivity of DNA–Au-NP probes, combined with the fact that they are environmentally benign, make them good candidates for use in bioassays where combinatorial labeling is needed. Size-dependent light scattering of Au-NPs is one way to accomplish this, but as described above it is limited in large part due to overlapping signals and the complexity of the scattering elements (probes, probe orientation, probe aggregation).

An extremely powerful new method for carrying out multiplexed detection in a chip-based microarray format takes advantage of Raman spectroscopy and the ability of Au-NPs to promote the reduction of silver in the presence of Raman-active spectroscopic labels (Figure 19.8). In this type of assay, a large number of probes can be designed and synthesized by functionalizing the nanoparticles with Raman dyes, each with a unique Raman spectrum. In a typical assay, a microarray spotted with different capture strands is used to capture one or more DNA target strands in a solution. If the target strand is captured, particles with unique signatures are hybridized to the appropriate spots. Silver development followed by Raman analysis allows one to quickly analyze which strands were present in the sample in a sensitive and highly selective manner (Figure 19.9). Further, one can correlate the intensity of the Raman spectrum with the relative amount of target present when comparing two different samples. In proof-of-concept experiments, this detection scheme was shown to be effective for simultaneously detecting six specific targets and ratioing RNA strands with single-base mismatches [8].

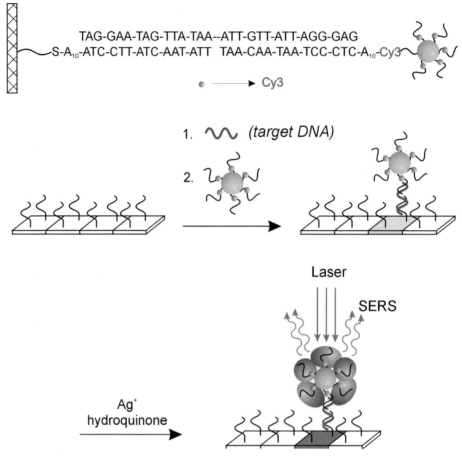

Figure 19.8 Scheme for surface-enhanced Raman spectroscopic (SERS) detection of DNA targets of interest. In a sandwich assay similar to those above, DNA-nanoparticles encoded with Raman-active dyes (e. g., Cy3) are hybridized to the surface -immobilized capture/target hybrid and silver enhancement is performed. Upon single wavelength laser excitation, the particles emit a strong and reproducible Raman spectrum specific to the Raman-active dye chosen.

There are numerous advantages that this system offers when compared to conventional fluorophore labeling of microarrays. Some of these advantages have already been discussed with respect to the DNA–Au-NP's increased sensitivity (this assay has an unoptimized detection limit of ~2 fM) and selectivity (easily able to distinguish single-base mismatched targets using either thermal or salt stringency washes). In addition, the number of labels one can prepare using this SERS approach is, in principle, much greater than can be prepared with organic fluorophores, due to the relatively narrow line widths in Raman spectroscopy. Furthermore, detecting and ratioing Raman label spectroscopic intensities can be carried out with single-source laser excitation; likewise, photobleaching is not a significant problem with the nanoparticle materials, but can be with fluorophore labels [8].

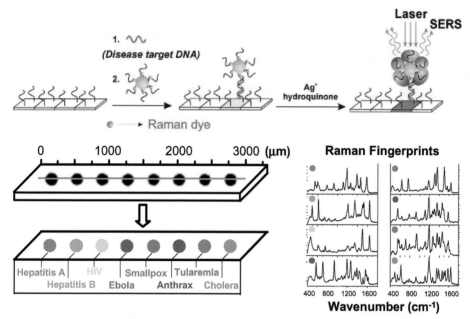

Figure 19.9 Raman-encoded DNA-nanoparticle detection of multiple DNA targets of interest where each Raman spectrum, or "color" corresponds to a specific target of interest. Here, eight targets were chosen and each assigned a Raman-encoded dye. Note that after silver staining, all of the spots appear black and it is impossible to discern which spot corresponds to which target. However, by using SERS, one is able to scan the spots with a single wavelength excitation laser and observe dye- (and thus, target) specific Raman spectra.

19.1.5
DNA-Nanoparticle Detection of Proteins: Biobarcodes

The previous sections focused on the use of DNA-nanoparticle conjugates used for DNA/RNA detection. However, one can also take advantage of DNA–Au-NPs for the detection of proteins. This is similar to Au-NP–protein conjugates which, as discussed previously, are commercially available for both DNA and protein detection. However, there are inherent differences as pointed out below when using DNA–Au-NPs.

Like DNA, proteins have "complementary" counterparts. Protein–protein and protein–small molecule interactions regulate the homeostatic balance of entire organisms. For instance, the human immune system is built upon the ability of immunoglobulins to recognize foreign proteins so that invading organisms can be recognized and destroyed. Analogous to using a DNA target sequence as a linker, protein targets can also be used to link Au-NP probes for their subsequent detection. An ideal detection scheme should enable one to detect numerous protein targets without substantial sample processing.

We have designed an assay which takes advantage of the DNA–Au-NP probes and a "biobarcode" specific to a protein of interest. A "biobarcode" is a synthetic DNA sequence of choice with a unique base sequence corresponding to a *protein* target of interest. Thus, by changing the DNA sequence, one effectively changes the biobarcode and its corresponding protein target. Therefore, for protein detection, one takes advantage of DNA as decoding agents and proteins as recognition agents. The identification of the protein target is made using the DNA biobarcode. DNA–Au-NP probes are fabricated such that half of the biobarcode sequence is complementary to the DNA–Au-NP probe sequence. The other half of the biobarcode complement is present on a DNA oligomer terminally functionalized with the target protein's counterpart. In this case, the target protein was a monoclonal antibody specific for either biotin or dinitrophenol (DNP). Therefore, in solution, and in the presence of the DNA–Au-NP probes, the specific biobarcode, a DNA-oligomer terminated with biotin or DNP, and the anti-biotin or anti-DNP antibody detection target, aggregation of Au-NPs takes place due to the polymeric network structure which results (Figure 19.10, see p. 302). As with the homogeneous DNA target detection scheme detailed above, a melting analysis can be performed on the aggregate structures such that specific melting temperatures, pre-chosen specifically for each biobarcode/protein, correspond to the presence of specific protein targets in solution. One limitation of this approach is that, while an infinite number of protein specific biobarcodes are available, their melting temperatures must be chosen such that one can discriminate individual biobarcode melting profiles in the presence of numerous others during the DNA melting analysis. Therefore, in practical terms this method is limited to five to seven analytes.

As an alternative approach, and in order to increase both the number of analytes possible and the sensitivity of the system, biobarcode–protein target aggregates can be centrifuged such that the aggregates are collected as a pellet. The supernatant containing unreacted biobarcodes is removed and the particles are resuspended in water such that the hybridization between the complementary strands is lost. Filtering this mixture allows for the isolation of single-stranded biobarcodes which can then be sensitively detected in chip-based format using appropriate DNA–Au-NPs and capture strands with subsequent silver enhancement [39]. The use of the target-specific biobarcodes is important as proteins are largely dependent on their three-dimensional (3D) structure for function. As 3D protein structure is easily disrupted, one depends on the protein target only for target recognition and aggregate formation. The more stable DNA biobarcode then becomes responsible for the identification of the protein either through a melting analysis or in a chip-based assay. Furthermore, as the number of biobarcode DNA sequences is almost limitless, one can detect large numbers of proteins in solution simultaneously using the chip-based assay. Finally, the detection of a target protein can be made as sensitive as the hybridization of DNA–Au-NPs to biobarcode sequences in a chip-based format, with subsequent silver signal amplification.

19.1.6
Conclusion

The preceding sections provide an overview of the investigations carried out with respect to DNA-nanoparticle conjugates and their use as DNA and protein detection materials. The use of such novel materials in genomic and proteomic research, as well as for the prevention, diagnosis, and treatment of human disease is imminent. Using such materials allows one to realize that such endeavors will be carried out with increased sensitivity and selectivity such that answers to difficult and complex queries can be made more timely and with greater certainty, as can the direction of treatment. The following is a discussion of essential methods and protocols.

19.2
The Essentials: Methods and Protocols

19.2.1
Nanoparticle Synthesis

Gold nanoparticles (13 nm diameter) are prepared by the citrate reduction of $HAuCl_4$ gold precursor salt [30]. Briefly, all glassware used in the synthesis of Au-NPs should be cleaned with aqua regia (3 parts HCl, 1 part HNO_3), rinsed with 18.1 MΩ nano-pure water (NP-H_2O), and oven-dried prior to use. An aqueous solution of $HAuCl_4$ (1 mM, 500 mL) is brought to reflux while stirring, and then 50 mL of 38.8 mM trisodium citrate solution is *rapidly* added such that the solution quickly changes color from pale yellow to deep red. After waiting 15 minutes, the mixture is allowed to cool to room temperature and subsequently filtered through a 0.45 µm filter. The colloid is characterized using UV-Visible spectroscopy where the characteristic surface plasmon resonance band of mono-disperse particles is located at 518–520 nm. Particles with more uniform size distributions have narrower absorption bands. Note that aggregated 13 nm colloid, as discussed above, displays a flattened and red-shifted absorption peak at approximately 570 nm. The preparation of Au-NPs of different sizes is accomplished using different ratios of gold precursor to reducing agent as dictated by the growth versus nucleation properties of nanoparticle synthesis [30]. In general, the higher the ratio, the larger the particles as more gold precursor allows for increased colloid growth around fewer nucleation centers due to the relatively low concentration of reducing agent. Accordingly, the increased colloid nucleation that takes place when relatively more reducing agent is present results in smaller particle diameters. Transmission electron microscopy (TEM) is the best means for confirming particle size and monodispersity.

19.2.2
DNA-functionalized Au-NP Probe Synthesis

The fabrication of DNA–Au-NP probes is the single most important process in either homogeneous or chip-based detection assays, and it starts with the synthesis of gold colloid as described above. The next step is the DNA-functionalization of the appropriately sized Au-NP colloid as described in two key publications [7, 40]. Usually, 13 nm colloid is used and has been shown to be quite stable, but Au-NPs as large as 100 nm in diameter can be functionalized with DNA [40]. In short, DNA probe strands are typically 12- to 15-mers (can be longer) that are complementary to a DNA target of interest. As a general rule, stringency washes are more efficient at removing mismatched labels with shorter oligonucleotide lengths. Often, a poly-adenine (poly-A) stabilization strand of 10–20 bases is added on the Au-NP side of the DNA probe strand as this has been shown to increase hybridization efficiency [41]. Probe strands are synthesized using solid-state DNA synthetic methods taking advantage of standard phosphoramidite chemistry. Importantly, synthesized DNA probe strands are functionalized at either the 3'- or 5'-end with terminal sulfur-containing groups either in the form of a mono-thiol, cyclic disulfide, or trithiol. Probe strands are linked to the Au-NP by taking advantage of the strong coordination chemistry that exists between gold and sulfur. It was found that DNA–Au-NPs functionalized with probe strands terminated in a trihexylthiol group were stable at higher temperatures and in the presence of reducing agents such as dithiothreitol (DTT) and mercaptoethanol for longer periods of time when compared to mono- or cyclic dithiane species [40]. In addition, DNA–Au-NPs made with a trithiol linker had higher DNA-probe surface coverages (approximately 129 versus 84 for mono-thiol) than those made with either the mono-thiol or cyclic dithiane species [40, 41]. This is thought to be due to the increased binding affinity of the trithiane to the gold surface when compared to the cyclic dithiane or monothiol species. Finally, either the cyclic di-thiane or the trithiol is necessary to stabilize Au-NPs larger than 30 nm in diameter [40]. Indeed, in general the smaller particles (5–13 nm) can be stabilized better than the larger particles (> 30 nm).

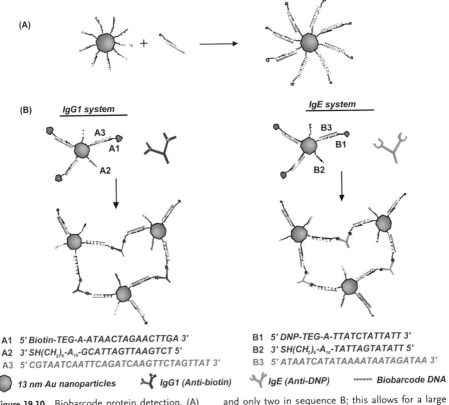

(A)

(B) *IgG1 system*   *IgE system*

A3   B3
A1   B1
A2   B2

A1 5' *Biotin-TEG-A-ATAACTAGAACTTGA 3'*
A2 3' *SH(CH$_2$)$_6$-A$_{10}$-GCATTAGTTAAGTCT 5'*
A3 5' *CGTAATCAATTCAGATCAAGTTCTAGTTAT 3'*

B1 5' *DNP-TEG-A-TTATCTATTATT 3'*
B2 3' *SH(CH$_2$)$_6$-A$_{10}$-TATTAGTATATT 5'*
B3 5' *ATAATCATATAAAATAATAGATAA 3'*

⬤ **13 nm Au nanoparticles**   Y **IgG1 (Anti-biotin)**   Y **IgE (Anti-DNP)**   ⸺ **Biobarcode DNA**

Figure 19.10 Biobarcode protein detection. (A) Preparation of hapten-modified DNA–Au-NP probes and (B) protein detection using protein binding. Note that there are nine G,C pairs in sequence A, and only two in sequence B; this allows for a large difference in the melting temperature signature that is unique for each protein analyte present.

The salt stabilization of DNA–Au-NPs is extremely important and is usually an indicator of whether or not a given DNA–Au-NP probe is DNA-functionalized and stable. In a typical synthesis, thiol-modified DNA-probe oligomers are added to gold colloid (usually in a ratio of 0.5–1 OD DNA to 1 mL Au-NP colloid which is at a concentration of ~15 nM), covered in tin foil and placed on a shaker for 16 hours. This solution is then brought to a phosphate buffer (PB) concentration of 0.01 M (pH = 7.0) and a sodium chloride (NaCl) concentration of 0.05 M. The solution is then re-covered with tin foil and allowed to shake for another 40 hours during which time the salt concentration is increased gradually from 0.05 M to 0.3 M NaCl by adding concentrated (2 M) NaCl. The 2 M NaCl is added drop-wise, shaking the solution gently between drops, and increasing by 0.05 M every 8 hours (i.e., 0.05 to 0.1 to 0.15, etc. with 8 hours between additions to 0.3 M NaCl). Usually, DNA–Au-NP probes are then stored at room temperature in the final phosphate-buffered saline (PBS) concentration of 0.3 M NaCl, 0.01 M PB. Sodium azide (NaN$_3$) is added to a concentration of 0.01–0.05 % if one plans to use and store the DNA–Au-NP probe over a length of time. The DNA–Au-NP probes should be monitored at all stages of DNA-functionalization and salt addition for aggregation using UV-Visible spectroscopy such that monodisperse probes absorb at a wavelength of 520 nm, while aggregated or "crashed" probes demonstrate a flattening and red-shift of the absorbance peak to 574 nm. Aggregated DNA–Au-NP probes are the result of either self-dimerization between the DNA probe strands or the failure to functionalize the Au-NPs with DNA either due to faulty synthesis (no terminal thiol) or rapid increases in salt concentration. Probes that are aggregated due to self-dimerization can be heated to effect their dispersion. Probe strands with increased G–C content are more likely to show this behavior due to their increased degree of hydrogen bonding. In order to remove unreacted DNA and Au-NP colloid from salt-stabilized DNA–Au-NP probes, centrifugation at 13 000 g for 25–30 minutes must be carried out to pellet the DNA–Au-NP probes. The supernatant containing unreacted DNA and gold colloid is subsequently removed. DNA–Au-NP probes are then re-dispersed in 0.3 M PBS, 0.01 M PB (pH = 7.0) ± 0.01–0.05 % NaN$_3$. Probe stability at assay-specific salt concentrations and temperature must be checked using UV-Visible spectroscopy in a typical melting experiment, monitoring the spectra at 520 nm for particle aggregation.

19.2.3
Chip Functionalization with DNA Target "Capture" Strands

For chip-based "microarray" or "gene chip" assays, glass slides must be modified with the appropriate DNA "capture" strands for hybridization and detection. As noted previously, it is more convenient to describe the DNA–Au-NP "probes" as such in our detection schemes; however, the immobilized DNA strand on the chip surface is often referred to as the "probe" strand in microarray terminology. As this chapter is not a review of either microarray technology or the fabrication of the chips used in such arrays, readers are referred to the following sources. Excellent reviews of this subject can be found in recent supplements to *Nature Genetics* [42]. Further, Cold Spring Harbor recently published, *DNA Microarrays*, which is a comprehensive manual with a copious discussion of the fabrication techniques and applications of microarrays [43]. In addition, Chrisey et al. have published an extremely useful report describing the fabrication and use of chips functionalized with heterobifunctional crosslinker molecules for covalent DNA attachment to glass slides that we often employ [44]. Note that crosslinking molecules are often light-sensitive, and so correct synthesis and storage methods should be adhered to in order to maintain functional chips.

19.2.4
Typical Assay Design

As described above, homogeneous detection assays are essentially preformed as detailed by Storhoff et al. [45], while chip-based assays are performed as detailed by Taton et al. [28], Park et al. [6], and Cao et al. [8]. All assays are carried out in appropriate strength biological buffer systems, typically PBS. Chip-based hybridizations are carried out in a humidity chamber at the appropriate temperature, either under coverslips or in Grace BioLabs on-chip hybridization vessels (GraceBio, OR, USA). For homogeneous assays, both the probes and the target are mixed together, frozen in liquid nitrogen or dry ice to expedite hybridization, left to thaw at room temperature, after which a melting analysis is performed. For chip-based assays, a mixture of the DNA–Au-NP probe and target strand is usually prehybridized at the appropriate temperature and then added to the chip surface to effect hybridization. Silver developing and enhancement is carried out with commercial silver-developing solutions after the chip has been rinsed with appropriate strength phosphate-buffered sodium nitrate ($NaNO_3$) in order to remove excess chloride ions. Scanometric array readout is performed using a conventional flatbed scanner, while Raman detection assays require a Raman spectrometer, as detailed in Cao et al. [8]. Overall, successful assays depend on the careful selection of DNA probe(s) and capture strand sequences, appropriate reaction conditions based on calculated and empirically determined melting information, and appropriate strength buffer solutions and temperatures for hybridization and washes.

19.3
Outlook

19.3.1
Challenges Ahead

DNA–nanoparticle conjugate systems for research and diagnostic purposes are currently the cutting edge in DNA and protein labeling technology. That said, there are numerous steps that must be taken from the demonstrated proof-of-concept experiments detailed above to their common place use in the clinical and diagnostic arenas. Most importantly, research demonstrating that DNA–Au-NP probes are useful for labeling numerous types of nucleic acid targets (e. g., PCR products, RNA, genomic DNA, etc.) in practical detection assays is imperative. Testing DNA–Au-NPs on biological samples (e. g., urine, saliva, blood), whether for the diagnosis of an infectious agent in a blood sample or for the identification of a single nucleotide mutation in growing cancer cells, must be carried out so that the question of how the sample should be handled to ensure probe stability and functionality in complex environments can be answered. Further, direct comparisons with current biomolecule detection technologies (e. g., ELISA immunoassays, organic fluorophore probes) must be made to ensure that the superior selectivity, sensitivity, and stability of DNA–Au-NP probes demonstrated in the discussed proof-of-concept experiments holds true for real samples.

Further, convenient high-throughput chip-reading instrumentation is necessary (Raman, electric, light scattering, or Scanometric) so that meaningful data can be derived from chip-based detection or diagnostic assays based on DNA–Au-NP probes. The ability to ratio and compare nanoparticle generated hybridization signals will ultimately require computer software and instrumentation capable of rapid analysis and distribution of data specific to the nanoparticle probe technology being used. In addition, nanoparticle probe technology itself will no doubt move forward with novel chemistries formulated for the synthesis of more highly uniform nanoparticles with desirable physical properties and for the surface ligands imparting chemical functionality. Researchers are currently working both in academia and industry to address these challenges [46].

19.3.2
Academic and Commercial Applications

Commercial applications for protein– and DNA–Au-nanoparticle conjugate systems are already being realized. Numerous companies are taking advantage of gold nanoparticle conjugates and the novel characteristics that they display to commercialize high sensitivity and selectivity assays for a wide range of analytes. Specific to DNA–Au-NPs, Nanosphere – a company which is based on much of the research and technology reported in this chapter – aims to perfect and commercialize DNA–Au-NP conjugate biomolecule sensing systems, along with the appropriate readout technology, for basic science research as well as for the clinical diagnosis of human disease [46]. In addition, Genicon, also referred to in this chapter, offers gold nanoparticle conjugates used for a wide range of biomolecule sensing applications [13]. Genicon uses protein–Au-NP probes for molecular sensing which have been commercialized by numerous companies, notably Nanoprobes, which offers Au-NP probes with a wide range of surface ligands [47]. For a more thorough review of protein–NP conjugates, please see the appropriate chapters in this volume.

Currently, nanotechnology is a field that is rapidly advancing and having a major impact on the research conducted both in academic settings and in industry. There is no doubt that fundamentally new information will result from the research conducted, methods for the synthesis and characterization of highly uniform nanoparticles will be improved and, ultimately, the future of numerous industries will be dictated by the successes enjoyed in the exciting and growing field of nanotechnology.

References

[1] Subramanian, G., et al. Implications of the human genome for understanding human biology and medicine. *JAMA* **2001**, *286*, 2296–2307.

[2] Myers, E. W., et al. On the sequencing and assembly of the human genome (vol. 99, p. 4145, 2002). *Proc. Natl. Acad. Sci. USA* **2002**, *99*, 9081–9081.

[3] Venter, J. C., The sequence of the human genome (vol. 291, p. 1304, 2001). *Science* **2002**, *295*, 1466–1466.

[4] Mirkin, C. A., et al. A DNA-based method for rationally assembling nanoparticles into macroscopic materials. *Nature* **1996**, *382*, 607–609.

[5] Taton, T. A., G. Lu, C. A. Mirkin. Two-color labeling of oligonucleotide arrays via size-selective scattering of nanoparticle probes. *J. Am. Chem. Soc.* **2001**, *123*, 5164–5165.

[6] Park, S. J., T. A. Taton, C. A. Mirkin. Array-based electrical detection of DNA with nanoparticle probes. *Science* **2002**, *295*, 1503–1506.

[7] Storhoff, J. J., et al., One-pot colorimetric differentiation of polynucleotides with single base imperfections using gold nanoparticle probes. *J. Am. Chem. Soc.* **1998**, *120*, 1959–1964.

[8] Cao, Y. W. C., R. C. Jin, C. A. Mirkin. Nanoparticles with Raman spectroscopic fingerprints for DNA and RNA detection. *Science* **2002**, *297*, 1536–1540.

[9] Bruchez Jr., M., et al. Semiconductor nanocrystals as fluorescent biological labels. *Science* **1998**, *281*, 2013–2015.

[10] Elghanian, R., et al. Selective colorimetric detection of polynucleotides based on the distance-dependent optical properties of gold nanoparticles. *Science* **1997**, *277*, 1078–1081.

[11] Niemeyer, C. M., B. Ceyhan, D. Blohm. Functionalization of covalent DNA-streptavidin conjugates by means of biotinylated modulator components. *Bioconjug. Chem.* **1999**, *10*, 708–719.

[12] Niemeyer, C. M., B. Ceyhan. DNA-directed functionalization of colloidal gold with proteins. *Angew. Chem. Int. Ed.* **2001**, *40*, 3685.

[13] Yguerabide, J., E. E. Yguerabide. Resonance light scattering particles as ultrasensitive labels for detection of analytes in a wide range of applications. *J. Cell. Biochem.* **2001**, *00*, 71–81.

[14] Bendayan, M. Worth its weight in gold. *Science* **2001**, *291*, 1363.

[15] Keating, C. D., K. M. Kovaleski, M. J. Natan. Protein:colloid conjugates for surface enhanced Raman scattering: stability and control of protein orientation. *J. Phys. Chem. B.* **1998**, *102*, 9404–9413.

[16] He, L., et al. Colloidal Au-enhanced surface plasmon resonance for ultrasensitive detection of DNA hybridization. *J. Am. Chem. Soc.* **2000**, *122*, 9071–9077.

[17] Hainfeld, J. F., R. D. Powell. New frontiers in gold labeling. *J. Histochem. Cytochem.* **2000**, *48*, 471–480.

[18] Grabar, K. C., et al. Preparation and characterization of Au colloid monolayers. *Anal. Chem.* **1995**, *67*, 735–743.

[19] Frens, G. Controlled nucleation for regulation of particle-size in monodisperse gold suspensions. *Nature-Physical Science* **1973**, *241*, 20–22.

[20] Buckle, P. E., et al. The Resonant Mirror – a Novel Optical Sensor for Direct Sensing of Biomolecular Interactions. 2. Applications. *Biosensors Bioelectronics* **1993**, *8*, 355–363.

[21] He, L., et al. Colloidal Au-enhanced surface plasmon resonance for ultrasensitive detection of DNA hybridization. *J. Am. Chem. Soc.* **2000**, *122*, 9071–9077.

[22] Goodrich, G. P., et al. Strategies for optimizing particle-amplified SPR. *Abstr. Am. Chem. Soc.* **2000**, *220*, U246–U246.

[23] Musick, M. D., et al. Colloidal Au amplified immunoassays: enhanced surface plasmon resonance biosensors. *Abstr. Am. Chem. Soc.* **1998**, *216*, U626–U626.

[24] Lyon, L. A., M. D. Musick, M. J. Natan. Colloidal Au-enhanced surface plasmon resonance immunosensing. *Anal. Chem.* **1998**, *70*, 5177–5183.

[25] Biacore. http://www.biacore.com.

[26] Nice, E. C., B. Catimel. Instrumental biosensors: new perspectives for the analysis

of biomolecular interactions. *BioEssays* **1999**, *21*, 339–352.

[27] Silin, V., A. Plant. Biotechnological applications of surface plasmon resonance. *Trends Biotechnol.* **1997**, *15*, 353–359.

[28] Taton, T. A., C. A. Mirkin, R. L. Letsinger. Scanometric DNA array detection with nanoparticle probes. *Science* **2000**, *289*, 1757–1760.

[29] Zimmermann, N., et al., Immunocytochemical investigations of the membrane of experimentally altered and physiologically aged erythrocytes. *Acta Histochem.* **1986**, 61–67.

[30] Hayat, M. A. *Colloidal gold : principles, methods, and applications.* Academic Press, San Diego, **1989**.

[31] Patolsky, F., A. Lichtenstein, I. Willner, Electrochemical transduction of liposome-amplified DNA sensing. *Angew. Chem. Int. Ed.* **2000**, *39*, 940.

[32] Patolsky, F., et al., Electronic transduction of polymerase or reverse transcriptase induced replication processes on surfaces : highly sensitive and specific detection of viral genomes. *Angew. Chem. Int. Ed.* **2001**, *40*, 2261.

[33] Storhoff, J. J., et al. What controls the optical properties of DNA-linked gold nanoparticle assemblies? *J. Am. Chem. Soc.* **2000**, *122*, 4640–4650.

[34] Schena, M., et al. Quantitative monitoring of gene-expression patterns with a complementary-DNA microarray. *Science* **1995**, *270*, 467–470.

[35] Yguerabide, J., E. E. Yguerabide. Light-scattering submicroscopic particles as highly fluorescent analogs and their use as tracer labels in clinical and biological applications – II. Experimental characterization. *Anal. Biochem.* **1998**, *262*, 157–176.

[36] Patolsky, F., et al. Dendritic amplification of DNA analysis by oligonucleotide-functionalized Au-nanoparticles. *Chem. Commun.* **2000**, 1025–1026.

[37] Han, M. Y., et al. Quantum-dot-tagged microbeads for multiplexed optical coding of biomolecules. *Nature Biotechnol.* **2001**, *19*, 631–635.

[38] Chan, W. C. W., S. Nie. Quantum dot bioconjugated for ultrasensitive nonisotopic detection. *Science* **1998**, *281*, 2016.

[39] Nam, J. M., S. J. Park, C. A. Mirkin. Bio-barcodes based on oligonucleotide-modified nanoparticles. *J. Am. Chem. Soc.* **2002**, *124*, 3820–3821.

[40] Li, Z., et al. Oligonucleotide-nanoparticle conjugates with multiple anchor groups. *Abst. Am. Chem. Soc.* **2001**, *222*, U342–U342.

[41] Demers, L. M., et al. A fluorescence-based method for determining the surface coverage and hybridization efficiency of thiol-capped oligonucleotides bound to gold thin films and nanoparticles. *Anal. Chem.* **2000**, *72*, 5535–5541.

[42] Duyk, G. M. Sharper tools and simpler methods. *Nature Genet.* **2002**, *32*, 465–468.

[43] Bowtell, D., J. Sambrook. *DNA microarrays : a molecular cloning manual.* Cold Spring Harbor Laboratory Press, Cold Spring Harbor, NY, vol. xxiii, p. 712, **2003**.

[44] Chrisey, L. A., G. U. Lee, C. E. O'Ferrall. Covalent attachment of synthetic DNA to self-assembled monolayer films. *Nucleic Acids Res.* **1996**, *24*, 3031–3039.

[45] Storhoff, J. J., et al. One-pot colorimetric differentiation of polynucleotides with single base imperfections using gold nanoparticle probes. *J. Am. Chem. Soc.* **1998**, *120*, 1959–1964.

[46] Nanosphere. http://www.nanosphere-inc.com.

[47] Nanoprobes. http://www.nanoprobes.com/index.html.

20
DNA Nanostructures for Mechanics and Computing: Nonlinear Thinking with Life's Central Molecule

Nadrian C. Seeman

20.1
Overview

The chemistry of DNA is not restricted to use in biological systems. The generalization of the biological process of reciprocal exchange leads to stable branched motifs that can be used for the construction of DNA-based geometrical and topological objects, arrays, and nanomechanical devices. In addition, it is possible to use these systems for algorithmic assembly, both to compute logical quantities and to define complex parameters for the shapes and sizes of nanotechnological systems. The information in DNA is the basis of life, but it can also be used to control the physical states of a variety of systems, leading ultimately to nanorobotics. We expect ultimately to be able to use the dynamic information-based architectural properties of nucleic acids as a basis for advanced materials with applications from nanoelectronics to biomedical devices on the nanometer scale.

20.2
Introduction

DNA is the genetic material of all living organisms; it serves as the repository for all of the information needed for the organism to grow, to replicate, to respond to its environment, and (for eukaryotes) to develop from a zygote to an adult organism. The molecular properties of DNA that allow it serve so well in this role also can be exploited for chemical ends on the nanometer scale. The DNA molecule is inherently a nanoscale species: the diameter of the double helix is about 2 nm and its helical repeat is about 3.5 nm. However, such dimensions could easily describe a variety of proteins. Hence, the question remains – what is so special about DNA?

The key feature of DNA that makes it useful as a nanoscale building block is its specificity and programmability in intermolecular interactions. We are all familiar with the notion of hydrogen bond-mediated Watson–Crick base pairing, wherein adenine (A) pairs specifically with thymine (T) and guanine (G) pairs specifically with cytosine (C) [1]. These interactions are responsible for the cohesion of the two strands of the double helix. However, it is also possible to make two different double helices cohere by having

Nanobiotechnology. Edited by Christof Niemeyer, Chad Mirkin
Copyright © 2004 WILEY-VCH Verlag GmbH & Co. KgaA, Weinheim
ISBN 3-527-30658-7

short overhangs at the ends of the strand [2]; these overhangs are called "sticky ends", and the way that they bring helices together is illustrated in Figure 20.1a. The upper drawing in Figure 20.1a shows two double helices with four-residue sticky ends that are complementary to each other. The middle drawing shows that these two molecules can cohere by hydrogen bonding to form a single molecular complex. The bottom drawing illustrates that it is possible to ligate these molecules so that they consist of one double helical complex.

The predictable specificity of sticky-ended affinity is only one half of the story of sticky ends. The other part is structure. It is known that sticky ends cohere to form the conventional structure of DNA, known as B-DNA [3]. To understand the importance of this fact,

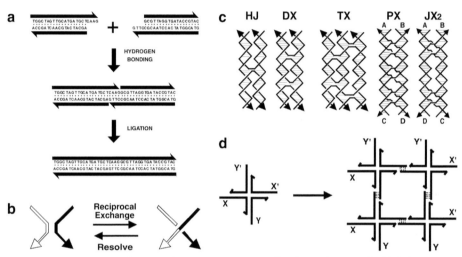

Figure 20.1 Components of Structural DNA Nanotechnology. (a) Sticky ended cohesion. Two linear double helical molecules of DNA are shown at the top of panel (a). The antiparallel backbones are indicated by the black lines terminating in half-arrows. The half-arrows indicate the 5′ → 3′ directions of the backbones. The right end of the left molecule and the left end of the right molecule have single-stranded extensions ("sticky ends") that are complementary to each other. The middle portion shows that, under the correct conditions, these bind to each other specifically by hydrogen bonding. The bottom of panel (a) shows that they can be ligated to covalency by the proper enzymes and cofactors. (b) Reciprocal exchange of DNA backbones. Two strands are shown on the left, one filled, and one unfilled. Following reciprocal exchange, one strand is filled-unfilled, and the other strand is unfilled-filled. (c) Key motifs in structural DNA nanotechnology. On the left is a Holliday junction (HJ), a four-arm junction that results from a single reciprocal exchange between double helices. To its right is a double crossover (DX) molecule, resulting from a double exchange. To the right of the DX is a triple crossover (TX) molecule, that results from two successive double reciprocal exchanges. The HJ, the DX and the TX molecules all contain exchanges between strands of opposite polarity. To the right of the TX molecule is a paranemic crossover (PX) molecule, where two double helices exchange strands at every possible point where the helices come into proximity. To the right of the PX molecule is a JX₂ molecule that lacks two of the crossovers of the PX molecule. The exchanges in the PX and JX₂ molecule are between strands of the same polarity. (d) The combination of branched motifs and sticky ends. At the left is a four-arm branched junction with sticky ends. On the right, four such molecules are combined to produce a quadrilateral. The sticky ends on the outside of the quadrilateral are available so that the structure can be extended to form a 2D lattice.

let us consider another system from which specific affinity could also be derived, say, an antigen–antibody complex. Although the structures of antibodies are well known, the geometrical relationship between the components of any given antigen–antibody pair would have to be determined individually, perhaps by a crystal structure. By contrast, the structure of sticky-ended DNA complexes is well defined, at least locally. Given that the persistence length of DNA is about 50 nm [4], the local structure in the vicinity of a combining site is well defined. Thus, sticky ends can be used to join DNA molecules together. One might think of an analogy to Velcro, but Velcro is floppy. The strongly coherent pieces of a jigsaw puzzle or, perhaps of Lego blocks, are a better analogy. Recently, other types of cohesive nucleic acid interactions have been described, so-called Tecto-RNA [5], which is cohesion through loops, and PX cohesion [6], which joins topologically closed double helices. However, the pertinent three-dimensional (3D) structures of these cohesive systems have yet to be determined.

So far, of course, I have described the ways to assemble linear DNA molecules. It is not terribly interesting to join linear DNA molecules for nanotechnological purposes, because they cannot produce precise complex systems in two or three dimensions. To do that, it is necessary to work with branched DNA molecules, rather than linear DNA molecules. Fortunately, branching is a concept familiar in nucleic acid chemistry. Ephemeral branch points are found as Holliday junction [7] intermediates in the process of genetic recombination. In general, branching can occur when exchange occurs between two DNA strands, as illustrated in Figure 20.1b [8]. Two strands are shown, one outlined, and one filled. When reciprocal exchange occurs between such molecules, new species are generated; in Figure 20.1b, the products are a filled-outlined molecule, and an outlined-filled molecule.

This simple protocol can lead to a variety of branched species, some of which are illustrated in Figure 20.1c. A single reciprocal exchange event can produce a Holliday junction-like molecule (HJ), shown at the left of Figure 20.1c. This is a four-arm branch, which is stabilized by ensuring that there is no two-fold symmetry around its branch point; minimizing sequence symmetry is key to the design of all unusual DNA motifs [9, 10]. In addition to four-arm junctions [11], three-arm junctions [12] and five-arm and six-arm junctions have been reported [13]. A double exchange between DNA double helices produces double crossover (DX) molecules [14], and double exchanges between three successive helices produces triple crossover (TX) molecules [15]. These molecules are all shown in Figure 20.1c to have undergone reciprocal exchange between strands of opposite polarity. Strands with the same polarity can also undergo reciprocal exchange. An example of this, shown with exchange events at every possible position is the paranemic crossover (PX) molecule [8]. A PX molecule lacking two exchanges, called JX_2, is shown next to the PX molecule. These last two motifs are used in a nanomechanical device described below.

Figure 20.1d illustrates the combination of unusual motifs with sticky-ended association. A four-arm junction, shown in a cruciform arrangement of arms has its arms tailed in sticky ends, X and its complement X', and Y and its complement Y'. The panel on the right illustrates four of these molecules arranged parallel to each other to form a quadrilateral. It is important to realize that in this arrangement all of the sticky ends are not satisfied. Those on the outside of the quadrilateral can participate in further interactions so that a two-dimensional (2D) periodic array can be formed. A variety of 2D DNA arrays have been produced, and they are described below. The helical nature of the DNA molecule means that constructs are not restricted to two dimensions. In principle, motifs can be rotated out of the plane to produce 3D designed periodic arrangements, although this goal has yet to be achieved in the laboratory. Nevertheless, a number of objects have been generated that are certainly incompatible with a planar graph. These include trefoil and figure-8 knots [16, 17], Borromean rings [18], and polyhedral catenanes whose helix axes have the connectivities of a cube [19] and of a truncated octahedron [20].

20.3
DNA Arrays

Being able to produce a series of topological targets, such as the knots, Borromean rings, and polyhedral catenanes is of little value unless functionality can be included in the system. The essence of nanotechnological goals is to place specific functional species at particular loci, using the architectural properties of DNA. Functionality includes the use of periodic DNA arrays to scaffold molecular arrangements in other species, algorithmic assemblies that perform computations, and the development of DNA-based nanomechanical devices. In addition to the intermolecular specificity described above, the key architectural property that is needed to build and demonstrate these arrays and devices is high structural integrity in the components; even if their associations are precise, the assembly from marshmallow-like components will not produce well-structured materials. It turns out that single-branched junctions, such as the HJ structure in Figure 20.1c are relatively flexible [12, 21]. Fortunately, the DX molecule is considerably more rigid, possibly stiffer than even double helical DNA [22], and the TX and PX molecules appear to share this rigidity [15, 23]. Consequently, it has been possible to use these molecules as the building blocks of both arrays and devices. Arrays are useful both as the basis of DNA computation by self-assembly and as frameworks to mount DNA nanomechanical devices.

Figure 20.2 illustrates the 2D arrangements that entail the use of DX molecules to produce periodic patterns. Figure 20.2a illustrates a two-component array that can tile the plane. One tile is a DX molecule labeled A, and the second is a DX molecule labeled B*. B* contains another DNA domain that projects out of the plane of the helix axes. This other domain can serve as a topographic marker for the atomic force microscope when the AB* array is deposited onto the surface of mica. The dimensions of the two DX tiles in Figure 20.2a are about 4 nm tall × 16 nm wide × 2 nm thick. Thus, the B* markers in the 2D array shown should appear as stripe-like features separated by ~32 nm, which has been confirmed experimentally. Figure 20.2b shows a four-tile arrangement that should produce stripes separated by ~64 nm, also confirmed by experiment [24]. Thus, it is possible to design and produce patterns using DNA components; these patterns contain predictable features, based on the sticky-ended cohesion of individual motifs. In addition to forming arrays from DX molecules, it is also possible to produce periodic arrays from TX molecules [14]. A variety of DNA parallelograms have also been used to produce arrays [25–27], these motifs being produced by combining four HJ-like branched junctions. Unlike the DX, TX, and PX molecules, the two domains of the HJ molecule are not parallel to each other. As a function of the sequence and backbone

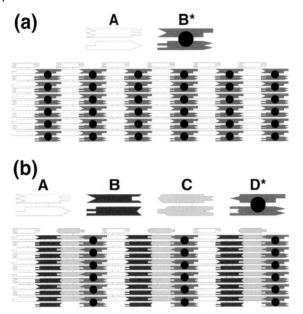

Figure 20.2 Tiling the plane with DX molecules. (a) A two-tile pattern. The two helices of the DX molecule are represented schematically as rectangular shapes that terminate in a variety of shapes. The terminal shapes are a geometrical representation of sticky ends. The individual tiles are shown at the top of the drawing; the way tiles fit together using complementary sticky ends to tile the plane is shown at the bottom. The molecule labeled A is a conventional DX molecule, but the molecule labeled B* contains a short helical domain that protrudes from the plane of the helix axes; this protrusion is shown as a black dot. The black dots form a stripe-like feature in the array. The dimensions of the tiles are 4 × 16 nm in this projection. Thus, the stripe-like features should be about 32 nm apart. (b) A four-tile pattern. The same conventions apply as in (a). The four tiles form an array in which the stripes should be separated by about 64 nm, as confirmed by atomic force microscopy (AFM).

connections at the crossover point, they adopt angles of ∼60° [25], ∼−70° [26], or ∼40° [27, 28], thus producing a diversity of parallelogram angles.

Programmed arrays bear on several aspects of DNA nanotechnology. First, they appear to be appropriate for the scaffolding of other molecules. One suggestion is that periodic arrays could be used to arrange biological macromolecules into crystalline arrangements [9] (Figure 20.3a). A second suggestion is that they could be used to arrange the components of nanocircuitry [29] (Figure 20.3b). Similarly, arrays may be used as a supporting surface to organize nanomechanical switches and devices. In addition to periodic arrays, one can also organize algorithmically-ordered DNA arrays. These will be discussed below in the section of DNA-based computation.

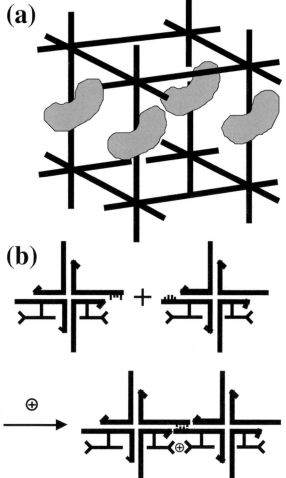

Figure 20.3 Applications of DNA periodic arrays. (a) Biological macromolecules organized into a crystalline array. A cube-like box motif is shown, with sticky ends protruding from each vertex. Attached to the vertical edges are biological macromolecules that have been aligned to form a crystalline arrangement. The idea is that the boxes are to be organized into a host lattice by sticky ends, thereby arranging the macromolecular guests into a crystalline array, amenable to diffraction analysis. (b) Nanoelectronic circuit components organized by DNA. Two DNA branched junctions are shown, with complementary sticky ends. Pendent from the DNA are molecules that can act like molecular wires. The architectural properties of the DNA are seen to organize the wire-like molecules, with the help of a cation, which forms a molecular synapse.

20.4
DNA Nanomechanical Devices

Nanomechanical action is a central target of nanotechnology. The first DNA-based devices were predicated on structural transitions of DNA driven by small molecules. The first deliberate DNA device entailed the extrusion of a DNA cruciform structure from a cyclic molecule [30]. The position of the branch point was controlled by the addition or removal of an intercalating dye to the solution. This system was not very convenient to operate, and the large size of the DNA circle made it unwieldy to handle. Nevertheless, it demonstrated that DNA could form the basis of a two-state system.

(a)

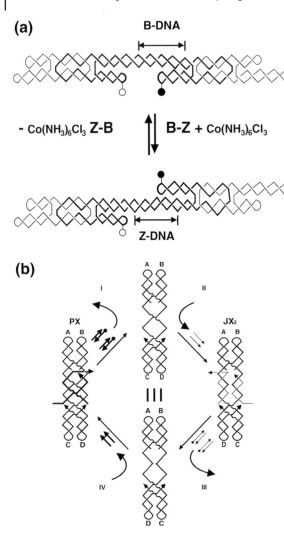

B-DNA

- Co(NH$_3$)$_6$Cl$_3$ **Z-B** ↕ **B-Z** + Co(NH$_3$)$_6$Cl$_3$

Z-DNA

(b)

PX   JX$_2$

Figure 20.4 DNA-based nanomechanical devices. (a) A device predicated on the B-Z transition. The molecule consists of two DX molecules, connected by a segment containing proto-Z-DNA. The molecule consists of three cyclic strands, two on the ends drawn with a thin line, and one in the middle, drawn with a thick line. The molecule contains a pair of fluorescent dyes to report their separation by FRET. One is drawn as a filled circle, and the other as an empty circle. In the upper molecule, the proto-Z segment is in the B conformation, and the dyes are on the same side of the central double helix. In the lower molecule, the proto-Z segment is in the Z conformation, and the dyes are on opposite sides of the central double helix. The length of the proto-Z-DNA and its conformation are indicated at top and bottom by the two vertical lines flanking the conformation descriptor. (b) A sequence-dependent device. This device uses two motifs, PX and JX$_2$. The labels A, B, C, and D on both show that there is a 180 °C difference between the wrappings of the two molecules. There are two strands drawn as thick lines at the center of the PX motif, and two strands drawn with thin lines at the center of the JX$_2$ motif; in addition to the parts pairing to the larger motifs, each has an unpaired segment. These strands can be removed and inserted by the addition of their total complements (including the segments unpaired in the larger motifs) to the solution; these complements are shown in processes I and III as strands with black dots (representing biotins) on their ends. The biotins can be bound to magnetic streptavidin beads so that these species can be removed from solution. Starting with the PX, one can add the complement strands (process I), to produce an unstructured intermediate. Adding the set strands in process II leads to the JX$_2$ structure. Removing them (III) and adding the PX set strands (IV) completes the machine cycle. Many different devices could be made by changing the sequences to which the set strands bind.

The second DNA-based device (Figure 20.4a) was a marked advance. It was relatively small, and included two rigid components, DX molecules like those used to make the arrays of Figure 20.2. It, too, relied on the addition or removal of a small molecule. The basis of the device was the transition between right-handed (conventional) B-DNA and left-handed Z-DNA. There are two requirements for the formation of Z-DNA, a "'proto-Z" sequence capable of forming Z-DNA readily (typically a $(CG)_n$ sequence), and conditions (typically high salt or molecules like $Co(NH_3)_6^{3+}$ that emulate the presence of high salt) to promote the transition [31]. The sequence requirement enables us to control the transition in space, and the requirement for special conditions allows us to control the transition, and hence the device, in time. As shown in Figure 20.4a, the device consists of two DX molecules connected by a shaft containing a proto-Z sequence that consists of $(CG)_{10}$. In B-promoting conditions, both of the DX helices not collinear with the shaft are on the same side of the shaft. In Z-promoting conditions, one of these helices winds up on the other side of the shaft. This difference is the result of converting a portion of the shaft to Z-DNA, which rotates one DX motif relative to the other by 3.5 turns. The motion is demonstrated by fluorescence resonance energy transfer (FRET) measurements that monitor the difference between dye separations on the DX motifs [32].

The problem with the B-Z device is that it is activated by an unspecific molecule, $Co(NH_3)_6^{3+}$. Thus, a number of such devices, embedded in an array, would all respond similarly, at least within the limits of a small amount of chemical nuance [17]. Thus, N two-state devices would result in essentially two structural states; clearly it would be of much greater value to have N distinct 2-state devices capable of producing 2^N structural states. It is evident that a sequence-dependent device would be an appropriate vehicle for the goal of achieving multiple states. The method for devising sequence-dependent devices was worked out by Yurke and his colleagues [33]. It entails setting the state of a device by the addition of a "set" strand that contains an unpaired tail. When the full complement to the set strand is added, the set strand is removed, and a different set strand may be added. This system has been adapted to the PX and JX₂ motifs (see Figure 20.1) [23]. Figure 20.4b shows how these two states can be interconverted by the removal of one pair of set strands (processes I and III) and the addition of the opposite pair (processes II and IV). The tops and bottoms of the two states differ by a half-rotation, as seen by comparing the A and B labels at the tops of the molecules, and the C and D labels at the bottoms of the molecules. A variety of devices can be produced by changing the sequences of the regions where the set strands bind.

20.5
DNA-based Computation

Experimental DNA-based computation was founded by Leonard Adleman in 1994 [34]. His approach is different from the use of DNA to scaffold nanoelectronic component assembly, as suggested in Figure 20.3b. Instead, Adleman combined the information in DNA molecules themselves, using standard biotechnological operations (ligation, PCR, gel electrophoresis, and sequence-specific binding) to solve a Hamiltonian path problem. The idea is that there exist certain classes of computational problems for which the parallelism of molecular assembly overcomes the slow speed of the required macroscopic manipulations. Many varieties of DNA-based computation have been proposed, and a number of them have been executed experimentally for relatively small cases. Limited space within this chapter does not permit discussion of all of them.

However, there is one approach to DNA-based computation that is relevant to our discussion of structural DNA nanotechnology. This was a method suggested by Winfree, who noticed that the system described above, branched junctions with sticky ends, could be a way to implement computation by "Wang tiles" on the molecular scale [35]. This is a system of tiles whose edges may contain one or more different markings; the tiles self-assemble into a mosaic according to the local rule that all edges in the mosaic are flanked by the same color. Such a form of assembly can be shown to emulate the operation of a Turing machine, a general-purpose computer [36]. The relationship between the sticky ends of a branched junction and the markings on a Wang tile is shown in Figure 20.5a.

This form of DNA-based computation has been prototyped successfully in a 4-bit cumulative XOR calculation [37]. The XOR calculation yields a 1 if the two inputs are different, and a 0 if they are the same. Figure 20.5b shows the components of this calculation. Each component is a TX molecule, schematized as three rectangles with geometrical shapes on their ends to represent complementarity. The input bits are 'x' tiles (upper left), and the output bits are 'y' tiles (bottom), and there are two initiator tiles, C1 and C2, as well (upper right). The upper left corner of Figure 20.5c shows the strand structure of the TX tiles;

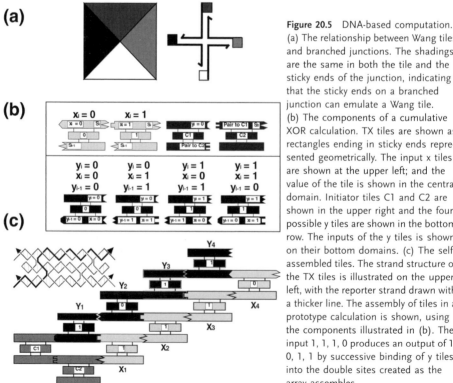

(a)

(b)

(c)

Figure 20.5 DNA-based computation. (a) The relationship between Wang tiles and branched junctions. The shadings are the same in both the tile and the sticky ends of the junction, indicating that the sticky ends on a branched junction can emulate a Wang tile. (b) The components of a cumulative XOR calculation. TX tiles are shown as rectangles ending in sticky ends represented geometrically. The input x tiles are shown at the upper left; and the value of the tile is shown in the central domain. Initiator tiles C1 and C2 are shown in the upper right and the four possible y tiles are shown in the bottom row. The inputs of the y tiles is shown on their bottom domains. (c) The self-assembled tiles. The strand structure of the TX tiles is illustrated on the upper left, with the reporter strand drawn with a thicker line. The assembly of tiles in a prototype calculation is shown, using the components illustrated in (b). The input 1, 1, 1, 0 produces an output of 1, 0, 1, 1 by successive binding of y tiles into the double sites created as the array assembles.

each strand contains a "reporter strand" (drawn with a thicker line); the value of x and y tiles is set to 0 or 1 depending whether it contains a *Pvu*II or *Eco*RV restriction site, respectively. The y_i tiles perform the gating function; there are four of them, corresponding to the four possible combinations of 0 and 1 inputs. The input involves the bottom domain (Figure 20.5b). The assembly of periodic arrays discussed above entails competition between correct and incorrect tiles for particular positions; by contrast, the competition here is between correct and partially correct tiles. For example, the $y_{i-1} = 0$ sticky end on the leftmost tile is the same as the y_{i-1} sticky end on the rightmost tile. In the cumulative XOR calculation, $y_i = \text{XOR} \, (x_i, y_{i-1})$. The implementation of this formula is shown in Figure 20.5c. The x_i tiles and the initiators are given longer sticky ends than the y_i tiles, so they assemble a template first when the tiles are cooled. This creates a double site where the y_1 tile can bind. This binding creates the double site where the y_2 tile can bind, and so on. When the assembly is complete, the reporter strands are ligated together, creating a long strand that connects the input to the output through the initiator tiles. Partial restriction analysis of the resulting strand reveals that the correct answer is obtained almost exclusively.

20.6
Summary and Outlook

This chapter has discussed the current state of structural DNA nanotechnology, emphasizing the individual components, their assembly into periodic and algorithmic arrays, and their manipulation as nanomechanical devices. Where is this area going? The achievement of several key near-term goals will move structural DNA nanotechnology from an elegant structural curiosity to a system with practical capabilities. First among these goals is the extension of array-making capabilities from two to three dimensions, particularly with high order. Likewise, heterologous molecules must be incorporated into DNA arrays, so that the goals both of orienting biological macromolecules for diffraction purposes and of organizing nanoelectronic circuits may be met. A DNA nanorobotics awaits the incorporation of the PX-JX$_2$ device into arrays. Algorithmic assembly in three dimensions will lead ultimately to very smart materials, particularly if combined with nanodevices. The development of self-replicating systems using branched DNA appears today to be somewhat oblique [38, 39], but it nevertheless represents an exciting challenge that will significantly economize on the preparation of these systems. Currently, structural DNA nanotechnology is a biokleptic pursuit, stealing genetic molecules from biological systems; ultimately, it must advance from biokleptic to biomimetic, not just using the central molecules of life, but improving on them, without losing their inherent power as central elements of self-assembled systems.

Acknowledgments

This research has been supported by grants GM-29554 from the National Institute of General Medical Sciences, N00014-98-1-0093 from the Office of Naval Research, grants DMI-0210844, EIA-0086015, DMR-01138790, and CTS-0103002 from the National Science Foundation, and F30602-01-2-0561 from DARPA/AFSOR.

References

[1] J. D. Watson, F. H. C. Crick, *Nature* **1953**, *171*, 737–738.

[2] S. N. Cohen, A. C. Y. Chang, H. W. Boyer, R. B. Helling, *Proc. Natl. Acad. Sci. USA* **1973**, *70*, 3240–3244.

[3] H. Qiu, J. C. Dewan, N. C. Seeman, *J. Mol. Biol.* **1997**, *267*, 881–898.

[4] P. J. Hagerman, *Annu. Rev. Biophys. Biophys. Chem.* **1988**, *17*, 265–286.

[5] L. Jaeger, E. Westhof, N. B. Leontis, *Nucleic Acids Res.* **2001**, *29*, 455–463.

[6] X. Zhang, H. Yan, Z. Shen, N. C. Seeman, *J. Am. Chem. Soc.* **2002**, *124*, 12940–12941.

[7] R. Holliday, *Genet. Res.* **1964**, *5*, 282–304.

[8] N. C. Seeman, *Nano Lett.* **2001**, *1*, 22–26.

[9] N. C. Seeman, *J. Theoret. Biol.* **1982**, *99*, 237–247.

[10] N. C. Seeman, *J. Biomol. Struct. Dyn.* **1990**, *8*, 573–581.

[11] N. R. Kallenbach, R.-I. Ma, N. C. Seeman, *Nature* **1983**, *305*, 829–831.

[12] R.-I. Ma, N. R. Kallenbach, R. D. Sheardy, M. L. Petrillo, N. C. Seeman, *Nucleic Acids Res.* **1986**, *14*, 9745–9753.

[13] Y. Wang, J. E. Mueller, B. Kemper, N. C. Seeman, *Biochemistry* **1991**, *30*, 5667–5674.

[14] T.-J. Fu, N. C. Seeman, *Biochemistry* **1993**, *32*, 3211–3220.

[15] T. LaBean, H. Yan, J. Kopatsch, F. Liu, E. Winfree, J. H. Reif, N. C. Seeman, *J. Am. Chem. Soc.* **2000**, *122*, 1848–1860.

[16] J. E. Mueller, S. M. Du, N. C. Seeman, *J. Am. Chem. Soc.* **1991**, *113*, 6306–6308.

[17] S. M. Du, B. D. Stollar, N. C. Seeman, *J. Am. Chem. Soc.* **1995**, *117*, 1194–1200.

[18] C. Mao, W. Sun, N. C. Seeman, *Nature* **1997**, *386*, 137–138.

[19] J. Chen, N. C. Seeman, *Nature* **1991**, *350*, 631–633.

[20] Y. Zhang, N. C. Seeman, *J. Am. Chem. Soc.* **1994**, *116*, 1661–1669.

[21] M. L. Petrillo, C. J. Newton, R. P. Cunningham, R.-I. Ma, N. R. Kallenbach, N. C. Seeman, *Biopolymers* **1988**, *27*, 1337–1352.

[22] X. Li, X. Yang, J. Qi, N. C. Seeman, *J. Am. Chem. Soc.* **1996**, *118*, 6131–6140.

[23] H. Yan, X. Zhang, Z. Shen, N. C. Seeman, *Nature* **2002**, *415*, 62–65.

[24] E. Winfree, F. Liu, L. A. Wenzler, N. C. Seeman, *Nature* **1998**, *394*, 539–544.

[25] C. Mao, W. Sun, N. C. Seeman, *J. Am. Chem. Soc.* **1999**, *121*, 5437–5443.

[26] R. Sha, F. Liu, D. P. Millar, N. C. Seeman, *Chem. Biol.* **2000**, *7*, 743–751.

[27] R. Sha, F. Liu, N. C. Seeman, *Biochemistry* **2002**, *41*, 5950–5955.

[28] B. F. Eichman, J. M. Vargason, B. H. M. Mooers, P. S. Ho, *Proc. Natl. Acad. Sci. USA* **2000**, *97*, 3971–3976.

[29] B. H. Robinson, N. C. Seeman, *Prot. Eng.* **1987**, *1*, 295–300.

[30] X. Yang, A. Vologodskii, B. Liu, B. Kemper, N. C. Seeman, *Biopolymers* **1998**, *45*, 69–83.

[31] A. Rich, A. Nordheim, A. H.-J. Wang, *Annu. Rev. Biochem.* **1984**, *53*, 791–846.

[32] C. Mao, W. Sun, Z. Shen, N. C. Seeman, *Nature* **1999**, *397*, 144–146.

[33] B. Yurke, A. J. Turberfield, A. P. Mills, Jr., F. C. Simmel, F. C., J. L. Neumann, *Nature* **2000**, *406*, 605–608.

[34] L. Adleman, *Science* **1994**, *266*, 1021–1024.

[35] E. Winfree, in: E. J. Lipton, E. B. Baum (eds), *DNA Based Computing*. Am. Math. Soc., Providence, **1996**, pp. 199–219.

[36] B. Grünbaum, G. C. Shephard, *Tilings Patterns*. W. H. Freeman Co., New York, **1987**, pp. 583–608.

[37] C. Mao, T. LaBean, J. H. Reif, N. C. Seeman, *Nature* **2000**, *407*, 493–496.

[38] N. C. Seeman, N. C., *DNA Cell Biol.* **1991**, *10*, 475–486.

[39] L. H. Eckardt, K. Naumann, W. M. Pankau, M. Rein, M. Schweitzer, N. Windhab, G. von Kiedrowski, *Nature* **2002**, *420*, 286–286.

21
Nanoparticles as Non-Viral Transfection Agents

M. N. V. Ravi Kumar, Udo Bakowsky, and Claus-Michael Lehr

21.1
Introduction to Gene Delivery

Direct injection of naked DNA plasmids is possible, but relatively few cells take up the DNA (1–3%), leading to a small production of the encoded protein. The most important use of naked DNA plasmids is in vaccine development, as the small amount of protein produced can elicit a protective immune response. However, in most of the cases, it is also well established that naked DNA is not suitable for in-vivo transport of genetic material into selected cell body types due to its degradation by serum nucleases; hence, the use of a carrier system is suggested. Unfortunately, there is no single system universally applicable in vivo. An ideal system should be biocompatible, nonimmunogenic, and stable in the bloodstream, protect DNA during transport, avoid entrapment by components of the reticuloendothelial system (RES), be small enough to extravagate, and should be cell- or tissue-specific to reach selected cells in the body. The delivery system is expected

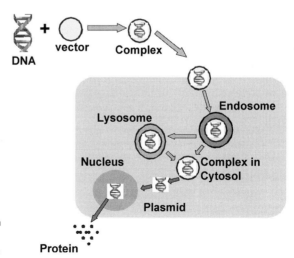

Figure 21.1 Schematic representation of biological barriers that need to be overcome by the gene transfer vector.

Nanobiotechnology. Edited by Christof Niemeyer, Chad Mirkin
Copyright © 2004 WILEY-VCH Verlag GmbH & Co. KgaA, Weinheim
ISBN 3-527-30658-7

to enter the cell via endocytosis, thereby avoiding its interaction with lysosomal enzymes, and to facilitate endosomal escape, resulting in DNA delivery to the nucleus, as shown in Figure 21.1. Thus, it is quite obvious that a successful gene delivery system must contain a variety of structural elements responsible for the specific behavior.

In gene therapy, plasmid DNA is introduced into cells of patients to express the therapeutic proteins. On the other hand, an oligonucleotide is used to suppress the expression of a disease-causing gene in antisense therapy. The clinical application of these new types of gene-drugs is severely hindered by their instability in biological fluids and the low cellular uptake efficiency due to the high molecular weight and polyanionic nature of the nucleic acids. Thus, it is necessary to develop an efficient delivery system that can transfer the gene/DNA in to the target site. So far, two major approaches have been tried and are in use for gene delivery: these are viral vectors and nonviral vectors.

In spite of their relatively high efficacy, several major problems are associated with the use of viral delivery systems in clinical treatment, particularly in relation to the risk of an immune response against viral particles, and also to the risk of random integration mediated by viruses or their recombination with wild-type viruses [1]. In an efficient gene therapy, plasmid DNA is introduced into target cells, transcribed and the genetic information ultimately translated into the corresponding protein (see Figure 21.1).

Successful/efficient transfection is hampered by: (i) targeting the delivery systems to the target cell; (ii) transport through the cell membrane; (iii) uptake and degradation in endolysomes; and (iv) intracellular trafficking of plasmid DNA to the nucleus. Although, viral vectors yield high transfection efficiency over a wide range of cell targets [2, 3], they present major drawbacks, such as virally induced inflammatory responses and oncogenic effects [4].

There is need for the development of safer and more effective gene delivery vehicles. These should offer freedom to manipulate the complex stoichiometry, surface charge density, and hydrophobicity needed for interaction with the cellular lipid components. Cationic phospholipids and cationic polymers are the two major types of nonviral gene delivery vectors currently being investigated. Due to their permanent cationic charge, both types interact electrostatically with negatively charged DNA and form complexes (lipo- or poly-plexes). Despite the ease of fabrication of the lipoplexes, their low transfection efficiency and toxicity has limited its success. However, polyplexes involving cationic polymers, on the other hand are more stable than cationic lipids [5], although the transfection is relatively low when compared to viral vectors. Cationic polymers have been used to condense and deliver DNA both in vitro and in vivo. Several cationic polymers have been investigated that lead to higher transfection efficiencies when compared to the other nonviral vectors in use [5, 6]. They form polyelectrolyte complexes with plasmid DNA, in which the DNA becomes better protected against nuclease degradation [7]. They also show structural variability and versatility, including the possibility of linking the targeting moieties for gene expression mediated through specific receptors [5].

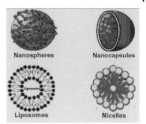

Figure 21.2 Structures of various nanoparticles in use for pharmaceutical applications.

21.2
Nanoparticles for Drug and Gene Targeting

Pharmaceutical nanoparticles were first developed by Speiser and co-workers [8] during the 1970s, and are defined as solid colloidal particles, less then 1 μm in size, that consist of macromolecular compounds. Since then, a considerable amount of work has been carried out on nanoparticles worldwide in the field of drug/gene delivery. Nanoparticles were initially devised as carriers for vaccines and anticancer drugs [9], but their use for ophthalmic and oral delivery has also been investigated [10]. Drugs or other biologically active molecules are dissolved, entrapped or encapsulated in the nanoparticles, or are chemically attached to the polymers or adsorbed to their surface. The selection of an appropriate method for preparing drug-loaded nanoparticles depends on the physico-chemical properties of the polymer and the drug. On the other hand, the procedure and the formulation conditions will determine the inner structure of these polymeric colloidal systems. Two types of systems with different inner structures are possible:

- A matrix-type system composed of an entanglement of oligomer or polymer units, defined here as a nanoparticle or nanosphere.
- A reservoir type system, consisting of an oily core surrounded by a polymer wall, defined here as a nanocapsule.

Various colloidal nanoparticulate systems in use for drug/gene delivery are shown in Figure 21.2. In this chapter, we will discuss various nanoparticulate systems which are used as gene carriers. Great care has been taken in compiling the literature; however, any omission in the references is purely inadvertent and is highly regretted.

21.3
Nonviral Nanomaterials in Development and Testing

21.3.1
Chitosan

Chitin is the most abundant natural aminopolysaccharide, and its annual production is estimated to be almost as much as that of cellulose (Figure 21.3). Chitin has become of major interest not only as an underutilized resource, but also as a new functional material of high potential in various fields, and recent progress in chitin chemistry is quite noteworthy. Chitosan is an aminopolysaccharide obtained by the alkaline deacetylation of

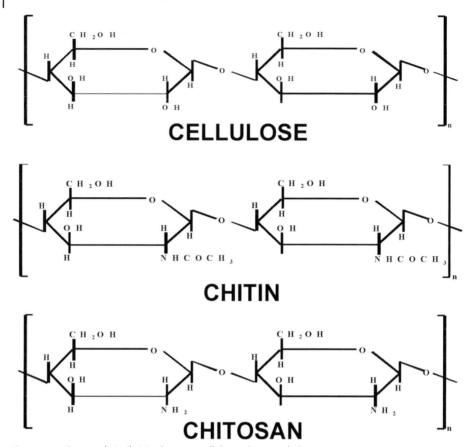

Figure 21.3 Structural similarities between cellulose, chitin, and chitosan.

chitin, a cellulose-like polymer which is present in fungal cell walls and the exoskeletons of arthropods such as insects, crabs, shrimps, lobsters, and other vertebrates [11]. Chitosan is a nontoxic, biocompatible polymer that has found a number of applications in drug delivery [12], and recently has emerged as an alternative nonviral gene delivery system [13]. Borchard [14] recently described chitosans as efficient gene delivery systems in his review.

MacLaughlin et al. [15] studied chitosan and depolymerized chitosan oligomers as condensing carriers for in-vivo plasmid delivery. For forming the complexes with plasmid, each chitosan oligomer or polymer (2 %, w/v) was dissolved in acetic acid by sonication and the final concentration of the solution was brought to 0.4 % (w/v) by adding water; the solution was sterile-filtered to remove particles above 200 nm. This chitosan solution is then added to aqueous suspension containing 25–400 mg of plasmid in a total volume of 1 mL by gentle pipetting to form nanocomplexes of a selected charge ratio [14]. The solution was vortexed rapidly for 3–5 minutes and left for 30 minutes at room temperature to ensure complete complexation [15]. The in-vitro transfection was tested in Cos-1 cells, and in-vivo expression in intestinal tissues. In vivo, higher levels of expression were measured with plasmid/chitosan/GM225.1 formulation over naked plasmid in the upper small intestine, the overall expression levels achieved using the DOTMA:DOPE based formulation were lower than those achieved with plasmid/chitosan/GM225.1 complexes, and there were a lower number of tissue extracts showing positive chloramphenicol acetyltransferase (CAT) gene expression (Table 21.1). It was interesting however to note that the in-vitro results were the reverse of these findings. The parameters which influenced

Table 21.1 CAT gene expression in tissue extracts of rabbits dosed in the upper small intestine or colon with plasmid/chitosan complexes

| Formulation | Region | Rabbit number | | | | CAT expression[*] |
		#1	#2	#3	#4	[pg mg^{-1}]
Dosed in upper small intestine	PP1	0	0	0	0	0 ± 0
	PP2	0	0	0	0	0 ± 0
Plasmid/CT/GM225.1	PP3	6.12	5.01	0	5.83	4.24 ± 2.87
(n = 4)	ENT1	1.45	0	8.10	6.62	4.04 ± 3.92
	ENT2	0	4.50	0	9.60	3.52 ± 4.57
	ENT3	6.16	0	11.41	4.32	5.47 ± 4.72
	Col	0	0	0	0	–
	MLN	0	0	3.68	10.96	3.66 ± 5.17
Plasmid in 10% lactose	PP3	0	0	–	–	0 ± 0
(n = 2)	ENT1	0	0	–	–	0 ± 0
	MLN	0	0	–	–	0 ± 0
Plasmid/DOTMA:DOPE	PP1	0	0	0	6.75	1.69 ± 3.37
(n = 4)	PP2	0	0	0	0	0 ± 0
	PP3	0	8.85	0	6.83	3.9 ± 2.46
	ENT1	0	0	0	10.13	2.53 ± 5.07
	ENT2	0	0	0	0	0 ± 0
	ENT3	0	0	0	0	0 ± 0
	Col	0	0	0	4.74	1.19 ± 2.37
	MLN	0	0	0	4.60	1.15 ± 2.30
Dosed in the colon						
Plasmid/CT/GM225.1	Col1	0	0	0	6.30	1.58 ± 3.20
(n = 4)	Col2	0	0	7.47	6.30	3.57 ± 4.13
	Col3	0	0	0	0	0 ± 0
	MLN	5.97	10.11	12.14	0.00	7.06 ± 5.36
Plasmid in 10% lactose	PP3	0	0	–	–	0 ± 0
(n = 2)	ENT1	0	0	–	–	0 ± 0
	MLN	0	0	–	–	0 ± 0
Plasmid/DOTMA:DOPE	Col1	0	0	0	0	0 ± 0
(n = 4)	Col2	0	0	0	0	0 ± 0
	Col3	0	0	0	0	0 ± 0
	MLN	0	0	0	0	0 ± 0

[*]Values ae mean ± SD.

particle size and stability included chitosan molecular weight, plasmid concentration, and the charge ratio. Plasmid/chitosan complexes made of higher molecular-weight chitosan were more stable to salt and serum challenge. Complexes of a 1:2 (−/+) charge ratio were shown to be most stable. In vitro, the highest level of expression in the absence of serum was obtained using a 1:2 (−/+) complex made with 102 kDa chitosan, and was approximately 250-fold lower than that observed with a positive control Lipofectamine™. Surprisingly, particle size was found not to influence the expression, but inclusion of the pH-sensitive endoosomolytic peptide GM227.3 in the formulation enhanced the levels of expression with the 1:2:0.25 (−/+/−) complex (200 µg mL^{-1}), though expression levels were very low (100-fold less) when compared to that of Lipofectamine.

Detailed investigations on chitosans as efficient gene transfection agents in vitro and in vivo by Leong and co-workers [16–20] resulted in a series of papers describing various modifications. They investigated the important parameters for the preparation of nanoparticles and characterized the physico-chemical properties of the system. The protection for encapsulated DNA by chitosan particles was confirmed [19]. These authors have also investigated the effects of co-encapsulating a lysoosomolytic agent chloroquine, on the transfection efficiency [19]. They have also proposed plausible schemes (Figure 21.4) for transferrin and KNOB protein conjugation in an attempt to improve the surface property resulting in improved transfection efficiency [19]. Furthermore, the chitosan-DNA particles were PEGylated (Figure 21.5) to improve their storage stability and to yield a formulation that could be lyophilized without loss of the transfection ability [19]. Tissue distribution of the chitosan-DNA nanoparticles and PEGylated nanoparticles following intravenous administration was investigated [19], and clearance of the PEGylated nanoparticles was found to be slower than that of the unmodified particles at 15 minutes, but with higher deposition in the kidney and liver, though no difference was seen at the 1-hour time point [19]. It was noted that the transfection efficiency was cell type-dependent, with three to four orders of magnitude (in relative light units) higher than background level in HEK293 cells, and two- to ten-fold lower than that achieved by Lipofectamine [19]. Mohapatra and co-workers [20] used these chitosan nanoparticles [19] to demonstrate the protection factor of the particles towards acute respiratory syncytial virus infection in BALB/c mice following intranasal gene transfer.

Hoggard et al. [21] used chitosan nanoparticles particles made from different molecular weight chitosans (Table 21.2) for intratracheal administration in mice, and compared the results obtained with those reported for polyethyleimine (PEI) in vitro. Following intratracheal administration, both polyplexes were seen to be distributed to the mid-airways, where transgene expression was observed in virtually every epithelial cell, using a sensitive pLacZ reported containing a translational enhancer element [21]. However, a similar kind of result as others was observed, where PEI polyplexes induced more rapid onset of gene expression than that of chitosans [21].

Sato and co-workers [22], in their studies, optimized the transfection conditions for chitosan-mediated gene delivery. The transfection studies were carried out using a tumor cell line (Human-lung Carcinoma A549 cells, HeLa cells and B16 melanoma cells; 1×10^5 cells/well). The transfection efficiency of the chitosan complexes was seen to depend on the pH of the culture medium, the stoichiometry of pGL3:chitosan, serum and the molecular mass of chitosan [22]. Transfection efficiency at pH 6.9 was higher than that

at pH 7.6, and the optimum charge ratio of the pGL3 :chitosan was 1 :5. Chitosan polymers of 15 and 52 kDa largely promoted luciferase activities. Transfection efficiency
mediated by chitosan of >100 kDa was less than that by chitosan of 15 and 52 kDa [22].

Kim et al. [23] reported self-aggregates of deoxycholic acid-modified chitosan as being
DNA carriers. In order to control the size of the self-aggregates, chitosan was depolymerized with various amounts of sodium nitrite, and hydrophobically modified with deoxycholic acid to form self-aggregates in aqueous media (Figure 21.6). These authors observed that the size of the self-aggregates varied in the range of 130 to 300 nm in diameter,
where the chitosan molecular mass has influence on the sizes.

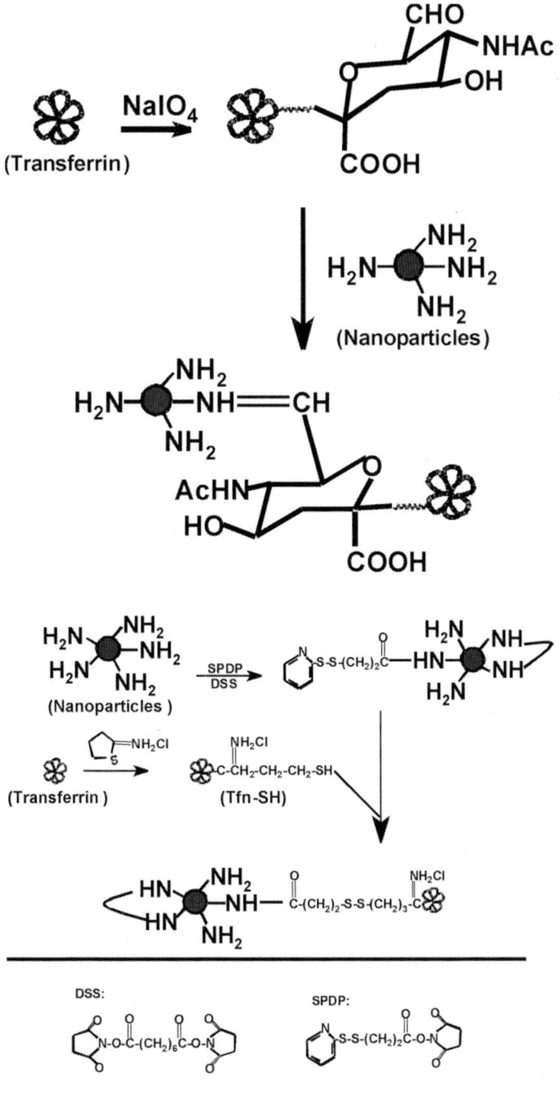

Figure 21.4 (a) Conjugation of transferrin through periodate oxidation; (b) conjugation of transferrin through a reversible disulfide linkage.

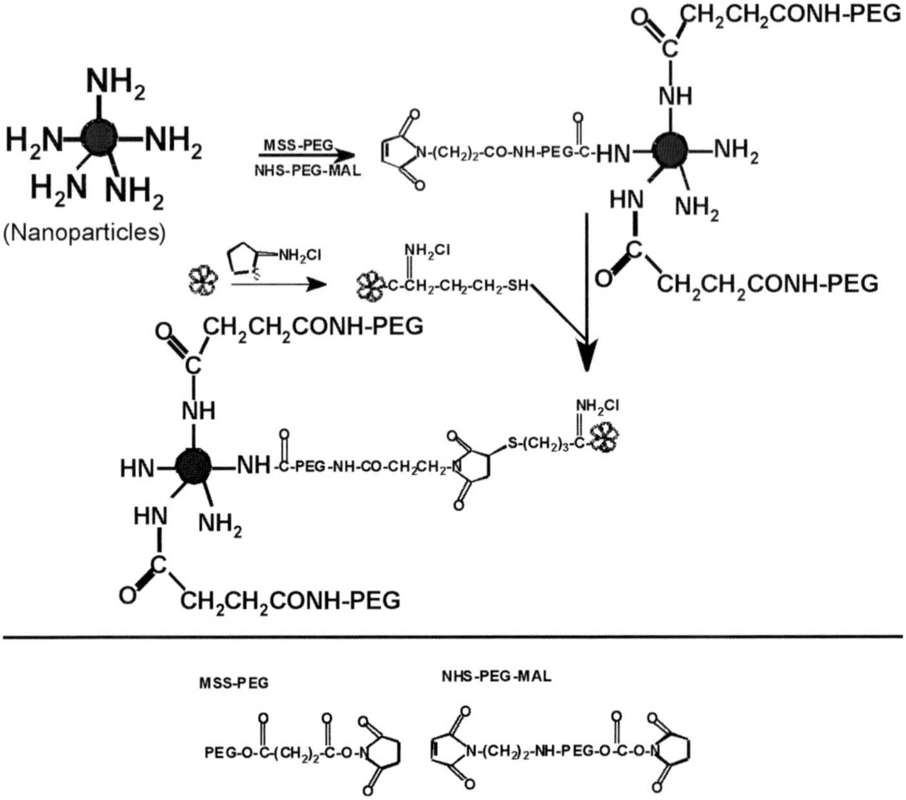

Figure 21.5 PEGylation of chitosan–DNA nanoparticles and conjugation of transferrin through a PEG spacer.

Table 21.2 Transfection efficiency in 292 cells and particle size of different chitosan/pDNA complexes at their optimal charge ratios

Chitosan [–][$^{-1}$][]	Charge ratio [+/–]	Gene expression [pg mg^{-1} protein]	Particle size [nm]
C(1;31)	3.6:1	6.0 ± 2.8	131 ± 9
C(1;170)	3.6:1	5.0 ± 2.0	174 ± 23
C(15;190)	3.0:1	7.2 ± 1.5	144 ± 12
C(35;170)	3.0:1	0.2 ± 0.04	195 ± 15
C(49;98)	3.6:1	0.1 ± 0.04	229 ± 2

Charge ratios covering the range 0.6:1–4.2:1 (+/–) were investigated.
Cells were analyzed for CAT gene expression 48 hours after transfection. Data are expressed as mean values ± SD from one representative experiment (n = 4) of three performed.
Chitosan represented by previously published nomenclature in which C is followed by two numbers: the first is the degree of deacetylation (in %); the second is the molecular weight in kDa.

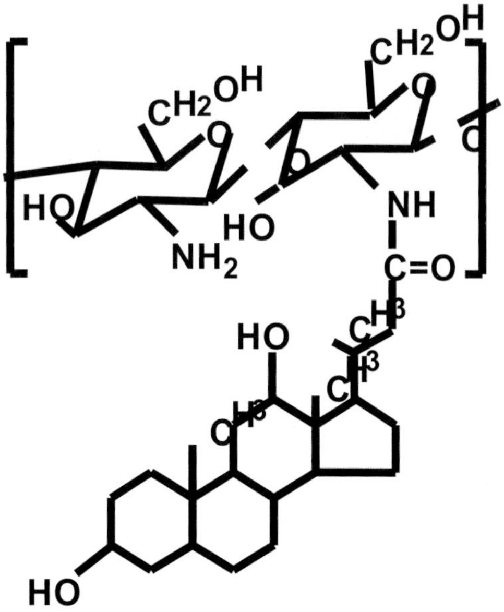

Figure 21.6 Chemical structure of deoxycholic acid-modified chitosan (DMAC).

Thanou et al. [24] reported quaternized chitosan oligomers as novel gene delivery vectors in epithelial cell lines. They synthesized trimethylated chitosan (TMO) using oligomeric chitosan (<20 monomer units), and observed spontaneous complex formation of TMOs with RSV-α3 luciferase plasmid DNA with a size ranging from 200 to 500 nm [24]. They also found the transfection efficiencies of chitoplexes to be lower than that of DOTAP (N-[1-(2, 3-dioleoyloxy)propyl]-N,N,N-trimethylammonium sulfate)-DNA lipoplexes, when tested in Cos-1, whereas, both these complexes showed lower transfection in Caco2 cells, with TMOs taking lead over DOTAP [24].

21.3.2
Liposomes and Solid Lipids

Felgner and co-workers [25] were the first to use cationic lipids dioleyltrimethylammonium chloride (DOTMA) in a 1:1 molar ratio with the neutral lipid dioleoylphosphatidyl-ethanolamine (DOPE) to condense and transfect DNA. Since then, a variety of cationic lipids have been synthesized and evaluated for gene transfection. Because of their amphiphilic nature, they easily form vesicular structures termed liposomes, when suspended in water; given their cationic charge, these liposomes readily and efficiently interact with DNA, thus forming so-called "lipoplexes". Cationic liposomes are prepared by evaporation of the organic solvent in which the cationic lipid (mixture) is dissolved, followed by hydration of the lipid film in aqueous buffer, and subsequent vortexing, which results in the formation of multilamellar vesicles (MLVs). Small unilamellar vesicles are generated by sonication or extrusion of the MLVs. Addition of DNA to performed cationic liposomes triggers significant structural changes in both the liposomes and the DNA, whereby the

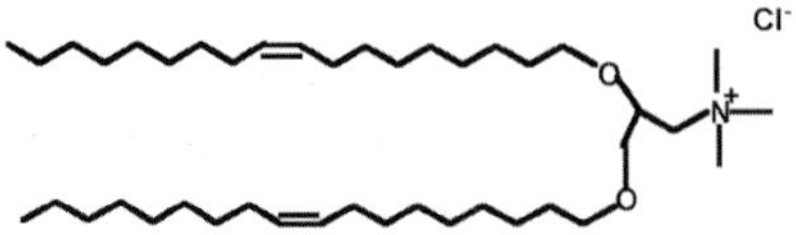

2,3-bis(oleyl)oxipropyl-trimethylammonium chloride (DOTMA)

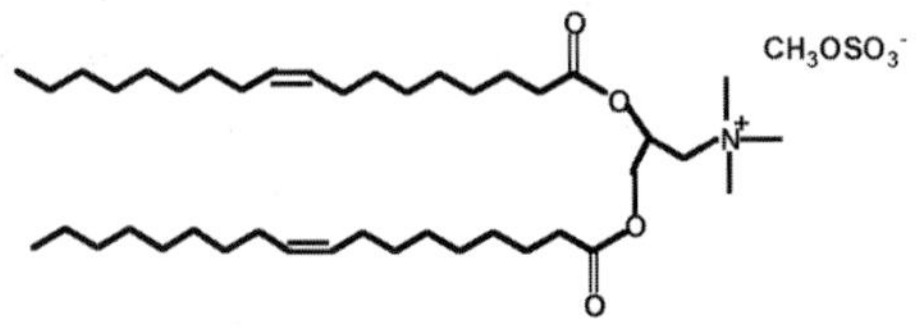

1,2-dioleoyl-3-trimethylammonium propane (DOTAP)

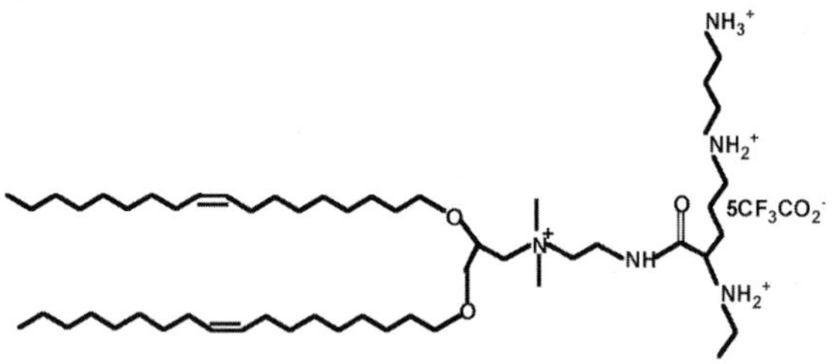

2,3 dioleyloxy-*N*-[2(spermine carboxamino)ethyl]-*N,N*-dimethyl-1-propanaminium trifluoroacetate (DOSPA)

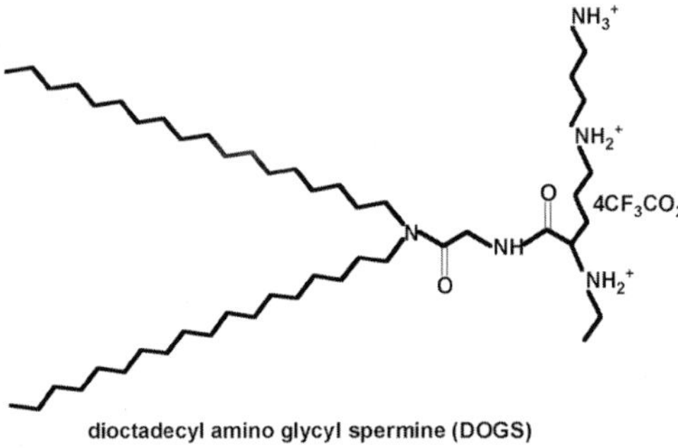

dioctadecyl amino glycyl spermine (DOGS)

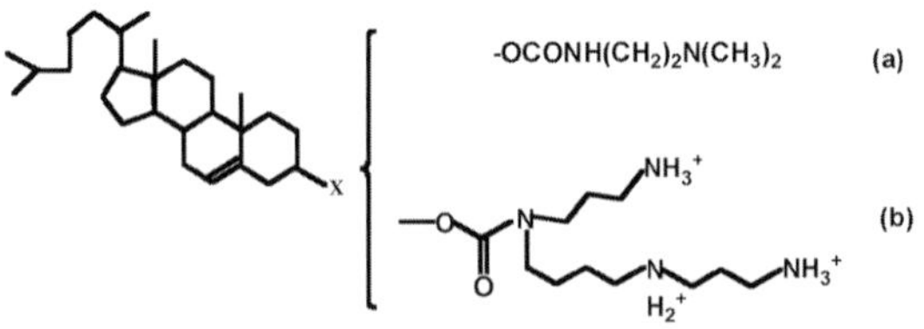

(a) 3 b[*N,N'*-dimethylaminoethane)-carbamoyl]cholesterol (DC-chol);
(b) Lipid 67

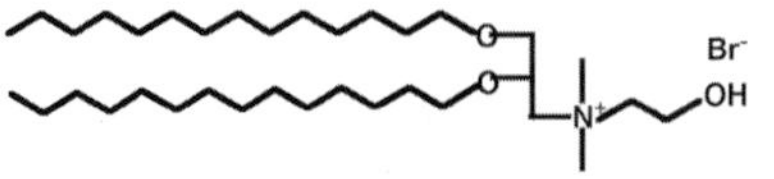

N-(2-hydroxyethyl)-*N,N*-dimethyl-2,3-bis(tetradecyloxy)-1-propanaminium bromide (DMRIE)

Figure 21.7 Structures of some cationic lipids commonly used in gene therapy.

initial liposomal structure is lost and new structural organization, the lipoplex, is adopted [26, 27]. The structures of some cationic lipids commonly used in gene therapy are shown in Figure 21.7. Pedroso de Lima et al. [28] recently reviewed cationic lipid–DNA complexes in gene delivery, where they discussed various aspects starting from biophysics to biological applications.

Ochiya et al. [29] evaluated gene transfection in pregnant animals by testing several cationic liposomes conjugated with plasmid DNA carrying the β-galactosidase gene through intravenous injection. These authors identified DMRIE-C reagent, which was a liposome formulation of the cationic lipid DMRIE (1, 2-dimyristyloxypropyl-3-dimethyl-hydroxy ethyl ammonium bromide) and cholesterol, as being suitable for their purpose. When plasmid DNA/DMRIE-C complexes were administered intravenously to pregnant mice at day 11.5 post coitus (p.c.), transferred genes were observed in several organs in dams and were expressed (Table 21.3). Furthermore, gene expression was observed in the progeny, although the copy numbers transferred into embryos were low [29].

Ishiwata et al. [30] developed a novel liposome formulation for gene transfection, consisting of *O,O′*-ditetradecanoyl-*N*-(a-trimethyl ammonioacetyl) diethanolamine chloride (DC-6-14) as a cationic lipid (Figure 21.8), phospholipid and cholesterol in a molar ratio of 4:3:3. The DC-6-14 liposome–DNA complexes were usually thought to have an overall positive surface charge, but it was found to be DNA ratio-dependent, and the diameter of the complex was also dependent on DNA concentration. The complexes had a maximum diameter when the surface charge was neutral. These formulations showed effective gene transfection activity in cultured cells with serum and, when administered intraperitoneally to mice, the positively charged particles showed immediate accumulation in lung tissues. By contrast, the negatively charged particles did not show any accumulation [30]. The authors therefore concluded that surface modification of the liposome

Table 21.3 Distribution of the transfected gene in vivo

Organ	Number of	Plasmid form copies/cell Open circle	Linear	Supercoil
Brain	0	–	–	–
Heart	2–5	–	+	–
Lung	20–50	+	+	+
Liver	10–40	+	+	+
Kidney	2–10	–	+	–
Spleen	2–5	–	+	–
Pancreas	2–10	–	+	–
Stomach	1–2	–	+	–
Colon	0–2	–	+	–
Small intestine	0–2	–	+	–
Testis	0	–	–	–
Ovary	0	–	–	–

Note: Seven-week-old ICR male and female mice (n = 12) were injected i.v. with 100 μg pCMVβ. SPORT βgal plasmid DNA conjugated with 500 μg DMRIE-C reagent. Two days later, several organs were obtained, and the genomic DNAs from them were subjected to Southern blot analysis for the presence of transferred gene.

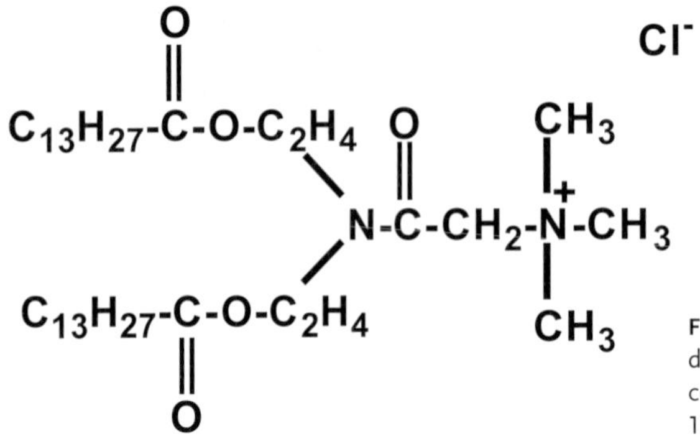

$$C_{13}H_{27}\text{-}\overset{\displaystyle O}{\overset{\|}{C}}\text{-O-C}_2H_4$$

Figure 21.8 Chemical structure of *O,O'*-ditetradecanoyl-*N*-(a-trimethylammonioacetyl) diethanolamine chloride (DC-6-14).

would improve the biodistribution and hence the targetability of their DNA complexes [30].

No single system thus far seems to transfect DNA efficiently into the nucleus, and attempts are ongoing to achieve this. Oku et al. [31] reported a novel nonviral gene delivery system which they named as the polycation liposome system (PCL). Basically, this approach was to combine the favorable properties of liposomes with the polycation, PEI. The interesting point about PCL was that it did not require phosphatidylethanolamine or cholesterol as a component, unlike most conventional liposomes [31]. Egg yolk phosphatidylcholine- and dipalmitoylphosphatidylcholine-based PCLs were found to be as effective as dioleoylphosphatidylethanolamine-based PCL for gene transfer [31]. These authors examined the effect of molecular weight of the PEI on PCL-mediated gene transfer and found that PEIs with molecular weights of 600 and 1800 Da to be quite effective, whereas PEI of 25 000 Da was much less effective. Furthermore, they demonstrated an increased transfection efficacy in the presence of serum, with effective gene transfer being observed in all eight malignant and two normal cell lines tested, as well as in Cos-1 cells [31]. Interestingly, the same authors also demonstrated gene transfer in vivo: GFP and luciferase genes were expressed in mouse lung when the animals were given tail vein injections, whereas gene expression occurred in the liver after portal vein injection [31].

Huang and co-workers made very significant contributions to the liposomal approach in gene delivery [32–40]. Li and Huang [41] have recently reviewed their developments in nonviral delivery systems, with a major focus on cationic liposome entrapped, polycation-condensed DNA, which they termed LPD1. The original complex was composed of poly-L-lysine (PLL), DNA, and cationic liposomes, which resulted in a high interaction between PLL and DNA, while liposomes led to the formation of a lipid shell on the surface of the particles [33]. A recent review also highlighted the developments of LPD1, systemic gene delivery, and mechanisms involved in cationic mediated gene delivery [41].

The other leading group in liposome research for nonviral gene delivery is that of Szoka and co-workers [42–52]. These authors have contributed extensively to the fascinating area of liposome research, and have applied cationic liposomes for targeting drugs to various lesions as well as the delivery of genes. They suggested that intravenously administered lipoplexes might serve as a depot for the extracellular release of naked DNA, and initially considered that the naked DNA mediated gene delivery in the lung [49]. They introduced a lung inflation-fixation protocol to examine the distribution and gene transfer efficiency of fluorescently tagged lipoplexes using confocal microscopy within thick lung tissue sections [52]. Infusion of plasmid DNA at a rate of 80 µg min^{-1} into the tail vein of a mouse resulted in a DNA serum concentration of 800 µg mL^{-1}. In spite of this high level of transcriptionally active DNA, there was no significant gene expression in the lung or any other organ tested [49]. In addition, when lipoplex containing a reporter gene was injected, followed by an infusion of noncoding plasmid DNA as a potential competing molecule for DNA released from the lipoplex, there was no effect on gene expression. Thus, it was concluded that the cationic lipid component of the lipoplex functions in an active capacity beyond that of a simple passive matrix for plasmid DNA [49]. Baraldo et al. [53] recently reported sphingosine-based liposomes as DNA vectors for intramuscular gene delivery; these liposomes were formulated in a range of solutions with phosphatidylcholine, and then complexed with DNA. The physico-chemical characteristics of the sphingosine/EPC liposomes and sphingosine/EPC/DNA lipoplexes were determined. It was found that, by increasing the charge ratios, colloidally stable sphingosine/DNA particles with a 170 nm average diameter and a positive zeta potential were obtained [53]. These authors concluded that the cationic sphingosine/EPC/DNA complexes formed a weakly compacted structure which potentially was labile in vivo and which might be useful for in-vivo gene transfer [53].

Oberle et al. [54] showed that, under equilibrium conditions, lipoplex formation involves a three-step mechanism, the interaction between plasmid and cationic liposomes being investigated using atomic force microscopy (AFM) [54]. In the first step, the plasmids – when interacting with the monolayer – display a strong tendency for orientational ordering. Subsequently, individual plasmids enwrap themselves with amphiphile molecules in a multilamellar fashion. The size of the complex formed is determined by the supercoiled

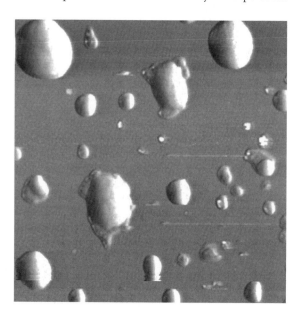

Figure 21.9 Lipoplex (d = 300 nm) composed of SAINT 2/DOPE 50/50 pCMVβ plasmid.

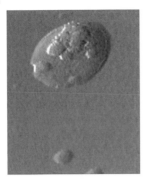

Figure 21.10 Solid lipid nanoparticles (SLNs) (d ≤100 nm) composed of compritol ATO, Tween 80, Span 85, and distearoylethyl dimethyl ammonium chloride form discrete, submicron particles with plasmid DNA (upper half). Coalescence of the individual SLNs following interaction with DNA stabilize the complex. A single SLN can be seen in the lower half of the image.

size of the plasmid, and calculations reveal that the plasmid can be surrounded by three to five bilayers of the amphiphile. The eventual size of the transfecting complex is finally governed by fusion events between individually wrapped amphiphile/DNA complexes. The AFM images of the liposome complexes are as shown in Figure 21.9.

Müller and co-workers [55–58] carried out extensive research on solid lipid nanoparticles for drug targeting as well as gene delivery. Müller et al. [59] have discussed the state of art of solid lipid nanoparticles for pharmaceutical applications, and in recent investigations found that cationic solid-lipid nanoparticles (SLNs) could bind efficiently and transfect plasmid DNA [60]. They produced highly cationic charged SLNs with a zeta potential of up to +40 mV at pH 7.4 and 100 nm size, by using a hot homogenization technique. They also characterized the SLN–DNA complexes using AFM, and suggested fusion of the SLNs after binding to the DNA, resulting in near-spherical lipid–DNA particles with sizes between 300 and 800 nm (Figure 21.10). These investigations also demonstrated that efficient transfection occurs in Cos-1 cells [60].

21.3.3
Poly-L-Lysine and Polyethylenimines

Poly-L-lysine (PLL) – one of the first polymers to be used in nonviral gene delivery [61, 62] – is biodegradable but has a high toxicity which prevents its use in vivo. Initial reports stated that, if prepared with PLL of suitable molecular weight and N/P ratio, 100 nm-sized complexes are formed which are easily taken up by the cells, although the transfection remained lower [61]. The lack of amino groups was thought to prevent PLL complexes from undergoing endoosomolysis, leading to low levels of transgene expression [63]. Nevertheless, inclusion of targeting moieties or co-application of endoosomolytic agents such as chloroquine [64] or fusogenic peptides [61] might improve reporter gene expression.

Parker et al. [65] recently discussed methods for monitoring nanoparticle formation by self-assembly of DNA with PLL. They compared three different methods viz., light scattering, inhibition of ethidium bromide fluorescence, and modified electrophoretic mobility of DNA. Results obtained with the first two methods indicated that stable nanoparticles form over the lysine/phosphate ratio range 0.6 to 1.0 [65], though no transfection studies were reported.

In an attempt to improve transfection efficiency and reduce cytotoxicity, Lee et al. [66] utilized PEG grafted polylysine with fusogenic peptide. The method is initiated by complexing the PEG-grafted PLL with DNA, and subsequently coating the positively charged fusogenic peptide by ionic interaction onto the surface. This results in a net positively charged complex [66]. The authors showed that the use of PEG-grafted PLL for peptide coating significantly suppressed aggregation of the complexes via peptide mediated interparticulate crosslinking, and also enhanced transfection [66]. Hence, both PEG and peptides play key roles in reducing cytotoxicity and improving transfection [74].

Gonzalez Ferreiro et al. [67] described the use of alginate/PLL particles for antisense oligonucleotides wherein they utilized the gel-forming abilities of the alginate in the presence of calcium ions and further crosslinked it with PLL. Particle formation was found to occur when the alginate concentration was 0.04–0.08 % (w/v) of the total formulation; moreover, the particle size increased from nm to µm when the alginate concentration was 0.055 % (w/v) [67]. These authors found that with increasing PLL content, both the size and encapsulation efficiency of the particle increased, but the release rate was decreased [67].

Polyethylenimines (PEIs) were the first synthetic transfection agents to be discovered, and have been widely studied, although the toxicity of PEI is a major drawback. Kircheis et al. [68] reviewed the design and gene delivery of modified PEIs, and emphasized the synthesis of modified PEIs, DNA condensation, particle size, cellular uptake of the particles, endosomal release, in-vivo gene delivery and many other features.

Sagara and Kim [69] more recently reported galactose-polyethylene glycol-PEI (Gal-PEG-PEI) for gene delivery to hepatocytes. They found that transfection efficiency with 1 % Gal-PEG-PEI in human hepatocyte-derived cell lines (HepG2) – a model of parenchymal cells in the liver – was superior to PEI at corresponding optimal ratios of polymer to plasmid DNA [69]. In HepGe cells, luciferase activity with 1 % Gal-PEG-PEI at an N/P ratio of 20 was 2.1-fold higher than that of PEI at an N/P ratio of 5. However, in mouse fibroblasts (NIH-3T3) that had no asialoglycoprotein (ASGP) receptor, the transfection efficiency with 1 % Gal-PEG-PEI fell drastically to one-fortieth of that with PEI. These studies suggest that Gal-PEG-PEI is more suited for targeting specific genes to the liver [69].

Rudolph et al. [70] reported modified PEIs for gene transfer to the lungs via either direct intratracheal instillation or nebulization. In this study, PEG-coated PEI polyplexes were investigated for their stability and interaction with human plasma and bronchoalveolar lavage fluid (BALF); their potential for gene delivery to the mouse lungs in vivo was also examined. Gene transfer efficiency of the PEG-coated PEI polyplexes decreased as compared with uncoated PEI polyplexes when administered intratracheally to the lung. The higher the molecular weight of the copolymerized PEG, the stronger was the observed gene transfer reduction. These authors speculated that the gene transfer was decreased due to a reduced interaction of the coated gene vectors with the cell surface. To circumvent this problem, transferrin was combined with PEI/DNA polyplexes for specific binding to the cell surface, but this resulted in a decrease in gene transfer efficiency. Gene transfer of the copolymer-protected and transferrin-modified gene vectors increased as compared with the copolymer-protected gene vectors alone, but did not reach the level of uncoated gene vectors. These data show that copolymers could be used effectively to shield polyplexes from interaction with components of the airway surface liquid. The increase in gene expression upon transferrin modification of the coated PEI polyplexes still suggests a targeting effect [70].

21.3.4
Poly(lactide-*co*-glycolide)

Various biodegradable polymers are being investigated for the formulations of nanoparticles aimed at drug and gene delivery applications [71]. Poly(D,L lactide-*co*-glycolide) (PLGA) and poly(lactic acid) (PLA) – both of which are FDA-approved biocompatible polymers – have been the most extensively studied [72, 73]. Recently, Jain [72] reviewed the PLGA nanoparticle preparation techniques, and reported an emulsion-solvent evaporation technique to be the most widely used method to formulate PLA and PLGA nanoparticles, while poly(vinyl alcohol) (PVA) is the most commonly used stabilizer.

Labhasetwar and co-workers [74–77] made several reports on aspects of PLGA nanoparticles for gene delivery, viz., characterization of nanoparticle uptake by endothelial cells [74], size dependency of nanoparticle-mediated gene transfection [75], effects on nanoparticle properties and their cellular uptake associated with the residual PVA [76], and rapid endolysosomal escape of the PLGA nanoparticles [77]. They followed the standard emulsion-solvent evaporation technique, which resulted in particles with a heterogeneous size distribution. Moreover, they investigated the relative transfectivity of the smaller and the larger-sized particles in cell cultures [75]. Those particles which passed through a membrane of pore size 100 nm (mean diameter = 70 ± 2 nm) were designated as smaller particles, whereas those retained on the membrane were deemed to be larger-sized nanoparticles (202 ± 9 nm). The smaller-sized nanoparticles showed a 27-fold higher transfection than the larger-sized nanoparticles in a Cos-7 cell line, and a 4-fold higher transfection in an HEK-293 cell line [75]. The surface charge, cellular uptake, and DNA release were similar for the two fractions of nanoparticles, and suggests that other unknown factor(s) might be responsible for the observed differences in transfection levels [75].

The same authors also studied the effects of residual PVA associated with PLGA nanoparticles on cellular uptake [76]. PVA removal during nanoparticle preparation is difficult by water washing, and small amounts are inevitably associated with the particles. This occurs despite repeated washings as PVA forms an interconnected network with the polymer at the interface. The amount of residual PVA was found to depend on the initial PVA concentration and type of organic solvent used in the emulsion, and affected particle size, zeta potential, polydispersity index, and surface hydrophobicity of the nanoparticle surface [76]. Particles with more residual PVA, despite their lower size, were seen to have the lowest cellular uptake, but this may be due to the higher hydrophilicity of the nanoparticle surface [76]. These authors also studied endolysosomal escape of the PLGA nanoparticles [77], and showed the mechanism to be by selective reversal of the surface charge of NPs (from anionic to cationic) in the acidic endolysosomal compartment, which in turn causes the NPs to interact with the endolysosomal membrane and escape into cytosol [77].

21.3.5
Silica

A major limitation of the polymers is the difficulty of including additional bioactive molecules (in order to enhance or modify the DNA activity) during the self-assembly process,
unless these are covalently linked to the polymer backbone. Recently, it was shown in the
present author's laboratory that the colloidal silica particles with cationic surface modifications interact with plasmid DNA and transfect in vitro [78, 79]. Surface-modified silica
nanoparticles were synthesized by modification of commercially available silica particles
(IPAST; Nissan Chemical Industries, Tokyo, Japan) with various aminoalkylsilanes, viz.,
N-(2-aminoethyl)-3-aminopropyltrimethoxysilane (AEAPS) and N-(6-aminohexyl)-amino-
propyltrimethoxysilane (AHAPS). The resulting particles have a hydrodynamic diameter
of 26 nm and a zeta potential up to +31 mV [78, 79]. The scheme for modification of
the silica particles is shown in Figure 21.11. In analogy to the terms lipoplex and polyplex,
we propose that be the nanoparticle–DNA complexes be described by the term "nanoplex".
Transfection was strongly increased in the presence of 100 µM chloroquine in the incubation medium, and reached approximately 30 % of the efficiency of a 60-kDa PEI. In contrast to PEI, no toxicity was observed at the concentrations required. AFM of Si_26H–DNA
complexes revealed a spaghetti-meatball-like structure (Figure 21.12).

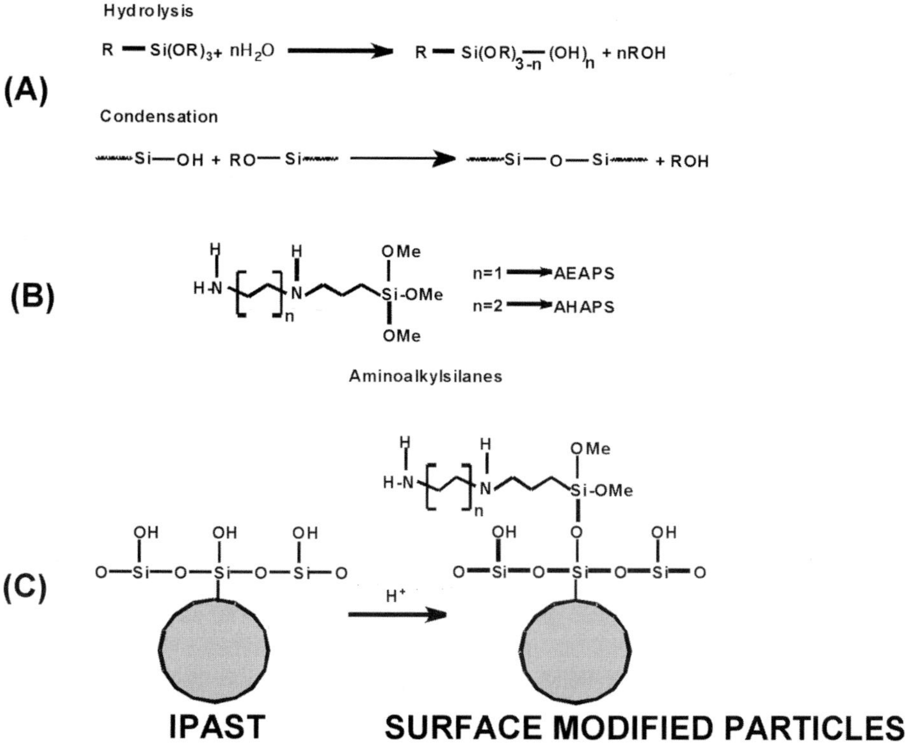

Figure 21.11 (A) Hydrolysis and condensation of unmodified particles. (B) Alkylaminoalkanes. (C)
Modification scheme.

Figure 21.12 Cationic silica particles (d ≤300 nm) self-associate with plasmid DNA into submicron complexes that protect, release, and transfect the DNA in vitro.

We also investigated the stability of freeze-drying for the conservation of zwitterionic nanoparticles and usefulness of different lyoprotective agents (LPAs) for the minimization of particle aggregation [80]. The activity of the nanoparticles was measured as DNA-binding capacity and transfection efficiency in Cos-1 cells before and after lyophilization. Massive aggregation was found to occur in the absence of any LPA. Of the various LPAs screened in the present investigations, trehalose and glycerol were found to provide exceptional conservation of cationically modified silica nanoparticles with simultaneous preservation of their DNA-binding and transfection activity in Cos-1 cells [80].

21.3.6
Block Copolymers

Block copolymers composed of the cationic segment and hydrophilic segments are expected spontaneously to associate with polyanionic DNA to form block copolymer micelles. Kataoka and co-workers [81–85] pioneered this area of research, and recently provided an excellent review of block copolymer micelles for gene delivery and related compounds [86]. They discussed various aspects of the synthesis of cationic block and graft copolymers, the physico-chemical properties and the biological aspects including cellular uptake and lysosomal escape [86]. Kataoka's group is actively involved in block copolymers of poly(ethylene glycol)-*block*-poly(L-lysine) (PEG-*b*-PLL) [82, 83], poly(ethylene glycol)-*block*-poly(ethylenimine) (PEG-*b*-PEI) [87], and poly(ethylene glycol)-*block*-poly(dimethylaminoethyl methacrylate) (PEG-*b*-PPAMA) [84], all of which are synthesized from their respective monomers using PEG with terminal functional groups as a macroinitiator (Figure 21.13).

The other active group involved in block copolymers is that of Kabanov et al. [88–92], who carried out extensive investigations on Pluronic® block copolymers as depots for drugs as well as genes. In a recent review, these authors described Pluronic block copolymers as novel functional molecules for gene therapy, with special focus on gene delivery to skeletal muscle [93].

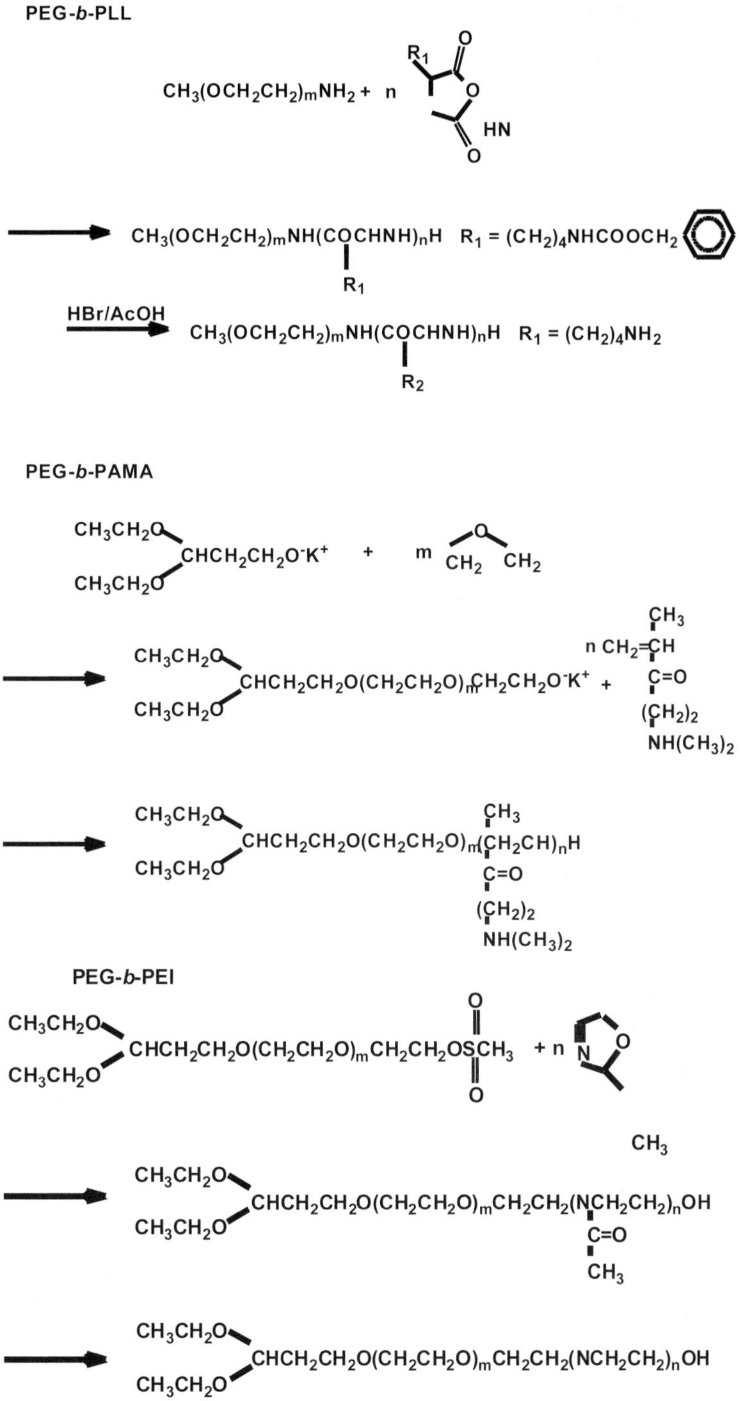

PEG-b-PLL

$CH_3(OCH_2CH_2)_mNH_2 +$ n

$CH_3(OCH_2CH_2)_mNH(COCHNH)_nH$ $R_1 = (CH_2)_4NHCOOCH_2$ ⬡
             R_1

HBr/AcOH $CH_3(OCH_2CH_2)_mNH(COCHNH)_nH$ $R_1 = (CH_2)_4NH_2$
             R_2

PEG-b-PAMA

CH_3CH_2O — $CHCH_2CH_2O^-K^+$ +   m $\begin{smallmatrix}O\\CH_2\ CH_2\end{smallmatrix}$
CH_3CH_2O

CH_3CH_2O — $CHCH_2CH_2O(CH_2CH_2O)_mCH_2CH_2O^-K^+$ +   n $CH_2{=}CH$
CH_3CH_2O CH_3 / $C{=}O$ / $(CH_2)_2$ / $NH(CH_3)_2$

CH_3CH_2O — $CHCH_2CH_2O(CH_2CH_2O)_m(CH_2CH)_nH$
CH_3CH_2O CH_3 / $C{=}O$ / $(CH_2)_2$ / $NH(CH_3)_2$

PEG-b-PEI

CH_3CH_2O — $CHCH_2CH_2O(CH_2CH_2O)_mCH_2CH_2OSCH_3$ + n
CH_3CH_2O

CH_3CH_2O — $CHCH_2CH_2O(CH_2CH_2O)_mCH_2CH_2(NCH_2CH_2)_nOH$
CH_3CH_2O $C{=}O$ / CH_3

CH_3CH_2O — $CHCH_2CH_2O(CH_2CH_2O)_mCH_2CH_2(NCH_2CH_2)_nOH$
CH_3CH_2O

Figure 21.13 Synthetic procedures for cationic copolymers.

21.4
Setbacks and Strategies to Improve Specific Cell Uptake of Nonviral Systems

Nonviral gene delivery has been used to express various genes in animal tissues, including muscle, lung, skin, the central nervous system, and liver. For in-vivo gene delivery, nonviral systems have the advantage of being more similar to pharmaceutical drugs in terms of safety, uniformity, and administration. However, the transduction frequency of the target cells is often barely adequate, and the persistence of cells expressing the transduced gene in vivo is poor. Further development of nonviral gene delivery technology will expand its application and provide a viable alternative to the use of viral vectors.

For in-vivo therapy it is necessary to address gene delivery vehicles to specific cell types in order to avoid unwanted effects in nontarget cells. Targeting is achieved actively by incorporating structures, which facilitate exclusive uptake of the vector in certain tissues of cell types. However, in some cases this can be achieved passively by taking advantage of particular physiological conditions of the target tissue; for example, irregular endothelial fenestration in tumors in conjugation with certain complex properties. To achieve an efficient active targeting in vivo, the vector must fulfill two main requirements: (i) unspecific interactions must be reduced by shielding net positive surface charges of the complexes by using rather low nitrogen to phosphate content; and (ii) the targeting moiety which enables uptake into a specific cell type needs to be incorporated. Various ligands are in use depending upon the target structure, and these include folate, transferrin, lactose, galactose, mannose, low-density lipoprotein (LDL), fibroblast growth factor (FGF), endothelial growth factor (EGF), and several antibodies. An early review of the strategies was provided by Nettelbeck et al. [94], and more recently by Merdan et al. [95].

21.5
Prospects for Nonviral Nanomaterials

Currently, gene therapy researchers are attempting to improve each vector, as well as to match vector characteristics with diseases that they will target most successfully. It is likely that, in the future, there will be many specialized vectors rather than one universal vector. Some vectors may be required in situations where doctors require short term expression, for example in the expression of a toxic gene product in cancer cells. However, in situations where long-term expression is required (e. g., for most genetic diseases), vectors that deliver large and medium-sized sections of DNA would be required.

Gene insertion is not the only problem encountered, however, as the process is not automatic and the vector must contain a mechanism to activate the therapeutic gene. Hence, the vector must include a timing and regulatory "device" which allows the gene to be turned on and off and change levels of the therapeutic protein over time. Such devices are termed the gene's "promoters"; they are complex and sometimes quite large, and placing them into a therapeutic vector is difficult.

It is clear from the foregoing sections that polymers display striking advantages as vectors for gene delivery. They can be specifically tailored for a proposed application by choosing an appropriate molecular weight, by coupling the cell- or tissue-specific targeting moieties, and/or performing other modifications that confer certain physiological or physico-chemical properties. After identifying a suitable polymer structure, a scale-up to the production of large quantities is somewhat more straightforward. It is understood that all these modifications/manipulations are possible only by using nonviral vectors.

The bottom line in any type of biomedical research is its relevance. Despite the excitement that gene therapy might cause, the field is still in its infancy. Several clinical trials of gene therapy have been completed or are under way, and all have provided information that cannot be derived from tests conducted in animals. Although in theory this treatment is good in principle, its efficacy has proved difficult to demonstrate in practice, and success rates in clinical trials have been relatively low. Poor results may reduce the enthusiasm for genetic approaches, but the field is still new and the pace and surprises of new discoveries continue to amaze.

Although in this chapter we have attempted to outline the most significant contributions on nonviral nanoparticulate gene delivery systems, it is very difficult to discuss each and every reported system within the confines of the text. Nonetheless, an attempt has been made to provide updated information on nonviral nanoparticulate gene delivery systems, and in consideration of the reader's interest we have included bibliographic information of those reports not discussed in the text [96–125].

Acknowledgments

M. N. V. R. K. is grateful to the Alexander Von Humboldt Foundation, Germany for providing personal fellowship. U. Bakowsky wishes to thank "Stiftung Deutscher Naturforscher Leopoldina" (BMBF/LPD-9901/8-6).

References

[1] V. Kabanov, V. A. Kabanov, *Adv. Drug Del. Rev.* **1998**, *30*, 49–60.

[2] Z. Q. Xiang, Y. Yang, J. M. Wilson, H. C. Ertl, *Virology* **1996**, *219*, 220–227.

[3] M. R. Knowles, K. W. Hohneker, Z. Zhou, J. C. Olsen, T. L. Noah, P. C. Hu, M. W. Leigh, et al. *N. Engl. J. Med.* **1995**, *333*, 823–831.

[4] R. H. Simon, J. F. Engelhardt, Y. Yang, M. Zepeda, S. Weber-Pendelton, M. Grossman, J. M. Wilson, *Hum. Gene Ther.* **1993**, *4*, 771–778.

[5] S. C. De Smedt, J. Demeester, W. E. Hennink, *Pharm. Res.* **2000**, *17*, 113–126.

[6] M. C. Garnett, *Crit. Rev. Ther. Drug Carrier Syst.* **1999**, *16*, 147–207.

[7] K. Minagawa, Y. Matzusawa, K. Yoshikawa, M. Matsumoto, M. Doi, *FEBS Lett.* **1991**, *295*, 67–69.

[8] J. Kreuter, P. Speiser, *J. Pharm. Sci.* **1976**, *65*, 1624–1627.

[9] P. Couvreur, L. Grislain, V. Lenaert, F. Brasseur, P. Guiot, A. Biernacki, in: P. Guiot, P. Couvreur (eds), *Polymeric Nanoparticles and Microparticles.* CRC Press, Boca Raton, **1986**, pp. 27–93.

[10] V. Labhasetwar, C. Song, R. J. Levy, *Adv. Drug Del. Rev.* **1997**, *24*, 63–85.

[11] R. A. A. Muzzarelli, *Chitin*, Pergamon Press, Oxford, **1977**.

[12] M. N. V. Ravi Kumar, *Reactive and Functional Polymers* **2000**, *46*, 1–27.

[13] R. J. Mumper, J. J. Wang, J. M. Claspell, A. P. Rolland, *Proc. Int. Symp. Controll. Rel. Bioactive Mater.* **1995**, *22*, 178–179.

[14] G. Borchard, *Adv. Drug Del. Rev.* **2001**, *52*, 145–150.

[15] F. C. MacLaughlin, R. J. Mumper, J. Wang, J. M. Tagliaferri, I. Gill, M. Hinchcliffe, A. P. Rolland, *J. Controll. Rel.*, **1998**, *56*, 259–272.

[16] H.-Q. Mao, K. Roy, V. Truong-Le, et al., *Proc. Int. Symp. Controll. Rel. Bioact. Mater.* Stockholm, Sweden, **1997**, *24*, 671–672.

[17] K. W. Leong, H.-Q. Mao, V. L. Truong-Le, et al., *J. Controll. Rel.*, **1998**, *53*, 183–193.

[18] K. Roy, H.-Q. Mao, S. K. Huang, K. W. Leong, *Nature Med.* **1999**, *5*, 387–391.

[19] H.-Q. Mao, K. Roy, V. L. Troung-Le, et al., 2001, *J. Controll. Rel.* **2001**, *70*, 399–421.

[20] M. Kumar, A. K. Behera, R. F. Lockey, J. Zhang, et al., *Hum. Gene Ther.* **2002**, *13*, 1415–1425.

[21] M. Koping-Hoggard, I. T. Guan, K. Edwards, M. Nilsson, K. M. Varum, P. Artursson, *Gene Ther.* **2001**, *8*, 1108–1121.

[22] T. Sato, T. Ishii, Y. Okahata, *Biomaterials* **2001**, *22*, 2075–2080.

[23] Y. H. Kim, S. H. Gihm, C. R. Park, *Bioconjug. Chem.* **2001**, *12*, 932–938.

[24] M. Thanou, B. I. Florea, M. Geldof, H. E. Junginger, G. Borchard, *Biomaterials* **2002**, *23*, 153–159.

[25] P. L. Felgner, T. R. Gadek, M. Holm, R. Roman, et al., *Proc. Natl. Acad. Sci. USA* **1987**, *84*, 7413–7417.

[26] E. K. Wasan, A. Fairchild, M. B. Balay, *J. Pharm. Sci.* **1998**, *87*, 9–14.

[27] A. A. P. Meekel, A. Wagenaar, J. Smisterova, J. E. Kroeze, *Eur. J. Org. Chem.* **2000**, *2000*, 665–673.

[28] M. Pedroso de Lima, S. Simoes, P. Pires, H. Faneca, N. Duzgunes, *Adv. Drug Deliv. Rev.* **2001**, *47*, 277–294.

[29] T. Ochiya, Y. Takahama, H. Baba-Toriyama, M. Tsukamoto, Y. Yasuda, H. Kikuchi, M. Terada, *Biochem. Biophys. Res. Commun.* **1999**, *258*, 358–365.

[30] H. Ishiwata, N. Suzuki, S. Ando, H. Kikuchi, T. Kitagawa, *J. Controll. Rel.* **2000**, *69*, 139–148.

[31] N. Oku, Y. Yamazaki, M. Matsuura, M. Sugiyama, M. Hasegawa, M. Nango, *Adv. Drug Deliv. Rev.* **2001**, *52*, 209–218.

[32] C. Y. Wang, L. Huang, *Proc. Natl. Acad. Sci. USA* **1987**, *84*, 7851–7855.

[33] X. Gao, L. Huang, *Biochemistry* **1996**, *53*, 1027–1036.

[34] F. L. Sorgi, S. Bhattacharya, L. Huang, *Gene Ther.* **1997**, *4*, 961–968.

[35] S. Li, M. A. Rizzo, S. Bhattacharya, L. Huang, *Gene Ther.* **1998**, *5*, 930–937.

[36] S. Li, L. Huang, *Gene Ther.* **1997**, *4*, 891–900.

[37] S. Li, W. C. Tseng, D. B. Stolz, S. P. Wu, S. C. Watkins, L. Huang, *Gene Ther.* **1999**,*6*, 585–594.

[38] X. Zhou, L. Huang, *Biochim. Biophys. Acta* **1994**, *1189*, 195–203.

[39] F. Liu, H. Qi, L. Huang, D. Liu, *Gene Ther.* **1997**, *4*, 517–523.

[40] Y. Tan, F. Liu, Z. Li, L. Huang, *Mol. Ther.* **2001**, *3*, 673–682.

[41] F. Liu, L. Huang, *J. Controll. Rel.* **2002**, *78*, 259–266.

[42] L. G. Barron, K. B. Meyer, F. C. Szoka, Jr., *Hum. Gene Ther.* **1998**, *9*, 315–323.

[43] O. Zelphati, L. S. Uyechi, L. G. Barron, F. C. Szoka, Jr., *Biochim. Biophys. Acta* **1998**, *1390*, 119–133.

[44] S. C. Davis, F. C. Szoka, Jr., *Bioconjug. Chem.* **1998**, *9*, 783–792.

[45] J. Y. Legendre, S.-K. Huang, F. C. Szoka, Jr., *J. Liposome Res.* **1998**, *8*, 347–366.

[46] L. G. Barron, L. S. Uyechi, F. C. Szoka, Jr., *Gene Ther.* **1999**, *6*, 1179–1183.

[47] Y. Xu, S.-W. Hui, P. Frederik, F. C. Szoka, Jr., *Biophys. J.* **1999**, *77*, 341–353.

[48] S. Hirota, C. Tros de Llarduyu, L. Barron, F. C. Szoka, Jr., *Biotechniques* **1999**, *27*, 286–289.

[49] L. Barron, L. Gagne, F. C. Szoka, Jr., *Hum. Gene Ther.* **1999**, *10*, 1683–1894.

[50] F. Nicol, S. Nir, F. C. Szoka, Jr., *Biophys. J.* **2000**, *78*, 818–829.

[51] X. Guo, F. C. Szoka, Jr., *Bioconjug. Chem.* **2001**, *12*, 291–300.

[52] L. S. Uyechi, G. Thurston, L. Gagne, F. C. Szoka, Jr., 2001, *Gene Ther.*, *8*, 828–836.

[53] K. Baraldo, N. Leforestier, M. Bureau, N. Mignet, D. Scherman, *Pharm. Res.* **2002**, *19*, 1144–1149.

[54] V. Oberle, U. Bakowsky, I. S. Zuhorn, D. Hoekstra, *Biophys. J.* **2000**, *79*, 1447–1454.

[55] R. H. Muller, S. Massen, H. Weyhers, F. Specht, J. S. Lucks, *Int. J. Pharm.* **1996**, *138*, 85–94.

[56] R. H. Muller, W. Mehnert, J. S. Lucks, C. Schwarz, et al., *Eur. J. Pharm. Biopharm.* **1995**, *41*, 62–69.

[57] C. Freitas, R. H. Muller, *Eur. J. Pharm. Biopharm.* **1999**, *47*, 125–132.

[58] A. Dingler, R. P. Blum, H. Niehus, S. Gohla, R. H. Muller, *J. Microencapsulation* **1999**, *16*, 751–767.

[59] R. H. Muller, K. Mader, S. Gohla, *Eur. J. Pharm. Biopharm.* **2000**, *50*, 161–177.

[60] C. Olbrich, U. Bakowsky, C. M. Lehr, R. H. Muller, C. Kneuer, *J. Controll. Rel.* **2001**, *77*, 345–355.

[61] E. Wagner, C. Plank, K. Zatloukal, M. Cotton, M. L. Birnstiel, *Proc. Natl. Acad. Sci. USA* **1992**, *89*, 7934–7938.

[62] M. A. Wolfert, P. R. Dash, O. Nazarova, D. Oupicky, et al., *Bioconjug. Chem.* **1999**, *10*, 993–1004.

[63] T. Merdan, K. Kunath, D. Fischer, J. Kopecek, T. Kissel, *Pharm. Res.* **2002**, *19*, 140–146.

[64] C. W. Pouton, P. Lucas, B. J. Thomas, A. N. Uduehi, D. A. Milroy, S. H. Moss, *J. Controll. Rel.* **1998**, *53*, 289–299.

[65] A. L. Parker, D. Oupicky, P. R. Dash, L. W. Seymour, *Anal. Biochem.* **2002**, *302*, 75–80.

[66] H. Lee, J. H. Jeong, T. G. Park, *J. Controll. Rel.* **2002**, *79*, 283–291.

[67] M. Gonzalez Ferreiro, L. Tillman, G. Hardee, R. Bodmeier, *Int. J. Pharm.* **2002**, *239*, 47–59.

[68] R. Kircheis, L. Wightman, E. Wagner, *Adv. Drug Deliv. Rev.* **2001**, *53*, 341–358.

[69] S. Kazuyoshi, S. W. Kim, *J. Controll. Rel.* **2002**, *79*, 271–281.

[70] C. Rudolph, U. Schillinger, C. Plank, et al., *Biochim. Biophys. Acta* **2002**, *1573*, 75–83.

[71] M. N. V. Ravi Kumar, N. Kumar, A. J. Domb, M. Arora, *Adv. Polym. Sci.* **2002**, *160*, 1–73.

[72] R. A. Jain, *Biomaterials* **2000**, *21*, 2475–2490.

[73] R. Gref, Y. Minamitake, M. T. Peracchia, et al., *Science* **1994**, *263*, 1600–1603.

[74] J. Davda, V. Labhasetwar, *Int. J. Pharm.* **2002**, *233*, 51–59.

[75] S. Prabha, W.-Z. Zhou, J. Panyam, V. Labhasetwar, *Int. J. Pharm.* **2002**, *244*, 105–115.

[76] S. K. Sahoo, J. Panyam, S. Prabha, V. Labhasetwar, *J. Controll. Rel.* **2002**, *82*, 105–115.

[77] J. Panyam, W.-Z. Zhou, S. Prabha, S. K. Sahoo, V. Labhasetwar, *FASEB J.* **2002**, *16*,1217–1226.

[78] C. Kneuer, M. Sameti, U. Bakowsky, T. Schiestel, H. Schirra, H. Schmidt, C. M. Lehr, *Bioconjug. Chem.* **2000**, *11*, 926–932.

[79] C. Kneuer, M. Sameti, E. G. Haltner, T. Schiestel, H. Schirra, H. Schmidt and C. M. Lehr, *Int. J. Pharm.* **2000**, *196*, 257–261.

[80] M. Sameti, G. Bohr, M. N. V. Ravi Kumar, C. Kneuer, U. Bakowsky, M. Nacken, H. Schmidt, C.-M. Lehr, *Int. J. Pharm.* **2003**, *266*, 51–60.

[81] K. Kataoka, A. Harada, Y. Nagasaki, *Adv. Drug Deliv. Rev.* **2001**, *47*, 113–131.

[82] A. Harada, K. Kataoka, *Macromolecules* **1995**, *28*, 5294–5299.

[83] K. Kataoka, H. Togawa, A. Harada, K. Yasugi, T. Matsumoto, S. Katayose, *Macromolecules* **1996**, *29*, 8556–8557.

[84] K. Kataoka, A. Harada, D. Wakebayashi, Y. Nagasaki, *Macromolecules* **1999**, *32*, 6892–6894.

[85] A. Harada, K. Kataoka, *Science* **1999**, *283*, 65–67.

[86] Y. Kakizawa, K. Kataoka, *Adv. Drug Deliv. Rev.* **2002**, *54*, 203–222.

[87] Y. Akiyama, A. Harada, Y. Nagasaki, K. Kataoka, *Macromolecules* **2000**, *33*, 5841–5845.

[88] V. Y. Alakhov, E. Y. Moskaleva, E. V. Batrakova, A. V. Kabanov, *Bioconjug. Chem.* **1996**, *7*, 209–216.

[89] E. V. Batrakova, S. Li, V. Y. Alakhov, A. V. Kabanov, *Polym. Prep.* **2000**, *41*, 1639–1640.

[90] P. Lemieux, N. Guerin, G. Paradis, R. Proulx, et al., *Gene Ther.* **2000**, *7*, 986–991.

[91] A. V. Kabanov, S. V. Vinogradov, Y. G. Suzdaltseva, V. Y. Alakhov, *Bioconjug. Chem.* **1995**, *6*, 639–643.

[92] S. V. Vinogradov, T. K. Bronich, A. V. Kabanov, *Bioconjug. Chem.* **1998**, *9*, 805–812.

[93] A. V. Kabanov, P. Lemieux, S. Vinogradov, V. Alakhov, *Adv. Drug Deliv. Rev.* **2002**, *54*, 223–233.

[94] D. M. Nettelbeck, V. Jerome, R. Muller, *Trends Genet.* **2000**, *16*, 174–181.

[95] T. Merden, J. Kopecek, T. Kissel, *Adv. Drug Deliv. Rev.* **2002**, *54*, 715–758.

[96] A. Prokop, E. Kozlov, W. Moore, J. M. Davidson, *J. Pharm. Sci.* **2002**, *91*, 67–76.

[97] I. Koltover, T. Salditt, J. O. Radler, C. R. Safinya, *Science* **1998**, *281*, 78–81.

[98] S. Simoes, V. Slepushkin, P. Pires, et al., *Gene Ther.* **1999**, *6*, 1798–1807.

[99] M. O. Hotiger, T. N. Dam, B. J. Nickoloff, T. M. Johnson, G. J. Nabel, *Gene Ther.* **1999**, *6*, 1929–1935.

[100] Y. Yamazaki, M. Nango, M. Matsuura, Y. Hasegawa, M. Hasegawa, N. Oku, *Gene Ther.* **2000**, *7*, 1148–1155.

[101] A. Chaudhuri, *Pharmatechnology* **2002**, 1–4.

[102] R. Chakraborty, D. Dasgupta, S. Adhya, M. K. Basu, *Biochem. J.* **1999**, *340*, 393–396.

[103] R. Weiskirchen, J. Kneifel, S. Weiskirchen, E. Van de Leur, D. Kunz, A. M. Gressner, *BMC Cell Biol.* **2000**, *1*, 4–12.

[104] G. Gregoriadis, R. Saffie, J. Brian de Souza, *FEBS Lett.* **1997**, *402*, 107–110.

[105] J. Wang, P. C. Zhang, H. Q. Mao, K. W. Leong, *Gene Ther.* **2002**, *9*, 1254–1261.

[106] U. Z. Stammberger, A. N. Uduehi, et al., *Ann. Thorac. Surg.* **2002**, *73*, 432–436.

[107] L. S. Siddall, L. C. Barcroft, A. J. Andrew, *Mol. Reprod. Dev.* **2002**, *63*, 413–421.

[108] K. Rittner, A. Benavente, S. Bompard, et al., *Mol. Ther.* **2002**, *5*, 104–114.

[109] H. Petersen, K. Holger, M. Klaus, et al., *Biomacromolecules* **2002**, *3*, 926–936.

[110] F. M. Orson, L. Song, A. Gautam, et al., *Gene Ther.* **2002**, *9*, 463–471.

[111] B. A. Lobo, S. A. Rogers, S. Choosakoonkroang, et al., *J. Pharm. Sci.* **2002**, *91*, 454–466.

[112] D. Y., Furgeson, R. N. Cohen, R. I. Mahato, S. W. Kim, *Pharm. Res.* **2002**, *19*, 382–390.

[113] M. Benns, R. I. Mahato, S. W. Kim, *J. Controll. Rel.* **2002**, *79*, 255–269.

[114] R. I. Mahato, S. W. Kim (eds), *Pharmaceutical Perspectives of Nucleic Acid-Based Therapeutics*, Taylor and Francis, London, **2002**.

[115] R. I. Mahato, O. D. Monera, L. C. Smith, A. Rolland, *Curr. Opin. Mol. Ther.* **1999**, *2*, 226–243.

[116] R. I. Mahato, L. C. Smith, A. Rolland, *Adv. Genet.* **1999**, *41*, 95–156.

[117] C. W. Pouton, L. W. Seymour, *Adv. Drug Deliv. Rev.* **2001**, *46*, 187–203.

[118] C. Perez, A. Samchez, D. Putnam, D. Ting, R. Langer, M. J. Alonso, *J. Controll. Rel.* **2001**, *75*, 211–224.

[119] J. Liaw, S.-F., F.-C. Hsiao, *Gene Ther.* **2001**, *8*, 999–1004.

[120] M. Hashida, M. Nishikawa, F. Yamashita, Y. Takakura, *Adv. Drug Deliv. Rev.* **2001**, *52*, 187–196.

[121] P. K. Yadava, *Molecular Biology Today*, **2000**, *1*, 1–6.

[122] M. Kurisawa, M. Yokoyama, T. Okano, *J. Controll. Rel.* **2000**, *69*, 127–137.

[123] V. L. Truong-Le, S. M. Walsh, E. Schweibert, H.-Q. Mao, W. B. Guggino, J. T. August, K. W. Leong, *Arch. Biochem. Biophys.* **1999**, *361*, 47–56.

[124] H. Cohen, R. J. Levy, J. Gao, et al., *Gene Ther.* **2000**, *7*, 1896–1905.

[125] H. Cohen-Sacks, Y. Najajreh, V. Tchaikovski, et al., *Gene Ther.* **2002**, *9*, 1607–1616.

22
Luminescent Quantum Dots for Biological Labeling

Xiaohu Gao and Shuming Nie

22.1
Overview

The integration of nanotechnology with biology and medicine is expected to produce major advances in medical diagnostics, therapeutics, molecular biology, and bioengineering [1, 2]. Recent advances have led to the development of functional nanoparticles (electronic, optical, magnetic, or structural) that are covalently linked to biological molecules such as peptides, proteins, and nucleic acids [3–14]. Due to their size-dependent properties and dimensional similarities to biomacromolecules, these bioconjugates are well suited as contrast agents for in-vivo magnetic resonance imaging (MRI) [15–17], as nanoscale carriers for drug delivery, and as nanostructured coatings and scaffolds for medical implants and tissue engineering [18, 19].

In this chapter, we discuss semiconductor quantum dots (QDs) and their applications in biological labeling. In comparison with organic dyes and fluorescent proteins, semi-

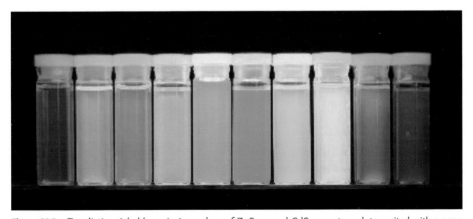

Figure 22.1 Ten distinguishable emission colors of ZnS-capped CdSe quantum dots excited with a near-UV lamp. From left to right (blue to red), the emission maxima are located at 443, 473, 481, 500, 518, 543, 565, 587, 610, and 655 nm.

Nanobiotechnology. Edited by Christof Niemeyer, Chad Mirkin
Copyright © 2004 WILEY-VCH Verlag GmbH & Co. KgaA, Weinheim
ISBN 3-527-30658-7

conductor QDs represent a new class of fluorescent labels with unique advantages and applications. For example, the fluorescence emission spectra of QDs can be continuously tuned by changing the particle size, and a single wavelength can be used for simultaneous excitation of all different-sized QDs (Figure 22.1, see p. 343). Surface-passivated QDs are highly stable against photobleaching and have narrow, symmetric emission peaks (25–30 nm full width at half maximum). It has been estimated that CdSe quantum dots are about 20 times brighter and 100 times more stable than single rhodamine 6G molecules [5].

Semiconductor QDs (e. g., CdSe, CdTe, CdS, ZnSe, InP, and InAs) are most often composed of atoms from groups I–VII, II–VI, or III–V elements. Earlier attempts to synthesize QDs were conducted in aqueous environments with stabilizing agents such as thioglycerol and polyphosphate. However, the resulting QDs showed poor quantum yields (<10%) and broad size distributions (relative standard deviation RSD >15%). In 1993, Bawendi and coworkers reported a high-temperature organometallic procedure for QD synthesis [20]. This method was later improved by three independent research groups [21–23], yielding near-perfect nanocrystals with quantum yields as high as 50% at room temperature, and a particle size distribution as narrow as 5%.

To prepare type II–VI QDs, a metal precursor (such as dimethyl cadmium) and a chalcogenide compound (such as selenium) are first dissolved in tri-*n*-butylphosphine (TBP) or tri-*n*-octylphosphine (TOP), and are then injected into a hot coordinating solvent such as tri-*n*-octylphosphine oxide (TOPO) at 340–360 °C. Recent studies conducted by Peng and coworkers have shown that high-quality nanocrystals could also be prepared by using CdO as an inexpensive starting material [24, 25]. The nanocrystal size can be tuned by heating QDs in TOPO at 300 °C for an extended period of time (ranging from seconds to days, depending on the desired particle size), in which the QDs grow by Ostwald ripening. In this process, smaller nanocrystals are broken down, and the dissolved atoms are transferred to larger nanocrystals. The rate of growth is dependent upon temperature and the amount of limiting reagents [26, 27]. Alternately, continuous injection of organometal/chalcogenide precursors at 300 °C can be used to increase the size of QDs [28].

For improved optical properties, the QDs are often coated and passivated by a thin layer of a higher bandgap material. For example, the fluorescence quantum yields of CdSe QDs increase from 5% to 50% with one to two monolayers of ZnS capping [21–23]. At present, ZnS and CdS are most commonly used to cap CdSe QDs. The bandgap energy of bulk CdS is about 0.9 eV higher than that of CdSe, while the ZnS and CdSe bond lengths are similar; these conditions lead to the epitaxial growth of a smooth ZnS layer on the surface of CdSe core particles. Similar procedures have been used to synthesize group III–V nanocrystals such as InP and InAs [29–32].

Semiconductor QDs absorb photons when the energy of excitation exceeds the bandgap energy. During this process, electrons are promoted from the valence band to the conduction band. Measurements of UV-Visible spectra reveal a large number of energy states in QDs. The lowest excited energy state is shown by the first observable peak (also known as the quantum-confinement peak), at a shorter wavelength than the fluorescence emission peak. Excitation at shorter wavelengths is possible because multiple electronic states are present at higher energy levels. In fact, the molar extinction coefficient gradually increases toward shorter wavelengths (Figure 22.2). This is an important feature for biological applications because it allows simultaneous excitation of multicolor QDs with a single light source.

Light emission arises from the recombination of mobile or trapped charge carriers. The emission from mobile carriers is called "excitonic fluorescence", and is observed as a sharp peak. The emission spectra of single ZnS-capped CdSe QDs are as narrow as 13 nm (full width at half maximum or FWHM) at room temperature [5]. Defect states in the crystal interior or on its surface can trap the mobile charge carriers (electrons or holes), leading to a broad emission peak that is red-shifted from the excitonic peak. Nanocrystals with a large number of trap states generally have low quantum yields, but surface capping or passivation can remove these defect sites and improve the fluorescence quantum yields.

The excitonic fluorescence is dependent on the nanocrystal size. Research conducted by several groups has demonstrated an approximately linear relationship between the particle size and the bandgap energy [21, 33]. This quantum–size effect is similar to that observed for a "particle in a box." Outside of the box, the potential energy is considered to be infinitely high. Thus, mobile carriers (similar to the particle) are confined within the di-

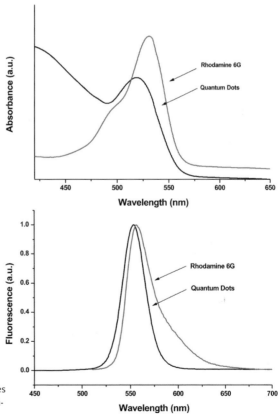

Figure 22.2 Comparison of the excitation (top) and emission (bottom) profiles between rhodamine 6G and CdSe quantum dots.

mensions of the nanocrystal (similar to the box) with discrete wavefunctions and energy levels. As the physical dimensions of the box become smaller, the bandgap energy becomes higher. For CdSe nanocrystals, the sizes of 2.5 nm and 5.5 nm correspond to fluorescence emission at 500 nm and 620 nm, respectively. In addition to size, the emission wavelength can be varied by changing the semiconductor material. For example, InP and InAs QDs usually emit in the far-red and near-infrared [29–32], while CdS and ZnSe dots often emit in the blue or near-UV [34]. It is also interesting to note that elongated QDs (called quantum rods) show linearly polarized emission [35], whereas the fluorescence emission from spherical CdSe dots is either circularly polarized or not polarized [36, 37].

In comparison to organic dyes such as rhodamine 6G and fluorescein, CdSe nanocrystals show similar or slightly lower quantum yields at room temperature. The lower quantum yields of nanocrystals are compensated by their larger absorption cross-sections and much reduced photobleaching rates. Bawendi and coworkers estimated that the molar extinction coefficients of CdSe QDs are about 10^5 to 10^6 M^{-1} cm^{-1}, depending on the particle size and the excitation wavelength [20, 21]. These values are 10- to 100-fold larger than those of organic dyes, but are similar to the absorption cross-sections of phycoerytherin, a multi-chromophore fluorescent protein. Chan and Nie have estimated that single ZnS-capped CdSe QDs are ~20 times brighter than single rhodamine 6G molecules [5]. Similarly, phycoerytherin is estimated to be 20 times brighter than fluorescein [38].

Another attractive feature of using QDs as biological labels is their high photostability. Gerion et al. examined the photobleaching rate of silica-coated ZnS-capped CdSe QDs against that of rhodamine 6G [39]. The QD emission stayed constant for 4 hours, while rhodamine 6G was photobleached after only 10 minutes. It has been suggested that capped CdSe nanocrystals are 100- to 200-fold more stable than organic dyes and fluorescent proteins [5]. Under intense UV excitation, single phycoerytherin molecules are found to photobleach after 70 seconds, while the fluorescence emission of quantum dots remain unchanged after 600 seconds [28]. The photobleaching of QDs is believed to arise from a slow process of photo-induced chemical decomposition. Henglein and coworkers speculated that CdS decomposition is initiated by the formation of S or SH radicals upon optical excitation [40, 41]. These radicals can react with O_2 from the air to form a SO_2 complex, resulting in slow particle degradation.

Single QDs have been shown to emit photons in an intermittent on-off fashion [42, 43], similar to a "blinking" behavior reported for single fluorescent dye molecules, proteins, polymers, and metal nanoparticles. The fluorescence of single QDs turns on and off at a rate that is dependent on the excitation power. This phenomenon has been suggested to arise from a light-induced process involving photoionization and slow charge neutralization of the nanocrystals [42]. When two or more electron-hole pairs are generated in a single nanocrystal, the energy released from the combination of one pair could be transferred to the remaining carriers, one of which is preferentially ejected into the surrounding matrix. Subsequent photogenerated electron-hole pairs transfer their energy to the resident, unpaired carrier, leading to nonradiative decay and dark periods. The luminescence is restored only when the ejected carrier returns to neutralize the particle. Banin et al. believe that thermal trapping of electrons and holes is also a contributing factor because they observed a dependence of the blinking rate on temperature [44]. A further find-

(a) Bifunctional linkage

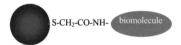

S-CH₂-CO-NH- biomolecule

(d) Electrostatic Attraction

(b) Hydrophobic Attraction

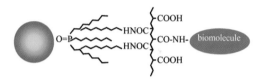

(c) Silanization

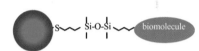

S ∿ Si-O-Si ∿ biomolecule

(e) Encoded Beads

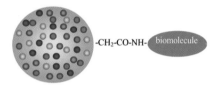

-CH₂-CO-NH- biomolecule

Figure 22.3 Schematic illustration of surface modification methods for linking quantum dots to biomolecules.

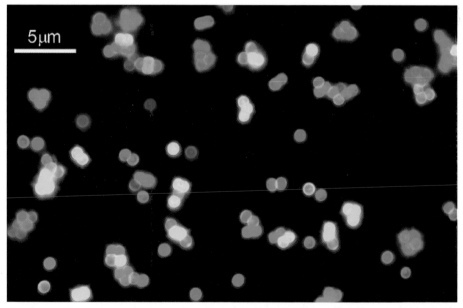

5 μm

Figure 22.4 Fluorescence micrograph of a mixture of CdSe/ZnS QD-tagged beads emitting single-color signals at 484, 508, 547, 575, and 611 nm. The beads were spread and immobilized on a polylysine-coated glass slide, which caused a slight clustering effect. (Reproduced with permission from Ref. [51].)

ing is that single dots exhibit random fluctuations in the emission wavelength (spectral wandering) over time [7, 45]. This effect is attributed to interactions between excitons with optically induced surface changes.

22.2
Methods

In order to exploit the novel optical properties of QDs for biological applications, a number of methods have been reported for converting hydrophobic QDs to water-soluble and biocompatible nanocrystals (Figure 22.3, see p. 347). In one approach, mercaptopropyl trimethoxysilane (MPS) adsorbs onto the QD surface, and displaces the surface-bound TOPO molecules [4]. A silica-shell is formed on the surface by introduction of a base and then hydrolysis of the MPS silanol groups. The polymerized silanol groups help stabilize nanocrystals against flocculation, and render the QDs soluble in intermediate polar solvents such as methanol and dimethylsulfoxide. Further reaction of bifunctional methoxy molecules, such as aminopropyltrimethoxysilane and trimethoxysilyl propyl urea, makes the QDs more polar and soluble in aqueous solution.

In another method, bifunctional molecules such as mercaptoacetic acid and dithiothreitol are directly adsorbed onto the QD surface [5]. Mercapto compounds and organic bases are added to TOPO-QDs dissolved in organic solvents. The base deprotonates the mercapto functional group and carboxylic acid (in the case of mercaptoacetic acid), which leads to a favorable electrostatic binding between negatively charged thiols and the positively charged metal atoms. The QDs precipitate out of solution and can be redissolved in aqueous solution (pH >5). The presence of highly polar functional groups, such as – COOH, –OH, or –SO$_3$Na (from bifunctional mercapto molecules) makes the nanocrystals soluble in water.

A third approach for linking biomolecules onto the particle's surface is to use an exchange reaction, in which mercapto-coated QDs are mixed with thiolated biomolecules (such as oligonucleotides and proteins). After overnight incubation at room temperature, a chemical equilibrium is reached between the thiolated molecules in solution and on the QD surface. This method has been used to adsorb oligonucleotides and biotinylated proteins onto the surface of QDs [7, 46].

Recent research has further improved the surface chemistry using a synthetic biopolymer coating. For example, the water-soluble QDs can be stabilized with a positively charged polymer or a layer of chemically denatured bovine serum albumin (BSA) [47, 48]. A key finding is that the polymer coating restores the optical properties of QDs nearly to that of the original QDs in chloroform. The polymer layer also provides functional groups (amines and carboxylic acids) for covalent conjugation with a variety of biological molecules. A similar approach has recently been used by Mattoussi and coworkers in which engineered proteins with a linear positively-charged peptide are directly adsorbed onto negatively charge nanocrystals through electrostatic interactions [6].

Most recently, Wu and coworkers used an amine-modified polyacrylic acid polymer to coat the surface of QDs [49]. The modified polymer was no longer soluble in water, and strongly adsorbed onto TOPO-capped QDs via hydrophobic interactions in chloroform. An important feature of this procedure is that QDs are solubilized without removing the surface ligands (TOPO), which maintains the optical properties of QDs in an aqueous environment. Similarly, Bubertret et al. encapsulated hydrophobic QDs in small micelles and demonstrated their use in in-vivo cellular imaging [50].

QDs have been used for multiplexed optical encoding and high-throughput analysis of genes and proteins, as reported by Nie and co-workers [51]. Polystyrene beads are embedded with multicolor CdSe QDs at various color and intensity combinations (Figure 22.4, see p. 347). The use of six colors and ten intensity levels can theoretically encode one million protein or nucleic acid sequences. Specific capturing molecules such as peptides, proteins, and oligonucleotides are covalently linked to the beads and are encoded by the bead's spectroscopic signature. A single light source is sufficient for reading all the QD-encoded beads. To determine whether an unknown analyte is captured or not, conventional assay methodologies (similar to direct or sandwich immunoassay) can be applied. This so-called "bar-coding technology" can be used for gene profiling and high-throughput drug and disease screening. Based on entirely different principles, Natan and coworkers reported a metallic nanobarcoding technology for multiplexed bioassays [52]. Together with QD-encoded beads, these "barcoding" technologies offer significant advantages over planar chip devices (e. g., improved binding kinetics and dynamic range), and are likely to find use in various biotechnological applications.

22.3
Outlook

A number of biological labeling applications have been demonstrated for QDs, including DNA hybridization, immunoassays, and receptor-mediated endocytosis. In particular, multicolor quantum dots are well-suited for the simultaneous labeling of multiple antigens on the surface of normal and diseased cells (Figure 22.5). The high photostability of QDs allows not only real-time monitoring or tracking of intracellular processes over long periods of time, but also quantitative measurements of fluorescent intensity. In fact, the QD labels are so bright that they allow target detection at the single-copy level, and are able to provide detailed structure information of biological specimens. Figure 22.6 shows a true-color fluorescent image of BT-474 cells labeled with Her-2/neo antibody (green color).

Far-red and near-infrared QDs are well-suited for applications in in-vivo molecular imaging and ultrasensitive biomarker detection. Visible light has been used for cellular imaging and tissue diagnosis, but optical imaging of deeper tissues (millimeters) requires the use of far-red or near-infrared light in the spectral range of 650–900 nm. This wavelength range provides a "clear" window for in-vivo optical imaging because it is separated from the major absorption peaks of blood and water. Under photon-limited in-vivo conditions (where light intensities are severely attenuated by scattering and absorption), the large absorption coefficients of QDs (on the order of 10^6 cm^{-1} M^{-1}, ca. 10–100 times larger than those of common organic dyes) will be essential for efficient probe excitation. Unlike current single-color molecular imaging, multi-wavelength optical imaging with QDs will allow intensity ratioing, spatial colocalization, and quantitative target measurements at single metastasized tumor sites and for single anatomical structures.

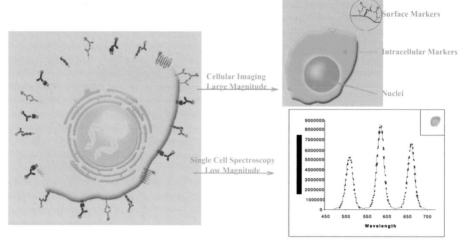

Figure 22.5 Schematic illustration of cell staining using biomolecules attached with multicolor QDs. Small color particles represent bioconjugated quantum dots.

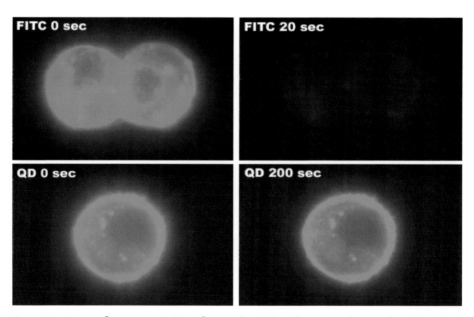

Figure 22.6 Immunofluorescence images of human breast tumor cells (BT-474) stained with organic dye (fluoroscein isothiocyanate; FITC) and green quantum dots. Within only 20 seconds of il- lumination, the organic dye was almost completely photobleached (top panel), while the QD fluores- cence image was stable

With an inert layer of surface coating, the nanocrystals are less toxic than organic dyes, similar to magnetic iron oxide nanoparticles. In preliminary studies, we have conjugated luminescent QDs to transferrin (an iron-transport protein), to antibodies that recognize cancer biomarkers, and to folic acid (a small vitamin molecule which is recognized by many cancer cells). In each case, we found that receptor-mediated endocytosis occurred and the nanocrystals were transported into the cell. Single QDs as well as clusters of dots trapped in vesicles were clearly visible inside living cells.

In conclusion, semiconductor QDs have been developed as a new class of biological labels with unique advantages and applications that are not possible with organic dyes or fluorescent proteins. When conjugated with fully functional biomolecules such as peptides, protein, and oligonucleotides, this class of fluorescent tags is well-suited for ultrasensitive imaging and detection. We envision that the design and construction of multifunctional QDs will allow molecular imaging and diagnostics of single diseased cells.

Acknowledgments

These studies were supported by grants from the National Institutes of Health (R01 GM58173 and R01 GM60562) and the Department of Energy (DOE FG02-98ER14873).

References

[1] W. C. W. Chan, D. J. Maxwell, X. H. Gao, R. E. Bailey, M. Y. Han, S. M. Nie, *Curr. Opin. Biotech.* **2002**, *13*, 40–46.

[2] C. M. Niemeyer, *Angew. Chem. Int. Ed.* **2001**, *40*, 4128–4158.

[3] S. R. Whaley, D. S. English, E. L. Hu, P. F. Barbara, A. M. Belcher, *Nature* **2000**, *405*, 665–668.

[4] M. Bruchez, M. Moronne, P. Gin, S. Weiss, A. P. Alivisatos, *Science* **1998**, *281*, 2013–2016.

[5] W. C. W. Chan, S. M. Nie. *Science*, **1998**, *281*, 2016–2018.

[6] H. Mattoussi, J. M. Mauro, E. R. Goldman, G. P. Anderson, V. C. Sundar, F. V. Mikulec, M. G. Bawendi, *J. Am. Chem. Soc.* **2000**, *122*, 12142–12150.

[7] G. P. Mitchell, C. A. Mirkin, R. L. Letsinger, *J. Am. Chem. Soc.* **1999**, *121*, 8122–8123.

[8] S. Pathak, S. K. Choi, N. Arnheim, M. E. Thompson, *J. Am. Chem. Soc.* **2001**, *123*, 4103–4104.

[9] R. Elghanian, J. J. Storhoff, R. C. Mucic, R. L. Letsinger, C. A. Mirkin *Science* **1997**, *277*, 1078–1081.

[10] R. A. Reynolds, C. A. Mirkin, R. L. Letsinger, *J. Am. Chem. Soc.* **2000**, *122*, 3795–3796.

[11] C. A. Mirkin, R. L. Letsinger, R. C. Mucic, J. J. Storhoff, *Nature* **1996**, *382*, 607–609.

[12] J. J. Storhoff, C. A. Mirkin, *Chem. Rev.* **1999**, *99*, 1849–1862.

[13] A. P. Alivisatos, K. P. Johnsson, X. G. Peng, T. E. Wilson, C. J. Loweth, M. P. Bruchez, P. G. Schultz, *Nature* **1996**, *382*, 609–611.

[14] B. Dubertret, M. Calame, A. J. Libchaber, *Nature Biotechnol.* **2001**, *19*, 365–370.

[15] L. Josephson, C. H. Tung, A. Moore, R. Weissleder, *Bioconjug. Chem.* **1999**, *10*, 186–191.

[16] J. W. M. Bulte, R. A. Brooks, Magnetic nanoparticles as contrast agents for MR imaging. in: *Scientific and clinical applications of magnetic carriers.* Plenum Press, New York, USA **1997**.

[17] J. W. M. Bulte, T. Douglas, B. Witwer, S. C. Zhang, E. Strable, B. K. Lewis, H. Zywicke, B. Miller, P. Van Gelderen, B. M. Moskowitz, I. D. Duncan, J. A. Frank, *Nature Biotechnol.* **2001**, *19*, 1141–1147.

[18] A. Curtis, C. Wilkinson, *Trends Biotechnol.* **2001**, *19*, 97–101.

[19] R. Gref, Y. Minamitake, M. T. Peracchia, V. Trubetskoy, V. Torchilin, R. Langer, *Science* **1994**, *263*, 1600–1603.

[20] C. B. Murray, D. J. Norris, M. G. Bawendi, *J. Am. Chem. Soc.* **1993**, *115*, 8706–8715.

[21] B. O. Dabbousi, J. Rodriguez-Viejo, F. V. Mikulec, J. R. Heine, H. Mattoussi, R. Ober, K. F. Jensen, M. G. Bawendi, *J. Phys. Chem. B* **1997**, *101*, 94639475.

[22] M. A. Hines, P. Guyot-Sionnest, *J. Phys. Chem. B* **1996**, *100*, 468471.

[23] X. G. Peng, M. C. Schlamp, A. V. Kadava-nich, A. P. Alivisatos, *J. Am. Chem. Soc.* **1997**, *119*, 7019–7029.

[24] L. Qu, Z. A. Peng, X. Peng, *Nano Lett.* **2001**, *1*, 333–337.

[25] L. Qu, X. Peng, *J. Am. Chem. Soc.* **2002**, *124*, 2049–2055.

[26] Y. DeSmet, L. Deriemaeker, E. Parloo, R. Finsy, *Langmuir* **1999**, *15*, 2327–2332.

[27] Y. DeSmet, L. Deriemaeker, R. Finsy, *Langmuir* **1997**, *13*, 6884–6888.

[28] W. C. W. Chan, PhD thesis, Semiconductor Quantum Dots for Biological Detection and Imaging, Bloomington, IN: Indiana University, **2001**.

[29] J. A. Prieto, G. Armeeles, J. Groenin, R. Cales, *Appl. Phys. Lett.* **1999**, *74*, 99–101.

[30] O. I. Micic, H. M. Cheong, H. Fu, A. Zunger, J. R. Sprague, A. Mascarenhas, A. J. Nozik, *J. Phys. Chem. B* **1997**, *101*, 4904–4912.

[31] B. Schreder, T. Schmidt, V. Ptatschek, U. Winkler, A. Materny, E. Umbach, M. Lerch, G. Muller, W. Kiefer, L. Spanhel, *J. Phys. Chem. B* **2000**, *104*, 1677–1685.

[32] J. Z. Shi, K. Zhu, Q. Zheng, L. Zhang, L. Ye, J. Wu, J. Zuo, *Appl. Phys. Lett.* **1997**, *70*, 2586–2588.

[33] X. G. Peng, J. Wickham, A. P. Alivisatos, *J. Am. Chem. Soc.* **1998**, *120*, 5343–5344.

[34] M. A. Hines, P. Guyot-Sionnest, *J. Phys. Chem. B* **1998**, *102*, 3655–3657.

[35] J. T. Hu, L. Li, W. Yang, L. Manna, L. Wang, A. P. Alivisatos, *Science* **2001**, *292*, 2060–2063.

[36] A. L. Efros, *Phys. Rev. B* **1992**, *46*, 7448–7458.

[37] S. A. Empedocles, R. Neuhauser, M. G. Bawendi, *Nature* **1999**, *399*, 126–130.

[38] R. Mathies, L. Stryer, Single-molecule fluorescence detection, in: *Applications of Fluroescence in the Biomedical Sciences*, Alan R. Liss, Inc., USA, **1986**.

[39] D. Gerion, F. Pinaud, S. C. Willimas, W. J. Parak, D. Zanchet, S. Weiss, A. P. Alivisatos, *J. Phys. Chem. B* **2001**, *105*, 8861–8871.

[40] A. Henglein, *Ber. Bunsenges. Phys. Chem.* **1982**, *86*, 301–305.

[41] S. Baral, A. Fojtik, H. Weller, A. Henglein, *J. Am. Chem. Soc.* **1986**, *108*, 375–378.

[42] M. Nirmal, B. O. Dabbousi, M. G. Bawendi, J. J. Macklin, L. E. Brus, *Nature* **1996**, *383*, 802–804.

[43] S. A. Empedocles, M. G. Bawendi, *Acc. Chem. Res.* **1999**, *32*, 389–396.

[44] U. Banin, M. Bruchez, A. P. Alivisatos, T. Ha, S. Weiss, D. S. Chemla, *J. Chem. Phys.* **1999**, *110*, 1195–1201.

[45] S. A. Blanton, M. A. Hines, P. Guyot-Sion-nest, *Appl. Phys. Lett.* **1996**, *69*, 3905–3907.

[46] D. M. Willard, L. L. Carillo, J. Jung, A. Van Orden, *Nano Lett.* **2001**, *1*, 469–474.

[47] I. Potapova, R. Mruk, S. Prehl, R. Zentel, T. Basche, A. Mews, *J. Am. Chem. Soc.* **2003**, *125*, 320–321.

[48] X. H. Gao, W. C. W. Chan, S. M. Nie, *J. Biomed. Opt.* **2002**, *7*, 532–537.

[49] X. Y. Wu, H. J. Liu, J. Q. Liu, K. N. Haley, J. A. Treadway, J. P. Larson, N. F. Ge, F. Peale, M. P. Bruchez, *Nature Biotechnol.* **2003**, *21*, 41–46.

[50] B. Dubertret, P. Skourides, D. J. Norris, V. Noireaux, A. H. Brivanlou, A. Libchaber, *Science* **2002**, *298*, 1759–1762.

[51] M. Y. Han, X. H. Gao, J. Z. Su, S. M. Nie, *Nature Biotechnol.* **2001**, *19*, 631–635.

[52] S. R. Nicewarner-Pena, R. G. Freeman, B. D. Reiss, L. He, D. J. Pena, I. D. Walton, R. Cromer, C. D. Keating, M. J. Natan, *Science*, **2001**, *294*, 137–141.

23
Nanoparticle Molecular Labels

James F. Hainfeld, Richard D. Powell, and Gerhard W. Hacker

23.1
Introduction

Nanotechnology means many things to many people, from *Fantastic Voyage* miniature submarines traveling through a person's arteries, to a ribosome that is a cellular "factory" which synthesizes proteins, while material scientists believe that it means making better materials from molecular building blocks.

The common theme seems to be "Nano" – that is, small – in the nanometer size range. In a sense, nanotechnology has been around a long time, as both chemistry and biochemistry rely on molecules and complexes in this size range or smaller. For example, nylon was first made in 1935 as a silk substitute, and was patterned after silk by making an amide-bonded polymer, similar to silk, but with a slightly different repeating unit [1]. Nanoparticles have also been around for a long time; presumably, the first nanoparticle was recognized in 1570 with aurum potable (potable gold) and luna potable (potable silver) which alchemists used as elixirs. Unfortunately, they did not make the consumer live forever, as evidenced by the high incidence (100 %) of dead alchemists. As early as 1595, gold colloids were incorporated into glass to make various colors, such as red, purple, violet, brown, or black. Faraday was the first to recognize that the colors were related to particle sizes [2].

This chapter deals with the biological application of metal nanoparticles to label biomolecules. A number of uses benefit from this combination: the metal nanoparticles make detection possible, easier or more sensitive. Colloidal gold particles, when adsorbed to antibodies [3] or to other targeting agents such as proteins [4] or peptides [5–7], are widely used as labels for the detection or microscopic localization of molecular and macromolecular targets. One popular pregnancy test kit develops a pink line which is in fact a 40-nm gold particle adsorbed to an antibody, anti-human chorionic gonadotropin (hCG). This "pregnancy hormone" keeps the corpus luteum producing progesterone after conception occurs. The urine sample flows over a capture stripe where anti-hCG has been previously adsorbed to nitrocellulose; if hCG is present, it is bound, or "captured". The gold with anti-hCG then flows over, and if hCG is bound, the gold-antibody binds, leaving a pink color on the stripe. The very high extinction coefficient of the 40-nm

Nanobiotechnology. Edited by Christof Niemeyer, Chad Mirkin
Copyright © 2004 WILEY-VCH Verlag GmbH & Co. KgaA, Weinheim
ISBN 3-527-30658-7

gold, $\sim 2 \times 10^9$ M^{-1} cm^{-1} compared with that of a fluorescent molecule (fluorescein: $\sim 7 \times 10^4$), increases detection sensitivity by a factor of about 30 000. The gold nanoparticle gives sufficient sensitivity to be perceived with the unaided eye, thereby producing an inexpensive assay [8].

Nanobiotechnology implies both some degree of supramolecular organization or cooperation, and the incorporation of a desirable function, ability or property into a supramolecular construct. The entry of metal particle bioconjugates as players in nanobiotechnology has been facilitated by the development of molecular control over the site, nature and formation of the link between the biological molecule and metal particle. This enables both the conjugation of metal particles to molecules with potential applications in nanotechnology, and the selective attachment of metal nanoparticles to specific sites within biological structures where the properties of the metal particles impart potentially useful functionality to the construct. This chapter will focus on the covalently linkable metal cluster labels, principally Nanogold [9, 10] and undecagold [11–13], which have been used for a number of such applications. Other types of particles, including unstabilized gold [14–16], clusters of other metals, and larger gold particles with controlled chemical reactivity [17], have also been utilized. In this chapter, both the use of metal particle labels to detect and localize biological targets will be discussed, together with the preparation and potential applications of metal cluster bioconjugates with novel properties and functionality.

23.2
Immunogold-Silver Staining: A History

For more than 30 years, nanometer-sized gold particles (mostly colloidal gold, with diameters ranging from 1 to 40 nm) have been the label of choice to demonstrate proteins and peptides (and other substances against which specific antibodies can be made) by transmission electron microscopy (TEM) [3]. Not only for TEM, but also for light microscopy (LM) and for scanning electron microscopy (SEM), immunogold-staining (IGS) methods show various advantages over other, nonparticulate immunostaining techniques, including enhanced visibility and increased detection efficiency and sensitivity. Accumulations of colloidal gold can be made visible with LM by the application of autometallography (AMG) [18–28], a group of techniques whereby silver or gold ions are reduced in situ to metal atoms and precipitated onto the surface of the gold particle nuclei. In the electron microscope, it can be shown that gold particles thereby grow considerably in size, until they conglomerate [29–31]. Combination of this reaction with colloidal gold-labeled enzyme histochemistry [21] and immunohistochemistry (IHC) [32, 33], both in 1983, led to the introduction of immunogold-silver staining (IGSS), and this was a major breakthrough in sensitivity and detection efficiency in the early 1980s. Initial attempts to apply Holgate's original IGSS method with a broader spectrum of antibodies used in general immunohistopathology often resulted in very high levels of unwanted background staining, and often the overall appearance of the stained section was "dirty". Successful attempts to modify IGSS to facilitate the highly sensitive demonstration of various kinds of substances in routine paraffin sections were published by a number of research teams. With the years, the methodology became increasingly sophisticated, and a variety of protocols were used [34–46]. Modifications included the use of gold particle sizes smal-

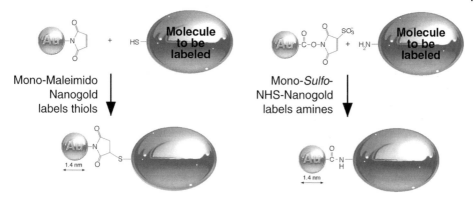

Figure 23.1 Schematic showing reactions of Mono-Maleimido-Nanogold and Mono-*Sulfo*-NHS-Nanogold.

ler than the ones originally used (1–5 nm in diameter), the use of fish gelatin, a number of new AMG developers, the use of gold-labeled protein-A [47–50], utilization of the streptavidin–biotin complex (S-ABC) principle [51, 52], and the introduction of antigen retrieval techniques. For review and update see Ref. [53].

The introduction of covalently linked gold cluster labels rather than colloidal gold marked another major advance in the methodology, giving researchers the ability to direct the attachment of metal nanoparticles to biological molecules with submolecular precision. The undecagold [12] and Nanogold [9, 10] labels have been conjugated to many different molecules that cannot be labeled with colloidal gold, including proteins, peptides, oligonucleotides, lipids, and small molecules, many of which have potential nanotechnology applications, which are discussed in more detail below. Conjugation reactions are shown in Figure 23.1. Antibody and protein conjugates give improved performance over colloidal gold probes, including increased cellular penetration, labeling density, and access to hindered antigens [54, 55].

For silver or gold amplification, numerous protocols had been described and are being marketed by a number of companies. Danscher's original protocols relied on the use of silver lactate [18–20], and had to be applied in a darkened room or under red light to give background-free preparations. Hacker et al. described a less light-sensitive modification using silver acetate as the silver ion source [27], thereby introducing the possibility to amplify gold-labeled LM preparations in normal laboratory daylight. It became possible now to optimize staining under microscopic control. Comparisons of different silver salts were reported and showed marked differences [46]. Most recently, AMG based on gold ions has been introduced commercially (*GoldEnhance*, Nanoprobes, Inc.), a new technique that further improved the spatial staining resolution and signal-to-noise ratio [56–59].

23.3
Combined Fluorescent and Gold Probes

Antibodies, or even antibody fragments, are sufficiently large that both gold clusters and fluorescent labels may be attached, spaced sufficiently far apart that quenching by Förster energy transfer [60] is minimal, fluorescence is largely maintained, and the probes are ef-

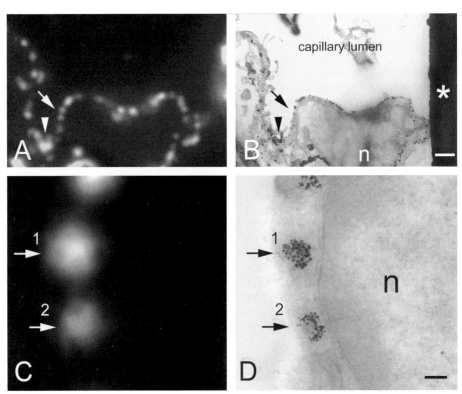

Figure 23.2 Demonstration of caveolin-1a (CAV-1a) localization with FluoroNanogold (FNG) using correlative immunofluorescence and immunoelectron microscopy on the same ultrathin cryosection. (A) Immunofluorescence localization of CAV-1a in a portion of a capillary endothelial cell from a terminal villus of the placenta; individual punctate structures can be observed (arrow). (B) Electron micrograph of the same region. This section has been subjected to silver enhancement so that the FNG could be visualized. Silver-enhanced gold labeling can be detected in the same structures observed by fluorescence microscopy (A) (see arrow). It should be noted that where individual caveolae are spaced very closely together, the fluorescence and silver-enhanced FNG signals appear to be "fused together" (arrowheads). The nucleus (n) of the endothelial cell is evident. A portion of the electron opaque grid bar is seen (*). Scale bar = 1 mm; (A) and (B) at the same magnification. (C) Enlargement of the region of panel (A) indicated with the white arrow. Two fluorescent structures are indicated (arrows). (D) Electron micrograph showing enlargement of the region of panel (B) indicated with the black arrow. The same structures shown by fluorescence microscopy in (C) (1 and 2) are heavily labeled with silver-enhanced gold particles. The characteristic omega-shaped morphology of caveolae is evident in structure 2. A portion of the nucleus (n) is evident. Scale bar = 100 nm; (C) and (D) at the same magnification. (Figure courtesy of T. Takizawa, Ohio State University, Columbus, OH) [69].)

fective for both immunofluorescence and immunogold labeling [61]. Initially, secondary Fab' antibody probes labeled with both Nanogold and fluorescein were prepared by the sequential conjugation of Monomaleimido Nanogold and fluorescein N-hydroxysuccinimide ester. These were used to label the SC35 pre-mRNA splicing factor in HeLa cells [62, 63]. The same probe was also used to label human lymphocytes, and labeling was visualized using fluorescence microscopy, different modalities of LM, and TEM [64]. Proof of principle was subsequently demonstrated using correlative light and electron microscopy of specimens with features localized on indexed grids [65]. Combined Cy3 and Nanogold probes have also been developed, and demonstrate high specificity and sensitivity both for immunocytochemistry and for in-situ hybridization [66, 67]; using Tyramide Signal Amplification (TSA) followed by combined Cy3/Nanogold-labeled streptavidin, the fluorescence signal was sufficiently bright that staining was observed for HPV 16/18 in SiHa cells, known to contain only one or two copies of the target. More recently, combined Alexa Fluor 488 and 594 and Nanogold probes have been prepared and found to give higher brightness and improved pH compatibility for fluorescent staining [68]. An example of correlative fluorescence and electron microscopic localization of caveolin is shown in Figure 23.2 [69]. It has also been reported that fluorescence is maintained even after a brief period of silver enhancement, enabling observation by epifluorescence microscopy [65]; preliminary results have also suggested that fluorescence may be sufficiently preserved to allow the preparation of dual-function antibody probes in which fluorescent labels are combined with larger platinum clusters [70].

23.4
Methodology

23.4.1
Choice of Gold and AMG Type

Originally, colloidal gold particles were mainly applied adsorbed to second layer antibodies and used in an indirect, two-step IHC method [32–35]. Later on, it was found that the use of covalently bound Nanogold–antibody and streptavidin conjugates, instead of colloidal gold electrochemically adsorbed to antibodies, gave a further boost in signal-to-noise ratio [9, 10]. A three-step S-ABC Nanogold technique was born that could be successfully applied in IHC, and also for in-situ hybridization (ISH) [38, 57]. Conglomerations of clustered gold particles amplified by silver- or gold-salt-based AMG now appeared as jet-black precipitates with a distinctly sharper appearance than the reaction products of most enzyme-labeled preparations. Also, when compared to the "classical" indirect and silver-enhanced colloidal gold techniques, the staining results were often much clearer than those achieved before (Figure 23.3). The new GoldEnhance technology (Nanoprobes, Inc.), based on the catalytic deposition of gold rather than silver [58, 59], allows for a more "metallic" appearance seen with LM – that is, the edges of staining appear clear-cut and the staining itself is completely black, provided that the amount of the substance to be detected is high enough and that the antigen–antibody reaction is adequate. A similar result is also most often obtained in ISH (Figure 23.4).

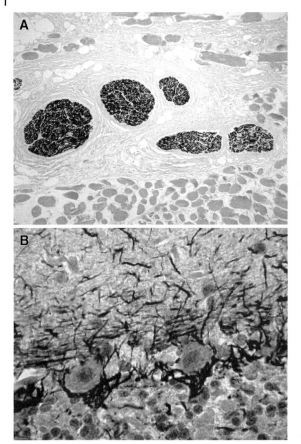

Figure 23.3 (A) S-100 immunostained Schwann cell sheets in cross-sectioned nerve bundles of human skin, embedded in connective tissue and muscle fibers. Jet-black specific staining of very high resolution is obtained with Nanogold-IGSS (staining with 1.4 nm-Nanogold–streptavidin). (B) High-power photomicrograph (objective magnification ×100) of cerebellum, immunostained for neurofilament protein triplet. Indirect IGSS with conventional colloidal gold (5 nm diameter) shows basket cell nerve fibers within the molecular and the white layers, partly surrounding large unstained Purkinje cells. (A, B) : HE counterstained, formalin-fixed 5 μm-thick paraffin sections. Both preparations were silver-amplified with silver acetate autometallography.

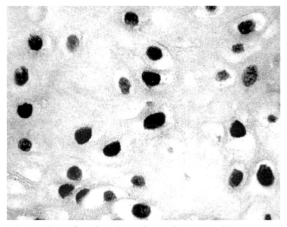

Figure 23.4 High-power (objective magnification ×100) photomicrograph of condyloma accuminatum, stained with a simple and straight-forward non-tyramide Nanogold-silver in-situ hybridization protocol. A biotinylated cDNA-probe (Enzo, NY) recognizing human papillomavirus (HPV) subtypes 6/11 was used, followed by only one step with streptavidin–Nanogold and GoldEnhance™ autometallography. Metallic-black labeling within infected nuclei is obtained. Most nuclei are stained as a whole, and some of them are more spotty. The latter usually contain fewer copies of HPV-DNA, or concentrations of HPV-DNA within certain nuclear areas. HE counterstained, formalin-fixed 5 μm-thick paraffin section.

23.4.2
Iodinization

One important fact also should be re-addressed here: it had previously been reported by the original authors of IGSS [32, 33] that pre-treating the sections with Lugol's iodine is essential for obtaining a high detection efficiency and sensitivity. Since then, most authors have confirmed this finding, although the exact process yet still seems not clearly understood. It had been suggested that the weak oxidizing activity of the halogen is responsible for the fact that many or most antigens can only be demonstrated if this step is used. Although some authors also suggested protocols without the Lugol's iodine steps, we highly recommend using it, for IHC as well as for ISH; in our experience, staining sensitivity and signal-to-noise ratio is far better in most experiments. The effect sometimes, however, appears to be less significant or even diverse in cryostat or in resin sections. Most protocols referenced above do rely on the use of iodinization.

23.4.3
Sensitivity

In comparison to most other IHC techniques, IGSS methods are extremely sensitive and detection efficient. Often, antigens can be detected with IGSS where other methods failed or gave equivocal reactions [35]. Although in IGSS, colloidal gold most often produced surprisingly good results, it was noted that often it could not clearly demonstrate intranuclear structures such as steroid receptors, proliferation markers (e. g., Ki-67), or certain tumor suppressor gene proteins (e. g., p53). For such applications, the use of streptavidin–Nanogold appears to be a very practical solution. Most likely due to its nonionizing characteristics at near-neutral pH values, and in contrast to the isoelectric point of colloidal gold (~8.4), Nanogold appears to produce much better labeling of proteins localized near DNA [37, 71].

23.5
Applications for the Microscopical Detection of Antigens

Immunocytochemistry or histochemistry using gold particles as the label can be recommended for many applications in routine and scientific detection experiments, especially those where a high sensitivity is needed, or when there is a need for spectacular photomicrographs to be produced. One of the major advantages of the resulting black stain is that conventional hematoxylin and eosin (HE) counterstaining is possible, thereby also enlarging the diagnostic potential of routine pathology [35, 72]. Multiple immunostaining reactions can be achieved which are of outstanding visual impact (e. g. [36]; Hacker in Refs. [41, 73, 74]). The combination of Nanogold with a fluorescent label in one-and-the-same preparation provides the fascinating new possibility of subsequently using fluorescence microscopy and transmitted light microscopy [61–64, 75]. Very recent experiments (D. Schwertner and G. W. Hacker, unpublished results) have shown fascinating new possibilities when applying gold-silver techniques for three-dimensional full-color computer light microscopy; for example, when using desmin-antibodies, the hexamer structure can be readily experienced at the LM level. For uses at the EM level (not within the scope of this chapter), the reader is referred to a few key articles in the literature (see Refs. [53, 54, 72, 75–81]).

Nanogold-silver staining has been used with conventional enzyme IHC for double-staining experiments. In one study, m4R and ChAT or -opiate receptors were localized for EM with silver-enhanced Nanogold-labeled secondary antibodies and peroxidase-DAB respectively in rat brain sections [82]; in another example, cytokeratin-19 and γ-tubulin were "simultaneously" localized at the EM level in CACO-2 cells. Both primary antibodies (mAb anti-CK19 and rabbit anti-γ-tubulin) were added together. To avoid interactions between the peroxidase and the silver enhancer of Nanogold, the Nanogold-silver enhancement procedure was completed first; the cells were then incubated with Fab anti-rabbit IgG-peroxidase followed by diaminobenzidine development [83].

23.6
Detection of Nucleic Acid Sequences

A number of protocols using colloidal gold with silver enhancement for nucleic acid sequence detection were described during the 1980s [84-87] but have not been widely adopted. ISH staining results at that time were often better when a peroxidase-based system was used, but with the introduction of gold cluster labels this situation has slowly changed. The nearly pH-neutral behavior of Nanogold also allowed a greatly improved applicability to the detection of gene sequences by ISH : Nucleic acid sequences present only in minute amounts could be readily demonstrated using a simple, two-step indirect ISH technique (Figure 23.5) [88].

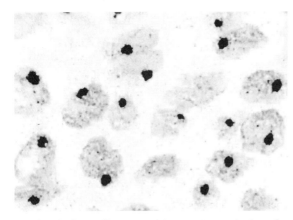

Figure 23.5 Single spots of human papillomavirus (HPV) 16/18 in cervical carcinoma cell nuclei. Tyramide signal-amplified streptavidin–Nanogold in-situ hybridization, intensified with GoldEnhance™. It is very likely that the black intranuclear spots are single copies of HPV-16/18-DNA. SiHa cell culture sections, stained in parallel, known for their content of only one to two copies of HPV-16-DNA per nuclei, showed comparable staining. Undifferentiated cervical squamous cell carcinomas usually contain only very few or only one HPV-DNA-copy integrated in the host cell genome. Other staining methods performed on serial sections of the one shown here, including peroxidase-based or colloidal-gold-based in-situ hybridization, did not give reproducible positive results on this particular paraffin block, whereas the use of biotinylated tyramides (TSA, tyramide signal-amplification; PerkinElmer; CSA, catalyzed signal amplification; DakoCytomation) in combination with streptavidin–Nanogold yielded reproducible and distinctly recognizable staining in nearly all epithelial carcinoma cells of this case. High-power photomicrograph (objective magnification ×100), HE counterstained, formalin-fixed 5 µm-thick paraffin section.

A

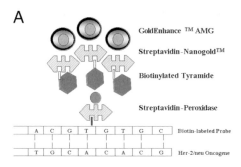

B

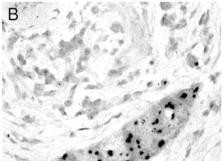

Figure 23.6 (A) Final configuration for GOLDFISH in-situ hybridization assay. (B) A non-neoplastic duct displaying normal endogenous (nonamplified) HER-2/neu gene copy (left) is surrounded by infil- trating duct carcinoma demonstrating HER-2/neu gene amplification (right). GOLDFISH with nuclear fast red counterstains; original magnification, ×400. (From Ref. [107].)

Applications of gold and silver for in-situ polymerase chain reaction (IS-PCR) have also been described [89–91], and in this still-emerging field numerous advantages of this highly sensitive detection method are clear [88, 92–94]. Due not only to the occasionally relatively low reproducibility and major specificity problems encountered with IS-PCR, but also to the relatively higher costs, the authors have during more recent years placed higher emphasis of label amplification techniques, rather than target (DNA/RNA) ampli- fication. Another application of gold-silver for super-sensitive RNA-detection which is still awaiting broader investigation is that of in-situ self-sustained sequence replication-based amplification (3SR) [95].

The introduction of catalyzed reporter deposition (CARD), now commercially termed "tyramide-signal-amplification" (TSA; PerkinElmer Life Sciences) or "catalyzed signal amplification" (CSA; DakoCytomation, Glostrup, Denmark, and Carpinteria, CA, USA) [96, 97], for the first time allowed the detection of single molecules of gene sequences in the light microscope by using a relatively straightforward ISH and applying the Nano- gold technique [56, 57, 98–104]. Today, the TSA protocol can be carried out fully automa- tically [105, 106]. A bright-field (LM) method, GOLDFISH (gold-facilitated in-situ hybridi- zation) has been developed for the detection of Her-2/neu gene amplification in paraffin- embedded sections of invasive ductal carcinoma [107], and showed both high reproduci- bility and excellent concordance with fluorescence in-situ hybridization (FISH) methods [108] (Figure 23.6).

23.7
Applications for Microscopical Detection of Nucleic Acids

For ISH, a large spectrum of uses is applicable. The in-situ detection of specific DNA- or RNA-sequences with molecular sensitivity gives rise to the bulk of routine uses for diag- nostic pathology, for example, in the detection of tumor-associated viruses or of tumor suppressor genes. Gold-silver-based ISH is also a very elegant way to demonstrate cancer gene amplification by LM, with numerous advantages when compared to FISH, for example, the direct applicability of automated computer-based image analysis in permanent preparations without the need for special equipment. The GOLDFISH method yields a dense, punctate staining pattern, which readily allows visualization of the underlying ultrastructure; this is important for a complete diagnosis. It also allows the use of other stains. In transmission LM, conventional histochemical counterstains can be used (e. g., HE); although counterstains such as pontamine sky blue may also be used with FISH, they do not provide as useful an interpretation as HE or Nuclear Fast Red counterstains. In LM or transmission EM, Nanogold-silver/gold ISH or ICC may be combined with a chromogenic second specific target stain, such as peroxidase-DAB-H_2O_2, to detect antigens or other nucleic acid sequences on the same section; this is most effective if the enzymatic chromogen is sufficiently different in color. With LM, in addition, a light nuclear counterstain such as hematoxylin may be applied, whilst with the TEM a light conventional contrasting stain should be used. This is more convenient for the practicing pathologist as it uses the standard bright-field light microscope rather than requiring expensive or less-accessible fluorescence optics. Furthermore, the interpretation is simpler because it is based on the overall pattern rather than requiring spot counting [107].

23.8
Technical Guidelines and Laboratory Protocols

Most recently, a book was published which was dedicated solely to gold and silver staining techniques [53] wherein a state-of-the-art review and exact technical guidelines on the most promising molecular morphological technologies related to gold labels and autometallography is provided. Constantly updated staining protocols are available on the internet, under the web address http://www.frontierquestions.com/labprotocols.htm

23.9
Gold Derivatives of Other Biomolecules

As gold nanoparticles can be covalently attached to antibodies, it is likely that they could also be attached to other biomolecules such as proteins, peptides, drugs, viruses, carbohydrates, lipids, and nucleic acids. The introduction of chemically selective reactivity has enabled conjugation of metal particle labels to almost any biological molecule containing an appropriate reactive group, in a similar manner to the conjugation of fluorescent labels. In practice, the relatively larger size of the gold particle compared with many biomolecules of interest means that conjugation should be approached with consideration of how the properties and biological activity of the conjugate might be modified by the attached gold particle. This is particularly appropriate with small molecules in which the gold can perturb binding. For example, while undecagold-conjugated phalloidin can be used to map the topography of actin filaments [109], a phalloidin conjugate prepared with the larger Nanogold was reported not to show comparable activity [110].

23.9.1
Protein Labeling

Because gold cluster labeling proceeds through specific chemical reactions, it is selective towards specific groups in the conjugated biomolecule. If the gold particle reacts with a unique group in the biomolecule, the reaction is site-specific, and can be chosen so that it does not interfere with biological function. As the reaction is no longer dependent on the charge properties of the conjugate protein, many proteins that are not amenable to conventional colloidal gold labeling may be conjugated with gold clusters.

This site-specificity implies higher resolution when the bound probe is microscopically localized. This in turn enables a higher level of resolution in EM studies; while colloidal gold labeling might typically localize targets within tissues, site-specific Nanogold labeling can be used to localize specific functional elements at the macromolecular level, enabling the localization or differentiation of different binding sites within multi-subunit protein complexes, structures, or organelles. Examples include the determination of the quaternary structure of the insulin-insulin receptor by cryoelectron microscopy using insulin labeled with Mono-*Sulfo*-NHS-Nanogold [111, 112], and the use of Nanogold to localize the two dimers of L7/L12 within the structure of the 70S ribosome. Protein L7/L12 was reduced with 1 % mercaptoethanol and labeled with Monomaleimido Nanogold; two reconstitution approaches, together with cryoelectron microscopy and single particle reconstruction, were used to determine the structure [113].

Subtle changes in binding can be differentiated by gold labeling and modifications in the complex assembly and binding procedures. In a recent example, undecagold was used to localize the site of microtubule-associated protein 2 (MAP2) and tau protein binding on the surface of pre-assembled microtubule protofilaments; Cryo-EM and helical image analysis showed that both the IR and MAP2 elements lie along the exterior ridges of microtubules [114]. In a subsequent study, tau protein, labeled with Nanogold at a repeat motif in the microtubule-binding domain, was used to study tau binding during microtubule assembly. Three-dimensional electron cryomicroscopy indicated that a repeat motif occupies a similar site to taxol on the inner surface of the microtubules, supporting the conclusion that one of the tau repeat loops is the natural substrate that occupies the taxol-binding pocket in beta-tubulin [115]. The reactivity of the gold labeling reagent may be used to localize a target chemical group. Monomaleimido Nanogold has been used to localize and quantitate interprotamine disulfide bonds during spermiogenesis : the disappearance of Nanogold labeling was an indicator of the formation of disulfide bonds by cysteine residues [116]. Nanogold may be prepared in a positively or negatively ionizing form by incorporation of synthetically modified ligands, bearing aliphatic amines or carboxyls respectively, into its surface; the resulting charge can provide a method for labeling. Prescianotto-Baschong and co-workers have used positively charged Nanogold to label elements of the yeast endocytic pathway [117, 118]. Negatively charged Nanogold was used to map the distribution of electrical charges over the surface of *Plasmodium falciparum* merozoites and erythrocytes. Atomic force microscopy with surface potential spectroscopy were used to map the surface charge directly; this was followed up by incubation with negatively charged Nanogold, silver enhancement and gold toning, localized with TEM [119].

The extreme insolubility of the aberrantly folded isoform (PrPSc) of the prion protein (PrP) responsible for Creutzfeldt–Jakob disease (CJD), bovine spongiform encephalopathy (BSE) and other spongiform encephalopathies has prevented structural determination by X-ray diffraction or nuclear magnetic resonance (NMR) imaging. However, 2-D electron crystallography and Nanogold labeling of two truncated but still infectious variants, N-terminally truncated PrPSc (PrP 27-30) and a miniprion (PrPSc106), yielded sufficient structural information to construct models for the PrPSc structure. N-linked sugars were oxidized with periodate, then selectively labeled using Monoamino Nanogold. Negative-stain EM and image processing allowed the extraction of limited structural information to 7 Å resolution. The dimensions of the monomer and the locations of the deleted segment and sugars, used as constraints in the construction of models for PrPSc, were satisfied only by structures featuring parallel beta-helices as the key element – a significant finding that will help derive an understanding of prion propagation and the process of neurodegeneration associated with these prion diseases [120].

23.9.2
Gold Cluster-labeled Peptides

Conjugation of gold cluster labels to peptides has been described by a number of researchers, and these findings have been reviewed previously [10]. Two groups have used Nanogold to label antibody Fv fragments, thus generating a probe smaller than a Fab' fragment that, as the gold particle is linked at a site where it does not affect binding, retains the immunoreactivity of the native antibody [121, 122]. Segond von Banchet has conjugated Nanogold to the undecapeptide Substance P [123], and the tetradecapeptide somatostatin [124] – the same peptide earlier conjugated with colloidal gold [5–7] – and used the labeled peptides to localize substance P binding sites and somatostatin receptors in the rat spinal cord. The larger peptide calmodulin, a 17 kDa protein that regulates the calcium release channel (ryanodine receptor) in the sarcoplasmic reticulum of skeletal and cardiac muscle, has been labeled with Nanogold at a cysteine residue (Cys27) and used as a high-resolution probe to directly visualize the binding and control site of calmodulin on isolated single molecules of the calcium release channel cryoembedded in amorphous ice [125]. More recently, Nanogold-labeled calmodulins have been used, exchanged for delta, to enable the localization of the delta subunit within the bridged, bilobal phosphorylase b kinase holoenzyme complex by scanning TEM [126].

23.9.3
Gold Cluster Conjugates of Other Small Molecules

Even small molecules, appropriately functionalized, may be labeled with gold clusters. The use of undecagold-conjugated phalloidin to map the topography of actin filaments by STEM has been mentioned previously; the phalloidin was synthetically modified to incorporate an aliphatic amino-group, which was then reacted with bis-(4-nitrophenyl) adipate followed by amino- undecagold [109]. In addition, a snake venom toxin, toxin-α from *Naja nigricollis*, has been derivatized with maleimido-undecagold to produce a small probe with a high affinity for the cholinergic binding site of the *Torpedo marmorata* nicotinic receptor [127]; the use of small molecule probes such as these may enable further improvements in the resolution achievable in EM labeling.

23.9.4
Gold–Lipids: Metallosomes

Fatty acids or phospholipids that bear an appropriate chemical group may also be conjugated with gold nanoparticles. Two such Nanogold conjugates, palmitoyl Nanogold and diphosphatidyl ethanolamino- (DPPE) Nanogold, are shown in Figure 23.7. Lipids in aqueous solutions form emulsions similar to mayonnaise. These possess the ability to self-organize to give a wide variety of nanoscale structures and morphologies with potentially useful applications: upon sonication, lipids form micelles, hollow liposome vesicles, multilayer liposomes, sheets and tubes depending on the phase state.

Lipid–gold conjugates have applications both as probes for the microscopic localization and tracking of liposomes, and as components for the templated assembly of supramolecular arrays of gold particles with a rich variety of morphologies. Liposomes are used to encapsulate drugs: they both keep the drug separate from metabolic activity, and deliver it to the target cells. These gold–lipid conjugates have been incorporated into antifungal drug liposomes to aid in visualization of the delivery process [128], and also to demonstrate the targeting of cationic liposomes to endothelial cells in tumors and chronic in-

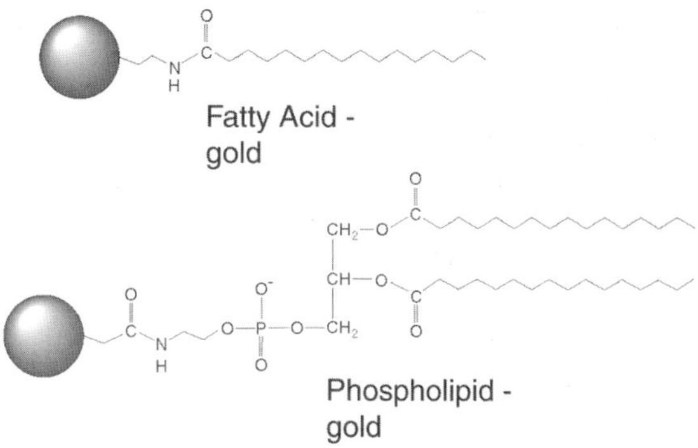

**Fatty Acid -
gold**

Figure 23.7 Gold-lipid diagram showing the chemical structures of two major lipid classes covalently bound to gold clusters. The top diagram shows the C_{15} palmitoyl fatty acid–gold conjugate; the bottom diagram shows the phospholipid dipalmitoyl phosphatidylethanolamine–gold conjugate.

**Phospholipid -
gold**

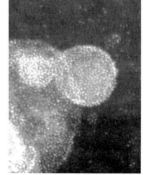

Figure 23.8 Metallosomes. These liposomes composed of 100 % gold-conjugated phospholipids (as shown in Figure 23.7) were formed by dissolving the gold-lipid in chloroform:methanol, evaporating the solvent, adding water, and sonicating. They were then placed on a carbon-coated grid and observed by STEM. Full width 90 nm. (From Ref. [131].)

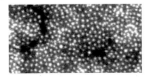

Figure 23.9 Gold–lipid monolayer. DPPE-Nanogold was dissolved in chloroform:methanol and applied to a water surface. After the volatile solvent had evaporated, a monolayer of lipid–gold particles formed, which was picked up on a carbon-coated electron microscope grid. Imaging was performed in the Brookhaven high-resolution STEM. Bright dots are the 1.4 nm-diameter Nanogold clusters. Full width 65 nm. (From Ref. [131].)

flammation [129]. Sonication of gold–lipids in aqueous solution leads to formation of gold–liposomes, which have been named "metallosomes" [130, 131] (Figure 23.8). Interestingly, each lipid molecule has one 1.4-nm Nanogold cluster attached, such that a single lipid molecule is then visualized. Lipids also form a monolayer when placed at an air–water interface, and this can be picked up on an electron microscope grid and viewed by EM (Figure 23.9).

23.10
Larger Covalent Particle Labels

Research is also being directed towards the bioconjugate chemistry of larger gold particles, functionalized in a similar manner to Nanogold, by synthetic modification of small organic molecules coordinated to the gold surface. A preliminary report has described the preparation and use of a covalent 10 nm gold–Fab′ conjugate for blotting and immunoelectron microscopy (Figure 23.10) [17]. Fundamental changes are observed in the chemical and electronic properties of gold particles from the small gold clusters such as undecagold and Nanogold to even slightly larger gold nanoparticles 3 and 5 nm in diameter, and this has important implications for the roles that these particles might serve in nanobiotechnology applications. Gold particles with controlled cross-linking functionality, in a range of precisely defined sizes with a variety of properties, would be a valuable component of this developing field.

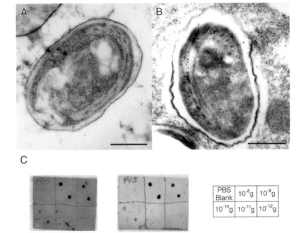

Figure 23.10 Electron micrographs of *G. americanus* spores incubated with monoclonal anti-PTP 43 GA primary antibody with (A) 12 nm colloidal gold anti-mouse secondary (Jackson), and (B) 10 nm covalent gold-Fab′ anti-mouse. Polar tube labeling shown by arrows (scale bar = 0.5 μm). (C) Immunoblot of 10 nm colloidal gold-IgG (left) and 10 nm covalent gold-Fab′ (right) anti-mouse conjugate against serial dilutions of mouse IgG spotted onto nitrocellulose membrane, with key showing the amounts of mouse IgG in each spot for the corresponding divisions of the blots [17].

23.11
Gold Targeted to His Tags

A popular molecular biology technique is to transfect a cell so that an engineered DNA is incorporated and the new protein expressed. Frequently, overexpression is an objective, so that the new protein can be produced in quantity. Purification of the protein from the other cell materials is then a challenge. It was discovered that histidine residues have an affinity for metals, and by coding a sequence of six histidines (usually at the amino or carboxyl end of the protein), the whole cell contents could be poured over a metal-functionalized column, and only the his-tagged protein would bind. The bound protein can then be eluted under more stringent conditions to yield highly purified bulk protein in one step. Although other chelators and metals are sometimes used [132], the optimum chromatography media was one derivatized with the nitrilotriacetic acid group (NTA) [133]: this binds a hexacoordinate nickel(II) atom in a configuration that leaves two adjacent coordination sites unoccupied (Figure 23.11), creating a binding site with a strong, selective affinity towards polyhistidine. Dissociation constants for NTA-Ni(II) binding to polyhistidine are thought to be about 10^{-7} M [134, 135]. This rivals that of antibodies, and means that NTA-Ni(II) functionalized gold nanoparticles may be used as probes to localize polyhistidine-tagged targets. Because they are much smaller than antibody conjugates, NTA-Ni(II) derivatives may be better able to penetrate and access sterically hindered targets, and form more closely spaced supramolecular structures such as nanowires.

For biomolecular labeling of His-tagged proteins (one containing the $6 \times$ His tag), the Ni-NTA group was incorporated synthetically into the organic shell of a gold particle [136] (Figure 23.12). The nickel(II)-charged derivative has been used to quantitatively label the three $6 \times$ His-tagged subunits of the 64-kDa adenovirus A12 knob protein for STEM observation [136, 137] (Figure 23.13), and also to label N-terminal $6 \times$ His-tagged PsbH protein which was then located within the Photosystem II multisubunit complex by EM and image analysis [138].

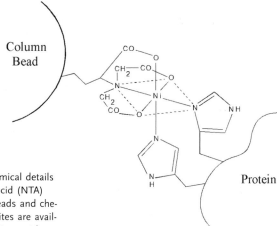

Figure 23.11 Diagram showing the chemical details of a nickel column. The nitrilotriacetic acid (NTA) group is covalently bound to column beads and chelates a nickel atom. Two coordination sites are available for bonding with two protein histidine residues.

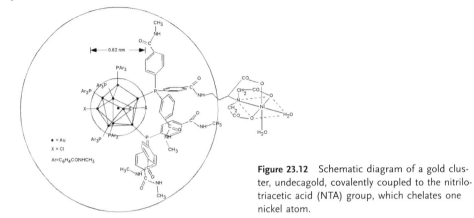

Figure 23.12 Schematic diagram of a gold cluster, undecagold, covalently coupled to the nitrilotriacetic acid (NTA) group, which chelates one nickel atom.

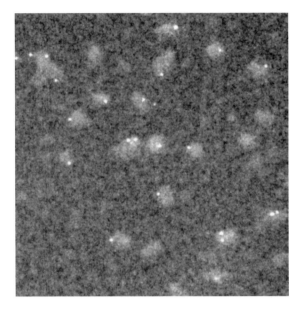

Figure 23.13 Knob protein from adenovirus 12 cloned with 6×-His tag, labeled with Ni-NTA-Nanogold, column purified from excess gold, and viewed in the Brookhaven STEM unstained; 128 nm full width.

23.12
Enzyme Metallography

Enzymes play a variety of roles in the sequestration, transport, and metabolic activity of metals within organisms. Some bacterial cells accumulate magnetic particles that allow them to navigate without a global positioning system (GPS), and some cells perform metal pumping and sequestration, for growth in toxic environments. In similar manner, our bodies store iron in ferritin, which consists of a protein shell with portals whereby about 5000 iron atoms can enter and be stored in an insoluble Fe(III) oxidation state. An interesting recent discovery was that enzymes can be controlled to deposit metal in the zero oxidation state [139]. The beauty of enzyme reactions is that the enzymes are cat-

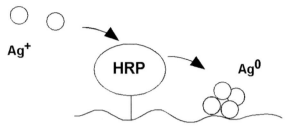

Figure 23.14 Enzyme metallography diagram showing the reduction of silver ions in solution by the biological enzyme horseradish peroxidase (HRP), producing a local deposit of silver metal.

alysts, and can cycle a very large number of times. If such an enzyme is targeted to a gene or antigen, it can then be used to locally deposit metal in the nanoscale range (Figure 23.14). If the enzyme is kept supplied with metal ion substrate, then macroscopic deposits will result, making sensitive detection possible even when using all methods that might be used to detect and localize metal deposits, including light scattering, reflectance, SERS, absorption, density, and conductivity.

23.13
Gold Cluster Nanocrystals

Gold cluster compounds may be prepared with sufficient monodispersity that they can deposit to form highly ordered 2-D arrays, or "nanocrystals". The self-assembly of larger colloidal gold particles has been documented [140], but we have observed similar behavior in a gold cluster, similar to Nanogold but slightly larger, known as "Greengold" [141]. Greengold is a highly regular compound, thought to contain 73–75 gold atoms based on MALDI mass spectroscopy data. Upon standing, it can form small microcrystals (found by STEM to be usually 2-D planar sheets), but occasionally it may form thin 3-D crystals. The center-to-center spacing between gold clusters is 2.6 nm, which is consistent with a gold core about 1.4 nm in diameter and an organic ligand shell of 0.6 nm thickness [142].

23.14
Gold Cluster–Oligonucleotide Conjugates: Nanotechnology Applications

The use of the supramolecular organizing properties of biomolecules to arrange metal nanoparticles into extended arrays offers one of the most promising approaches to developing their nanotechnology applications. In this respect, nucleic acids occupy a unique place in nanobiotechnology. The vast number of unique combinations available for their hybridization allows the "programming" of many simultaneous, unique interactions, which can be used to assemble complex structures, or even carry out complex processes such as DNA computing. Combined with the variety of physical, optical and electronic properties imparted by metal nanoparticles, this promises to give rise to a highly diverse set of nanotechnological applications for metal nanoparticle–nucleic acid conjugates.

Figure 23.15 1.4 nm Nanogold clusters (bright spots) bound to double-stranded bacteriophage T7 DNA (rope-like strands). Dark field, unstained STEM image on a thin carbon substrate. Full width 128 nm [145].

23.14.1
DNA Nanowires

DNA has some interesting properties for use in wiring nanocircuitry. It is tiny in width (2 nm), flexible, and may be easily synthesized in well-defined lengths. With other polymers, the lengths are difficult to control, whereas with DNA each base is programmed so that exact lengths can be mass-produced. The ends may be synthesized with unique nucleotide sequences that will bind to (hybridize) with a complementary sequence target. This means that the ends of such a DNA nanowire will self-assemble and connect to complementary target pads, even in three dimensions (Figure 23.15). Current computers rely on lithography and use wiring that is ~0.3 μm, so DNA would be a factor of 150 times smaller. Packing density in two dimensions could then be increased by $150^2 = 22\,500$, or in three dimensions by 3 375 000. While this is optimistic, even several orders of magnitude improvement would be significant.

A number of methods are available for attaching metal nanoparticles to DNA, including covalent attachment to specific bases [143, 144], photoreaction, intercalation, and charge binding. One method we have used is shown in Figure 23.15, which shows a positively charged Nanogold cluster bound to the negatively charged DNA [145]. The average spacing of gold quantum dots is ~2 nm. Although conduction can occur through tunneling or electron hopping, the dots may be processed by autometallography, using them as nuclei to deposit additional metal, growing them to confluence if desired to make a continuous wire.

23.14.2
3-D Nanostructured Mineralized Biomaterials

The assembly of complementary oligonucleotides with multiple hybridization sites into complex, 3-D structures with multiple unique sites and potential functionalities was first pursued by Mirkin, who used thiolated oligonucleotides coordinated directly to unstabilized colloidal gold particles to construct large oligomers containing semi-regular arrays of 13-nm gold particles [146]. The optical properties of the gold particles were quickly applied in a method for detecting DNA hybridization, using the color change that occurred when labeled strands hybridized and brought the gold particles into proximity [16, 147].

Further developments and applications of this technology are described in other chapters of this book; here, we will limit our discussion to the use of gold cluster labels.

The chemical selectivity of covalent gold cluster labeling affords an additional dimension in the design of hybrid DNA–metal nanoparticle materials, by letting the researcher choose a reaction to attach the gold particles directly at any suitably modified point within the structure. Programmed self-assembly of Nanogold-labeled oligonucleotides has been demonstrated by Liu and co-workers, who used Mono-*Sulfo*-NHS-Nanogold to label nano-tube-forming oligonucleotides before assembly [148]. Kiehl and co-workers recently assembled metallic nanoparticle arrays using DNA crystals, labeled site-specifically with Nanogold, as a programmable molecular scaffolding [149]. 2D DNA crystals as a scaffolding potentially offers fundamental advantages over other self-assembly approaches for the precision, rigidity, and programmability of the assembled nanostructures, and this represents a critical step toward the realization of DNA nanotechnology and its nanoelectronic applications. 2-D arrays were constructed by tiling together rigid DNA motifs composed of double-crossover (DX) molecules containing DNA hairpins. The nanoparticles formed precisely integrated components, covalently bonded to the DNA scaffolding. STEM showed that the gold particles formed 2-D arrays with interparticle spacings of 4 and 64 nm (Figure 23.16). DNA–Nanogold conjugates were prepared from 5′-thiol-modified C6 oligonucleotides, which were reacted with Monomaleimido Nanogold. DNA:Nanogold labeling stoichiometry of the purified conjugate was estimated spectroscopically to be very close to the desired 1:1 product.

Another important development is the use of autometallography to further modify such materials after assembly, and introduce new and useful functionality. This has been postulated as a method for forming metallic nanowires [145] and nanospheres or other structures [131] using gold-decorated DNA and liposomes respectively. Mirkin and co-workers have reported the fabrication of a conductimetric biosensor based on these processes. The device comprises lithographically prepared microelectrode pairs (separation 20 μm), a shorter "capture" oligonucleotide strand located in the gap between them, and a longer "target" oligonucleotide in solution. The target oligonucleotide has contiguous recognition elements complementary to the capture strand on one end and on the other to oligonu-

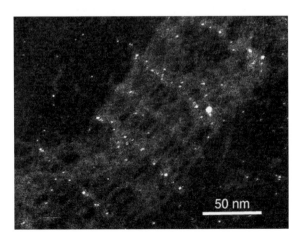

Figure 23.16 Dark field unstained STEM image of a DNA crystal incorporating the DNA–Au conjugate, showing 64 and 4 nm interparticle spacings [149].

cleotides attached to gold nanoparticles: upon immersion of the electrode pair device in a solution containing the appropriate probe and target, gold nanoparticle probes fill the gap, and when these are silver-enhanced, the gap becomes conductive [150].

23.14.3
Gold-quenched Molecular Beacons

Gold particles absorb strongly in the visible spectrum, and hence, when brought within Förster distance, are effective quenchers of fluorescent groups by energy transfer [60, 151]. This property has been explored for molecular beacons, a novel class of DNA hybridization probes comprising hairpin loops of DNA with a fluorescent group at one end and a quencher at the other. When they bind to their target, they open, the fluorophore and the quencher move apart, and fluorescent signal appears. These probes have an important advantage over conventional probes in that the presence of unbound probe does not generate background signal; therefore unbound probe need no longer be removed. This makes beacons useful for homogeneous real-time PCR detection, and may also enable monitoring of processes within living cells or tissues using microprobes equipped with immobilized beacon biosensors. The critical quantity in determining the utility of a quencher is the "signal-to-noise ratio": this is the ratio of the fluorescence intensity when the beacon is in the open configuration (i. e., hybridized to target) to the fluorescence intensity when it is closed. Molecular beacons are conventionally prepared with the organic quencher, 4-((4'-(dimethylamino)phenyl)azo)benzoic acid (DABCYL), which can give signal-to-noise ratios of up to 100; however, when Nanogold was used instead, the signal-to-noise ratio was significantly increased, in one case to more than 2000 using a Nanogold-quenched rhodamine 6G probe. This improved level of sensitivity easily permitted the detection of single mismatches. Probes were prepared by the conjugation of Monomaleimido Nanogold to a 5'-thiol-functionalized probe that had previously been labeled at the 3' end with the fluorophore [152].

Consideration of the other mechanisms by which quenching might proceed suggests that higher levels of quenching are possible than those suggested by Förster energy transfer alone. Dulkeith and co-workers investigated the effect on radiative and nonradiative fluorescence lifetimes for systems in which fluorophores were linked to metal nanoparticles of varying sizes from 1 to 30 nm. They found both an increase in the radiative lifetime, and a decrease in the nonradiative lifetime, both of which contribute to quenching; this implies that fluorescence quenching by attached gold particles is greater than predicted by Förster theory. With a gold-fluorophore separation of 1 nm, about 99.8 % quenching was found, even with 1-nm particles [153].

23.15
Other Metal Cluster Labels

Other metals besides gold can form cross-linkable nanoparticles of a suitable size for biomolecular labeling. The use of different metals offers potential methods for discriminating between different sites spectromicroscopically. Moreover, since different metals possess different chemical and electronic properties, this provides a method for introducing a choice of different functional properties. By imparting multiple functionalities to such a conjugate, the use of different metal particles may provide an approach to developing a molecular assembly capable of carrying out a multi-step or complex process.

23.15.1
Platinum and Palladium

Large, ligand-stabilized cluster compounds of platinum [154] and palladium [155, 156] have been described. Originally, these were prepared by addition of molecular hydrogen to solutions of platinum or palladium chloride in acetic acid in the presence of 1, 10-phenanthrolines; this is a modification of a well-established method for the preparation of finely divided platinum and palladium for catalytic use, in which the ligands are used to

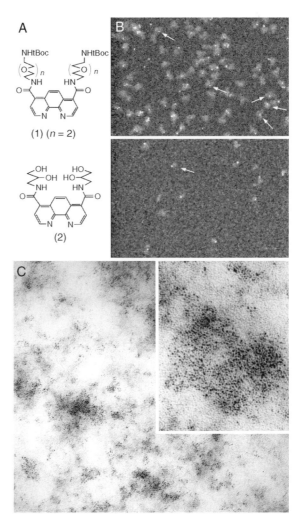

Figure 23.17 (A) Substituted 1, 10-phenanthroline ligands used for the preparation of water-soluble, cross-linkable large platinum clusters. (B) STEM micrographs of platinum cluster–IgG conjugate (upper) and platinum cluster–Fab' conjugate (lower) showing cluster conjugation localized to hinge region (full width 256 nm). (C) Transmission electron micrographs of labeled HeLa cells stained with monoclonal primary antibody against SC35 pre-mRNA splicing factor (1:30 dilution), followed by secondary platinum cluster-labeled Fab' goat anti-mouse secondary (1:20 dilution), then stained with 0.5 % uranyl acetate (original magnification ×35 000; inset, ×140 000). The staining pattern is consistent with distribution of SC35 found in previous immunofluorescence and immunogold studies [158].

trap smaller intermediates [157]. This reaction was adapted to prepare platinum clusters for biological labeling by using modified 1, 10-phenanthroline ligands to impart water-solubility and cross-linking functionality to the clusters; cluster formation was carried out by reducing platinum(II) acetate in ethanol with sodium borohydride. Platinum clusters 0.8 to 2.0 nm in size (mean size close to 1.6 nm) and functionalized with amines were isolated by gel filtration chromatography. Maleimide derivatives were then conjugated selectively to IgG molecules and Fab' fragments at a hinge thiol site in the same manner as the gold clusters. Fab' secondary antibodies labeled with these platinum clusters were used to label the SC35 pre-mRNA splicing site in HeLa cells (Figure 23.17) [158]. While size variation in these preliminary preparations was more than that found with similarly sized gold clusters, the use of two different metals may offer methods both to differentiate sites spectromicroscopically, and to incorporate differing functionalities or properties at different sites within a biological nanostructure.

23.15.2
Tungstates

Tungstates possess very different chemical and electronic properties to gold clusters, including catalytic and oxidative activity, and can be prepared in a number of different sizes [159]. Since tungsten is also a heavy atom, these clusters exhibit high density in the electron microscope, which gives them potential applications as biomolecule labels for EM. Unlike the similarly sized undecagold cluster, they suffer virtually no beam damage in the electron microscope, as measured by beam loss. Although, like undecagold, these are too small for routine use in a standard TEM, they are therefore potentially useful biomolecule labels for higher resolution methods such as STEM, or with image analysis.

Replacement of one of the tungsten atoms with titanium, tin, or, silicon creates a unique site for the incorporation of selective reactivity towards biomolecules. Keana and co-workers modified Dawson- ($[W_{18}PO_{62}]^{6-}$) type polytungstates by incorporating a cyclopentadienyl-titanium entity in which the cyclopentadienyl moiety had been previously functionalized with 1, 3-diene; after incorporation, this was converted to maleimide, bromoacetamide, and biotin derivatives. A benzaldehyde derivative was also prepared using a second modified cyclopentadienyl compound, and reacted smoothly with N6-[(aminohexyl)-carbamoyl] methyladenosine 5'-triphosphate to give an ATP derivative [160–162]. Keggin- ($[W_{12}PO_{40}]^{3-}$) polytungstates were functionalized using the same chemistry to give undecatungstate clusters $[RC_5H_4TiPW_{11}O_{39}]^{4-}$ [161, 162]. These compounds were prepared as organic-soluble tetraalkylammonium salts, which were converted to water-soluble potassium analogs by ion-exchange chromatography. They proved chemically stable in the electron beam. Although the Keggin-type compounds were not expected to be stable above pH 7, a larger Dawson-type derivative was found by NMR to be stable up to pH 8.2 [162]. The use of these compounds as biological labels is limited by their anionic charge, which results in precipitation with cationic biomolecules; however, a salt prepared using a novel tetraalkylammonium cation bearing oxygenated (ether and ester-containing) substituents was found to prevent precipitation with the basic proteins lysozyme, poly-ʟ-lysine and concanavalin A, raising the possibility that these compounds may be used for staining cationic proteins [163].

An alternative route to incorporating selective reactivity into the Keggin-type tungstates is replacement of a tungsten atom by an alkylsilicon derivative. Treatment of the organic-soluble "defect" tungstate salt, $\{[nBu_4N]^+\}_7[W_{11}PO_{39}]^{7-}$ with $Cl_3Si(CH_2)_3Br$ in acetonitrile, followed by $NaSSO_2CH_3$, yielded the thiolsulfonate derivative $[W_{11}PO_{39}Si(CH_2)_3 SSO_2CH_3]^{4-}$, which selectively reacts with thiols and therefore may be used to label thiol sites such as cysteine residues. Reaction of $\{[nBu_4N]^+\}_7[W_{11}PO_{39}]^{7-}$ with $Cl_3Si(CH_2)_2 (C_6H_4)SO_2Cl$ yielded the amine- and thiol- reactive sulfonate, $[W_{11}PO_{39}Si(CH_2)_2(C_6H_4) SO_2Cl]^{4-}$. Both were effectively conjugated to BSA, and conjugation of the thiol-sulfonate compound under conditions carefully controlled to exclude other labeling reactions resulted in labeling of close to 70% of the albumin molecules, consistent with the known free thiol content of 0.7 per molecule [164]. In a second series of experiments, the same Keggin-type tungstate was derivatized using an aliphatic organotin compound with chain lengths C_4, C_8, C_{12}, C_{18}, and C_{22}; the resulting alkylated clusters were effective as membrane labels, inserting into synthetic phospholipid vesicles and human erythrocyte cell membranes [165].

23.15.3
Iridium

Tetrairidium is a yet smaller heavy atom cluster label than undecagold, and this helps to give it unprecedented resolution as a label. Although not usually resolved in a standard TEM, it is clearly resolved by STEM, and has been used to measure the lengths of rigid organic molecules using this technique [166]. A phosphine compound, modified with a reactive aryl substituent that forms the terminal unit of the rigid organic molecule, was treated with $[Et_4N]^+[Ir_4(CO)_{11}I]^-$; the phosphine selectively displaces the coordinated iodide, and the resulting metal-tagged phosphine was incorporated into the rigid organic molecule. The separation of the two tetrairidium units was measured by STEM and found to be consistent with the calculated length of the rigid molecule.

Heavy atom compounds are used in the crystallography of biological macromolecules, as heavy atom derivatization reagents. When introduced at specific sites within the unit cell in biomolecule crystals, they help to enable phasing of the diffraction data. For very large molecules, supramolecular complexes, and organelles, small heavy atom clusters may be required to achieve sufficient phasing power. Jahn used solubilizing tris (amidoalkyl) phosphines to prepare a water-soluble, singly amino-substituted tetrairidium complex; following activation with a heterobifunctional cross-linker to prepare the maleimide, both this cluster and a similarly activated undecagold cluster were used to derivatize crystallized ribosomes with heavy atoms to enable phasing during crystallographic structure determination [167].

Although it is generally not directly visualized with TEM, tetrairidium may be located using image analysis. If it is used to label a specific site in a large, regular structure such as a virus, averaging of the data by image analysis of many particles may be used to obtain structural information at higher resolution than direct visualization; moreover, because of its smaller size the tetrairidium particle offers higher resolution in this application than the larger gold clusters. A water-soluble, maleimido tetrairidium cluster was used to label the C-terminus of the hepatitis B viral capsid protein; the cluster was clearly

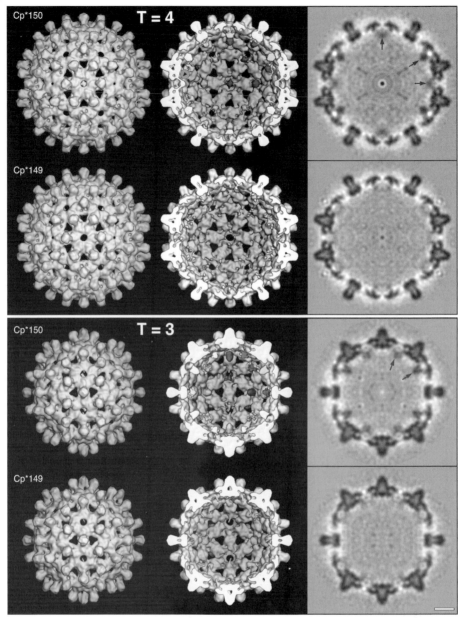

Figure 23.18 Density maps of [Ir$_4$]-labeled and unlabeled HBV capsids. Top: T = 4 capsids of the unlabeled Cp*150 construct (tetrairidium cluster conjugated to cysteine appended at position 150) and the control Cp*149 construct (unlabeled). The small black arrowhead (middle panel, top row) points to the organic cross-linker connecting the [Ir$_4$] cluster to the capsid protein. Bottom: A similar comparison for the corresponding T = 3 capsids. For each density map, isodensity renderings of the outer and inner surfaces are shown, together with a central section depicting local variations in density. The capsids are viewed along a two-fold axis symmetry axis and are contoured to enclose 100% of expected mass. For both capsids, labels are visible at the corresponding sites on their inner surfaces. Scale bar = 50 (From Ref. [168].)

visible in 3-D electron density maps calculated from cryoelectron micrographs even at partial occupancy, and remained visible down to 25 Å resolution. Furthermore, unlike the larger undecagold, the tetrairidium label was able to enter the assembled viral capsid and label these sites on the inner surface (Figure 23.18) [168].

Acknowledgments

R. D. Powell and J. F. Hainfeld acknowledge the generous support of the National Institute of General Medical Sciences through Small Business Innovation Research (SBIR) grants GM60067, GM56090, GM53900, GM62100, and GM64257, and the National Cancer Institute through Small Business Technology Transfer (STTR) grant CA83618. The contents of this chapter are solely the responsibility of the authors and do not necessarily represent the official views of the National Institute of General Medical Sciences or National Cancer Institute. G. W. Hacker would like to sincerely acknowledge the generous support by Professor Gernot Pauser (Medical Director of the St. Johanns-Hospital Salzburg) and Dr. Alois Gruener (Land Salzburg, Department of Health), as well as of the government of Land Salzburg.

References

[1] Carrothers, W. H. (to E. I. DuPont de Nemours and Co.), U. S. Patent 2, 130, 523, **1938**.

[2] Faraday, M., Experimental Relations of Gold (and other Metals) to Light, *Experimental Researches in Chemistry and Physics*, **1857** p. 391 ff. London, 1859.

[3] Faulk, W. P., G. M. Taylor, An immunocolloid method for the electron microscope, *Immunochemisty* **1971**, *8*, 1081–1083.

[4] Wang, B. L., L. Scopsi, M. Hartvig Nielsen, L. I. Larsson, Simplified purification and testing of colloidal gold probes, *Histochemistry* **1985**, *83*, 109–115.

[5] Krisch., B., C. Buchholz, R. Mentlein, A. Turzynski, Visualization of neuropeptide-binding sites on individual telencephalic neurons of the rat, *Cell Tissue Res.* **1993**, *272*, 523–531.

[6] Krisch, B., J. Feindt, R. Mentlein, Immunoelectronmicroscopic analysis of the ligand-induced internalization of the somatostatin receptor subtype 2 in cultured human glioma cells, *J. Histochem. Cytochem.* **1998**, *46*, 1233–1242.

[7] Mentlein, R., C. Buchholz, B. Krisch, Somatostatin-binding sites on rat telencephalic astrocytes. Light- and electron-microscopic studies in vitro and in vivo, *Cell Tissue Res.* **1990**, *262*, 431–443.

[8] Gribnau, T. C., J. H. Leuvering, H. van Hell, Particle-labelled immunoassays: a review, *J. Chromatogr.* **1986**, *376*, 175–189

[9] Hainfeld J. F., F. R. Furuya, A 1.4-nm gold cluster covalently attached to antibodies improves immunolabeling, *J. Histochem. Cytochem.* **1992**, *30*:177–184.

[10] Hainfeld, J. F., R. D. Powell, Nanogold technology: new frontiers in gold labeling, *Cell Vision* **1997**, *4*, 408–432.

[11] Hainfeld, J. F., A small gold-conjugated antibody label: Improved resolution for electron microscopy, *Science* **1987**, *236*, 450–453.

[12] Hainfeld, J. F., Undecagold-Antibody Method, in: M. A. Hayat (ed.), *Colloidal Gold: Principles, Methods and Applications*, Vol. 2, Academic Press, San Diego, CA, **1989**, pp. 413–429.

[13] Reardon J. E., P. A. Frey, Synthesis of undecagold cluster molecules as biochemical

labeling reagents. 1. Monoacyl and mono[N-(succinimidooxy)succinyl] undeca-gold clusters, *Biochemistry* **1984**, *23*, 3849–3856.

[14] Li, Z., R. Jin, C. A. Mirkin, R. L. Letsinger, Multiple thiol-anchor capped DNA-gold nanoparticle conjugates, *Nucleic Acids Res.* **2002**, *30*, 1558–1562.

[15] Storhoff, J. J., C. A. Mirkin, Programmed Materials Synthesis with DNA, *Chem. Rev.* **1999**, *99*, 1849–1862.

[16] Taton, T. A., G. Lu, C. A. Mirkin, Two-color labeling of oligonucleotide arrays via size-selective scattering of nanoparticle probes, *J. Am. Chem. Soc.* **2001**, *123*, 5164–5165.

[17] Gutierrez, E., R. D. Powell, J. F. Hainfeld, P. M. Takvorian, A covalently linked 10 nm gold immunoprobe. *Microsc. Microanal.*, *5, Suppl. 2 (Proceedings)*, Bailey, G. W., W. G. Jerome, S. McKernan, J. F. Mansfield, and R. L. Price (eds); Springer-Verlag, New York, NY; **1999**, pp. 1324–1325.

[18] Danscher G., Histochemical demonstration of heavy metals. A revised version of the sulphide silver method suitable for both light and electron microscopy, *Histochemistry* **1981**, *71*, 1–16.

[19] Danscher G., Localization of gold biological tissue. A photochemical method for light and electron microscopy, *Histochemistry* **1981**, *71*, 81–88.

[20] Danscher G., Light and electron microscopical localisation of silver in biological tissue, *Histochemistry* **1981**, *71*, 177–186.

[21] Danscher G., J. O. Rytter Nörgaard, Light microscopic visualisation of colloidal gold on resin-embedded, *J. Histochem. Cytochem.* **1983**, *31*, 1394–1398.

[22] Danscher G., Autometallography. A new technique for light and electron microscopical visualization of metals in biological tissue (gold, silver, metal sulphides and metal selenides), *Histochemistry* **1984**, *81*, 331–335.

[23] Danscher, G., B. Möller-Madsen, Silver amplification of mercury sulfide and selenides: A histochemical method for light and electron microscopic localization of mercury in tissue, *Histochem. Cytochem.* **1985**, *33*, 219–228.

[24] Danscher, G., G. W. Hacker, J. O. Norgaard, L. Grimelius., Autometallographic silver amplification of colloid gold, *J. Histotechnol.* **1993**, *16*, 201–207.

[25] Danscher, G., G. W. Hacker, C. Hauser-Kronberger, L. Grimelius, Trends in auto-metallographic silver amplification of colloidal gold particles, in: M. A. Hayat (Ed.), *Immunogold-silver staining: methods and applications*, CRC Press, Boca Raton, FL, USA, **1995**, pp. 11–18.

[26] Hacker, G. W., G. Danscher, Immunogold-silver staining – autometallography: recent developments and protocols, in: J. Gu (ed.), *Analytical Morphology – Theory, Protocols and Applications*, Eaton Publishing, Natick, MA, **1997**, pp. 41–54.

[27] Hacker, G. W., L. Grimelius, G. Danscher, G. Bernatzky, W. Muss, H. Adam, J. Thurner, Silver acetate autometallography: an alternative enhancement technique for immunogold-silver staining (IGSS) and silver amplification of gold, silver, mercury and zinc tissues, *J. Histotechnol.* **1988**, *11*, 213–221.

[28] Danscher, G., G. W. Hacker, M. Stoltenberg, Autometallographic tracing of gold, silver, bismuth, mercury, and zinc, in: Hacker G. W., J. Gu (eds): *Gold and silver staining: techniques in molecular morphology*, CRC Press, Boca Raton, London, New York Washington, D. C., **2002**, pp. 13–27.

[29] Lackie, P. M., R. J. Hennessy, G. W. Hacker, J. M. Polak, Investigation of immunogold-silver staining by electron microscopy, *Histochemistry* **1985**, *83*, 545–550.

[30] Rufner, R., N. E. Carson, M. Forte, G. Danscher, J. Gu, G. W. Hacker, Autometallography for immunogold-silver staining. in light and electron microscopy, *Cell Vision* **1995**, *2*, 327–333.

[31] Gu, J., M. Forte, N. Carson, G. W. Hacker, M. D'Andrea, R. Rufner, Quantitative evaluation of immunogold-silver staining, in: M. A. Hayat (ed.), *Immunogold-Silver Staining: Methods and Applications*, CRC Press, Boca Raton, **1995**, pp. 119–136.

[32] Holgate, C. S., P. Jackson, P. N. Cowen, C. C. Bird, Immunogold-silver staining: New method of immunostaining with enhanced sensitivity, *J. Histochem. Cytochem.* **1983**, *31*, 938–944.

[33] Holgate, C. S., P. Jackson, I. Lauder, P. Cowen, C. Bird, Surface membrane staining of immunoglobulins in paraffin sections of non-Hodgkin's lymphomas using immunogold-silver staining technique, *J. Clin. Pathol.* **1983**, *36*, 742–746.

[34] Springall, D. R., G. W. Hacker, L. Grimelius, J. M. Polak, The potential of the immunogold-silver staining method for paraffin sections, *Histochemistry* **1984**, *81*, 603–608.

[35] Hacker, G. W., D. R. Springall, S. Van Noorden, A. E. Bishop, L. Grimelius, J. M. Polak, The immunogold-silver staining method. A powerful tool in histopathology, *Virchows Arch. A* **1985**, *406*, 449–461.

[36] Hacker, G. W., D. R. Springall, A. Cheung, S. Van Noorden, J. M. Polak, Immunogold-silver staining and its potential use in multiple staining and histopathology. *J. Histochem. Cytochem.* **1986**, *34*, 118.

[37] Hacker, G. W., A.-H. Graf, O. Dietze, Basic principles of immunohistopathology, in: Gu J., G. W. Hacker (eds), *Modern methods in analytical morphology*, Plenum Press, New York London, **1994**, pp. 81–112.

[38] Hacker G. W., C. Hauser-Kronberger, A.-H. Graf, G. Danscher, J. Gu, L. Grimelius, Immunogold-silver staining (IGSS) for detection of antigenic sites and DNA sequences, in: Gu J., G. W. Hacker (eds), *Modern methods in analytical morphology*, Plenum Press, New York London, **1994**, pp. 19–36.

[39] De Brabander, M., G. Geuens, R. Nuydens, M. Moeremans, J. De Mey. Probing microtubule-dependent intracellular motility with nanometer particle video ultramicroscopy (Nanovid ultramicroscopy), *Cytobios* **1985**, *43*, 273–283.

[40] De Mey, J., Colloidal gold probes in immunocytochemistry, in: J. M. Polak and S. Van Norden (eds), *Immunocytochemistry – Practical Applications in Pathology and Biology*, Wright, Bristol, **1983**, pp. 82–112.

[41] De Mey, J., G. W. Hacker, M. De Waele, D. Springall, Gold probes in light microscopy, in: Polak, J. M. and S. Van Norden (eds), *Immunocytochemistry—Modern Methods and Applications*, Wright, Bristol, **1986**, pp. 71–88.

[42] De Waele, M., J. De Mey, W. Renmans, C. Labeur, K. Jochmans, B. Van Camp, Potential of immunogold-silver staining for the study of leukocyte subpopulations as defined by monoclonal antibodies, *J. Histochem. Cytochem.* **1986**, *34*, 1257–1263.

[43] Scopsi, L., L.-I. Larsson, L. Bastholm, M. Hartvig Nielsen, Silver-enhanced-colloidal gold probes as markers for scanning electron microscopy, *Histochemistry* **1986**, *86*, 35–41.

[44] Teasdale, J., P. Jackson, C. S. Holgate, P. N. Cowen, Identification of oestrogen receptors in cells of paraffin-processed breast cancers by IGSS, *Histochemistry* **1987**, *87*, 185–187.

[45] Westermark, K., M. Lundqvist, G. W. Hacker, A. Karlsson, B. Westermark, Growth factor receptors in thyroid follicle cells, *Acta Endocrinol. (Copenhagen) Suppl.* **1987**, *281*, 252–255.

[46] Hacker G. W., G. Danscher, L. Grimelius, C. Hauser-Kronberger, W. H. Muss, J. Gu, O. Dietze, Silver staining techniques, with special references to the use of different silver salts in light- and electron microscopical immunogold-silver staining, in: M. A. Hayat (ed.), *Immunogold-silver staining: methods and applications*, CRC Press, Boca Raton, FL, USA, **1995**, pp. 20–45.

[47] Roth, J., Applications of immunocolloids in light microscopy. Preparation of protein A-silver and protein A-gold complexes and their application for the localisation of single and multiple antigens in paraffin sections, *J. Histochem. Cytochem.* **1982**, *30*, 691–696.

[48] Roth, J., The colloid gold marker system for light and electron microscopic cytochemistry, *Immunochemistry* **1983**, *2*, 217–284.

[49] Fujimori, O., M. Nakamura, Protein A gold-silver staining method for light microscopic immunohistochemistry, *Arch. Histol. Jpn.* **1985**, *48*, 449–452.

[50] Fujimori O., The protein A-gold-silver staining method, in: Hacker G. W., and J. Gu (eds), *Gold and silver staining: techniques in molecular morphology*, CRC Press, Boca Raton, London, New York Washington, D. C., **2002**, pp. 71–84.

[51] Guesdon, J. L., T. Ternynck, S. Avrameas, The use of avidin-biotin interaction in immunoenzymatic techniques, *J. Histochem. Cytochem.* **1979**, *27*, 1131–1139.

[52] Coggi, G., P. Dell'Orto and G. Viale, Avidin-Biotin methods, in: Polak, J. M., S. Van Norden (eds), *Immunocytochemistry – Modern Methods and Applications*, Wright, Bristol, UK, **1986**, pp. 54–70.

[53] Hacker, G. W., J. Gu, Gold and silver staining: techniques in molecular morphology, CRC Press, Boca Raton, London,

New York Washington, D. C., 246 pages, **2002**.

[54] Takizawa, T., J. M. Robinson, Use of 1.4-nm immunogold particles for immunocytochemistry on ultra-thin cryosections, *J. Histochem. Cytochem.* **1994**, *43*, 1615–1623.

[55] Sun, X. J., L. P. Tolbert, J. G. Hildebrand, Using laser scanning confocal microscopy as a guide for electron microscopic study: a simple method for correlation of light and electron microscopy, *J. Histochem. Cytochem.* **1995**, *43*, 329–335.

[56] Cheung A. L. M., A.-H. Graf, C. Hauser-Kronberger, O. Dietze, R. R. Tubbs, G. W. Hacker G. W., Detection of human papillomavirus in cervical carcinoma: comparison of peroxidase-, Nanogold- and CARD-Nanogold *in situ* hybridization, *Modern Pathology* **1999**, *12 (7)*, 689–696.

[57] Graf A.-H., A. L. M., Cheung, C. Hauser-Kronberger, N. Dandachi, R. R. Tubbs, O. Dietze, G. W. Hacker. Clinical relevance of HPV 16/18 testing methods in cervical squamous cell carcinoma. *Appl. Immunohistochem. Mol. Morphol.* **2000**, *8*, 300–309.

[58] Hainfeld, J. F., R. D. Powell, J. K. Stein, G. W. Hacker, C. Hauser-Kronberger, A. L. M. Cheung, C. Schofer, Gold-based autometallography, *Microsc. Microanal., 5, (Suppl. 2: Proceedings)*, Bailey, G. W., W. G. Jerome, S. McKernan, J. F. Mansfield, and R. L. Price (eds), Springer-Verlag, New York, NY; **1999**, pp. 486–487.

[59] Hainfeld, J. F., R. D. Powell, Silver- and gold-based autometallography of Nanogold, in: Hacker G. W., J. Gu (eds), *Gold and silver staining: techniques in molecular morphology*, CRC Press, Boca Raton, London, New York Washington, D. C., **2002**, pp. 29–46.

[60] Förster, Th., Zwisschenmolekulare Energiewandung und Fluoreszenz, *Ann. Physik.* **1948**, *2*, 55–75.

[61] Powell, R. D., C. M. R. Halsey, J. F. Hainfeld, Combined fluorescent and gold immunoprobes: Reagents and methods for correlative light and electron microscopy, *Microsc. Res. Tech.* **1998**, *42*, 2–12.

[62] Powell, R. D., C. M. R. Halsey, D. L. Spector, S. L. Kaurin, J. McCann, J. F. Hainfeld, A covalent fluorescent-gold immunoprobe: simultaneous detection of a pre-mRNA splicing factor by light and electron microscopy, *J. Histochem. Cytochem.* **1997**, *45*, 947–956.

[63] Powell, R. D., J. F. Hainfeld, Combined fluoresent and gold probes for microscopic and morphological investigations, in: Hacker G. W., J. Gu (eds), *Gold and silver staining: techniques in molecular morphology*, CRC Press, Boca Raton, London, New York Washington, D. C., **2002**, pp. 107–118.

[64] Robinson, J. M., D. D. Vandré, Efficient immunocytochemical labeling of leukocyte microtubules with FluoroNanogold: an important tool for correlative microscopy, *J. Histochem. Cytochem.* **1997**, *45*, 631–642.

[65] Takizawa, T., K. Suzuki, J. M. Robinson, Correlative microscopy using FluoroNanogold on ultrathin cryosections. Proof of principle, *J. Histochem. Cytochem.* **1998**, *46*, 1097–1102.

[66] Keohane, E. M., G. A. Orr, P. M. Takvorian, A. Cali, H. B. Tanowitz, M. Wittner, L. M. Weiss, Analysis of the major microsporidian polar tube proteins. *J. Euk. Microbiol.* **1999**, *46*, 1–5.

[67] Powell, R. D., V. N. Joshi, C. M. R. Halsey, J. F. Hainfeld, G. W. Hacker, C. Hauser-Kronberger, W. H. Muss, P. M. Takvorian, Combined Cy3/Nanogold conjugates for immunocytochemistry and in situ hybridization. *Microsc. Microanal., 5, (Suppl. 2: Proceedings)*, Bailey, G. W., W. G. Jerome, S. McKernan, J. F. Mansfield, R. L. Price (eds), Springer-Verlag, New York, NY; **1999**, pp. 478–479.

[68] Liu, W., J. F. Hainfeld, R. D. Powell, Combined ALEXA-488 and Nanogold Antibody Probes, *Microsc. Microanal., 8, (Suppl. 2: Proceedings)*, Voekl, E., D. Piston, R. Gauvin, A. J. Lockley, G. W. Bailey, S. McKernan (eds), Cambridge University Press, New York, NY, **2002**, pp. 1030–1031CD.

[69] Takizawa, T., J. M. Robinson, Correlative microscopy of ultrathin cryosections is a powerful tool for placental research, *Placenta* **2003**, *24*, 557–565.

[70] Powell, R. D., C. M. R. Halsey, E. Gutierrez, J. F. Hainfeld, F. R. Furuya, Dual-labeled probes for fluorescence and electron microscopy, *Microsc. Microanal., 4, (Suppl. 2: Proceedings)*, Bailey, G. W., K. B. Alexander, W. G. Jerome, M. G. Bond, J. J. McCarthy (eds), Springer, New York, NY, **1998**, pp. 992–993.

[71] Hacker, G. W., A. L. M. Cheung, R. R. Tubbs, L. Grimelius, G. Danscher, C. Hauser-Kronberger, Immunogold-silver staining for light microscopy using colloidal or clustered gold (Nanogold), in: Hacker G. W., J. Gu (eds), *Gold and silver staining: techniques in molecular morphology*, CRC Press, Boca Raton, London, New York Washington, D. C., **2002**, pp. 47–69.

[72] Hacker, G. W., W. H. Muss, C. Hauser-Kronberger, G. Danscher, R. Rufner, J. Gu, H. Su, A. Andreasen, M. Stoltenberg, O. Dietze, Electron microscopical autometallography: immunogold-silver staining (IGSS) and heavy-metal histochemistry, *Methods (Companion Meth. Enzymol.)* **1996**, *10*, 257–269.

[73] Krenács, T., L., Krenács, B. Bozóky, B. Iványi, Double and triple immunocytochemical labelling at the light microscope level in histopathology, *Histochem. J.* **1990**, *22*: 530–536.

[74] Krenács, T., Krenács L, Immunogold-silver staining (IGSS) for immunoelectron microscopy and in multiple detection affinity cytochemistry, in: Gu J., G. W. Hacker (eds), *Modern methods in analytical morphology*, Plenum Press, New York London, **1996**, pp. 225–252.

[75] Robinson, J. M., T. Takizawa, D. D. Vandré, Gold cluster immunoprobes: light and electron microscopy, in: Hacker G. W., J. Gu (eds), *Gold and silver staining: techniques in molecular morphology*, CRC Press, Boca Raton, London, New York Washington, D. C., **2002**, pp. 177–187.

[76] Robinson, J. M., T. Takizawa, Novel labeling methods for EM analysis of ultrathin cryosections, in: G. W. Bailey, J. M. Corbett, R. V. W. Dimlich, J. R. Michael, N. J. Zaluzec (eds), *Proc. Micros. Microanal.*, San Francisco Press, San Francisco, CA, USA, **1996**, pp. 894–895.

[77] Lossi, L., P. Aimar, E. Beltramo, A. Merighi, Immunogold labeling techniques for transmission electron microscopy: applications in cell and molecular biology, in: Hacker G. W., J. Gu (eds), Gold and silver staining: techniques in molecular morphology, CRC Press, Boca Raton, London, New York Washington, D. C., **2002**, pp. 145–167.

[78] Samaj, J., H.-J. Ensikat, W. Barthlott, D. Volkmann, Immunogold-silver scanning electron microscopy using glycerol liquid substitution, in: Hacker G. W., J. Gu (eds), *Gold and silver staining: techniques in molecular morphology*, CRC Press, Boca Raton, London, New York Washington, D. C., **2002**, pp. 223–233.

[79] Sawada, H., M. Esaki, Pro-embedding immunoelectron microscopy with Nanogold immunolabeling, silver enhancement, and its stabilization by gold, in: Hacker G. W., J. Gu (eds), *Gold and silver staining: techniques in molecular morphology*, CRC Press, Boca Raton, London, New York Washington, D. C., **2002**, pp. 169–176.

[80] Schöfer, C., K. Weipoltshammer, Highly sensitive ultrastructural immunogold detection using tyramide and gold enhancement, in: Hacker G. W., J. Gu (eds), *Gold and silver staining: techniques in molecular morphology*, CRC Press, Boca Raton, London, New York Washington, D. C., **2002**, pp. 189–197.

[81] Wang, B.-L., N. Flay, G. W. Hacker, Immunogold-silver staining for scanning electron microscopy in cancer research, in: Hacker G. W., J. Gu (eds), *Gold and silver staining: techniques in molecular morphology*, CRC Press, Boca Raton, London, New York Washington, D. C., **2002**, pp. 211–222.

[82] Bernard, V., A. I. Levey, B. Bloch, Regulation of the subcellular distribution of m4 muscarinic acetylcholine receptors in striatal neurons in vivo by the cholinergic environment: evidence for regulation of cell surface receptors by endogenous and exogenous stimulation. *J. Neurosci.* **1999**, *19*, 10237–10249.

[83] Salas, P. J. I., Insoluble gamma-tubulin-containing structures are anchored to the apical network of intermediate filaments in polarized CACO-2 epithelial cells, *J. Cell Biol.* **1999**, *146*, 645–657.

[84] Varndell, I. M., J. M. Polak, K. L. Sikri, C. D. Minth, S. R. Bloom, J. E. Dixon, Visualisation of messenger RNA directing peptide synthesis by *in situ* hybridisation using a novel single-stranded cDNA probe, *Histochemistry* **1984**, *81*, 597–601.

[85] Cremers, A. F. M., N. Jansen In De Wal, J. Wiegant, R. W. Dirks, P. Weisbeek, M. Van Der Ploeg, J. E. Landegent. Non-radioactive *in situ* hybridisation. A comparison of several immunocytochemical detection systems using reflection-contrast and electron

microscopy, *Histochemistry* **1987**, *86*, 609–615.

[86] Löning, T., R.-P. Henke, P. Reichart, J. Becker, *In situ* hybridisation to detect Epstein-Barr virus DNA in oral tissues of HIV-infected patients, *Virchows Arch. A* **1987**, *412*, 127–133.

[87] Hacker, G. W., A.-H. Graf, C. Hauser-Kronberger, G. Wirnsberger, A. Schiechl, G. Bernatzky, U. Wittauer, H. Su, H. Adam, J. Thurner, G. Danscher, L. Grimelius, Application of silver acetate autometallography and gold-silver staining methods for *in situ* DNA hybridization, *Chinese Med. J.* **1993**, *106*, 83–92.

[88] Hacker, G. W., I. Zehbe, J. Hainfeld, J. Sällström, C. Hauser-Kronberger, A.-H. Graf, H.Su, O. Dietze, O. Bagasra, High-Performance Nanogold *in situ* hybridisation and *in situ* PCR, *Cell Vision* **1996**, *3*, 209–214.

[89] Bagasra, O., Polymerase chain reaction *in situ*. *Amplifications*, editorial note, **1990**, March issue, 20–21.

[90] Haase A. T., E. F. Retzel, K. A. Staskus, Amplification and detection of lentiviral DNA inside cells, *Proc. Natl. Acad. Sci. USA* **1990**, *87*, 4971–4975.

[91] Nuovo, G. J., Detection of human papillomavirus DNA in formalin-fixed tissues by *in situ* hybridisation after amplification by polymerase chain reaction, *Am. J. Pathol.* **1991**, *139*, 847–854.

[92] Zehbe, I., G. W. Hacker, J. Sällström, E. Rylander, E. Wilander, *In situ* polymerase chain reaction (*in situ* PCR) combined with immunoperoxidase staining and immunogold-silver staining (IGSS) techniques. Detection of single copies of HPV in SiHa cells, *Anticancer Res.* **1992**, *12*, 2165–2168.

[93] Zehbe, I., G. W. Hacker, W. H. Muss, J. Sällström, E. Rylander, A.-H. Graf, H. Prömer, E. Wilander, An improved protocol of *in situ* polymerase chain reaction (PCR) for the detection of human papillomavirus (HPV), *J. Cancer Res. Clin. Oncol.* **1993**, *119 Suppl.*, 225.

[94] Zehbe, I., G. W. Hacker, C. Hauser-Kronberger, E. Rylander, E. Wilander, Indirect and direct *in situ* PCR for the detection of human papillomavirus. An evaluation of two methods and a double staining technique, *Cell Vision* **1994**, *1*, 163–167.

[95] Zehbe, I., G. W. Hacker, J. F. Sällström, E. Rylander, E. Wilander, Self-sustained sequence replication-based amplification (3SR) for the *in situ* detection of mRNA in cultured cells, *Cell Vision* **1994**, *1*, 20–24.

[96] Bobrow, M. N., D. Thomas, D. Harris, K. J. Shaughnessy, G. J. Litt, Catalyzed reporter deposition, a novel method of signal amplification. Application to immunoassays, *J. Immunol. Methods* **1989**, *124*, 279–285.

[97] Bobrow, M. N., K. J. Shaugnessy, G. J. Litt., Catalyzed reporter deposition, a novel method of signal amplification. II. Application to membrane immunoassays. *J. Immunol. Methods* **1991**, *137*, 103–112.

[98] Hacker, G. W., C. Hauser-Kronberger, I. Zehbe, H. Su, A. Schiechl, O. Dietze, R. R. Tubbs, *In situ* localization of DNA and RNA sequences: super-sensitive *in situ* hybridisation using streptavidin-Nanogold-silver staining: Minireview, protocols and possible applications, *Cell Vision* **1997**, *4*, 54–67.

[99] Zehbe, I., G. W. Hacker, H. Su, C. Hauser-Kronberger, J. F. Hainfeld, R. Tubbs, Single human papillomavirus copy detection in formalin-fixed samples: sensitive *in situ* hybridization with catalyzed reporter deposition, streptavidin-Nanogold, and silver acetate autometallography, *Am. J. Pathol.* **1997**, *150*, 1553–1561.

[100] Totos, G., A. Tbakhi, C. Hauser-Kronberger, R. R. Tubbs, Catalyzed reporter deposition: a new era in molecular immunomorphology. Nanogold-silver staining and colorimetric detection and protocols, *Cell Vision* **1997**, *4*, 433–442.

[101] Tbakhi, A., G. Totos, C. Hauser-Kronberger, J. Pettay, D. Baunnoch, G. W. Hacker, and R. R. Tubbs, Fixation conditions for DNA and RNA *in situ* hybridization: a reassessment of molecular morphology dogma, *Am. J. Pathol.* **1998**, *152*, 35–41.

[102] Hacker, G. W., I., Zehbe, R. R. Tubbs, Super sensitive *in situ* DNA and RNA localisation, in: Polak J. M., J. Wharton J., O'D. McGee (eds), *In situ Hybridization: Principles and Practice*, 2nd ed., Oxford University Press, Oxford, U. K., **1999**, pp. 179–206.

[103] Tubbs, R. R., J. Pettay, T. Grogan, A. L. M. Cheung, R. D. Powell, C. Hauser-Kronberger, G. W. Hacker, Supersensitive *in situ* hybridization by tyramide signal amplification and Nanogold silver staining: The

contribution of autometallography and catalyzed reporter deposition to the rejuvenation of *in situ* hybridization, in: Hacker G. W., J. Gu (eds), *Gold and silver staining: techniques in molecular morphology,* CRC Press, Boca Raton, London, New York Washington, D. C., **2002**, pp. 127–144.

[104] Schöfer, C., K. Weipoltshammer, *In situ* hybridization at the electron microscopic level, in: Hacker G. W., J. Gu (eds), *Gold and silver staining: techniques in molecular morphology,* CRC Press, Boca Raton, London, New York Washington, D. C., **2002**, pp. 199–209.

[105] Nitta, H., K. Christensen, T. M. Grogan, Automated colorimetric tyramide signal amplification in situ hybridization (TISH) detection of single HER-2/neu oncogene copies, *Acta Histochem. Cytochem.* **2002**, *35,* 224.

[106] Nitta H., J. Wong, T. M. Grogan, T. Koji, E. E. Vela, M. Bobrow, Development of automated in situ hybridization applications using oligoprobes and tyramide signal amplification on the ventana discovery slide processor, *Acta Histochem. Cytochem.* **2002**, *35,* 224.

[107] Tubbs, R., J. Pettay, M. Skacel, R. Powell, M. Stoler, P. Roche, J. Hainfeld, Gold-facilitated in situ hybridization: a bright-field autometallographic alternative to fluorescence in situ hybridization for detection of HER-2/neu gene amplification, *Am. J. Pathol.* **2002**, *160,* 1589–1595.

[108] Tubbs, R., M. Skacel, J. Pettay, R. Powell, J. Myles, D. Hicks, J. Sreenan, P. Roche, M. H. Stoler, J. Hainfeld. J., Interobserver interpretative reproducibility of GOLDFISH, a first generation gold-facilitated auto-metallographic bright field in situ hybridization assay for her-2/neu amplification in invasive mammary carcinoma, *Am. J. Surg. Pathol.* **2002**, *26,* 908–913.

[109] Steinmetz, M. O., D. Stoffler, S. A. Muller, W. Jahn, B. Wolpensinger, K. N. Goldie, A. Engel, H. Faulstich, U. Aebi, Evaluating atomic models of F-actin with an undecagold-tagged phalloidin derivative, *J. Mol. Biol.* **1998**, *276,* 1–6.

[110] Faulstich, H., private communication.

[111] Luo, R. Z.-T., D. R. Beniac, A. Fernandes, C. C. Yip, F. P. Ottensmeyer, Quaternary structure of the insulin-insulin receptor complex, *Science* **1999**, *285,* 1077–1080.

[112] Ottensmeyer, F. P., R. Z.-T. Luo, A. B. Fernandes, D. Beniac, C. C. Yip,: Insulin receptor: 3D reconstruction from darkfield STEM images, structural interpretation and functional model. *Microsc. Microanal.,* *5, (Suppl. 2: Proceedings)*, Bailey, G. W., W. G. Jerome, S. McKernan, J. F. Mansfield, R. L. Price (eds), Springer-Verlag, New York, NY; **1999**, pp. 408–409.

[113] Montesano-Roditis, L., D. G. Glitz, R. R. Traut, P. L. Stewart, Cryo-electron microscopic localization of protein L7/L12 within the *Escherichia coli* 70 S ribosome by difference mapping and Nanogold labeling, *J. Biol. Chem.* **2001**, *276,* 14117–14123.

[114] Al-Bassam, J. R. S. Ozer, D. Safer, S. Halpain, R. A. Milligan, MAP2 and tau bind longitudinally along the outer ridges of microtubule protofilaments, *J. Cell Biol.* **2002**, *157,* 1187–1196.

[115] Kar, S., J. Fan, M. J. Smith, M. Goedert, L. A. Amos, Repeat motifs of tau bind to the insides of microtubules in the absence of taxol, *EMBO J.* **2003**, *22,* 70–77.

[116] Gimenez-Bonafe, P., E. Ribes, P. Sautière, A. Gonzalez, H. Kasinsky, M. Kouach, P.-E. Sautière, J. Ausio, M. Chiva, Chromatin condensation, cysteine-rich protamine, and establishment of disulphide interprotamine bonds during spermiogenesis of *Eledone cirrhosa* (Cephalopoda). *Eur. J. Cell Biol.* **2002**, *81,* 341–349.

[117] Seron, K., V. Tieaho, C. Prescianotto-Baschong, T. Aust, M. O. Blondel, P. Guillard, G. Devilliers, O. W. Rossanese, B. S. Glick, H. Riezman, S. Keranen, R. Haguenauer-Tapis, A yeast t-SNARE involved in endocytosis, *Mol. Biol. Cell* **1998**, *9,* 2873.

[118] Prescianotto-Baschong, C., H. Riezman, Morphology of the yeast endocytic pathway. *Mol. Cell Biol.* **1998**, *9,* 173–189.

[119] Akaki, M., E. Nagayasu, Y. Nakano, M. Aikawa, Surface charge of *Plasmodium falciparum* merozoites as revealed by atomic force microscopy with surface potential spectroscopy. *Parasitol. Res.* **2002**, *88,* 16–20.

[120] Wille, H., M. D. Michelitsch, V. Guenebaut, S. Supattapone, A. Serban, F. E. Cohen, D. A. Agard, S. B. Prusiner, Structural studies of the scrapie prion protein by electron crystallography, *Proc. Natl. Acad. Sci. USA* **2002**, *99,* 3563–3568.

[121] Ribrioux, S., G. Kleymann, W. Haase, K. Heitmann, C. Ostermeier, H. Michel, Use of Nanogold- and fluorescent-labeled antibody Fv fragments in immunocytochemistry, *J. Histochem. Cytochem.* **1996**, *44*, 207–213.

[122] Malecki, M., A. Hsu, L. Truong, S. Sanchez, Molecular immunolabeling with recombinant single-chain variable fragment (scFv) antibodies designed with metal-binding domains, *Proc. Natl. Acad. Sci. USA* **2002**, *99*, 213–218.

[123] Segond von Banchet, G., B. Heppelman, Non-radioactive localization of substance P binding sites in rat brain and spinal cord using peptides labeled with 1.4 nm gold particles, *J. Histochem. Cytochem.* **1995**, *43*, 821–827.

[124] Segond von Banchet, G., M. Schindler, G. J. Hervieu, B. Beckmann, P. C. Emson, B. Heppelmann, Distribution of somatostain receptor subtypes in rat lumbar spinal cord examined with gold-labelled somatostatin and anti-receptor antibodies, *Brain Res.* **1999**, *816*, 254.

[125] Wagenknecht, T., J. Berkowitz, R. Grassucci, A. P. Timerman, S. Fleischer, Localization of calmodulin binding sites on the ryanodine receptor from skeletal muscle by electron microscopy, *Biophys. J.* **1994**, *67*, 2286–2295.

[126] Traxler, K. W., M. T. Norcum, J. F. Hainfeld, G. M. Carlson, Direct visualization of the calmodulin subunit of phosphorylase kinase via electron microscopy following subunit exchange, *J. Struct. Biol.* **2001**, *135*, 231–238.

[127] Kessler, P., F. Kotzyba-Hilbert, M. Leonetti, F. Bouet, P. Ringler, A. Brisson, A. Mendez, M. P. Goeldner, C. Hirth, Synthesis of an acetylcholine receptor-specific toxin derivative regioselectively labeled with an undecagold cluster, *Bioconjug. Chem.* **1994**, *5*, 199–204.

[128] Adler-Moore, J., AmBisome targeting to fungal infections. *Bone Marrow Transplant* **1994**, *14*, S3–S7.

[129] Thurston, G., J. W. McLean, M. Rizen, P. Baluk, A. Haskell, T. J. Murphy, D. Hanahan, D. M. McDonald, Cationic liposomes target angiogenic endothelial cells in tumors and chronic inflammation in mice, *J. Clin. Invest.* **1998**, *101*, 1401–1413.

[130] Hainfeld J. F., Gold Liposomes, *Microsc. Microanal., 2, Suppl. 2 (Proceedings)*, Bailey, G. W., J. M. Corbett, R. V. W. Dimlich, J. R. Michael, N. J., Zaluzec (eds), San Francisco Press, San Francisco, CA, **1996**, pp. 898–899.

[131] Hainfeld, J. F., F. R. Furuya, R. D. Powell, Metallosomes, *J. Struct. Biol.* **1999**, *127*, 152–160.

[132] Brena, B. M., L. G. Ryden, J. Porath, Immobilization of beta-galactosidase on metal-chelate-substituted gels, *Biotechnol. Appl. Biochem.* **1994**, *19*, 217–231.

[133] Hochuli, E, H. Dobeli H, A. Schacher, New metal chelate adsorbent selective for proteins and peptides containing neighbouring histidine residues, *J. Chromatogr.* **1987**, *411*, 177–184.

[134] Kröger, D., M. Liley, W. Schiwek, A. Skerra, H. Vogel, Immobilization of histidine-tagged proteins on gold surfaces using chelator thioalkanes, *Biosens. Bioelectronics* **1999**, *14*, 155–161.

[135] Nieba, L., S. E. Nieba-Axmann, A. Persson, M. Hämäläinen, F. Edebratt, A. Hansson, J. Lidholm, K. Magnusson, .F. Karlsson, A. Plückthun, BIACORE analysis of histidine-tagged proteins using a chelating NTA sensor chip, *Anal. Biochem.* **1997**, *252*, 217–228.

[136] Hainfeld, J. F., W. Liu, C. M. R. Halsey, P. Freimuth, R. D. Powell, Ni-NTA-Gold Clusters Target His-Tagged Proteins, *J. Struct. Biol.* **1999**, *127*, 185–198.

[137] Hainfeld, J. F., R. D. Powell, C. M. R. Halsey, P. Freimuth, Ni-NTA-Nanogold for binding His tags, in: *Proc. XIV Int. Congress on Electron Microscopy*, Calderon Benevides, H. A., M. Jose Yacaman (eds), Institute of Physics Publishing, Bristol, UK, **1998**, pp. 859–860.

[138] Buchel, C., E. Morris, E. Orlova, E., J. Barber, Localisation of the PsbH subunit in photosystem II: a new approach using labelling of His-tags with a Ni(2+)-NTA goldcluster and single particle analysis, *J. Mol. Biol.* **2001**, *312*, 371–379.

[139] Hainfeld, J. F., R. N. Eisen, R. R. Tubbs, R. D. Powell, Enzymatic Metallography: A Simple New Staining Method, *Microsc. Microanal., 8, (Suppl. 2: Proceedings)*, Lyman, C. E., R. M. Albrecht, C. B. Carter, V. P. Dravid, B. Herman, H. Schatten (eds),

Cambridge University Press, New York, NY, **2002**, pp. 916–917CD.

[140] Stowell, C., B. A. Korgel, Self-assembled honeycomb networks of gold nanocrystals, *Nano Lett.* **2001**, *1*, 595–600.

[141] Gutierrez, E., R. D. Powell, F. R. Furuya, J. F. Hainfeld, T. G. Schaaff, M. N. Shafigullin, P. W. Stephens, R. L. Whetten, Greengold, a giant cluster compound of unusual electronic structure, *Eur. Phys. J. D.* **1999**, *9*, 647–651.

[142] Hainfeld, J. F., R. D. Powell, F. R. Furuya, J. S. Wall, Gold Cluster Crystals, *Microsc. Microanal., 6, (Suppl. 2: Proceedings)*, Bailey, G. W., S. McKernan, R. L. Price, S. D. Walck, P.-M. Charest, R. Gauvin (eds), Springer-Verlag, New York, NY, **2000**, pp. 326–327.

[143] Alivisatos, A. P., K. P. Johnsson, X. Peng, T. E. Wilson, C. J. Loweth, M. P. Bruchez, Jr., P. G. Schultz, Organization of 'Nanocrystal Molecules' using DNA, *Nature* **1996**, *382*, 609.

[144] Hamad-Schifferli, K., J. J. Schwartz, A. T. Santos, S. Zhang, J. M. Jacobson, Remote electronic control of DNA hybridization through inductive coupling to an attached metal nanocrystal antenna, *Nature* **2002**, *415*, 152–155.

[145] Hainfeld, J. F., F. R. Furuya, R. D. Powell, W. Liu, DNA Nanowires, *Microsc. Microanal., 7, (Suppl. 2: Proceedings)*, Bailey, G. W., R. L. Price, E. Voelkl, and I. H. Musselman (eds), Springer-Verlag, New York, NY, **2001**, pp. 1034–1035.

[146] Mirkin, C. A., R. L. Letsinger, R. C. Mucic, J. J. Storhoff, A DNA-based method for rationally assembling nanoparticles into macroscopic materials, *Nature* **1996**, *382*, 607–609.

[147] Elghanian, R., J. J. Storhoff, R. C. Mucic, R. L. Letsinger, C. A. Mirkin, Selective colorimetric detection of polynucleotides based on the distance-dependent optical properties of gold nanoparticles, *Science* **1997**, *277*, 1078–1081.

[148] Liu, D., J. H. Reif, T. H. LaBean, DNA Nanotubes: construction and characterization of filaments composed of TX-tile lattice. Hagiya, M. and A. Ohuchi (eds), *Lecture Notes in Computer Science*, **2003**, *2568*, pp. 10–21.

[149] Xiao, S., F. Liu, A. E. Rosen, J. F. Hainfeld, N. C. Seeman, K. Musier-Forsyth, R. A.

Kiehl, Self-assembly of metallic nanoparticle arrays by DNA scaffolding. *J. Nanoparticle Res.* **2002**, *4*, 313–317.

[150] Park, S.-J., T. A. Taton, C. A. Mirkin, Array-based electrical detection of DNA with nanoparticle probes, *Science* **2002**, *295*, 1503.

[151] Wu, P., L. Brand, Resonance energy transfer: methods and applications, *Anal. Biochem.* **1994**, *218*, 1–13.

[152] Dubertret, B., M. Calame, A. Libchaber, Single-mismatch detection using gold-quenched fluorescent oligonucleotides, *Nature Biotechnol.* **2001**, *19*, 365–370.

[153] Dulkeith, E., et al., Fluorescence quenching of dye molecules near gold nanoparticles: radiative and nonradiative effects, *Phys. Rev. Lett.* **2002**, *89*, 203002.

[154] Schmid, G., B. Morun, J.-O. Malm, $Pt_{309}Phen*_{36}O_{30\pm10}$, a four-shell platinum cluster, *Angew. Chem. Int. Ed. Eng.* **1989**, *28*, 778.

[155] De Aguiar, J. A. O., H. B. Brom, L. J. de Jongh, G. Schmid, EPR on the high-nuclearity palladium cluster, $Pd_{561}Phen_{36}O_{200}$, *Z. Phys. D.:Atoms, Molecules and Clusters* **1989**, *12*, 457–459.

[156] Vorgaftik, M. M., et al., A novel giant palladium cluster, *J. Chem. Soc. Chem. Commun.* **1985**, 937–939.

[157] Schmid, G., M. Harms, J. O. Malm, J. O. Bovin, J. Van Ruitenbeck, H. W. Zandbergen, W. T. Fu, Ligand-stabilized giant palladium clusters: promising candidates in heterogeneous catalysis, *J. Am. Chem. Soc.* **1993**, *115*, 2046–2048.

[158] Powell, R. D., C. M. R. Halsey, W. Liu, V. N. Joshi, J. F. Hainfeld, Giant platinum clusters: 2 nm covalent metal cluster labels, *J. Struct. Biol.* **1999**, *127*, 177–184.

[159] Kepert, D. L., Isopolyanions and heteropolyanions, in: *Comprehensive Inorganic Chemistry*, Bailar, J. C., J. H. Emeléus, R. Nyholm, A. F. Trotman-Dickenson (eds), Pergamon Press, Oxford, UK, **1973**, pp. 607–672.

[160] Keana, J. F. W., M. D. Ogan, Y. Lu, M. Beer, J. Varkey, Functionalized heteropolytungstate anions possessing a modified Dawson structure: small, individually distinguishable labels for conventional transmission electron microscopy, *J. Am. Chem. Soc.* **1985**, *107*, 6714–6715.

[161] Keana, J. F. W., M. D. Ogan, Functionalized Keggin- and Dawson-type cyclopentadie-

nyltitanium heteropolytungstate anions: small, individually distinguishable labels for conventional transmission electron microscopy. 1. Synthesis, *J. Am. Chem. Soc.* **1986**, *108*, 7951–7957.

[162] Keana, J. F. W., M. D. Ogan, Y. Lu, M. Beer, J. Varkey, Functionalized Keggin- and Dawson-type cyclopentadienyltitanium heteropolytungstate anions: small, individually distinguishable labels for conventional transmission electron microscopy. 2. Reactions, *J. Am. Chem. Soc.* **1986**, *108*, 7957–7963.

[163] Keana, J. F. W., Y. Wu, G. Wu, Di-, tri-, tetra-, and pentacationic alkylammonium salts. ligands designed to prevent the nonspecific electrostatic precipitation of polyanionic, functionalized cyclopentadienyltitanium-substituted heteropolytungstate electron microscopy labels with cationic biomolecules. *J. Org. Chem.* **1987**, *52*, 2571–2576.

[164] Hainfeld, J. F., C. J. Foley, L. E. Maelia, J. J. Lipka, Eleven tungsten atom cluster labels: High-resolution, site-specific probes for

electron microscopy, *J. Histochem. Cytochem.* **1990**, *38*, 1787–1793.

[165] Hainfeld, J. F., J. J. Lipka, F. E. Quaite, A high-resolution tungstate membrane label, *J. Histochem. Cytochem.* **1990**, *38*, 1793–1803.

[166] Furuya, F. R., L. L. Miller, J. F. Hainfeld, W. C. Christopfel, P. W. Kenny, Use of $Ir_4(CO)_{11}$ to measure the lengths of organic molecules with a scanning transmission electron microscope, *J. Am. Chem. Soc.* **1988**, *110*, 641–643.

[167] Weinstein, S., W. Jahn, H. Hansen, H. G. Wittmann, A. Yonath, Novel procedures for derivatization of ribosomes for crystallographic studies, *J. Biol. Chem.* **1989**, *264*, 19138–19142.

[168] Cheng, N.; J. F. Conway, N. R. Watts, J. F. Hainfeld, V. Joshi, R. D. Powell, S. J. Stahl, P. E. Wingfield, A. C. Steven, Tetrairidium, a 4-atom cluster, is readily visible as a density label in 3D cryo-EM maps of proteins at 10–25 resolution, *J. Struct. Biol.* **1999**, *127*, 169–176.

24

Surface Biology: Analysis of Biomolecular Structure by Atomic Force Microscopy and Molecular Pulling

Emin Oroudjev, Signe Danielsen, and Helen G. Hansma

24.1
Introduction

During the past decade, the atomic force microscope has been developed from an exotic and sometimes home-made instrument into a relatively widespread surface-imaging and probing instrument. Atomic force microscopy (AFM) and its related techniques have proliferated successfully into many different fields, including biology. Today, the method is used to obtain high-resolution static and dynamic images in investigations of the physical and mechanical properties of biological macromolecules. As a result, the number of AFM-related articles listed in PubMed has mushroomed such that the recommended number of references for a minireview constitutes little more than 10 % of the total number referred to in PubMed for a single year. This review will therefore, of necessity, neglect many excellent reports on the subject. Reviews published during the past year on AFM include Refs. [1–12], but other recent reviews are available in Refs. [13–20].

The atomic force microscope and other scanning probe microscopes have been used to pioneer a new field, that of surface biology. This is a logical advance in biological methodology, as we now know that cells are not simply bags of cytoplasm but are bounded by surface-membranes which comprise incredible arrangements of macromolecular complexes for signaling, communicating, and regulating cellular functions. Furthermore, we now know that cells are filled with complex structures and membrane-bound organelles, such that most of life's processes actually occur on the cell surfaces. This new biology – surface biology – goes beyond the realms of test-tube biology which have been so successful in our learning about ourselves. Research into surface biology is indeed performed at surfaces which sometimes are physiologically relevant (e. g., lipid bilayers) and sometimes are physiologically irrelevant (e. g., mica). Mica serves as a flat support upon which biomaterials can be observed, either by direct imaging or by pulling or through the use of other techniques to probe the properties of the biomaterials.

AFM is now typically used in conjunction with other techniques to probe biomaterials. This is a continuation of the progression first noted in 1998, at which time a large proportion of biological reports included the term 'AFM' in their title, to a present-day situation where this is limited to 50 % or fewer cases [21]. Today, the use of AFM in conjunction

Nanobiotechnology. Edited by Christof Niemeyer, Chad Mirkin
Copyright © 2004 WILEY-VCH Verlag GmbH & Co. KgaA, Weinheim
ISBN 3-527-30658-7

with other techniques is a further indication of the technique having matured from a no-velty to one which generates significant amounts of scientific material. Indeed, nowadays AFM is used in conjunction with other analytical methods and with theoretical analyses on a highly frequent basis.

24.2
Recent Results

Research investigations into many biological macromolecules, including those for DNA, RNA, and proteins, as well as other biopolymers and their corresponding supramolecular structures such as viruses, living cells and cell colonies, have all benefited from recent developments in the field of AFM.

24.2.1
DNA

DNA was one of the first biological objects to be visualized by AFM, and the imaging tech-nique has been used not only for DNA mapping and sizing but and also examining the structural changes induced by interactions with enzymes, DNA-binding proteins and con-densing agents. By using AFM to image DNA–protein interactions, the location of pro-tein-binding sites and the conformation of DNA at these sites have been identified. In fact, in some cases it has been possible to determine the stoichiometry of these interac-tions. For a recent review of AFM studies of DNA, see Ref. [18].

Detailed information can often be obtained about the conformations and organization of DNA and RNA molecules, as well as the native and artificial nanostructures con-structed from these nucleic acids. DNA sizing, fingerprinting and/or mapping can be per-formed by imaging DNA restriction fragments and using automated software packages to measure their lengths (e. g., Ref. [22]). Some other methods of mapping sequences on large DNA molecules are based on the ability of these sequences to interact specifically with complementary oligonucleotides or with certain proteins. The hybridization area be-tween DNA and the probe will usually display an increased width and height, or some other substructure resulting from DNA–probe interactions (loops, hairpins, kinks, etc.), and these effects can be further enhanced by tagging the probe with nanoparticles (gold nanospheres, biotin–avidin complex, etc). In one example of DNA mapping, a 300 nucleotide-long RNA probe was hybridized to double-stranded DNA (ds-DNA), where-upon the site of hybridization had a distinctive appearance due to presence of the double-stranded RNA–DNA hybrid and unpaired single-stranded DNA [23]. Triple-helix-forming DNA oligonucleotides can also be used as probes in these experiments.

24.2.1.1   **2.1.1 DNA Condensation**
DNA condensation is another area of the DNA field where AFM plays a significant role. (Figure 24.1). Nonviral methods of condensing long DNA molecules, to a size that is ap-propriate for gene delivery into cells, are now becoming increasingly important as health complications and even human deaths have been reported recently as a result of viral-based gene therapy.

Figure 24.1 Sequence-dependent DNA condensation [86] (A) Poly (dG-dC)·(dC-dG) (GC-DNA) condenses in 1 mM NiCl$_2$. Times in Ni(II), top to bottom, = 5 sec, 1 min, 3–7 min, 2 months. (B) Poly (dA-dT)·(dT-dA) (AT-DNA) barely condenses after 3–5 min, even in 6 mM NiCl$_2$. Images are 1 μm wide; DNA concentration = 2.5 ng μL^{-1} in (A) and (B). (C) ds-DNA loops are typically seen on bundles of condensed GC-DNA: AFM image (left) and models (below) for formation of loops on bundles, progressing from stage 1 to 2 and 2'. (D) Electrostatic Zipper theory [87] extended to condensation of GC-DNA and AT-DNA in Ni(II). Graphs show force per unit length between parallel ds-DNA molecules versus intermolecular separation, at low ionic strength, as predicted by Eq. 2 in [sitko. Upper graph is for zero axial shift between parallel ds-DNA molecules; lower graph is for optimal axial shift, to maximize electrostatic attraction between parallel ds-DNA molecules. Curves a and a' correspond to AT-DNA, which is in the B-DNA conformation in Ni(II); curves b and b' correspond to

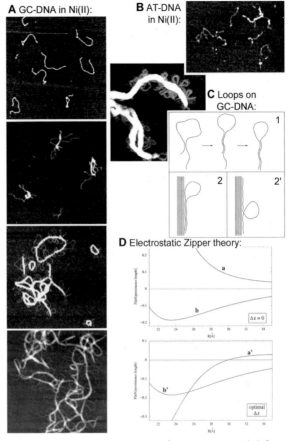

A GC-DNA in Ni(II):

B AT-DNA in Ni(II):

C Loops on GC-DNA:

D Electrostatic Zipper theory:

GC-DNA, which is in the Z-DNA conformation in Ni(II). Attraction (negative forces) at zero axial shift, as in curve b, is indicative of DNA condensation.

Of special interest has been the use of polycations for the development of nonviral gene delivery systems. One of the present authors (S. D.) has been using the linear polymer chitosan, a polycation derived from the polysaccharide chitin. Chitosan has emerged as a promising candidate for this purpose as it is nontoxic, biodegradable, and has been shown to yield high transfection efficiency [24]. The ability of various chitosans to compact DNA has been studied with AFM imaging (Danielsen and Stokke, in preparation), showing that chitosan effectively compacts DNA into toroidal, rod-like, and globular structures. This distribution of geometries was found to depend upon both the charge density and the degree of polymerization (DP) of the chitosan, with the low-DP chitosans resulting in less well-defined structures.

The AFM technique called chemical force microscopy was proposed to improve current high-throughput DNA micro-array screening analyses. DNA arrays are imaged in friction force mode with chemically modified probes that have complementary DNA probes attached to the tip. Increases in friction forces are detected as the tip reaches spots on the DNA array that contain DNA complementary to DNA on the tip. Enhanced sensitivity and spatial resolution of this method relative to current micro-array technology may make it possible to create significantly smaller and more densely packed micro-arrays or nano-arrays of DNA.

By using AFM to pull on single DNA molecules, information about single molecular mechanics and thermodynamics has been obtained. The forces required for DNA's B-S transition [25] and the attraction force between DNA molecules and cationic lipid bilayers have been measured [26].

24.2.1.2 DNA Sequences Recognized by Mica

Some intrinsically curved DNA sequences contain phased A-tracts, such that half of each helix turn has a run of As, paired with a run of Ts. Sequence-dependent DNA curvature such as this is an important element in specific DNA–protein interactions.

A-rich intrinsically curved DNA sequences were joined in head-to-head and head-to-tail palindromes to form 's' or backwards 's' conformations, depending on which side faced upwards. When these DNAs were deposited onto mica, the T-rich side bound preferentially to mica [27] – a finding which may have potential application(s) in the field of nanobiotechnology.

24.2.1.3 Drug-binding to Single ds-DNA Molecules

Molecular pulling may also become a valuable technique for monitoring the binding of drugs to DNA, and preliminary results in this area have recently been published [28]. Recently, a minor-groove binding drug (berenil), a cross-linking drug (cisplatin) and an intercalator (ethidium bromide) were each tested for their effects on the DNA pulling curves. At high levels, all three drugs completely or mostly abolished the B-to-S transition, but cisplatin also reduced the hysteresis between pulling and relaxing the DNA, even at short incubation times [28].

24.2.2
Proteins

Proteins are the second largest class of biological macromolecules to have benefited from developments in AFM methods [9]. Methods to image both separate protein molecules and protein layers with different degree of organization have been developed, and these utilize native or modified mica surfaces (see section 24.3.2). Isolated protein molecules, when applied to a surface, often are too mobile and too soft to be imaged with resolution sufficient for observing any submolecular features. Immunoglobulin G (IgG) and IgM are examples of submolecular resolution in the AFM imaging of isolated protein molecules. With IgG, the three-lobed shape (as identified by X-ray crystallography) was also observed by Anafi using AFM (see Ref. [29]) and imaging in aqueous buffer (see Ref. [30]). New information about IgM substructure was obtained by using cryo-AFM, which showed a tendency for IgM molecules to adopt a compact conformation with a raised center [31].

On the other hand, proteins deposited as a densely packed layer often produce images with very high resolution [32, 33]. Some such protein arrays occur naturally in specialized membranes such as the bacteriorhodopsin-containing purple membranes. To create these layers with other types of proteins, the proteins are a deposited onto artificial lipid bilayers [34], the latter being prepared by vesicle fusion onto mica or silicon surfaces or by using a Langmuir trough. The lipid monolayer that faces mica or silicon is usually made from phosphatidylcholine or related lipids. The opposite monolayer, which is used to bind and/or incorporate macromolecules of interest, can be constructed from different lipids that have the desired chemistry at their headgroups. Biological macromolecules (proteins and DNA) adsorbed onto these membrane-like surfaces are typically imaged with a resolution that is superior to that achieved on conventional surfaces such as native or modified mica.

AFM can be used to study different protein functions. For example, it was used to measure DNase I, DNA polymerase and RNA polymerase enzymes kinetics on surface-bound substrate (DNA) [35–37]. The dynamics of interactions between large and small subunits GroE chaperonin protein from *Escherichia coli* were studied using AFM at the level of individual protein molecules [38].

24.2.2.1 Prion Proteins

AFM analysis of a yeast prion protein, combined with Fourier transform infrared (FTIR) spectroscopy, indicates that fibrils can form from the prion protein in its native helical conformation and do not contain the crossed-beta structures typical of amyloid fibrils [39].

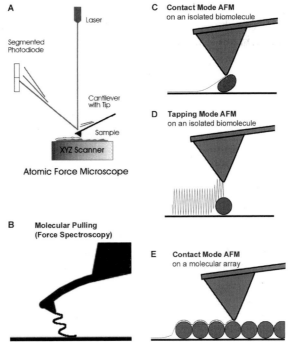

Figure 24.2 Atomic force microscopy (AFM) and some of its uses for biomolecular imaging and probing. (A) AFM schematic; see, for example, Ref. [20] for details. (B) Molecular pulling, or (Single Molecule) Force Spectroscopy is a method of obtaining information about the mechanics and folding of one or a few biomolecules. It is also used to measure unbinding forces and thermodynamic parameters of ligand–receptor interactions. (C) Contact mode AFM tends to deform, move or damage isolated biomolecules. (D) Tapping mode AFM reduces lateral forces on isolated biomolecules. (E) Contact mode AFM is often used for imaging biomolecular arrays, such as this "purple membrane protein" array. Resolution is typically higher on closely packed arrays of macromolecules than on isolated macromolecules. There are at least two explanations for this. First, the packed macromolecules have less freedom to move on the substrate. Second, less of the tip interacts with the surface of the arrays; many tips have an asperity at the apex, which can provide high-resolution imaging of relatively flat surfaces, such as those in protein arrays.

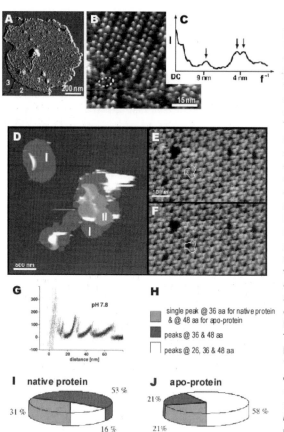

Figure 24.3 Membrane protein arrays [40, 41]. (A–C) Native eye disc membrane from the rod outer segment of a vertebrate retina, adsorbed on mica [40]. (A) Deflection image, showing that three different surface types are evident: 1, the cytoplasmic side of the disc membrane; 2, lipid; and 3, mica. To avoid the formation of opsin, the chromophore-depleted form of rhodopsin, membrane samples were never exposed to light. After adsorption of osmotically shocked disc membranes onto mica, their topography was measured in buffer solution (20 mM Tris–HCl (pH 7.8), 150 mM KCl and 25 mM MgCl$_2$). (B) Height image of the cytoplasmic surface of the disc membrane, showing rows of rhodopsin dimers, as well as individual dimers (inside dashed ellipse), presumably broken away from one of the rows. The rhodopsin molecules protrude from the lipid bilayer by 1.4 ± 0.2 nm (n = 111). This topograph is shown in relief, tilted by 5°. Vertical brightness range is 1.6 nm. (C) Angularly averaged powder-diffraction pattern, showing peaks at (8.4 nm)$^{-1}$, (4.2 nm)$^{-1}$ and (3.8 nm)$^{-1}$. (D–J) Purple membranes from *Halobacterium salinarum*: imaging and manipulation of bacteriorhodopsin (BR)

molecules [41]. (D) Purple membrane patches (I) adsorbed onto freshly cleaved mica and imaged in buffer solution (pH 7.8, 20 mM Tris-HCl, 300 mM KCl). In some areas, purple membrane patches overlap with other membranes, forming double layers (II). (E) High-resolution image of the cytoplasmic purple membrane surface showing BR trimers (outline) arranged into a hexagonal lattice. The topograph was recorded at minimum force allowing the longest cytoplasmic loops of the individual BR molecules (loop EF) to protrude fully from the membrane surface [88]. Individual defects show single or multiple BR monomers missing. After imaging, the AFM tip was brought into contact to the membrane surface (circle). This allowed the polypeptides of individual BR molecules to adsorb to the tip. During separation of tip and sample, this molecular bridge was used to pull on the protein, and the force spectrum was recorded. (F) Same purple membrane area imaged after the manipulation shows one individual BR monomer missing (outline, now enclosing dimer instead of trimer). Vertical brightness range corresponds to 50 nm (D) and 1.2 nm (E, F). (G) Force spectrum for unfolding of BR at pH 7.8 (n = 32). All molecules were unfolded by grabbing the C-terminus at the cytoplasmic surface [46]. (H–J) Probability distribution of pathways detected upon unfolding BR helices G and F. Probability of the unfolding pathways are shown for native BR and for the apoprotein. Although 58% of the unfolding events of apoprotein showed an additional unfolding barrier at 26 aa, this barrier was observed in only 16% of the unfolding traces of wild-type BR.

24.2.2.2 Membrane Proteins

Protein arrays in membranes can typically be imaged at higher resolution than isolated proteins, as shown diagrammatically for the "purple membrane" in Figure 24.2E and the isolated "protein" in Figure 24.2C and D. This permits the visualization submolecular structures and sometimes even substructural changes induced by such variables as pH and imaging force. Gap-junction membranes, for example, show Ca(II)-dependent reduction in the diameter of the gap-junction pore at the extracellular surface and a force-dependent collapse in pore structures at the cytoplasmic surface [5]. Figure 24.3 illustrates two recent research accomplishments on membrane protein arrays. First, rhodopsin dimers in a vertebrate retina were observed in a vertebrate retina (Figure 24.3A–C; see Ref. [40]). Second, the structure of bacteriorhodopsin has been analyzed by pulling apart individual bacteriorhodopsin molecules (Figure 24.3D–J; see Ref., [41]). The basic principle of molecular pulling is shown diagrammatically in Figure 24.2B.

Figure 24.4 The molecular structure of spider dragline silk nanofibers by AFM and molecular pulling [42, 89]. (A) Nanofibers form from molecules of a soluble bioengineered silk protein. These tapping-mode AFM images in air show segmented substructure of nanofibers and nanofiber aggregates; similar images are seen in aqueous fluid. (B) Two pulling curves (single-molecule force spectroscopy) on the silk molecules, with WLC fits to rupture peaks. In the first pull, much loosely adsorbed protein fell off the tip before the appearance of sawtooth rupture peaks. Rupture peaks are attributed to the unfolding of 38-aa sequence repeats from the silk protein. Peaks are typically 15 or 30 nm apart, which corresponds to the extended length of one or two polyA/GA+GGX repeats as shown in (C), below. (C) A model for the folding of single silk molecules. Length of the folded molecule is based on the mean length of the nanofiber segments (35 ± 9 nm) seen in (A) and (E). Amino-acid sequence is from [90]; zig-zags are polyA/GA repeats, and spirals are GGX repeats. D. Models for the stacking of silk molecules into nanofiber segments. (E) AFM image of a segmented silk nanofiber, showing proposed relationship between segments and nanofiber.

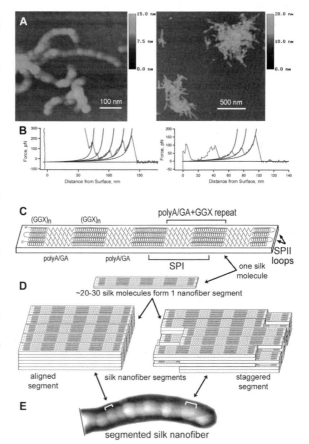

24.2.2.3 Spider Silk

Probe microscopy of a soluble bioengineered dragline silk protein provided results from both AFM imaging and molecular pulling approaches (Figure 24.4). These results were integrated into a molecular model for the folding pattern of individual silk molecules, plus a multi-molecular model for the association of these molecules into segmented silk nanofibers [42, 43].

This research involves the examination of mesostructures, which are more than single molecules but much less than bulk materials. Pulling on native capture silk also produced new information about spider silk at the mesostructure level [44].

24.2.3
Fossils

Today, even fossils are the subject of AFM research [45]. In the past, morphology has represented the major approach to characterizing fossils, although the earliest fossils – those of microorganisms – are too small for typical morphological observations to be made. AFM has been used to detect 200 nm-sized angular platelets stacked in arrays in the walls of petrified 650-million-year-old unicellular protists. In addition, Raman spectroscopy has been used to identify polycyclic aromatic kerogenous organic matter in the fossils. These results support the conclusion that the microscopic structures are true fossils, and not pseudo-fossil "look-alikes".

24.2.4
Science and Nature

Reports judged worthy of inclusion into *Nature* and *Science* [46–63] are worth citing in a review such as this, and news articles reported in both journals [64–68] are similarly valuable to the reader. The reports described in more detail below are representative of the breadth of topics covered in PubMed searches for AFM papers in *Science* and *Nature*.

Titin has been a popular molecule for single-molecule force spectroscopy or molecular pulling [69–71], and recent pulling research has focused on the relationship between the mechanical pulling of single molecules and the physiological milieu within the muscle [62, 72]. The tandem Ig domains of titin can withstand more force in vivo than was previously predicted, based on pulling experiments with mutant Ig domains [72]. In addition to the tandem Ig domains, titin in muscle has other types domains along its elastic region. Pulling experiments have been carried out on these other types of domains and, when combined with the results from Ig domains, the single-molecule pulls produced results which were similar to those observed with titin in situ in muscle sarcomere. In both cases, titin was seen not to be a simple entropic spring but rather to have a complex structure-dependent elasticity [62].

Protein mimics formed from the polymerization of isocyanopeptides fold into beta-helices that are different from the beta-helices found in proteins. The natural beta-helices contain arrays of large beta-sheets stacked in a helical fashion, while the beta-helices of the mimics have a central helical core with a beta-sheet-like arrangement of side arms [53].

The number-averaged molecular mass and polydispersity index were calculated from AFM images of the protein mimics.

Conductivity of DNA is an ongoing area of interest. DNA "combed" across a sub-micron slit between two rhenium/carbon superconducting electrodes showed ohmic conductance at temperatures between room temperature and 1 Kelvin, with a resistance per molecule of ca. 100 kΩ. Below 1 Kelvin, a proximity-induced superconductivity was observed. AFM was used in this research to characterize the topography of the samples, including DNA density and surface structures; the carbon film exhibited a "forest structure" [54].

Dip-pen nanolithography (DPN) was developed using DNA as the "ink", by dipping the tip into DNA and then tracing patterns [73]. This technique is now being used with protein "ink" to form arrays for AFM-based observation of specific protein–antigen interactions. There was virtually no nonspecific protein binding outside of the dots of the array. The array and background were formed by the deposition of two different alkane thiols on gold: a reactive (carboxy-terminated) alkane thiol in a patterned array, followed by a passive (ethylene glycol-terminated) alkane thiol deposited as a drop onto the patterned gold surface [60].

The nanoindentation of bulk amorphous metal induces crystallization at the indentations, at room temperature, as observed by both AFM and transmission electronic microscopy (TEM). The observed crystallites are similar to those formed by annealing at 783 Kelvin [59].

The interactions between biological surfaces and mineral surfaces are a novel and rapidly growing area of research. Genetically engineered bacterial viruses and ZnS quantum-dot solutions spontaneously formed ~72-μm ordered domains, arrays of which formed a complex hybrid film that was continuous over a 1 cm-long distance [61].

Force measurements between an iron-reducing bacterium and an iron oxide showed a several-fold increase in attraction when oxygen levels were reduced. In the absence of oxygen, the bacteria (*Shewanella*) transfer electrons to the iron oxide (goethite), near which they typically live. On the basis of Worm-Like Chain fits to the force curves, it appears that there is a 150-kDa iron reductase near the outer membrane of the bacterium [55].

Magnetoreceptors in rainbow trout formed the focus of one investigation, in which magnetic force microscopy was used to locate and characterize magnetic domains in receptor cells within olefactory lamellae. Magnetic crystals were arranged in a chain ~1 μm long in a single, multilobed cell [48].

24.3
Methodology

The basic design of an atomic force microscope and its main imaging modes are shown diagrammatically in Figure 24.2, and a collection of scanning-probe-microscopy methodologies has been published recently in *Methods in Cell Biology* [74].

AFM imaging can be carried out both in air and in a liquid environment. Imaging in liquid [75] is more demanding and more difficult to accomplish, but provides important advantages. Specifically, the imaging environment in the liquid cell can, ideally, be precisely controlled by using a buffer in which biological macromolecules are most stable and are held under native and optimal conditions for their functions. This also makes it possible to study the dynamics of behavior and function for many biological macromolecules. Two major scanning modes can be employed for image-producing purposes [17, 76] (Figure 24.2C–E). In contact mode, the AFM probe is kept in contact with the sample at all times during observation. In the second mode – the tapping or oscillating mode – the AFM probe oscillates at or near its resonant frequency. In this mode, the probe comes into contact with sample only at the lower end of each oscillation and, as a result, exerts much smaller lateral forces on the sample. The amplitude and phase of the oscillation signal recorded from the probe are very sensitive to interaction forces between probe and the sample. These changes are used to record sample topography (height imaging) plus amplitude and/or phase images.

24.3.1
The Probe

The resolving power of the AFM probe is one of the most critical issues, and silicon and silicon nitride probes are currently used for all AFM applications. The AFM probe consists of a cantilever (shaped usually like a diving board or as a "V") and a pyramidal or conical tip at the far end of the cantilever. The main characteristics of the cantilever are its softness (spring constant for cantilever) and resonant frequency. The softness and resonant frequency of each cantilever are related to each other such that softer cantilevers have lower resonant frequencies (provided that these cantilevers are of comparable size and shape). Soft cantilevers are preferable in contact mode on biological objects because they exert smaller forces and, thus, will not damage or dislocate the object from the surface during imaging. They are also often used for tapping in liquid. In contrast, imaging in tapping mode in air requires stiffer cantilevers that oscillate at higher resonant frequencies. This helps them overcome repulsive and attractive forces between tip and sample as well as meniscus forces at the water/air interface covering any surface in air [77]. To reduce meniscus forces, imaging in a constant flow of dry gas (e. g., nitrogen, helium) can be employed [78].

The tip of the AFM probe is characterized by its aspect ratio (ratio between width and height of the tip's pyramid or cone) and tip sharpness (expressed as tip radius of curvature). While the tip's aspect ratio is not critical for most applications (standard tips have cone or pyramid angles ~20–30°), for some samples higher aspect ratio tips can be crucial, as these will permit observation of features that otherwise are hidden by interactions between the sample and sides of the tip (deep pockets, etc).

An image obtained by AFM is the result of interactions between the very end of the tip and the biological macromolecules. The geometry of the tip end affects the final appearance of the image, including the dimensions of the sample structures and any artifacts and distortions of biological macromolecules [76]. For the majority of AFM samples, it is impossible to obtain an image resolution that exceeds the tip radius. Current silicon tips have radii of curvature of 2–10 nm, and silicon nitride tips of 20–60 nm. Silicon nitride tips can be further oxide sharpened to improve their sharpness up to 5 10 nm. Occasionally, resolution greater than tip radius is detected when imaging relatively hard samples or densely packed two-dimensional arrays of proteins. These unusually high-resolution images are attributed to the presence of small asperities on the end of the particular

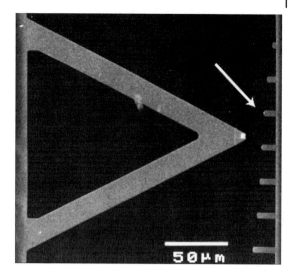

Figure 24.5 Small cantilevers (right) give better signal-to-noise ratio and permit higher imaging speeds than standard cantilevers (left). At the same spring constant, small cantilevers have higher resonant frequencies than large cantilevers. Arrow points to a cantilever similar to the one used to image protein–protein interactions in real time in Ref. [38]. Resonant frequency for small cantilever (arrow) is 30 times larger than that for large cantilever on left.

tip that effectively reduced the tip radius to a subnanometer range. In the case of imaging well-oriented or at least highly packed 2-D arrays of proteins and DNA, occasional high-resolution imaging can be attributed to the Fourier components from periodic nature of these arrays [29].

Hydronamic drag can be a problem for imaging in fluids, and can be increased by both large cantilevers and large movements. One promising improvement in AFM probe development is the development of small cantilevers [71, 79] with lengths of only 10–20 μm, unlike conventional cantilevers which have lengths of 100–400 μm (Figure 24.5). These small cantilevers have significantly higher resonant frequencies without the concomitant increase in spring constant, and this permits them to scan samples at much higher speeds but without increasing the forces applied to the samples by stiff cantilevers.

A recently emerging commercial system is that of nanotube AFM probes, in which the tip end is either a single- or multi-walled carbon nanotube with a tip radius as small as 0.5–2 nm and a 0° cone angle. Although to date these tips have seen only limited use, they show a significant improvement in spatial resolution and, due to their optimal aspect ratio, they may also be preferred for imaging objects with deep crevices.

24.3.2
The Sample

One of the first requirements for any object undergoing analysis by AFM is to be attached or adsorbed onto a hard (noncompressible) surface that is smooth and flat enough not to obscure the sample's topography. Freshly cleaved mica crystal (sometimes mounted on a support) is the most common mounting surface for imaging biomolecules by AFM, although some other materials (HOPG, glass, etc.) have also been used for this purpose.

Although mica has a negatively charged hydrophilic surface that is well-suited for mounting many positively charged biological macromolecules, the researcher often has to modify the mica surface in order to facilitate binding of the biological macromolecules to the surface. These modifications include treatment with ions of divalent metals (e. g., Ni(II) or Mg(II)) or long polymers carrying positive charges (e. g., poly-L-lysine, spermine, spermidine). This treatment leads to the formation of positively charged clusters on the mica surface and makes it possible for negatively charged molecules (such as DNA and RNA) to bind to the mica surface.

Another commonly used means of modifying mica is to treat it with derivatives of silane that can form monolayers on its surface. These monolayers are shown to be strongly attached to mica and to be smooth enough for imaging even relatively small objects. The outer ends of the silane molecules in the formed film can carry at their termini a number of different chemical groups (e. g., amino-, carboxyl-, epoxy-, mercapto-, alkyl-) to create a surface with the desired chemistry. For further details, the reader is referred to Refs. [34, 75]; for other articles, see Ref. [74].

24.4
The Future

Although some of the most exciting future developments in biological probe microscopy remain dreams in the minds of their inventors, two possible future uses are outlined in the following section.

24.4.1
Unity or Diversity?

Currently, single-molecule techniques are yielding information about the diversity seen in the fine structures of the individual molecules within a single population of molecules. One of the most exciting discoveries in single-molecule biophysics is that individual biomolecules of a particular type typically show qualitatively similar patterns of behavior, yet the quantitative behavior of the individual molecules often differs significantly – perhaps many-fold – between one molecule and the next. Xie observed this inter-molecular variability and defined static and dynamic forms of molecular variations [80]. Dynamic variations occur over time in a single molecule, while static variations occur between different molecules. Static variations in molecular behavior have also been noted in the interactions of the molecular chaperonins GroEL and GroES (Ref. [38] and P. Hansma, unpublished results), in lactate dehydrogenase by Xue [81], in the RedBCD enzyme by S. Kowalczykowski [82], and in a ribozyme by S. Chu [83], who named these variations "molecular memory".

These variations between one molecule and the next might be due to differences in the interactions between the molecule and its environment, such as surface effects for molecules attached to surfaces. While it is easy to label such variability as "surface artifacts", the cells in which these molecules reside are themselves filled with surfaces. In any case, this question is now at the forefront of single-molecule biophysics: namely, "Why are there such large differences between the individual molecules in a single so-called 'homogeneous' population?"

24.4.2
World-wide Research

Among the AFM-related reports listed in PubMed during this millennium, the vast majority were from the United States and other developed countries. Almost one-third were from the U.S., and one-third was from the United Kingdom, Germany, and Japan. The developed countries [84] together account for over 91 % of the AFM-related detailed papers in PubMed. One hope for the future is to see leading-edge research spreading to other countries around the world, and a related hope was expressed by Kofi Annan, Secretary-General of the United Nations, who extended such a challenge to the world's scientists [85].

References

[1] D. P. Allison, P. Hinterdorfer, W. Han, Biomolecular force measurements and the atomic force microscope, *Curr. Opin. Biotechnol.* 2002, *13*, 47–51.

[2] L. P. da Silva, Atomic force microscopy and proteins, *Protein Pept. Lett.* 2002, *9*, 117–126.

[3] Y. F. Dufrene, Atomic force microscopy, a powerful tool in microbiology, *J. Bacteriol.* 2002, *184*, 5205–5213.

[4] A. Engel, H. Stahlberg, Aquaglyceroporins: channel proteins with a conserved core, multiple functions, and variable surfaces, *Int. Rev. Cytol.* 2002, *215*, 75–104.

[5] D. Fotiadis, S. Scheuring, S. A. Muller, A. Engel, D. J. Muller, Imaging and manipulation of biological structures with the AFM, *Micron* 2002, *33*, 385–397.

[6] M. K. Higgins, H. T. McMahon, Snap-shots of clathrin-mediated endocytosis, *Trends Biochem. Sci.* 2002, *27*, 257–263.

[7] C. S. Hodges, Measuring forces with the AFM: polymeric surfaces in liquids, *Adv. Colloid Interface Sci.* 2002, *99*, 13–75.

[8] R. E. Marchant, I. Kang, P. S. Sit, Y. Zhou, B. A. Todd, S. J. Eppell, I. Lee, Molecular views and measurements of hemostatic processes using atomic force microscopy, *Curr. Protein Pept. Sci.* 2002, *3*, 249–274.

[9] D. J. Muller, H. Janovjak, T. Lehto, L. Kuerschner, K. Anderson, Observing structure, function and assembly of single proteins by AFM, *Prog. Biophys. Mol. Biol.* 2002, *79*, 1–43.

[10] D. J. Muller, K. Anderson, Biomolecular imaging using atomic force microscopy, *Trends Biotechnol.* 2002, *20*, S45–S49.

[11] P. J. Werten, H. W. Remigy, B. L. de Groot, D. Fotiadis, A. Philippsen, H. Stahlberg, H. Grubmuller, A. Engel, Progress in the analysis of membrane protein structure and function, *FEBS Lett.* 2002, *529*, 65–72.

[12] J. Zlatanova, S. M. Lindsay, S. H. Leuba, Single molecule force spectroscopy in biology using the atomic force microscope, *Prog. Biophys. Mol. Biol.* 2000, *74*, 37–61.

[13] H. G. Hansma, L. I. Pietrasanta, I. D. Auerbach, C. Sorenson, R. Golan, P. A. Holden, Probing biopolymers with the atomic force microscope: a review, *J. Biomater. Sci. Polymer Edition* 2000, *11*, 675–683.

[14] T. E. Fisher, M. Carrion-Vazquez, A. F. Oberhauser, H. Li, P. E. Marszalek, J. M. Fernandez, Single molecular force spectroscopy of modular proteins in the nervous system, *Neuron* 2000, *27*, 435–446.

[15] T. E. Fisher, P. E. Marszalek, J. M. Fernandez, Stretching single molecules into novel conformations using the atomic force microscope, *Nature Struct. Biol.* 2000, *7*, 719–724.

[16] A. Engel, D. J. Muller, Observing single biomolecules at work with the atomic force microscope, *Nature Struct. Biol.* 2000, *7*, 715–718.

[17] W. R. Bowen, R. W. Lovitt, C. J. Wright, Application of atomic force microscopy to

the study of micromechanical properties of biological materials [Review], *Biotechnol. Lett.* **2000**, *22*, 893–903.

[18] H. G. Hansma, Surface Biology of DNA by Atomic Force Microscopy, *Annu. Rev. Phys. Chem.* **2001**, *52*, 71–92.

[19] H. G. Hansma, L. I. Pietrasanta, R. Golan, J. C. Sitko, M. Viani, G. Paloczi, B. L. Smith, D. Thrower, P. K. Hansma, Recent Highlights from Atomic Force Microscopy of DNA, *Biol. Struct. Dynamics, Conversation* **2000**, *11*, 271–276.

[20] H. G. Hansma, D. O. Clegg, E. Kokkoli, E. Oroudjev, M. Tirrell, Analysis of matrix dynamics by atomic force microscopy, in: J. Adams (ed.), *Methods in Cell Biology*, vol. 69, San Diego, Academic Press, **2002**, pp. 163–193.

[21] H. G. Hansma, L. Pietrasanta, Atomic force microscopy and other scanning probe microscopies, *Curr. Opin. Chem. Biol.* **1998**, *2*, 579–584.

[22] N. Kaji, M. Ueda, Y. Baba, Direct measurement of conformational changes on DNA molecule intercalating with a fluorescence dye in an electrophoretic buffer solution by means of atomic force microscopy, *Electrophoresis* **2001**, *22*, 3357–3364.

[23] D. V. Klinov, I. V. Lagutina, V. V. Prokhorov, T. Neretina, P. P. Khil, Y. B. Lebedev, D. I. Cherny, V. V. Demin, E. D. Sverdlov, High resolution mapping DNAs by R-loop atomic force microscopy, *Nucleic Acids Res.* **1998**, *26*, 4603–4610.

[24] M. Koping-Hoggard, I. Tubulekas, H. Guan, K. Edwards, M. Nilsson, K. M. Varum, P. Artursson, Chitosan as a nonviral gene delivery system. Structure–property relationships and characteristics compared with polyethylenimine in vitro and after lung administration in vivo, *Gene Ther.* **2001**, *8*, 1108–1121.

[25] M. Rief, H. Clausen-Schaumann, H. E. Gaub, Sequence-dependent mechanics of single DNA molecules, *Nature Struct. Biol.* **1999**, *6*, 346–349.

[26] X. E. Cai, J. Yang, Molecular forces for the binding and condensation of DNA molecules, *Biophys. J.* **2002**, *82*, 357–365.

[27] B. Sampaolese, A. Bergia, A. Scipioni, G. Zuccheri, M. Savino, B. Samori, P. De Santis, Recognition of the DNA

sequence by an inorganic crystal surface, *Proc. Natl. Acad. Sci. USA* **2002**, *99*, 13566–13570.

[28] R. Krautbauer, L. H. Pope, T. E. Schrader, S. Allen, H. E. Gaub, Discriminating small molecule DNA binding modes by single molecule force spectroscopy, *FEBS Lett.* **2002**, *510*, 154–158.

[29] J. H. Hafner, C. L. Cheung, A. T. Woolley, C. M. Lieber, Structural and functional imaging with carbon nanotube AFM probes, *Prog. Biophys. Mol. Biol.* **2001**, *77*, 73–110.

[30] H. G. Hansma, Varieties of imaging with scanning probe microscopes, *Proc. Natl. Acad. Sci. USA* **1999**, *96*, 14678–14680.

[31] Z. Shao, Y. Zhang, Biological cryo atomic force microscopy: a brief review, *Ultramicroscopy* **1996**, *66*, 141–152.

[32] D. M. Czajkowsky, Z. Shao, Submolecular resolution of single macromolecules with atomic force microscopy, *FEBS Lett.* **1998**, *430*, 51–54.

[33] D. J. Muller, A. Engel, Conformations, flexibility, and interactions observed on individual membrane proteins by atomic force microscopy, *Methods Cell Biol.* **2002**, *68*, 257–299.

[34] D. M. Czajkowsky, Z. Shao, Supported lipid bilayers as effective substrates for atomic force microscopy, *Methods Cell Biol.* **2002**, *68*, 231–241.

[35] M. Guthold, X. Zhu, C. Rivetti, G. Yang, N. H. Thomson, S. Kasas, H. G. Hansma, B. Smith, P. K. Hansma, C. Bustamante, Direct observation of one-dimensional diffusion and transcription by *Escherichia coli* RNA polymerase, *Biophys. J.* **1999**, *77*, 2284–2294.

[36] S. Kasas, N. H. Thomson, B. L. Smith, H. G. Hansma, X. Zhu, M. Guthold, C. Bustamante, E. T. Kool, M. Kashlev, P. K. Hansma, *E. coli* RNA polymerase activity observed using atomic force microscopy, *Biochemistry* **1997**, *36*, 461–468.

[37] C. Bustamante, S. B. Smith, J. Liphardt, D. Smith, Single-molecule studies of DNA mechanics, *Curr. Opin. Struct. Biol.* **2000**, *10*, 279–285.

[38] M. B. Viani, L. I. Pietrasanta, J. B. Thompson, A. Chand, I. C. Gebeshuber, J. H. Kindt, M. Richter, H. G. Hansma, P. K. Hansma, Probing protein–protein

interactions in real time, *Nature Struct. Biol.* **2000**, *7*, 644–647.

[39] L. Bousset, N. H. Thomson, S. E. Radford, R. Melki, The yeast prion Ure2p retains its native alpha-helical conformation upon assembly into protein fibrils in vitro, *EMBO J.* **2002**, *21*, 2903–2911.

[40] D. Fotiadis, Y. Liang, S. Filipek, D. A. Saperstein, A. Engel, K. Palczewski, Atomic-force microscopy: Rhodopsin dimers in native disc membranes, *Nature* **2003**, *421*, 127–128.

[41] D. J. Muller, M. Kessler, F. Oesterhelt, C. Moller, D. Oesterhelt, H. Gaub, Stability of bacteriorhodopsin alpha-helices and loops analyzed by single-molecule force spectroscopy, *Biophys. J.* **2002**, *83*, 3578–3588.

[42] E. Oroudjev, J. Soares, S. Arcidiacono, J. B. Thompson, S. A. Fossey, H. G. Hansma, Segmented nanofibers of spider dragline silk: atomic force microscopy and single-molecule force spectroscopy, *Proc. Natl. Acad. Sci. USA* **2002**, *99*, 6460–6465.

[43] E. Oroudjev, C. Y. Hayashi, J. Soares, S. Arcidiacono, S. A. Fossey, H. G. Hansma, Nanofiber Formation in Spider Dragline-Silk as Probed by Atomic Force Microscopy and Molecular Pulling, in: D. A. Bonnell, J. Piqueras, A. P. Shreve, F. Zypman (eds), *Spatially Resolved Characterization of Local Phenomena in Materials and Nanostructures*, vol. 738, Boston, MA, **2003**, in press.

[44] N. Becker, E. Oroudjev, S. Mutz, J. P. Cleveland, P. K. Hansma, C. Y. Hayashi, D. E. Makarov, H. G. Hansma, Molecular nanosprings in spider capture silk threads, *Nature Mater.* **2003**, *2*, 278–283.

[45] A. Kempe, J. W. Schopf, W. Altermann, A. B. Kudryavtsev, W. M. Heckl, Atomic force microscopy of Precambrian microscopic fossils, *Proc. Natl. Acad. Sci. USA* **2002**, *99*, 9117–9120.

[46] F. Oesterhelt, D. Oesterhelt, M. Pfeiffer, A. Engel, H. E. Gaub, D. J. Muller, Unfolding pathways of individual bacteriorhodopsins, *Science* **2000**, *288*, 143–146.

[47] T. R. Serio, A. G. Cashikar, A. S. Kowal, G. J. Sawicki, J. J. Moslehi, L. Serpell, M. F. Arnsdorf, S. L. Lindquist, Nucleated conformational conversion and the replication of conformational information by a prion determinant, *Science* **2000**, *289*, 1317–1321.

[48] C. E. Diebel, R. Proksch, C. R. Green, P. Neilson, M. M. Walker, Magnetite defines a vertebrate magnetoreceptor, *Nature* **2000**, *406*, 299–302.

[49] L. K. Nielsen, T. Bjornholm, O. G. Mouritsen, Fluctuations caught in the act, *Nature* **2000**, *404*, 352.

[50] S. R. Whaley, D. S. English, E. L. Hu, P. F. Barbara, A. M. Belcher, Selection of peptides with semiconductor binding specificity for directed nanocrystal assembly, *Nature* **2000**, *405*, 665–668.

[51] S. T. Yau, P. G. Vekilov, Quasi-planar nucleus structure in apoferritin crystallization, *Nature* **2000**, *406*, 494–497.

[52] H. Seelert, A. Poetsch, N. A. Dencher, A. Engel, H. Stahlberg, D. J. Muller, Structural biology. Proton-powered turbine of a plant motor, *Nature* **2000**, *405*, 418–419.

[53] J. J. Cornelissen, J. J. Donners, R. de Gelder, W. S. Graswinckel, G. A. Metselaar, A. E. Rowan, N. A. Sommerdijk, R. J. Nolte, Beta-helical polymers from isocyanopeptides, *Science* **2001**, *293*, 676–680.

[54] A. Y. Kasumov, M. Kociak, S. Gueron, B. Reulet, V. T. Volkov, D. V. Klinov, H. Bouchiat, Proximity-induced superconductivity in DNA, *Science* **2001**, *291*, 280–282.

[55] S. K. Lower, M. F. Hochella, Jr., T. J. Beveridge, Bacterial recognition of mineral surfaces: nanoscale interactions between *Shewanella* and alpha-FeOOH, *Science* **2001**, *292*, 1360–1363.

[56] D. Y. Takamoto, E. Aydil, J. A. Zasadzinski, A. T. Ivanova, D. K. Schwartz, T. Yang, P. S. Cremer, Stable ordering in Langmuir-Blodgett films, *Science* **2001**, *293*, 1292–1295.

[57] C. A. Orme, A. Noy, A. Wierzbicki, M. T. McBride, M. Grantham, H. H. Teng, P. M. Dove, J. J. DeYoreo, Formation of chiral morphologies through selective binding of amino acids to calcite surface steps, *Nature* **2001**, *411*, 775–779.

[58] J. B. Thompson, J. H. Kindt, B. Drake, H. G. Hansma, D. E. Morse, P. K. Hansma, Bone indentation recovery time correlates with bond reforming time, *Nature* **2001**, *414*, 773–776.

[59] J. J. Kim, Y. Choi, S. Suresh, A. S. Argon, Nanocrystallization during nanoindentation of a bulk amorphous metal alloy at room temperature, *Science* **2002**, *295*, 654–657.

[60] K. B. Lee, S. J. Park, C. A. Mirkin, J. C. Smith, M. Mrksich, Protein nanoarrays generated by dip-pen nanolithography, *Science* **2002**, *295*, 1702–1705.

[61] S. W. Lee, C. Mao, C. E. Flynn, A. M. Belcher, Ordering of quantum dots using genetically engineered viruses, *Science* **2002**, *296*, 892–895.

[62] H. Li, W. A. Linke, A. F. Oberhauser, M. Carrion-Vazquez, J. G. Kerkvliet, H. Lu, P. E. Marszalek, J. M. Fernandez, Reverse engineering of the giant muscle protein titin, *Nature* **2002**, *418*, 998–1002.

[63] H. Yan, X. Zhang, Z. Shen, N. C. Seeman, A robust DNA mechanical device controlled by hybridization topology, *Nature* **2002**, *415*, 62–65.

[64] J. G. Forbes, G. H. Lorimer, Structural biology. Unraveling a membrane protein, *Science* **2000**, *288*, 63–64.

[65] R. F. Service, DNA imaging. Getting a feel for genetic variations, *Science* **2000**, *289*, 27–28.

[66] D. R. Meldrum, TechSight. Sequencing genomes and beyond, *Science* **2001**, *292*, 515–517.

[67] D. K. Newman, Microbiology. How bacteria respire minerals, *Science* **2001**, *292*, 1312–1313.

[68] S. Bunk, Better microscopes will be instrumental in nanotechnology development, *Nature* **2001**, *410*, 127–129.

[69] M. Rief, M. Gautel, F. Oesterhelt, J. M. Fernandez, H. E. Gaub, Reversible unfolding of individual titin immunoglobulin domains by AFM, *Science* **1997**, *276*, 1109–1112.

[70] P. E. Marszalek, H. Lu, H. B. Li, M. Carrion-Vazquez, A. F. Oberhauser, K. Schulten, J. M. Fernandez, Mechanical unfolding intermediates in titin modules, *Nature* **1999**, *402*, 100–103.

[71] M. B. Viani, T. E. Schaeffer, G. T. Paloczi, L. I. Pietrasanta, B. L. Smith, J. B. Thompson, M. Richter, M. Rief, H. E. Gaub, K. W. Plaxco, A. N. Cleland, H. G. Hansma, P. K. Hansma, Fast imaging and fast force spectroscopy of single biopolymers with a new atomic force microscope designed for

small cantilevers, *Rev. Sci. Instruments* **1999**, *70*, 4300–4303.

[72] P. M. Williams, S. B. Fowler, R. B. Best, J. L. Toca-Herrera, K. A. Scott, A. Steward, J. Clarke, Hidden complexity in the mechanical properties of titin, *Nature* **2003**, *422*, 446–449.

[73] L. M. Demers, D. S. Ginger, S. J. Park, Z. Li, S. W. Chung, C. A. Mirkin, Direct patterning of modified oligonucleotides on metals and insulators by dip-pen nanolithography, *Science* **2002**, *296*, 1836–1838.

[74] B. P. Jena, J. K. H. Horber, *Atomic Force Microscopy in Cell Biology*, vol. 68. Amsterdam: Academic Press, **2002**.

[75] J. H. Kindt, J. C. Sitko, L. I. Pietrasanta, E. Oroudjev, N. Becker, M. B. Viani, H. G. Hansma, Methods for biological probe microscopy in aqueous fluids, *Methods Cell Biol.* **2002**, *68*, 213–229.

[76] C. A. Siedlecki, R. E. Marchant, Atomic force microscopy for characterization of the biomaterial interface, *Biomaterials* **1998**, *19*, 441–454.

[77] B. Drake, C. B. Prater, A. L. Weisenhorn, S. A. Gould, T. R. Albrecht, C. F. Quate, D. S. Cannell, H. G. Hansma, P. K. Hansma, Imaging crystals, polymers, and processes in water with the atomic force microscope, *Science* **1989**, *243*, 1586–1589.

[78] H. G. Hansma, I. Revenko, K. Kim, D. E. Laney, Atomic force microscopy of long and short double-stranded, single-stranded and triple-stranded nucleic acids, *Nucleic Acids Res.* **1996**, *24*, 713–720.

[79] M. B. Viani, T. E. Schaffer, A. Chand, M. Rief, H. E. Gaub, P. K. Hansma, Small cantilevers for force spectroscopy of single molecules, *J. Appl. Physics* **1999**, *86*, 2258–2262.

[80] X. S. Xie, H. P. Lu, Single-molecule enzymology, *J. Biol. Chem.* **1999**, *274*, 15967–15970.

[81] Q. F. Xue, E. S. Yeung, Differences in the chemical reactivity of individual molecules of an enzyme, *Nature* **1995**, *373*, 681–683.

[82] P. R. Bianco, L. R. Brewer, M. Corzett, R. Balhorn, Y. Yeh, S. C. Kowalczykowski, R. J. Baskin, Processive translocation and DNA unwinding by individual RecBCD enzyme molecules, *Nature* **2001**, *409*, 374–378.

[83] X. Zhuang, H. Kim, M. J. Pereira, H. P. Babcock, N. G. Walter, S. Chu, Correlating structural dynamics and function in single

ribozyme molecules, *Science* **2002**, *296*, 1473–1476.

[84] http://millenniumindicators.un.org/unsd/ mi/mi_dict_xrxx.asp?def_code=491.

[85] K. Annan, A challenge to the world's scientists, *Science* **2003**, *299*, 1485.

[86] J. C. Sitko, E. M. Mateescu, H. G. Hansma, Sequence-dependent DNA condensation and the electrostatic zipper, *Biophys. J.* **2003**, *84*, 419–431.

[87] A. A. Kornyshev, S. Leikin, Electrostatic zipper motif for DNA aggregation, *Physical Rev. Lett.* **1999**, *82*, 4138–4141.

[88] D. J. Mueller, H.-J. Sass, S. Mueller, G. Bueldt, A. Engel, Surface structures of na-tive bacteriorhodopsin depend on the molecular packing arrangement in the membrane, *J. Mol. Biol.* **1999**, *285*, 1903–1909.

[89] E. Oroudjev, C. Y. Hayashi, J. Soares, S. Arcidiacono, S. A. Fossey, H. G. Hansma, Nanofiber Formation in Spider Dragline-Silk as Probed by Atomic Force Microscopy and Molecular Pulling, presented at MRS Fall Meeting, Boston, MA, **2002**.

[90] J. T. Prince, K. P. McGrath, C. M. Digiolamo, D. L. Kaplan, Construction, cloning, and expression of synthetic genes encoding spider dragline silk, *Biochemistry* **1995**, *34*, 10879–10885.

25
Force Spectroscopy

Markus Seitz

25.1
Overview

For many biological molecules, force is an important functional and structural parameter. The measurement of mechanical forces at the molecular level may thus provide detailed insights into the function and structure of many biological systems [1–6]. The binding potentials of receptor–ligand pairs involved in cell adhesion, protein folding pathways, DNA mechanics, and the function of molecular motors raise important questions which have stimulated the recent instrumental development of techniques for the precise application and measurement of minute forces. Fundamental intra- and intermolecular interactions can now be studied directly at the molecular level, and after its rapid evolution during the past decade, single molecule force spectroscopy has become a versatile analytical approach for the structural and functional investigation of single biomolecules in their native environments. By addressing individual molecules, it has become possible to go beyond the ensemble average, as one may now directly study transient intermediate states and resolve the individuality of reaction pathways. This has also stimulated new theoretical approaches for the understanding of the complex and dynamic interactions in biological processes.

Today, a number of techniques differing in force- and dynamical ranges are available, the most prominent of which are magnetic beads [7], optical tweezers [8], glass microneedles [9], the biomembrane force probe (BFP) [10], and the atomic force microscope (AFM) [11]. With the accessible force window, the whole range from entropic forces at several femtonewton (fN) [7] to the rupture of covalent bonds at a few nanonewton (nN) [12] can be investigated (see Figure 25.1 and Table 25.1).

One may – somewhat arbitrarily – distinguish between intra- and intermolecular forces acting within and between biomolecules. Hereby, "intramolecular forces" are considered to be the entropic and enthalpic elasticity of a biopolymer chain, the latter also including specific structural, for example, conformational, changes. The underlying molecular processes typically follow equilibrium pathways under the conditions of the stretching experiment. If the underlying molecular interactions are transmitted through the surrounding medium (at least, at a certain point of the process), as for unbinding and desorption pro-

Nanobiotechnology. Edited by Christof Niemeyer, Chad Mirkin
Copyright © 2004 WILEY-VCH Verlag GmbH & Co. KgaA, Weinheim
ISBN 3-527-30658-7

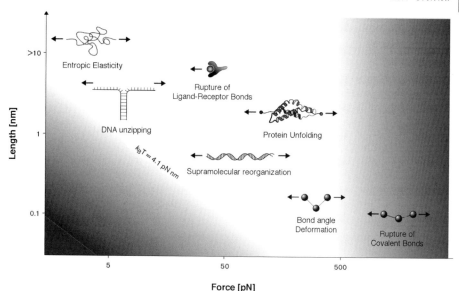

Figure 25.1 Typical forces and lengths scales in single-molecule force spectroscopy. The experimental accessibility of mechanical information is limited to the light areas of the plot. The shaded area in the lower left corner indicates the region of limited thermal stability of molecular structures (length multiplied by force = thermal energy, $k_BT = 4.1 \times 10^{-21}$ J = 4.1 pN nm at room temperature). The upper limit to the accessible experimental force range is determined by the rupture of covalent bonds at forces of a few nanonewton. (Figure adapted from Ref. [3].)

cesses, the resulting forces will be classified as "intermolecular". This includes the rupture of covalent chemical bonds, as well as the unbinding of different kinds of intermolecular aggregates based on noncovalent binding interactions as in ligand–receptor pairs, coordination complexes, hydrogen-bonded systems, ion pairs, or hydrophobically assembled structures. In the context of this definition, consequently, the unfolding of protein domains is an intermolecular process (although all the interactions are between segments of the same molecule). Unbinding processes proceed under highly nonequilibrium conditions whenever the natural off-rate of the binding interactions is much lower than the maximum force loading rates applicable in the experiment. This may lead to a pronounced time- (or rate-) dependence of measured unbinding forces, as will be discussed below.

25.1.1
Dynamic Force Spectroscopy of Specific Biomolecular Bonds

The specific binding of a ligand molecule to a receptor protein is a basic principle of cell adhesion and many other biomolecular recognition processes [13]. To name just two examples, the immunoresponse or the communication between nerve cells rely on highly specific molecular interactions, which allow each antibody or messenger substance to find their correct targets. However, in addition to chemical specificity, it is also the hier-

Table 25.1 Comparison of techniques used for the mechanical characterization of biomolecules. Typical values for accessible force-range and experimental time-window are given.

Method	Accessible range of forces [pN]	Spring-constants [N m^{-1}]; accessible time window	Minimum displacement [nm]	Practical advantages	Typical applications
Magnetic beads	0.01–100	NA; 0.1–1 Hz	10	Ability to induce torque	DNA stretching and twisting
Optical tweezers	0.1–150	10^{-3}–10^{-9} ≥ 10 ms	1	High force resolution	DNA, protein unfolding, molecular motors
Microneedles	>0.1	10^{-6}–10^{-1}; ≥ 100 ms	1	Good operator control, Soft spring constants	Actin, DNA stretching, unzipping and twisting
BFP	0.5–1000	10^{-4}–10^{-2}; ≥ 1 ms	5	Tunable spring constant, Broad dynamical range	Receptor–ligand pairs, membrane anchors
AFM cantilevers	>1	0.01–100 N m^{-1}; 0.01 ms–10 s	0.1	High spatial resolution, Commercially available	Bond strengths, (Bio-)polymer stretching
Computational	Full range	NA; ≤ 10 ps	Full range	atomic resolution	Bond strengths, (Bio-)polymer stretching

NA, Data not available.

archy of binding strengths that determines the proper function of many biological signaling sequences.

The strength of an isolated molecular bond can be represented by the maximum force, F_{max}, that the bond can withstand before it breaks. However, it has been predicted by Bell in his famous article on cell adhesion some 25 years ago that bond strength depends not only on the equilibrium binding energy but also on the temperature and the timescale of the measurement [14].

For any molecular bond, there is a finite probability that it acquires sufficient thermal energy to overcome the activation barrier for unbinding, even in the absence of an external pulling force. The lifetime of the bond t (or its inverse, the bond's off-rate, k_{off}) is a measure of the probability of the unbinding process. A constant stretching force acting along the unbinding pathway introduces additional potential energy, effectively tilting the potential landscape of the system along the stretching coordinate (Figure 25.2). This lowers the transition barrier to the unbound state, which is naturally located at larger separations than the bound state, and thus favored by pulling the binding partners apart. As a result, the rate of bond rupture, $k_{off}(F)$, increases with the applied force. In general,

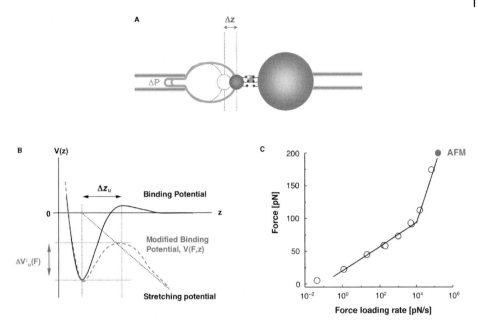

Figure 25.2 (A) The biomembrane force probe (BFP) technique. A microbead is attached to the apex of a red blood cell or a phospholipid vesicle, which is held in place by a micropipette. The suction pressure, ΔP, which directly relates to the membrane tension of the membrane capsule. Adhesive binding of the microbead to a second surface via interacting ligand and receptor molecules at the two opposing surfaces results in a deformation of the membrane capsule. This is utilized to measure the interaction force. (B) The potential energy curve of a single molecular bond. The binding potential is affected by a mechanical force field acting along the binding coordinate, z, which lowers the activation barrier for unbinding, $\Delta V^{\ddagger}_u$, by the mechanical stretching energy $F \cdot \Delta z_u$. (C) Dynamic strength spectrum of the biotin–streptavidin bond as measured by the BFP technique. In this particular case, the slopes of the solid lines correspond to activation barriers at $\Delta z_{u,1} = 0.5$ nm and $\Delta z_{u,2} = 0.12$ nm along the stretching coordinate. Data measured with the AFM are consistent with the high forces regime of this spectrum. (For details, cf. Ref. [19], from which the graph was adapted.)

any bond will break under any force if the measuring time is long enough, that is, approximately the bond lifetime, $\tau(F)$. Therefore, the apparent bond strength as given by the rupture force decreases with the time that the stretching force is applied. Bond rupture forces are therefore no equilibrium, or time-independent values, but depend on the intrinsic lifetime of the bond, the temperature, and on the measurement time.

The Bell model does not account for the stochastic nature of single bond rupture, and it cannot predict the actual distributions of measured bond strengths around the average value as they are experimentally observed. These statistical fluctuations arise from random fluctuations of the system in its equilibrium state and must be distinguished from the previously discussed time-dependence of the average value. Recent theoretical models for single bond rupture, further take into account viscous dissipation due to damping by the surrounding fluid and include a more realistic description of the stretching experiments, in which the force is not applied instantaneously, but increases at some finite rate until bond rupture eventually occurs [15–17]. Thus, they predict probability distributions around a most probable bond rupture force as a function of a slowly increasing stretching force.

If the unbinding trajectory is characterized by a single barrier located at Dz_u from the bound state, the dissociation rate will increase linearly with the logarithm of the applied force. If the force is gradually increased in the experiment, the bond dissociation rate also continues to increase with time. Under these conditions, the most probable rupture force varies with the loading rate according to

$$F = \frac{k_B T}{\Delta z_u} \ln \left(\frac{R_F \cdot \Delta z_u \, \tau_o}{k_B T} \right) \tag{1}$$

in which τ_0 is the intrinsic lifetime of the unperturbed bond, and R_F is the force loading rate [15].

While early experiments investigating receptor–ligand interactions have reported bond-rupture forces for fixed pulling velocities and spring constants [18], it is thus now generally appreciated, that with the bond dissociation being a nonequilibrium dynamical process, the force loading rate must be considered. The experimental demonstration was given by a variation of the force loading rate applied to biotin–streptavidin (and avidin) pairs over six orders of magnitude employing the biomembrane force probe technique [19]. The corresponding variation in bond rupture force revealed a detailed picture of the binding potential. Similar experiments have been reported investigating the protein A–IgG bond [20, 21] and the extraction of single lipid anchors from biomembranes [22]. Following early work on the rupture of complementary DNA strands [23, 24], meanwhile, their unbinding forces have been determined as a function of the force loading rate as well as the number of base pairs [25, 26].

Equation (1) allows us to determine the position of the activation barrier, Δz_u, from the slope of a F versus $\ln(R_F)$ plot [19]. The given relationship between detachment force and activation barrier further suggests that the variation of the transducer's spring constants and force loading rates in dynamic force spectroscopy can be used even to reveal spatially resolved information required for the reconstruction of more complex potential energy surfaces of intermolecular bonds [27]. Kinetic processes that involve molecules such as proteins, nucleic acids, or ligand–receptor pairs may exhibit multiple local maxima and minima along the reaction pathway. However complex, there will be in most cases a unique way connecting the bound and unbound state along which the system crosses the lowest possible energy barriers. In such cases, F versus $\ln(R_F)$ plots will exhibit a sequence of lines with different slopes, each one mapping the position of a particular energy barrier in the unbinding path [17, 19] (Figure 25.2). Recent theoretical models go beyond this single path picture by analyzing the force-induced dissociation of molecular adhesion complexes along alternative trajectories in a multidimensional energy landscape [28].

Despite this tremendous progress, details of ligand–receptor bond separation at the atomic scale are far from being resolved experimentally. Molecular dynamics simulations have thus been employed to theoretically study the rate-dependent unbinding of ligand–receptor pairs. From comparison with experimental data for the most prominent example, the streptavidin–biotin pair, a very detailed microscopic model for the unbinding along the highly complex reaction path has evolved [29].

25.1.2
Force Spectroscopy and Force Microscopy of Cell Membranes

Cell adhesion measurements employing force probes [4, 13, 30–33] have recently evolved towards in-vivo measurements at the single-molecule level [34, 35]: in the first example, genetic controls have helped to regulate the adhesion protein contact site A [34]. But aside from the mechanical stability of ligand–receptor bonds, the localization of binding sites by a ligand via molecular recognition is of particular importance in nature. This situation may be simulated in force experiments by laterally scanning during force–distance cycles [36]. The simultaneous acquisition of information on topography and ligand–receptor interactions by the atomic force microscope has led to lateral force mapping tools of receptor-functionalized surfaces [37, 38] with the restriction that they were either slow in data acquisition or lacked high lateral resolution. The combination with dynamic force spectroscopy [39, 40] has led to two-dimensional recognition imaging, in which unbinding forces of individual receptor–ligand pairs are used to localize receptors on functionalized surfaces at a lateral resolution of a few nanometer and at frequencies about hundred times faster than in conventional force mapping [41]. Topography and recognition images can now be assigned directly to structures in the topography image. This presents new perspectives for nanometer-scale epitope mapping of biomolecules and for the localizing of receptor sites during biological or cellular processes [31].

For a quantitative and subsequent modeling of membrane-bound processes, the lateral interaction forces between various membrane components and their mobility must be known. Hereby, AFM technology also permits the measurement of viscoelastic properties of single cells or cell layers [42, 43]. However, even superior spatial and temporal resolution has been achieved by the development of the photonic force microscope (PFM), a particular application of the optical tweezers approach, which has been applied to image the membrane of developing neurons, to determine elasticity and viscosity of the plasma membrane as well as the rate of diffusion of single-membrane proteins within the membrane [44, 45].

25.1.3
Protein (Un-)folding

In contrast to the unbinding of specific bonds between two different biomolecules considered in the previous section, biopolymer chains may also backfold onto themselves as a result of "intermolecular" forces between their individual segments. Such intrachain interactions may turn out to be very complex, and may be directed by specific interactions between its monomer units (hydrophobic forces, ionic and hydrogen bonds), as is the case when a protein folds into its native conformation. The configurational space of proteins exhibits a large number of local energy minima, so that a large protein molecule may need up to minutes to find its global free-energy minimum in the course of folding. Not surprisingly, the theoretical prediction of native three-dimensional protein structures only from the knowledge of the amino acid sequence still remains one of the key problems in biophysics. During the past five years, mechanical protein unfolding experiments have provided new insights into the course of the (reversible) protein folding process [46–

51]. As the mechanical unfolding forces of protein domains typically lie in the range between a few ten to a few hundred piconewton (pN), the atomic force microscope has emerged as an ideal and most widely used instrument for this purpose [3, 49, 50, 52–54].

Apart from the general issue of how proteins fold into their native, functional conformations, force experiments have helped to investigate the mechanical function of proteins, namely muscle proteins, cytoskeletal proteins and proteins of the extracellular matrix. The first experiments on the modular protein titin [46–48], which consists of repeating globular domains, demonstrated this use of force spectroscopy for the determination of mechanical properties of individual proteins, and already to some extent, for the collection of structural information. They also indicated the tremendous potential of the technique for the study of protein-folding mechanisms, and hereby inspired new theoretical approaches to this problem [27, 55–57]. Some fundamental differences become apparent when comparing the forced mechanical unfolding described here and the "classical" experiment, in which a protein is unfolded by means of chemical denaturants or by exposure to heat. First of all, as may be expected from the previous discussion of receptor–ligand pairs above, the unfolding forces of protein domains also depend on the force-loading rate [48, 58, 59]. Thus, the folding free energies as obtained from denaturing experiments do not provide direct information on the mechanical stability of protein conformations under external force. In addition, a broad unfolding potential well with shallow slope will lead to a lower unfolding force than a narrow and steep potential well, even if the height of the activation barrier is the same. The relevant parameter is the unbinding length Δz_u (see Figure 25.2), which can be viewed as the length of a "lever arm" by which the stretching force is acting on the protein domain. This was shown experimentally for two protein constructs based on modular titin and spectrin repeats (Δz_u (spectrin) ≈ 1.5 nm; Δz_u (titin) ≈ 0.3 nm). While of comparable thermal stability, the forces needed for their mechanical unfolding at comparable pulling rates differed by an order of magnitude, $F_u \approx 200$–250 pN for the β-barrel structures of titin, and $F_u \approx 25$–35 pN for the triple α-helical bundle domains of spectrin [58, 60].

Finally, mechanically forced unfolding may proceed via a different path along the protein's configurational energy landscape than the thermal process. Force experiments may thus allow to control the folding pathway externally and explore regions of the energy landscape potentially distant from thermal folding pathways (Figure 25.3A). While a first comparison of chemical and mechanical unfolding results for a titin analogue suggested similar transition rates and the same transition states when extrapolated to comparable conditions [59], recent evidence has revealed that similarities of unfolding pathways may be coincidental [61, 62].

For the general use of force spectroscopy as a tool to study structurally more complex proteins (naturally not always consisting of a regular structure of modular domains, but of various subdomains), the most critical issue is the exact assignment of specific features in the recorded force traces to the individual secondary and/or tertiary structural elements in a folded protein. Hereby, genetic engineering of tailor-made polypeptides has contributed significantly to extending the scope of mechanical protein unfolding experiments [52, 63]. Specific proteins (or domains thereof), which may be too small or too unstable to be studied by themselves, can be introduced into a modular construct of multiple identical and well-characterized structures, such as the titin Ig domains. In the force

A

B

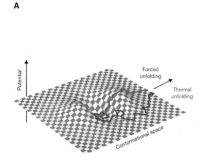

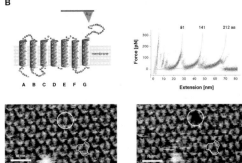

Figure 25.3 (A) Protein folding potential in conformational space. Although, in general, an externally applied force increases the unfolding rate of an individual protein domain, the forced unfolding pathway may deviate from the thermal pathway. This may add an extra contribution to the activation barrier along the forced unfolding pathway. (Figure reproduced from Ref. [3]; © Elsevier Science London, 2000). (B) Controlled extraction of an individual bacteriorhodopsin molecule from the purple membrane by a combination of AFM-based imaging and force spectroscopy [68]. Bacteriorhodopsin is a 248-amino acid membrane protein that consists of seven transmembrane α-helices (A–G), which are connected by loops. A terminal cysteine was introduced near the C-terminus to allow for specific attachment to a gold-coated AFM tip during the force spectroscopy experiment. The worm-like chain (WLC) analysis of several superimposed force curves allows for the correlation of the individual adhesive peaks with the unfolding of the protein's transmembrane helices. The comparison of two high-resolution images taken before and immediately after measuring a trace of adhesive force peaks proves the extraction of a single transmembrane protein from the membrane as indicated by the white circle. (Adapted from Ref. [68].)

traces, the unfolding characteristics of the smaller protein of interest can be distinguished easily within the known titin "fingerprint". By this approach, unfolding intermediates prior to the full unfolding of protein domains were first identified [64], and the introduction of specific mutant polypeptide sequences allowed for the determination of mechanical phenotypes that may play critical roles in protein unfolding diseases [65]. This strategy has also been employed for the study of the nonmechanical protein barnase, which apparently only resists much smaller forces than proteins with mechanical function [61].

Another elegant approach to the study of more complex tertiary structures is the extraction and unfolding of membrane proteins from their naturally ordered assemblies [66, 67]. It is then possible specifically to address and bind only those regions of the proteins, which are exposed at the cytoplasmic membrane surface, allowing for a most direct access to structural assignment of the individual protein domains to the measured unfolding peaks [68]. The ordering of proteins into an ordered 2-D structure within a membrane has also allowed to combine force spectroscopy on individual protein molecules with high-resolution AFM-imaging [67]. In a first example, individual hexagonally packed protomers were sequentially stretched, unfolded and removed from a bacterial surface layer until an entire bacterial pore formed by six merely identical protomers was unzipped. Subsequent imaging provided an exact correlation of the recorded force-extension curves with the tip-induced structural alterations [66]. In an extension to structurally more complex proteins, the complete extraction of bacteriorhodopsin from purple membrane patches revealed the individuality of the folding pathway of the structurally different subdomains (Figure 25.3B). Cleaved loops provided evidence for the stabilization of secondary structure by interhelix interactions [68], and a multitude of possible unfolding pathways with intermediate states was observed [67, 69].

The large forces that are required for mechanical protein unfolding can be explained by the fact that forced unfolding occurs under nonequilibrium conditions. In contrast to the ideal elasticity of polypeptide chains themselves [70], most of the deposited energy is dissipated, for instance, when the Ig domains of titin are ruptured which makes it such a good shock absorber in muscles. In the same way, it has been suggested that the dissipation of mechanical stress into the rupture of globular intra- and intermolecular polypeptide chain aggregates could be the origin for the unique fracture resistance of biominerals [71, 72]. This notion holds independently of specific binding interactions or the formation of regular aggregates. All that is required for the dissipation of a maximum of mechanical energy is a strong binding interaction which ruptures at a considerably high force in a nonequilibrium process. On the other hand, it was recently observed that the unfolding and refolding of the two-stranded coiled-coil structures in myosin occurs at equilibrium, demonstrating that protein aggregates may also build up truly elastic structures that can produce forces of up to 25 pN [73]. These findings may also provide interesting conceptual approaches for the materials scientist.

25.1.4
Elasticity of Individual Polymer Molecules

Flexible polymers (that is, all linear polymer molecules if the length scale is chosen large enough) adopt a random coil conformation in solution, and Brownian motion causes a permanent fluctuation of the molecule around a mean equilibrium conformation. The problem is closely related to problems such as random walks, diffusion, etc., and the mean values describing the conformation of a polymer chain in solution can be derived by a statistical approach. The classical partition function of the system is set up, from which the probability of configurations with, for example, the same specified end–end-distance, R, can be deduced. Based on this, the well-accepted concept of entropic polymer elasticity manifests itself in an effective restoring force upon stretching (or compressing) a flexible polymer chain by an external force field as a result of a loss in conformational freedom. However, as the full range from thermal fluctuations to the strength of individual chemical bonds is explored by single molecule force spectroscopy, the elastic profile of a polymer chain can be dominated by other elastic (namely enthalpic) contributions, especially in the high force range above a few hundred pN.

The two most simple theoretical models which are most often used to describe the force–extension profiles of individual polymer chains are the "freely-jointed chain" (FJC) model [74], and the "worm-like chain" (WLC, or Kratky-Porod) model [75]. In the FJC model, a polymer is described as a chain of N segments of equal length, l_K. The segments are freely jointed, and there are no restrictions to their spatial distribution such that each segment can point in every direction with equal probability. The WLC model entirely neglects any discrete structure along the chain, and describes the polymer as a continuous "rod" of constant bending module. The characteristic length scale denoting the flexibility of the polymer coil is the persistence length, L_P, which is defined as the decay length of the directional correlation along the polymer chain, and is also directly related to the bending module. In the limit of flexible chains ($L \gg L_P$), and for small forces, the persistence length equals half the segment length, l_K, of the FJC model [75].

From both models, theoretical relations between restoring force, F, and the polymer's end–end-distance along the stretching axis, R_z, can be deduced:

FJC model:
$$F(R_z) = \frac{k_B T}{l_K} L^{-1} \left(\frac{R_z}{N \cdot l_K} \right) \tag{2}$$

WLC model:
$$F = \frac{k_B T}{L_P} \cdot \left[\frac{R_z}{L} + \frac{1}{4(1 - R_z/L)^2} \, min \, \frac{1}{4} \right] \tag{3}$$

(L^{-1} is the inverse of the Langevin function, $L(x) = \coth(x) - x^{-1}$; also note that the exact solution for the WLC model can only be given numerically, so that Eq. (3) is only one analytical approximation [76] most frequently employed.)

However, it must be borne in mind that both models only consider entropic contributions to polymer elasticity, which does not hold true for stretching forces exceeding a few ten of pN. Especially in AFM experiments, which reach up to nanonewton forces at which even covalent bond rupture can be observed [12], the deformation of bond angles and the stretching of covalent bonds will result in an effective increase of the segment length, l_K, or the contour length, L, respectively. In a similar sense, it has to be stressed that not only the "classical" conformational flexibility, but also the damping of internal chain dynamics such as rotational, vibrational and librational modes may contribute significantly to entropic elasticity [77, 78].

If a more refined and detailed understanding of the molecular origin for single polymer elasticity is desired, more information on the chemical structure of the polymer as well as the solvent needs to be introduced in the theoretical models. Strictly, this means that an individual model needs to be established for each particular system, which can, in principle, be done in a most complete way by quantum chemical ab-initio approaches. But despite the recent implementation of solvent water molecules [79], their use is still limited to rather simple polymers. Also, molecular dynamics simulations have been successfully applied to explain polysaccharide elasticity [80]. On the other hand, more generalized descriptions of single chain elasticity based on a freely-rotating chain model have been introduced recently, which consider specific common features of a group of polymer backbones, and thus allow to introduce enthalpic elastic contributions such as arising from bond angle deformations [81]. The investigation of single chain elasticity continues to stimulate advancements in this area.

A better correlation between polymer elasticity and structure is often possible when specific features in the measured force–extension curve are observed, that is, when it deviates from the simple FJC or WLC behavior. Such features are often caused by conformational transitions within a polymer chain (or a superstructure thereof), upon which its elastic properties undergo a marked change. If such a transition is also connected to a considerable length change in the molecular structure and occurs in equilibrium on the time scale of the experiment, the force–distance curves may show a transition plateau of constant

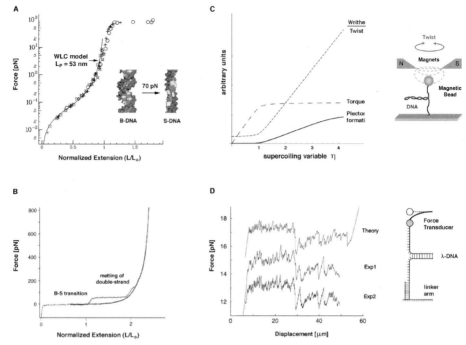

Figure 25.4 (A) Force versus extension data of a single double-stranded DNA molecule. At low forces the data can be fitted well with the WLC model, indicating entropic elasticity of DNA. Above 70 pN, the length of the molecule abruptly increases, corresponding to the structural transition between B-DNA and S-DNA. (Adapted from Ref. [4].) (B) At higher forces beyond the B–S transition, one finds mechanically induced melting of the DNA double strand [26]. Upon relaxation of the molecule, the "featureless" WLC curve of a single-stranded DNA molecule is measured. (C) Schematic view of the magnetic tweezers set-up for DNA twisting which is used for investigating plectoneme formation. The plot shows the dependence of plectoneme formation, the torque acting on the DNA, and the ratio of writhe to twist on the supercoiling in arbitrary units. (For details, see Ref. [4].) (D) Two force curves obtained during the mechanical unzipping of DNA employing microneedles obtained at two different unzipping velocities of 40 nm s^{-1} (Exp2) and 200 nm s^{-1} (Exp1), in comparison with a theoretical curve derived from the GC content of the molecule. For clarity, the curves 'Exp1' and 'Theory' were shifted by 2 and 4 pN, respectively. (Curves reproduced from Ref. [106].)

force over a certain distance range (see Figure 25.4B). Typical examples are polysaccharides and DNA [80, 82, 83].

25.1.5
DNA Mechanics

Double-stranded DNA (ds-DNA) was one of the first molecules investigated by single molecule force spectroscopy, and its elastic profile has been well-characterized in various experiments over a wide range of forces (Figure 25.4) [5, 7,82–88].

One of the most prominent mechanical features of the ds-DNA superstructure is a highly cooperative conformational transition from its natural form (B-DNA) to an overstretched and underwound conformation (S-DNA) at a force of approximately 70 pN, upon which the length of the molecule approximately doubles (Figure 25.4A and B) [82–84]. A number of theoretical models, as well as molecular dynamics simulations have investigated the molecular details of this overstretching transition [89–92]. At higher stretching forces beyond the B–S transition, a force-induced melting transition was found, during which double helical DNA splits up into two single strands [93]. The mechanical energy deposited in the DNA double helix before melting occurs agrees well with the base pairing free enthalpy of DNA under the given experimental conditions, and its variation with counter-ion concentration, temperature, and sequence agrees well with the base pairing free enthalpies under the given experimental conditions [26, 94, 95].

By employing the magnetic tweezers technique, it has even been possible to twist DNA molecules [96, 97], and to induce supercoiling and the formation of plectonemes (that is, twisted loops that branch laterally from the direct end-to-end path in the rod; Figure 25.4C). As more details about the mechanical properties of DNA have now been revealed, a complete picture has emerged which explains the coupling of stretch and twist, supercoiling, and overstretching mechanics, as well as base pairing forces, in a consistent way [5, 88, 94, 96–100]. Moreover, new detailed insights into the process of DNA condensation by multivalent cations have been obtained [101], and there have been first studies of the force loading rate dependence of RNA unfolding and refolding [102].

The accessibility of the full force range, including force-induced melting of DNA, has opened new perspectives in studying its interaction with anticancer drugs, after it was observed that cisplatin, which is known to form crosslinks by its binding to guanine, has a significant effect on DNA overstretching [103]. As more compounds are now being tested, force spectroscopy is used to discriminate between their different binding modes to DNA [104]. More experiments will be needed before a clear picture about the relationship between binding modes and pharmacological function will evolve, but there is justified hope that force spectroscopy may serve for the screening of DNA drugs.

Perhaps the most fascinating notion in this context is the mechanical sequencing of genomes [105]. The sequence-specific, base-pairing forces of DNA determined from the unzipping of lambda phage DNA with glass microneedles [106] and from the unzipping of synthetic DNA sequences with atomic force microscopy [93] and with laser tweezers [107], were found to be consistent and given as 9 pN for adenine–thymine base pairs and 20 pN for guanine–cytosine base pairs [93].

The first theoretical papers on this issue considered the difficulty of opening single pairs due to the elasticity of the single-strand segments created by the unzipping procedure, as well as to the influence of thermal agitation. However, neither of these models took into account the impact of a variable base sequence [105, 108]. When the corresponding experiment was carried out employing glass microneedles, it transpired that DNA opened in a stick-slip mechanism, with periods of only slight extension coupled to increasing molecular strain, interrupted by the sudden opening of segments containing tens to hundreds of base pairs [106, 109]. While DNA strands separated under an average force of 13 pN, force fluctuations on the order of 2 pN were found reproducible when strand separation was repeated for a single molecular construct. Furthermore, saw-tooth profiles along the force curve correlated with fluctuations in local GC content and agreed well with equilibrium statistical mechanics models of strand separation, into which sequence effects had been incorporated (Figure 25.4D) [109–111]. Meanwhile, DNA unzipping measurements have been performed with an optical trapping interferometer combining sub-pN force resolution and millisecond time resolution. With respect to the earlier studies, this resulted in a significant increase in basepair sensitivity so that, presently, sequence features appearing at a scale of ten basepairs can be resolved [107].

25.1.6
DNA–Protein Interactions

Following the mechanical investigation of the DNA molecule itself, dynamic studies of DNA and its interactions with proteins have recently come into the focus of biophysical research. An investigation of DNA interaction with the protein RecA, which is known to lengthen DNA up to a factor of 1.5 upon binding, has shown that binding of RecA to prestretched lambda phage DNA is largely accelerated [112]. As binding to AT-rich thus "softer" plasmid molecules occurs much faster than binding to stiffer sequences, it had been suggested that thermally induced stretching of DNA should be an important factor for RecA binding. More details about the kinetics of the RecA polymerization process were revealed using fragments of plasmid [113] as well as lambda phage DNA [114, 115].

Further, the activity of T7 DNA polymerase was determined as a function of the mechanical tension applied to the DNA template [116]. Since enzyme activity increased two-fold when the single-stranded DNA (ss-DNA) was prestretched to the length of ds-DNA, it was suggested that force needs to be generated by the enzyme during its rate-limiting step in order to adjust mechanically for the different lengths of ss-DNA and ds-DNA. Upon further stretching, the enzymatic DNA polymerization process slowed down until at high extensions of ss-DNA, the enzyme switched to its exonucleolysis mode of activity. In related studies, it was observed that *Escherichia coli* RNA polymerase exhibits pronounced variations in intrinsic transcription rates [117].

In another recent application of force spectroscopy on DNA–protein complexes, the HIV-1 nucleocapsid protein (NC) – a nucleic acid chaperone which is essential for HIV replication and facilitates the structural rearrangement of the nucleic acids to the lowest energy state – was proven to alter significantly the DNA overstretching transition. Recent results have now revealed that the zinc finger structures of the protein are essential to its capability to alter the helix-coil transition of nucleic acids, a crucial step in the reverse transcription process during HIV replication [118].

DNA compaction was investigated by addressing the mechanical properties of single chromatin fibers in different ionic environments [119]. Decondensation occurs at stretching forces of around 5 pN and removal of DNA from the core particles was found at forces above 20 pN. The internucleosomal attractive energy was determined as roughly 3.5 $k_B T$, suggesting a mechanism for the local access of trans-actin factors to chromatin in which two adjacent nucleosomes should be found in an open state for about 4 % of the time. The forces exerted by a bacteriophage portal motor when packaging DNA into a virus were found to be surprisingly high (up to 50 pN) [120]. It was thus suggested that internal pressure may provide the driving force for the injection of viral DNA into host cells.

Finally, the above-mentioned induction of plectonemic supercoils by winding a single DNA molecule with magnetic beads (Figure 25.4C) made it possible to study the relaxation of DNA supercoils induced by individual type II topoisomerase molecules. The enzyme's activity could be observed in real time, as each relaxation event was reflected by a 90-nm step in the extension of the DNA molecule, indicating the removal of two supercoils during a single enzyme turnover. The reaction rate as a function of external force was determined and revealed that increasing torque decreases turnover rates – an observation that was unprecedented in bulk [121]. Recently, this technique was applied to bacterial topoisomerases I and IV. For the latter, a chiral substrate specificity was found [122, 123]

25.1.7
Molecular Motors

Enzymes may be considered as small molecular machines that use chemical energy to perform specific tasks within a complex biochemical system such as our body, and often, these functions involve the creation of mechanical forces and motion. This is most obvious for motor molecules which are associated with intracellular trafficking, cell division, and muscle contraction. The diversity of biological functions of molecular motors and the complexity of molecular processes for energy conversion into active movement is still being explored. Most important are the microtubule-based kinesin motors and actin-based motors, which were previously considered to work by very different mechanisms, though it has now become clear that kinesin and myosin share a common core structure and convert energy from ATP into protein motion using a similar conformational change strategy [124]. With the instrumentational development towards the measurement and application of minute forces, it became possible to observe the stepwise motion of single kinesin molecules along microtubule tracks [125] The maximum force generated by the kinesin molecule to transport along a microtubule was determined between 5 and 7 pN [125, 126]. Further examples include the measurement of the displacement of actin filaments by single myosin molecules [127] and the discrete rotations of single F_1 subunits of the F_0F_1-ATP synthase [128, 129], as well as the torque–speed relationship of the flagellar rotary motor of *E. coli* [130, 131]. Hereby, microneedles and optical traps have been the techniques most successfully applied (Figure 25.5) [2, 6, 132]. Based on the experimental data, new general theories on the generation of mechanical force by molecular motors have evolved [133, 134].

A series of recent experiments on actomyosin motors have inspired a controversial debate about their mechanism of motion [135–141]. In the most widely accepted mechanistic model – the "lever-arm model" – a single cycle of actin attachment and power stroke for one myosin head involves the hydrolysis of one ATP molecule. Because of this straightforward relationship between energy input and mechanical action, the lever-arm theory is also called the "tight-coupling model" [127, 142, 143]. However, this model continues to be challenged as the existence of a "loose-coupling" model has also been suggested, in which the motor domain and/or the actomyosin interface enable myosins to produce larger processive steps during translocation along actin upon hydrolysis of only one ATP molecule [135, 141, 144]. The precise role of ATP in this model has not been fully clarified, but it might simply be responsible for the changing of the myosin's shape, and thus in-

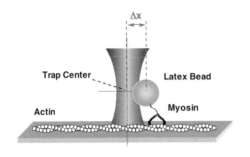

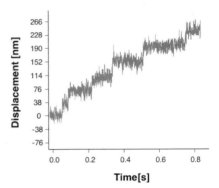

Figure 25.5 Experimental geometry used for the measurement of the movement of single myosin-V molecules along an actin filament. An optically trapped bead attached to the myosin molecule is moved near an actin filament immobilized on the surface of a glass slide. The myosin binds to the actin, and pulls the bead away from its trap center. A force feedback regulates the position of the optical trap, and the position of the pulled bead is plotted against time. Actin advances in steps of 38 nm, which are separated by dwell periods of approximately 100–200 ms. (Adapted from Ref. [136].)

itiate its motion. Much of the energy needed for the myosin movement against actin would then come from Brownian motion of the molecules. The biomolecular motors would effectively use thermal noise for their function, but it remains to be resolved how motion would be biased in one direction.

25.1.8
Synthetic Functional Polymers

After force spectroscopy has emerged from the study of biopolymers, its specific application to functional synthetic polymers has also opened a highly innovative field of research. While the specific questions being asked in materials research are often inspired by some specific application in the macroscopic world, and may thus be different from those in biophysical research, some of the experimental and theoretical understanding brought about by force spectroscopy on single biomolecules can be applied directly to "smart" synthetic materials. As mentioned above, the underlying physical concepts found in living organisms may stimulate the development of synthetic biomimetic polymers with unique elastic properties [71, 145], and the highly controversial issue of how biological molecular motors convert chemical energy into molecular movement is of fundamental importance. It will certainly inspire our designing of man-made nanoscopic machines, as related questions in materials science are just beginning to be explored at the single molecule level: What are the forces that drive structural transitions and how are they influenced by environmental changes? How can external stimuli be transformed into mechanical force at the single molecule level?

The latter question was recently addressed by investigating a synthetic photochromic polymer containing the bistable photosensitive azobenzene moiety. Individual polymer chains were found to lengthen and to contract by reversible optical switching of the azo groups between their trans and is configurations. The polymer was found to contract against an external force acting along the polymer backbone, thus delivering mechanical work. As a proof of principle, the polymer was operated in a periodic mode demonstrating for the first time opto-mechanical energy conversion in a single molecule device [146].

Further questions are: What determines the interaction between polymers and surfaces? How are colloidal assemblies held together by intermolecular (unspecific) interactions? What is the mechanical strength of covalent bonds and how could mechanical force be used as an energy source for chemical reactions? Again, for instance, the simple question of how a single polymer chain attaches and adheres to a solid surface is not only of fundamental basic interest, but related to many practical applications in materials science [54, 147–150].

25.2
Methods

In this section, the currently most prominent experimental tools for force spectroscopy will be summarized. In general, force spectroscopy experiments on single biomolecules should be performed under physiological conditions for which the molecules' functionality is retained, that is, in aqueous buffer solution and at temperatures between typically ~5 °C and ~50 °C. As seen in the previous section, the relevant forces range from ~10 femtonewton (fN) to a few hundred pN. Several probes capable of measuring such minute forces have been developed, of which here the mechanical transducers such as AFM cantilevers, microneedles, and the biomembrane force probe (micropipette aspiration technique), as well as small glass, polystyrene or metallic beads which are held and moved in space by external fields (optical and magnetic tweezers) will be discussed. All of these force probes are typically of micrometer size, so that the application and measurement of forces at the molecular level further requires the mechanical addressing ("grabbing") of individual molecules. In certain cases, this may require specific strategies for selective chemical or physical binding of the respective molecules to the force probe.

The relevant force ranges, minimum displacements, probe stiffness, typical applications and practical advantages of each technique are summarized in Table 25.1. The most important technical, experimental and historical aspects and current developments will be addressed in the brief sections below.

25.2.1
AFM Cantilevers

The first atomic force microscope was introduced in 1986 [11] as a tool for the imaging of surfaces with high resolution by employing a sharp tip (diameter approx. 3 μm, radius of curvature ~50 nm) at the end of a soft cantilever spring, which is typically 100–300 μm long and a few micrometers thick. During the imaging process, the vertical and lateral position of the cantilever with respect to the substrate is varied by the movement of a piezoelectric crystal supporting the sample or the tip, which allows for spatial control with sub-nanometer resolution. The interaction force between the cantilever tip and the surface is probed via optical detection of the cantilever deflection, that is, force is applied and sensed by the displacement of a bendable spring.

The first single-molecule force experiments using the AFM were performed on individual ligand–receptor pairs. A receptor-coated AFM-tip was brought into contact with a sample surface coated with the according ligand and then pulled off. Statistical analysis of the recorded force–distance traces allowed determination of the adhesion force of individual ligand–receptor pairs [18]. While these first experiments were performed with commercial instruments originally developed for image acquisition, custom-built instruments were soon optimized for single-molecule force spectroscopy by decoupling the z-movement of the cantilever from the other directions, and controlling the piezo's z-position by an interferometrically calibrated strain gauge. The major advantages of the AFM technique are its high spatial range and resolution, and the possibility of studying many different molecules immobilized on a single substrate surface. Typical applications lie in the high forces range (i. e., between ~10 pN and ~3 nN): the probing of ligand–receptor interactions, protein unfolding, as well as of single chain elasticity of biological and synthetic functional macromolecules with a particular emphasis on enthalpic contributions arising from conformational and configurational changes along the polymer backbone (e. g., DNA) [53, 54]. Meanwhile, the first highly accurate commercial instruments are now available which are designed specifically for force spectroscopy.

Nominal spring constants of most commonly used cantilevers are 10 to 500 mN m^{-1}, with typical resonance frequencies of 7 to 120 kHz in air (cantilevers for tapping mode imaging: ~300 kHz) and 1–30 kHz in water. The force resolution of the best instruments is only limited by thermal noise. For instance, for a cantilever with a spring constant of 10 mN m^{-1}, the root-mean-square force fluctuation is 6 pN at room temperature. Stiffer cantilevers have lower force sensitivity, while cantilever size determines the response times. A promising instrumental development was the recent fabrication of small, but soft cantilevers for AFM-based single-molecule force spectroscopy, which opens a broad dynamical range without increasing the spring constant. Using these cantilevers, the unfolding forces of titin domains were measured over a dynamical range of four orders of magnitude [151]. In order to decouple conservative and dissipative components of the forces associated with molecular stretching, a dynamic mode of force spectroscopy was developed which relies on actively controlling the quality factor of the cantilever in a buffer environment employing an appropriate feedback counteracting the hydrodynamic damping. This should, to some extent, allow the measurement of the effective viscosity of a single molecule [152].

Further important recent developments have been the implementation of a single-molecule force-clamp [153], as well as the coupling of optical excitation into the AFM experiment [146]. The first allows for the AFM measurement to be made under conditions of constant force, which can be used for studying the force dependence of the unfolding probability of protein modules. The latter has been introduced for the study of opto-mechanical energy transduction polymeric systems on the single-molecule level, but also widens the scope of the AFM technique towards the investigation of photoswitchable ligand–receptor systems, or the use of photoactivated compounds (such as caged calcium, protons, or ATP).

The particular power and versatility of atomic force microscopy is further exploited by combining force spectroscopy with its high precision in lateral scanning. High-resolution imaging of cell surfaces via ligand–receptor interactions at rapid acquisition rates was recently introduced [41]. In this technique, very gentle tip–surface interactions are provided when a magnetically coated and antibody functionalized tip is oscillated by an alternating magnetic field at an amplitude of 5 nm while being scanned along the surface. A polymeric tether connecting the antibody to the tip of 6 nm length provides an enhanced binding probability to an antigenic site. Commercial instruments with a 3-D closed loop feedback should allow for a precise spatial control in all three dimensions.

25.2.2
Microneedles

Glass microneedles are softer than AFM cantilevers as they are slightly greater in length (50–500 μm) but of smaller thickness (0.1–1 μm). They have a typical stiffness of $\sim$0.01 mN m^{-1}, so provided that the microneedle's displacement can be measured with a 10-nm resolution, the accessible forces lie in the fN range, and their softness makes them an excellent alternative for the probing of delicate biological samples. However, note that the large size of the glass fibers limits the experimental sampling rates. At higher frequencies such as 10–100 Hz, the force resolution is limited to a few pN. Also, microneedles are not commercially available, and the high-resolution displacement detection is usually not as straightforward as in the AFM technique. It may be achieved by imaging of the microneedle itself, or by employing optical fibers, and projecting the exiting light from their tips onto a photodiode. Applications of this technique are therefore less frequent. While there have been reports on their use in studying protein unfolding, they are more typically used in measuring molecular motor forces and DNA–protein interactions under mechanical stress [9, 99, 112, 135].

25.2.3
Optical Tweezers

Dielectric particles (or spheres) experience forces as light is scattered, emitted, or adsorbed by them, of which scattering and gradient forces are most important. They may be utilized to trap small objects in a focused laser beam, which are thus held by a spring-like force (applicable range: 0.1 to 100 pN) [8]. In early experiments, bacteria, yeast, or mammalian cells were directly mechanically manipulated, but in the recent investigation of DNA, DNA–protein complexes, and protein unfolding as well as biological molecular motors, the principal tools for the mechanical addressing of single molecules are small microspheres such as latex or glass beads, to which the individual molecules under investigation need to be chemically linked [6, 8,132].

The "spring constant" of an optical tweezer set-up depends on the bead size (usually comparable to the wavelength of light) and refractive index, as well as on the power of the laser beam (typically several tens of milliwatts) and the intensity distribution within the trap. A particular advantage of the optical tweezers technique is that the momentum transfer between the trapped object and the laser beam can be directly measured and correlated with displacement and force. On the other hand, laser damage of biologically active materials may prove disadvantageous, namely in the blue and ultraviolet range, and because water strongly adsorbs in the visible range, lasers employed in biophysical optical tweezers experiments operate in the near infrared (but even then, the active lifetime of enzymes may be significantly decreased).

If a single focused laser beam is used, light must be collected over a very large angular range so that the gradient force may overcome the scattering forces, which is only possible with lenses of high numerical aperture. As a result, particles can only be trapped in very close proximity to a glass slide, employing relatively high light intensities. A larger working distance and large size of the trapping zone can be achieved when two co-axial counter-propagating laser beams are used [154]. In such a dual-beam set-up the same trapping forces require a much lower light intensity, so that small transparent beads can be trapped with a force of ~100 pN under conditions that are gentle enough to avoid irradiation damage of the biomolecules. However, the need for elaborate optical alignment of the two laser beams is a disadvantage, so this method is only used when these benefits are of crucial importance to the experiments.

Various techniques may be utilized for the measurement of bead displacement. It can be done in a direct way by observing the bead movement with a camera or by projecting its image onto a four-quadrant photodiode; however, if ultra-fast displacement detection is desired, then interferometric techniques have proven most useful, where the trapping laser is also used as a displacement sensor. Hereby, frequencies of up to 100 kHz and a spatial resolution of better than 1 nm can be reached [155]. Recently, a trapping interferometer was introduced for determining DNA unzipping forces with high sequence sensitivity, which provides sub-pN resolution up to 100 pN [107].

One particular application of the optical tweezers technology is the recently developed photonic force microscopy (PFM) [44, 45]. Two-dimensional images are obtained by laterally scanning a trapped latex bead across biological samples while recording the bead's deflection from its resting position. Under experimental conditions, the maximum axial (imaging) force applied by the probe is well below 5 pN, and the lateral force is at maximum three-fold higher. It is possible to introduce molecular specificity to the sensor or even to use a single molecule as a sensor itself. Under appropriate conditions, molecular diffusion and mechanics can be studied at the scale of a few molecules.

25.2.4
Magnetic Tweezers

Magnetic fields, which rarely interact with biomolecules (an important requirement for minimizing artifacts), may be used to trap magnetic metallic particles. In an approach which is largely similar to that of optical tweezers, very stable and small forces can be applied to biomolecules in a magnetic field gradient when they are tethered to magnetic beads that are often 1–5 μm in size and consist of iron oxide microcrystals embedded in a polymer matrix. Small permanent magnets or electromagnets which can be moved and rotated are used to pull and rotate the microbead, and therefore to stretch and even twist biomolecules (the latter deformation mode is of particular importance during

DNA replication and transcription) [5, 96, 97, 156]. The forces applied in magnetic tweezers measurements are typically below 10 pN, but can be as small as a few fN, and are measured with a relative accuracy of ~10%. In contrast to the above-mentioned techniques, force measurements with magnetic tweezers do not require calibration of the sensor. They can be directly determined by analyzing the Brownian fluctuations of the tethered bead employing the equipartition theorem because, in contrast to optical tweezers, the variation of the trapping gradients occurs on the millimeter scale, so that the stiffness of the magnetic trap is much smaller than that of the molecule under investigation. Also, working at constant force is easily achieved with magnetic tweezers by keeping constant the position of the magnet (keeping constant the displacement of the force sensors is more elaborate in the previous techniques and typically requires an appropriate feedback). However, since stiffness depends on the force, spatial resolution of the magnetic tweezers technique is limited to ~10 nm below 1 pN.

25.2.5
Biomembrane Force Probe

In this technique, a red blood cell or a phospholipid vesicle is used as a mechanical transducer with tunable stiffness [10]. This is achieved by applying a suction pressure to a 10–20 μm vesicle with a micropipette (inner diameter 1–10 μm), which allows the vesicle to be positioned and a hydrostatic pressure difference to be set across its membrane. A microbead carrying the biomolecules of interest (e. g., ligand molecules) is glued to the apex of the biomembrane capsule opposing the pipette opening (see Figure 25.2A) and is brought into contact with a second surface carrying the binding partners (e. g., the corresponding receptor molecules). This second surface may either be the surface of a cell that is held by a second pipette, or a suitably functionalized plane glass substrate. A piezoelectric transducer is typically used to control the pipette position, but the determination of the bead distance from flat surfaces may also be made interferometrically with an accuracy of ~5 nm [13].

The suction pressure which is applied to the membrane capsule results in a mechanical surface tension, and this allows the capsule to resist any external forces that act on the microbead resulting from the ligand–receptor interactions under study. In-plane shear does not play a significant role for phospholipid membranes (blood cells relax by plastic flow at shear tensions >20 μN m^{-1}). The bending modules of biomembranes are also very low as long as the applied forces are distributed over a large membrane area by the microbead so that instabilities resulting from point forces are prevented (if pulled at moderate speeds, single lipid anchors are extracted from biomembranes at forces of a few tens of pN [22]).

The membrane capsule acts like a spring, for which it was shown that the force–extension relation is linear for extensions below 200 nm [157], and proportional to the suction pressure of the pipette. Its stiffness can be tuned during the experiment simply by changing the suction pressure, which can be easily applied and accurately measured in the range from 1 to 50 000 Pa. For red blood cells, the lower limit for the transducer stiffness is 45 μN m^{-1}. For phospholipid vesicles, there is practically no such lower limit, and binding forces of individual ligand–receptor bonds can be measured from below 1 pN to a few hundreds of pN [10, 157]. For extensions beyond 300 nm, the force–extension relationship becomes nonlinear, because membrane area and enclosed volume of the capsule are no longer preserved. While the convenient use of this technique is thus constrained to short extensions, a particular advantage is the high variability of force loading rates (approximately six orders of magnitude). It is thus ideally suited for the force-induced dissociation of single adhesive bonds [4, 19].

25.3
Outlook

Since its introduction about 15 years ago, single-molecule force spectroscopy has rapidly evolved, and is now becoming a standard technique for the structural and functional investigation of biomolecules in their native environments. As the field is gradually moving from the physics laboratories (where the instrumentation has largely been developed) to the life sciences laboratories, an increasing number of researchers with chemical, biological, medical, and pharmaceutical backgrounds are joining the field. Clearly, a trend towards the investigation of dynamical processes and the interaction of single molecules, not only with each other but also with their environment, can be observed. Dynamic force spectroscopy will continue to reveal more details about the binding potentials of receptor–ligand pairs, and hopefully force spectroscopy on genetically engineered protein domains will shed more light on the protein folding problem. The question of what makes molecular motors move also promises to provide further stimulating results in the future. As more DNA-binding proteins will be investigated, the role of mechanics in gene expression, regulation, and replication will be fully appreciated. Another promising approach will be the structural investigation of specific DNA sequences and its adducts with cancer drugs.

Taken together with new theoretical concepts of biomolecules under mechanical tension, single-molecule force spectroscopy will clearly generate new insights into force and its relation to structure and functions. It is a particularly fascinating goal to mimic these functions of biomolecular systems for the realization of artificial "smart" materials, such as new functional stimuli-responsive polymers capable of performing specific tasks. Such stimuli may be direct interactions with other molecules or external energy inputs. Also, based on our continuously improving understanding of the function of biomolecules, they will find increasing use in newly designed functional assemblies. The knowledge of molecular forces acting in such assemblies will be of central importance to nanobiotechnology, as we are looking forward to developing highly sensitive chemical and biological sensor devices as well as molecular switches and nanoscopic machines.

Acknowledgments

These studies were made possible by financial support from the Fonds der Chemischen Industrie. The author's own research is further funded by the Deutsche Forschungsgemeinschaft, the Volkswagen-Stiftung. Frequent discussions with Hermann Gaub, Matthias Rief, and Roland Netz have inspired the writing of this chapter.

References

[1] D. Bensimon, *Structure* **1996**, *4*, 885–889.

[2] C. Bustamante, J. C. Macosko, G. J. L. Wuite, *Nat. Rev. Mol. Cell. Biol.* **2000**, *1*, 130–136.

[3] H. Clausen-Schaumann, M. Seitz, R. Krautbauer, H. E. Gaub, *Curr. Opin. Chem. Biol.* **2000**, *4*, 524–530.

[4] R. Merkel, *Physics Reports* **2001**, *346*, 343–385.

[5] R. Lavery, A. Lebrun, J. F. Allemand, D. Bensimon, V. Croquette, *J. Phys.: Condens. Matter* **2002**, *14*, R383–R414.

[6] A. D. Mehta, M. Rief, J. A. Spudich, D. A. Smith, R. A. Simmons, *Science* **1999**, *283*, 1689–1695.

[7] S. B. Smith, L. Finzi, C. Bustamante, *Science* **1992**, *258*, 1122–1126.

[8] M. P. Sheetz, *Methods in Cell Biology, Vol. 55*, Academic Press, New York, **1997**.

[9] A. Kishino, T. Yanagida, *Nature* **1988**, *334*, 74–76.

[10] E. Evans, K. Ritchie, R. Merkel, *Biophys. J.* **1995**, *68*, 2580–2587.

[11] G. Binnig, C. F. Quate, C. Gerber, *Phys. Rev. Lett.* **1986**, *56*, 930–933.

[12] M. Grandbois, M. Beyer, M. Rief, H. Clausen-Schaumann, H. E. Gaub, *Science* **1999**, *283*, 1727–1730.

[13] P. Bongrand, *Rep. Prog. Phys.* **1999**, *62*, 921–968.

[14] G. I. Bell, *Science* **1978**, *200*, 618–627.

[15] E. Evans, K. Ritchie, *Biophys. J.* **1997**, *72*, 1541–1555.

[16] S. Izrailev, S. Stepaniants, M. Balsera, Y. Oono, K. Schulten, *Biophys. J.* **1997**, *72*, 1568.

[17] E. Evans, *Annu. Rev. Biophys. Biomol. Struct.* **2001**, *30*, 105–128.

[18] E.-L. Florin, V. T. Moy, H. E. Gaub, *Science* **1994**, *264*, 415–417.

[19] R. Merkel, P. Nassoy, A. Leung, K. Ritchie, E. Evans, *Nature* **1999**, *397*, 50–53.

[20] D. A. Simson, M. Strigl, M. Hohenadl, R. Merkel, *Phys. Rev. Lett* **1999**, *83*, 652–655.

[21] M. Strigl, D. A. Simson, C. M. Kacher, R. Merkel, *Langmuir* **1999**, *15*, 7316–7324.

[22] E. Evans, F. Ludwig, *J. Phys.: Condens. Matter* **2000**, *12*, A315–A320.

[23] G. U. Lee, L. A. Chris, R. J. Colton, *Science* **1994**, *266*, 771–773.

[24] T. Boland, B. D. Ratner, *Proc. Natl. Acad. Sci. USA* **1995**, *92*, 5297–5301.

[25] T. Strunz, K. Oroszlan, R. Schafer, H. J. Guntherodt, *Proc. Natl. Acad. Sci. USA* **1999**, *96*, 11277–11282.

[26] H. Clausen-Schaumann, M. Rief, C. Tolksdorf, H. E. Gaub, *Biophys. J.* **2000**, *78*, 1997–2007.

[27] B. Heymann, H. Grubmüller, *Phys. Rev. Lett.* **2000**, *84*, 6126–6129.

[28] D. Bartolo, I. Derény, A. Ajdari, *Phys. Rev. E* **2002**, *65*, 051910.

[29] H. Grubmüller, B. Heymann, P. Tavan, *Science* **1996**, *271*, 997–999.

[30] M. Benoit, in: B. P. Jena, H. J. K. Hörber (eds), *Atomic Force Microscopy in Cell Biology, Vol. 68*, Academic Press, San Diego, **2002**, pp. 91–114.

[31] P. Hinterdorfer, in: B. P. Jena, H. J. K. Hörber (eds), *Atomic Force Microscopy in Cell Biology, Vol. 68*, Academic Press, San Diego, **2002**, pp. 115–139.

[32] A. Razatos, Y. L. Ong, M. M. Sharma, G. Georgiou, *Proc. Natl. Acad. Sci. USA* **1998**, *95*, 11059–11064.

[33] G. Sagvolden, I. Giaver, E. O. Pettersen, J. Feder, *Proc. Natl. Acad. Sci. USA* **1999**, *96*, 471–475.

[34] M. Benoit, D. Gabriel, G. Gehrisch, H. E. Gaub, *Nature Cell Biol.* **2000**, *2*, 313–317.

[35] X. Zhang, E. Wojcikiewicz, V. T. Moy, *Biophys. J.* **2002**, *83*, 2270–2279.

[36] P. Hinterdorfer, W. Baumgartner, H. J. Gruber, K. Schilcher, H. Schindler, *Proc. Natl. Acad. Sci. USA* **1996**, *93*, 3477–3481.

[37] M. Ludwig, W. Dettmann, H. E. Gaub, *Biophys. J.* **1997**, *72*, 445–448.

[38] O. H. Willemsen, M. M. E. Snel, K. O. van der Werf, B. G. de Grooth, J. Greve, P. Hinterdorfer, H. J. Gruber, H. Schindler, Y. van Kooyk, C. G. Figdor, *Biophys. J.* **1998**, *75*, 2220–2228.

[39] W. Han, S. M. Lindsay, T. Jing, *Appl. Phys. Lett.* **1996**, *69*, 1–3.

[40] W. Han, S. M. Lindsay, M. Dlakic, R. E. Harrington, *Nature* **1997**, *386*, 563.

[41] A. Raab, W. H. Han, D. Badt, S. J. Smith-Gill, S. M. Lindsay, H. Schindler, P. Hin-

terdorfer, *Nature Biotechnol.* **1999**, *17*, 902–905.

[42] M. Radmacher, M. Fritz, C. M. Kacher, J. Cleveland, P. K. Hansma, *Biophys. J.* **1996**, *70*, 556–567.

[43] M. Radmacher, in: B. P. Jena, H. J. K. Hörber (eds), *Atomic Force Microscopy in Cell Biology, Vol. 68*, Academic Press, San Diego, **2002**, pp. 67–90.

[44] E. L. Florin, A. Pralle, J. K. H. Hörber, E. H. K. Stelzer, *J. Struct. Biol.* **1997**, *119*, 202–211.

[45] A. Pralle, E.-L. Florin, in: B. P. Jena, H. J. K. Hörber (eds), *Atomic Force Microscopy in Cell Biology, Vol. 68*, Academic Press, San Diego, **2002**, pp. 193–212.

[46] M. S. Kellermayer, S. B. Smith, H. L. Granzier, C. Bustamante, *Science* **1997**, *276*, 1112–1116.

[47] L. Tskhovrebova, J. Trinick, J. A. Sleep, R. M. Simmons, *Nature* **1997**, *387*, 308–312.

[48] M. Rief, M. Gautel, F. Oesterhelt, J. M. Fernandez, H. E. Gaub, *Science* **1997**, *276*, 1109–1112.

[49] S. M. Altmann, P. F. Lenne, in: B. P. Jena, H. J. K. Hörber (eds), *Atomic Force Microscopy in Cell Biology, Vol. 68*, Academic Press, San Diego, **2002**, pp. 311–335.

[50] M. Rief, H. Grubmüller, *ChemPhysChem* **2002**, *3*, 255–261.

[51] X. Zhuang, M. Rief, *Curr. Opin. Struct. Biol.* **2003**, *13*, 88–97.

[52] T. E. Fisher, A. F. Oberhauser, M. Carrion-Vazquez, P. E. Marszalek, J. M. Fernandez, *Trends Biochem. Sci.* **1999**, *24*, 379–384.

[53] A. Janshoff, M. Neitzert, Y. Oberdörfer, H. Fuchs, *Angew. Chem. Intl. Ed.* **2000**, *39*, 3212–3237.

[54] T. Hugel, M. Seitz, *Macromol. Rapid Commun.* **2001**, *22*, 989–1016.

[55] H. Lu, B. Isralewitz, A. Krammer, V. Vogel, K. Schulten, *Biophys. J.* **1998**, *75*, 662–671.

[56] D. K. Klimov, D. Thirumalai, *Proc. Natl. Acad. Sci. USA* **1999**, *96*, 6166–6170.

[57] N. D. Socci, J. N. Onuchic, *Proc. Natl. Acad. Sci. USA* **1999**, *96*, 2031–2035.

[58] M. Rief, J. Pascual, M. Saraste, H. E. Gaub, *J. Mol. Biol.* **1999**, *286*, 553–561.

[59] M. Carrion-Vazquez, A. F. Oberhauser, S. B. Fowler, P. E. Marszalek, S. E. Broedel, J. Clarke, J. M. Fernandez, *Proc. Natl. Acad. Sci. USA* **1999**, *96*, 3694–3699.

[60] S. M. Altmann, R. G. Grunberg, P. F. Lenne, J. Ylanne, A. Raae, K. Herbert, M. Saraste, M. Nilges, J. K. H. Horber, *Structure (Camb)* **2002**, *10*, 1085–1096.

[61] R. B. Best, B. Li, A. Steward, V. Daggett, J. Clarke, *Biophys. J.* **2001**, *81*, 2344–2356.

[62] D. J. Brockwell, G. S. Beddard, J. Clarkson, R. C. Zinober, A. W. Blake, J. Trinick, P. D. Olmsted, A. Smith, S. E. Radford, *Biophys. J.* **2002**, *2002*, 458–472.

[63] M. Rief, M. Gautel, A. Schemmel, H. E. Gaub, *Biophys. J.* **1998**, *75*, 3008–3014.

[64] P. E. Marszalek, H. Lu, H. B. Li, M. Carrion-Vazquez, A. F. Oberhauser, K. Schulten, J. M. Fernandez, *Nature* **1999**, *402*, 100–103.

[65] M. Carrion-Vazquez, P. E. Marszalek, A. F. Oberhauser, J. M. Fernandez, *Proc. Natl. Acad. Sci. USA* **1999**, *96*, 11288–11292.

[66] D. J. Müller, W. Baumeister, A. Engel, *Proc. Natl. Acad. Sci. USA* **1999**, *96*, 13170–13174.

[67] D. J. Müller, A. Engel, in: B. P. Jena, H. J. K. Hörber (eds), *Atomic Force Microscopy in Cell Biology, Vol. 68*, Academic Press, San Diego, **2002**, pp. 257–299.

[68] F. Oesterhelt, D. Oesterhelt, M. Pfeiffer, A. Engel, H. E. Gaub, D. J. Müller, *Science* **2000**, *288*, 143–146.

[69] D. J. Müller, M. Kessler, F. Oesterhelt, C. Möller, D. Oesterhelt, H. Gaub, *Biophys. J.* **2002**, *83*, 3578–3588.

[70] D. W. Urry, T. Hugel, M. Seitz, H. E. Gaub, L. Sheiba, J. Dea, J. Xu, T. Parker, *Proc. R. Soc. London, Phil. Trans. B* **2002**, *357*, 169–184.

[71] B. L. Smith, T. E. Schaffer, M. Viani, J. B. Thompson, N. A. Frederick, J. Kindt, A. Belcher, G. D. Stucky, D. E. Morse, P. K. Hansma, *Nature* **1999**, *399*, 761–763.

[72] J. B. Thompson, J. H. Kindt, B. Drake, H. G. Hansma, D. E. Morse, P. K. Hansma, *Nature* **2001**, *414*, 773–776.

[73] I. Schwaiger, C. Sattler, D. R. Hostetter, M. Rief, *Nature Mater.* **2002**, *1*, 232–235.

[74] P. J. Flory, *Statistical Mechanics of Chain Molecules*, Hanser, München, **1988**.

[75] O. Kratky, G. Porod, *Rec. Trav. Chim.* **1949**, *68*, 1106–1123.

[76] T. Odijk, *Macromolecules* **1995**, *28*, 7016–7018.

[77] D. K. Chang, D. W. Urry, *J. Comput. Chem.* **1989**, *10*, 850–855.

[78] Z. R. Wasserman, F. R. Salemme, *Biopolymers* **1990**, *29*, 1613–1631.

[79] H. J. Kreuzer, R. L. C. Wang, M. Grunze, *New J. Phys.* **1999**, *1*, 21.1.

[80] M. Rief, F. Oesterhelt, B. Heymann, H. E. Gaub, *Science* **1997**, *275*, 1295–1298.

[81] R. R. Netz, *Macromolecules* **2001**, *34*, 7522–7529.

[82] S. B. Smith, Y. Cui, C. Bustamante, *Science* **1996**, *271*, 795–798.

[83] P. Cluzel, A. Lebrun, C. Heller, R. Lavery, J.-L. Viovy, D. Chatenay, F. Caron, *Science* **1996**, *271*, 792–794.

[84] D. Bensimon, A. J. Simon, V. Croquette, A. Bensimon, *Phys. Rev. Lett.* **1995**, *74*, 4754–4757.

[85] T. T. Perkins, D. E. Smith, R. G. Larson, S. Chu, *Science* **1995**, *268*, 83–87.

[86] M. D. Wang, H. Yin, R. Landick, J. Gelles, S. M. Block, *Biophys. J.* **1997**, *72*, 1335–1346.

[87] M. C. Williams, I. Rouzina, *Curr. Opin. Struct. Biol.* **2002**, *12*, 330–336.

[88] C. Bustamante, Z. Bryant, S. B. Smith, *Nature* **2003**, *421*, 423–427.

[89] M. W. Konrad, J. I. Bolonick, *J. Am. Chem. Soc.* **1996**, *118*, 10989.

[90] A. Lebrun, R. Lavery, *Nucleic Acids Res.* **1996**, *24*, 2260–2267.

[91] A. Ahsan, J. Rudnick, R. Bruinsma, *Biophys. J.* **1998**, *74*, 132–137.

[92] W. K. Olson, V. B. Zhurkin, *Curr. Opin. Struct. Biol.* **2000**, *10*, 286–297.

[93] M. Rief, H. Clausen-Schaumann, H. E. Gaub, *Nature Struct. Biol.* **1999**, *6*, 346–349.

[94] M. C. Williams, I. Rouzina, V. A. Bloomfield, *Acc. Chem. Res.* **2002**, *35*, 159–166.

[95] J. R. Wenner, M. C. Williams, I. Rouzina, V. A. Bloomfield, *Biophys. J.* **2002**, *82*, 3160–3169.

[96] T. R. Strick, J. F. Allemand, D. Bensimon, A. Bensimon, V. Croquette, *Science* **1996**, *271*, 1835–1837.

[97] T. R. Strick, J.-F. Allemand, D. Bensimon, V. Croquette, *Biophys. J.* **1998**, *74*, 2016–2028.

[98] J. F. Allemand, D. Bensimon, R. Lavery, V. Croquette, *Proc. Natl. Acad. Sci. USA* **1998**, *95*, 14152–14157.

[99] J. F. Léger, G. Romano, A. Sarkar, J. Robert, L. Bourdieu, D. Chatenay, J. F. Marko, *Phys. Rev. Lett.* **1999**, *83*, 1066–1069.

[100] J. F. Marko, *Phys. Rev. E* **1998**, *57*, 2134–2149.

[101] C. G. Baumann, V. A. Bloomfield, S. B. Smith, C. Bustamante, M. D. Wang, S. M. Block, *Biophys. J.* **2000**, *78*, 1965–1978.

[102] J. Liphardt, B. Onoa, S. B. Smith, I. J. Tinoco, C. Bustamante, *Science* **2001**, *292*, 733–737.

[103] R. Krautbauer, H. Clausen-Schaumann, H. E. Gaub, *Angew. Chem. Intl. Ed.* **2000**, *39*, 3912–3915.

[104] R. Krautbauer, L. H. Pope, T. E. Schrader, S. Allen, H. E. Gaub, *FEBS Lett.* **2002**, *510*, 154–158.

[105] J. L. Viovy, C. Heller, F. Caron, P. Cluzel, D. Chatenay, *C. R. Acad. Sci., Paris* **1994**, *317*, 795–800.

[106] B. Essevaz-Roulet, U. Bockelmann, F. Heslot, *Proc. Natl. Acad. Sci. USA* **1997**, *94*, 11935–11940.

[107] U. Bockelmann, P. Thomen, B. Essevaz-Roulet, V. Viasnoff, F. Heslot, *Biophys. J.* **2002**, *82*, 1537–1553.

[108] R. E. Thompson, E. D. Siggia, *Europhys. Lett.* **1995**, *31*, 335–340.

[109] U. Bockelmann, B. Essevaz-Roulet, F. Heslot, *Phys. Rev. Lett.* **1997**, *79*, 4489–4492.

[110] U. Bockelmann, B. Essevaz-Roulet, F. Heslot, *Phys. Rev. E* **1998**, *58*, 2386–2394.

[111] D. K. Lubensky, D. R. Nelson, *Phys. Rev. Lett.* **2000**, *85*, 1572–1575.

[112] J. F. Léger, J. Robert, L. Bourdieu, D. Chatenay, J. F. Marko, *Proc. Natl. Acad. Sci. USA* **1998**, *95*, 12295–12299.

[113] M. Hegner, S. B. Smith, C. Bustamante, *Proc. Natl. Acad. Sci. USA* **1999**, *96*, 10109–10114.

[114] G. V. Shivashankar, M. Feingold, O. Krichevsky, A. Libchaber, *Proc. Natl. Acad. Sci. USA* **1999**, *96*, 7916–7921.

[115] M. L. Bennink, O. D. Scharer, R. Kanaar, K. Sakata-Sogawa, J. M. Schins, J. S. Kanger, B. G. de Groth, J. Greve, *Cytometry* **1999**, *36*, 200–208.

[116] G. J. L. Wuite, S. B. Smith, M. Young, D. Keller, C. Bustamante, *Nature* **2000**, *404*, 103–106.

[117] R. J. Davenport, G. J. L. Wuite, R. Landick, C. Bustamante, *Science* **2000**, *287*, 2497–2500.

[118] M. C. Williams, I. Rouzina, J. R. Wenner, R. J. Gorelick, K. Musier-Forsyth, V. Bloomfield, *Proc. Natl. Acad. Sci. USA* **2001**, *98*, 6121–6126.

[119] Y. Cui, C. Bustamante, *Proc. Natl. Acad. Sci. USA* **2000**, *97*, 127–132.

[120] D. E. Smith, S. J. Tans, S. B. Smith, S. Grimes, D. L. Anderson, *Nature* **2001**, *413*, 748–752.

[121] T. R. Strick, V. Croquette, D. Bensimon, *Nature* **2000**, *404*, 901–904.

[122] N. H. Dekker, V. V. Rybenkov, M. Duguet, N. J. Crisona, N. R. Cozzarelli, D. Bensimon, V. Croquette, *Proc. Natl. Acad. Sci. USA* **2002**, *99*, 12126–12131.

[123] N. J. Crisona, T. R. Strick, D. Bensimon, V. Croquette, N. R. Cozzarelli, *Genes Dev.* **2000**, *14*, 2881–2892.

[124] R. D. Vale, R. A. Milligan, *Science* **2000**, *288*, 88–95.

[125] K. Svoboda, C. F. Schmidt, B. J. Schnapp, S. M. Block, *Nature* **1993**, *365*, 721–727.

[126] H. Kojima, E. Muto, H. Higuchi, T. Yanagida, *Biophys. J.* **1997**, *73*, 2012–2022.

[127] J. T. Finer, R. M. Simmons, J. Spudich, *Nature* **1994**, *368*, 113–119.

[128] R. Yasuda, H. Noji, K. Kinosita, Jr., M. Yoshida, *Cell* **1998**, *93*, 1117–1124.

[129] W. Junge, *Proc. Natl. Adad. Sci. USA* **1999**, *96*, 4735–4737.

[130] S. M. Block, D. F. Blair, H. C. Berg, *Nature* **1989**, *338*, 514–518.

[131] X. Chen, H. C. Berg, *Biophys. J.* **2000**, *78*, 1036–1041.

[132] Y. Ishii, T. Yanagida, *Single Mol.* **2000**, *1*, 5–16.

[133] D. Keller, C. Bustamante, *Biophys. J.* **2000**, *78*, 541–556.

[134] M. E. Fisher, A. B. Kolomeisky, *Proc. Natl. Acad. Sci. USA* **1999**, *96*, 6597–6602.

[135] K. Kitamura, M. Tokunaga, A. H. Iwane, T. Yanagida, *Nature* **1999**, *397*, 129–134.

[136] M. Rief, R. S. Rock, A. D. Mehta, M. S. Mooseker, R. E. Cheney, J. A. Spudich, *Proc. Natl. Acad. Sci. USA* **2000**, *97*, 9482–9486.

[137] T. Yanagida, K. Kitamura, H. Tanaka, A. H. Iwane, S. Esaki, *Curr. Opin. Cell Biol.* **2000**, *12*, 20–25.

[138] K. Visscher, M. J. Schnitzer, S. M. Block, *Nature* **1999**, *400*, 184–189.

[139] C. Veigel, L. M. Coluccio, J. D. Jontes, J. C. Sparrow, R. A. Milligan, J. E. Molloy, *Nature* **1999**, *398*, 530–533.

[140] M. J. Tsyka, D. E. Dupuis, W. H. Guilford, J. B. Patlak, G. S. Waller, K. M. Trybus, D. M. Warshaw, S. Lowey, *Proc. Natl. Acad. Sci. USA* **1999**, *96*, 4402–4407.

[141] H. Tanaka, K. Homma, A. H. Iwane, E. Katayama, R. Ikebe, J. Saito, T. Yanagida, M. Ikebe, *Nature* **2002**, *415*, 192–195.

[142] J. E. Molloy, J. E. Burns, J. Kendrick-Jones, R. T. Treager, D. C. S. White, *Nature* **1995**, *378*, 209–212.

[143] A. D. Mehta, J. T. Finer, J. A. Spudich, *Proc. Natl. Acad. Sci. USA* **1997**, *94*, 7927–7931.

[144] T. Yanagida, T. Arata, F. Oosawa, *Nature* **1985**, *316*, 366–369.

[145] D. W. Urry, *J. Phys. Chem. B* **1997**, *101*, 11007–11028.

[146] T. Hugel, N. B. Holland, A. Cattani, L. Moroder, M. Seitz, H. E. Gaub, *Science* **2002**, *296*, 1103–1106.

[147] X. Châtellier, T. J. Senden, J. F. Joanny, J. M. di Meglio, *Europhys. Lett.* **1998**, *41*, 303–308.

[148] T. J. Senden, J. M. di Meglio, P. Auroy, *Eur. Phys. J. B* **1998**, *3*, 211–216.

[149] T. Hugel, M. Grosholz, H. Clausen-Schaumann, A. Pfau, H. Gaub, M. Seitz, *Macromolecules* **2001**, *34*, 1039–1047.

[150] T. J. Senden, *Curr. Opin. Coll. Interf. Sci.* **2001**, *6*, 95–101.

[151] M. B. Viani, T. E. Schaffer, G. T. Paloczi, L. I. Pietrasanta, B. L. Smith, J. B. Thompson, M. Richter, M. Rief, H. E. Gaub, K. W. Plaxco, A. N. Cleland, H. G. Hansma, P. K. Hansma, *Rev. Sci. Instr.* **1999**, *70*, 4300–4303.

[152] A. D. L. Humphris, J. Tamayo, M. J. Miles, *Langmuir* **2000**, *16*, 7891–7894.

[153] A. F. Oberhauser, P. K. Hansma, M. Carrion-Vazquez, J. M. Fernandez, *Proc. Natl. Acad. Sci. USA* **2001**, *98*, 468–472.

[154] A. Ashkin, *Phys. Rev. Lett.* **1970**, *24*, 156–159.

[155] F. Gittes, C. F. Schmidt, *Opt. Lett.* **1998**, *23*, 7–9.

[156] C. Gosse, V. Croquette, *Biophys. J.* **2002**, *82*, 3314–3329.

[157] D. A. Simson, F. Ziemann, M. Strigl, R. Merkel, *Biophys. J.* **1998**, *74*, 2080–2088.

26
Biofunctionalized Nanoparticles for Surface-Enhanced Raman Scattering and Surface Plasmon Resonance

Mahnaz El-Kouedi and Christine D. Keating

26.1
Overview

26.1.1
Introduction

Nano- and microparticle–biomolecule conjugates have been used as amplification tags in a wide variety of biosensing schemes [1]. The general format for many of these assays, which are analogous to fluorescently-detected sandwich immunoassays, is illustrated in Figure 26.1. A sensing surface is derivatized with a biorecognition element (such as an antibody or oligonucleotide) which is capable of selectively binding the desired analyte from solution. Detection is achieved by completing the "sandwich" with a second selective

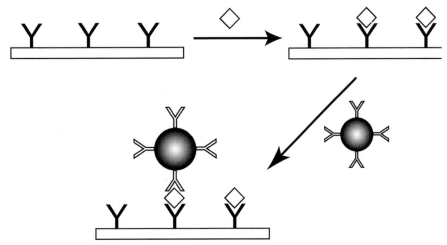

Figure 26.1 Typical bioconjugate-based sandwich bioassay. A surface functionalized with biorecognition chemistry (in this case an antibody) is exposed to analyte (diamond), after which a secondary antibody-labeled particle binds. The particle amplifies the sensor response, thereby increasing sensitivity.

Nanobiotechnology. Edited by Christof Niemeyer, Chad Mirkin
Copyright © 2004 WILEY-VCH Verlag GmbH & Co. KgaA, Weinheim
ISBN 3-527-30658-7

biorecognition molecule, this time tagged with a particle. This chapter will focus on metal nanoparticle bioconjugates in two surface-based optical sensing platforms: surface plasmon resonance (SPR) and surface-enhanced Raman scattering (SERS) [2–4]. In both techniques, the nanoscale structure at a metal–dielectric interface is critically important to successful sensing. We will discuss the application of metal nanoparticle bioconjugates in sensing strategies based on refractive index changes (optical extinction and surface plasmon resonance) and electromagnetic field enhancement of vibrational scattering (SERS).

26.1.2
Applications in SPR

In recent years, a wide variety of analytical techniques has been developed based on changes associated with the SPR of gold and silver nanoparticles and thin films. The SPR is a consequence of the collective oscillations of metal valence electrons resulting in a strong absorption peak that can be used to monitor changes in the surrounding medium such as biomolecule adsorption events at the metal surface, or nanoparticle aggregation [3, 4]. Analytical plasmon resonance techniques can be divided into two classes:

1. extinction-based methods, in which changes in the visible or near-infra-red (IR) transmission are monitored, typically for collections of nanoparticles; and
2. reflectivity-based methods, in which changes in the angle-dependent intensity of reflected light are monitored for planar metal films or gratings.

26.1.2.1 Nanoparticle Substrates

An introduction to aggregation-based bioassays can be found in a review by Englebienne [5]. Briefly, metal nanoparticle aggregation results in a red shift of the plasmon resonance absorption, accompanied by a decrease in peak intensity due to aggregate sedimentation. Aggregation of "bare", charge-stabilized nanoparticles can be induced by the simple addition of high concentrations of salt; bioconjugates can be selectively crosslinked by the presence of analyte. For example, Mirkin and coworkers reported a DNA sensor based on the dramatic redshift in absorbance upon hybridization-driven gold nanoparticle aggregation [6, 7] (Figure 26.2A). Recently, Englebienne has used colloidal nanoparticles for the high-throughout screening of proteins, and has determined ligand–protein interactions for 30 antibody–antigen pairs [8]. For Au nanoparticles, optical changes occur in the visible; to monitor bioassays in whole blood, Hirsch et al. have used SiO_2 core/Au shell bioconjugates, which absorb in the near-IR where interference from cell and tissue absorption is minimal [9].

Bioconjugate aggregation is not a requirement for analyte detection; it is also possible to monitor analyte binding through small shifts in the plasmon resonance absorbance due to changes in the refractive index surrounding the particle [3, 10, 11]. Indeed, Van Duyne and coworkers have taken advantage of changes in this localized SPR effect (LSPR) to monitor protein binding to triangular Ag nanoparticles bound to a glass substrate (Figure 26.2B) [12]. In this investigation, ~50 nm-high, 100 nm-wide nanoparticles were prepared by nanosphere lithography [13] and derivatized with biotin for streptavidin detection. The binding of 100 nM streptavidin to biotinylated prisms results in an impressive 27-nm red-

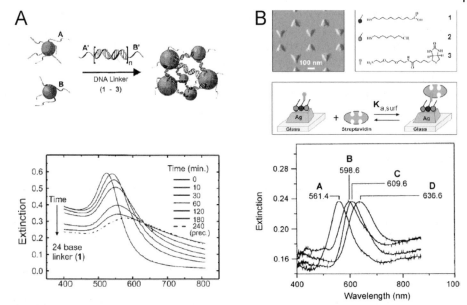

Figure 26.2 (A) Aggregation-based solution assay for the detection of DNA as utilized by Mirkin et al. [6]. A solution of DNA-coated nanoparticles changes color from red to blue upon the addition of a complementary DNA linker sequence, due to DNA duplex formation and subsequent particle aggregation (top). The aggregation results in the shift and broadening of the plasmon resonance peak, which can be monitored using absorbance spectroscopy (bottom). (Reprinted with permission from ACS; © 2000.) (B) Localized surface plasmon resonance (LSPR) biosensor utilized by Van Duyne et al. [12]. Triangular nanoparticles synthesized using nanosphere lithography techniques were functionalized with a biotinylated SAM layer. Streptavidin was then detected by looking at the shift in the Ag plasmon resonance band. In the lower panel, curve (A) unfunctionalized Ag; (B) Ag with aminated SAM; (C) nanoparticles after biotin attachment; (D) streptavidin detection shift. (Reprinted with permission from ACS; © 2002.)

shift of the plasmon resonance band. Low picomolar/high femtomolar limits of detection were reported for streptavidin binding; samples were measured dry, in a N_2 environment. These authors also demonstrated that the LSPR response could be amplified by using biotin-modified colloidal Au nanoparticles in a sandwich format [12]. Recent improvements have enabled real-time LSPR analysis of antibody binding in physiological buffer, with a detection limit of ~700 pM [14].

26.1.2.2 Planar Substrates

SPR techniques based on the planar Au substrate configuration have been utilized extensively for the detection of biomolecules, and have become increasingly popular since the development of the first commercial BIAcore instrument in 1990 [2, 15]. SPR detection is based on a refractive index change near a thin noble metal film upon analyte binding. The conventional SPR system is based on the Kretschmann configuration. In this configuration, a metal film (usually ~50 nm of Au) is evaporated either directly onto a hemispherical prism or evaporated onto a glass slide and index matched to the prism [2, 16]. Under conditions of total internal reflection, p-polarized light is used to illuminate the metal film through the prism. At the plasmon resonant angle, the light induces the collective oscillations of the valence electrons of the metal film, resulting in a decrease in the reflectivity. The position of this resonance angle depends strongly on the refractive index at the metal interface; changes in the resonant angle or in the intensity of reflected light at a fixed angle can be used to detect binding events at the substrate surface.

Although one advantage of SPR is the possibility of label-free detection, in many cases conventional SPR lacks the desired sensitivity for biosensing. SPR response can be amplified by using a sandwich geometry, where the change in refractive index due to analyte binding is enhanced by the binding of a macromolecule or particle with either large MW or high refractive index (see Figure 26.1). Examples include protein–DNA multilayers [17], liposomes [16], and polymer beads [18, 19]. We will focus on Au nanoparticle bioconjugates as amplification tags, an excellent introduction to which subject can be found in Ref. 20.

Large perturbations in the SPR response upon Au nanoparticle binding are thought to result largely from the particles acting as roughness features, enabling nonradiative surface plasmons propagating in the thin film to become radiative modes [20]. Au nanoparticle-amplified SPR has been used for detection of human serum albumin (HAS), using 30-nm diameter Au:protein bioconjugates [21], and also for the detection of immunoglobulins [22, 23]. A thousand-fold enhancement was observed for HAS detection as compared to the unamplified binding event [21]. Au:IgG bioconjugate binding was monitored in real time by Gu et al., who observed SPR response saturation in minutes [22].

Lyon et al. used 11-nm diameter colloidal Au for detection of human IgG using a sandwich immunoassay such as that described in Figure 26.1 [23]. For a 6.7 pM solution of IgG, a $0.3\bar{3}$ shift was observed in the resonance angle; given that angle shifts $\leq 0.00\bar{5}$ are detectable, these authors reasoned that actual detection limits may be much lower than picomolar [23]. The experiments showed a "quasi-linear" relationship between the particle coverage and the shift of the plasmon resonance angle, such that SPR shift can be related to antigen concentration.

Some experimental and theory papers have addressed the impact of nanoparticles on the SPR response of thin metal films. The electromagnetic coupling efficiency and hence the amplification efficiency of the Au tags typically decreases with increasing separation [3, 4,24]. This is demonstrated by the reduced SPR shift for the protein detection experiments as compared to the shifts observed for the binding of Au nanoparticles on a mercaptoethylamine self-assembled monolayer (SAM) [23]. Leung et al. stressed the importance of nanoparticle film morphology, predicting that clustering of the nanoparticles would lead to greater signal amplification [25]. Fendler and coworkers have examined the SPR response from Au and Ag thin films upon binding 10-nm Au or 30-nm diameter Ag nanoparticles with both experiments and modeling [26], while Lyon et al. have investigated the effect of Au nanoparticle size on SPR shift for particles 30–60 nm in diameter [27]. At constant particle coverages, greater shifting and broadening of the plasmon peak was observed with increasing particle size. From surface coverage data, these authors concluded that larger particles provide greater sensitivity, but at the same time decrease the maximum concentration that can be quantified [27].

Recently, Au particle-amplified SPR has been used to increase the sensitivity for the detection of short oligonucleotide sequences 1000-fold over the unamplified experiment [28].

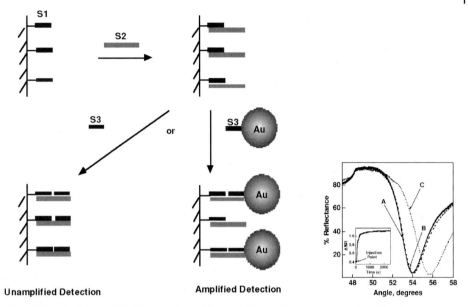

Unamplified Detection   **Amplified Detection**

Figure 26.3 Particle-amplified SPR for the detection of DNA as demonstrated by He et al. [28]. DNA sequence 1 (S1) is attached to a thin gold film evaporated onto a glass slide (curve A). The analyte sequence S2 binds to the surface, resulting in a very small shift in the SPR response (curve B). A large shift was observed upon the binding of sequence S3 attached to Au nanoparticle amplification tags, as shown in curve C. Inset shows real-time monitoring of the change in reflectivity at 53.2̄ after exposure to S3:Au particles. (Reprinted with permission from ACS; © 2002.)

In these experiments, a sandwich hybridization assay analogous to the immunoassay of Figure 26.1 was employed, in which three DNA strands were used (Figure 26.3). Submonolayer coverage of a DNA oligonucleotide that was half complementary to the analyte sequence, was attached to the sensor surface. The analyte was then introduced, followed by binding of the Au amplification tags. Upon binding of the nanoparticles to the surface, a large perturbation to the resonance angle was observed (Figure 26.3). Binding reversibility was demonstrated using thermal denaturation and enzyme cleavage to remove bound DNA and DNA:Au bioconjugates [28].

He et al. also demonstrated the feasibility of using imaging SPR and Au particle amplification to detect hybridization events in an array format with ~10 pM detection limits and a five orders of magnitude dynamic range [28]. These experiments show promise for particle amplified SPR in array-based DNA analysis. A key challenge will be reducing the background arising from nonspecific binding of bioconjugates to the metal film; it is this nonspecific background that currently limits the sensitivity of the technique. In the absence of nonspecific binding, substantial improvements in detection limits are expected.

26.1.3
Applications in SERS

SERS refers to the large signal enhancements observed for molecules adsorbed at roughened metal surfaces, particularly Ag and Au [4]. Two phenomena give rise to enhancements:

1. Local amplification of the electromagnetic field near roughness features in free electron metals. This portion of the SERS enhancement can be theoretically modeled, and has been reported to account for the bulk of the observed enhancements [4].
2. For molecules chemisorbed to the metal surface, an additional chemical effect is observed. This effect has been described by Campion and others [4, 29]. In combination, these effects can give rise to experimentally observed enhancements sufficient to permit single molecule detection [30].

An extensive literature exists describing SERS and resonant-SERS (or SERRS) of various biomolecules [4, 30]. In SERRS, the laser line used for excitation is resonant with an optical absorbance of the molecule, leading to an additional enhancement in scattering from vibrations coupled to the electronic transition. In nearly all cases, proteins are added to colloidal Ag sols to induce aggregation or are added immediately after aggregation has been induced by addition of salt. The resulting Ag nanoparticle–protein aggregates are ill-defined but often give rise to very intense SERS spectra. Due in part to the success of this approach, very few studies of well-defined nanoparticle bioconjugates have appeared. SERS studies of protein and nucleic acid bioconjugates are described separately, as these molecules have very different binding modalities to metal nanoparticles.

26.1.3.1 **Proteins**
Unlike many SPR and LSPR approaches, the goal of protein SERS has often been to elucidate binding chemistry and active site structure–function rather than to detect the protein of interest. The bioactivity of adsorbed protein molecules is of great importance, particularly given the propensity of proteins to denature on solid surfaces [31–33]. Several groups have addressed the issue of bioactivity for adsorbed enzymes in SERS studies. The characteristic vibrational frequencies of heme chromophores present in many enzymes enable spectroscopic determination of enzyme stability, orientation, and charge state. Several groups have used the heme spin state marker bands to monitor conformational changes in proteins adsorbed to aggregated Ag sols. Although early studies at Ag electrodes [32] and BH_4-reduced Ag sols [33] indicated some protein denaturation at the metal surface, later investigations for cytochrome c (Cc) and several cytochromes P450 showed the retention of native spin states, which suggested retention of native protein structure [34–36]. The principal difference in the latter studies was the use of Ag sols prepared by citrate reduction of $AgNO_3$ [37]; such particles have an adsorbed layer of citrate which seems to increase their biocompatibility. Citrate-reduced Ag nanoparticles were also used in a study by Broderick et al. which showed that the nonheme iron enzyme chlorocatechol dioxygenase retained 60–85 % of its activity when adsorbed to aggregated colloidal Ag nanoparticles. These authors were able to observe enzymatic turnover at the Ag surface via SERRS [38].

Well-defined, discrete protein–nanoparticle bioconjugates were not employed in any of the aforementioned SERS investigations, and indeed are rarely used for SERS. The principal reason for this is that it is simply not necessary to prepare discrete bioconjugates in order to acquire SERS spectra. In addition to the ease of acquiring SERS spectra for biomolecules adsorbed to colloidal Ag aggregates or other roughened Ag substrates, another important reason for the relative dearth of biomolecule–nanoparticle conjugate SERS studies is the relative difficulty of handling colloidal Ag sols, the substrate of choice for biomolecule SERS. To our knowledge, the first use of SERS to characterize well-defined bioconjugates was by Ahern and Garrell in 1991 [39]. For protein–Ag nanoparticle conjugates, these authors observed a time-dependent increase in SERS signal for proteins adsorbed to colloidal nanoparticles over the course of several days, consistent with conformational changes of the protein. No SERS signal was observed for Au nanoparticle bioconjugates [39]. Despite the poor SERS activity of isolated nanometer Au spheres, Au bioconjugates are more stable than their Ag counterparts. Several researchers have described methods for acquiring SERS for Au nanoparticle bioconjugates.

Keating et al. have demonstrated that preconjugation of Cc to colloidal Au nanoparticles provided greater stability and control over protein orientation as compared to direct adsorption onto aggregated Ag sols [40]. Well-defined Cc:Au nanoparticle conjugates were prepared and SERRS spectra for the Cc:Au bioconjugates were compared to spectra for Cc alone. The two geometries investigated in these studies are shown in Figure 26.4. Ag:Cc and Ag:Cc:Au samples were prepared by adsorption of Cc or Cc:Au conjugates, respectively, to aggregated colloidal Ag. When Cc adsorbs to negatively charged surfaces, such as colloidal metal nanoparticles, the heme group is located near the metal surface. Thus, in Cc:Au bioconjugates, the heme chromophore faces in towards the Au, and remains so upon adsorption of the bioconjugates to the surface of aggregated Ag sol. In contrast, Cc directly adsorbed to aggregated Ag nanoparticles binds with its heme close to the Ag surface (Figure 26.4). SERRS spectra for Ag:Cc:Au and Ag:Cc samples were closer in intensity than expected based on heme proximity to the Ag surface, due to the presence of the Au nanoparticles. In addition, preconjugation to colloidal Au led to reproducible protein orientation on the surface and increased resistance to conformational changes indicated by the spin state marker bands [40].

These Ag:Cc:Au sandwiches provided an opportunity to probe the heightened electromagnetic fields that had long been predicted between closely spaced metal particles [4, 30]. SERRS spectra were acquired under identical conditions for Ag:Cc:Au and Ag:Cc at several wavelengths in the visible spectrum [41]. Electromagnetic coupling between the Au nanoparticles and the aggregated Ag surface was evidenced as increased relative intensity for Ag:Cc:Au versus Ag:Cc at wavelengths matched to the optical absorbance maximum of the Ag:Cc:Au sandwich. Keating et al. also prepared Au:Cc:Au and Ag:Cc:Ag sandwiches; scattering intensities for the Ag:Cc:Ag sandwiches exceeded that for directly adsorbed Cc, despite the greater distance between the heme chromophore and the aggregated Ag surface for the sandwiches. These studies demonstrated strong electromagnetic coupling between the bioconjugated metal nanoparticles and the aggregated Ag surface, despite the intervening protein layer (the Cc diameter is ~34) [41].

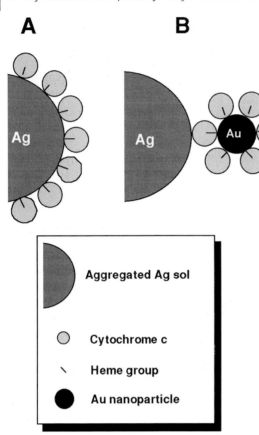

A **B**

Aggregated Ag sol

Cytochrome c

Heme group

Au nanoparticle

Figure 26.4 Schematic depiction of the two types of SERS samples investigated by Keating et al. [40]. In (A), Cc is directly adsorbed on aggregated Ag nanoparticles (Ag:Cc), while in (B) the Cc:Au bioconjugates are adsorbed to aggregated Ag, resulting in a greater separation between the SERS-active Ag surface and the heme chromophore (Ag:Cc:Au). (Reprinted with permission from ACS; ©1998.)

Porter and coworkers have reported using antibody–Au bioconjugates in a SERS-based immunoassay [42]. In these studies, a submonolayer of small-molecule Raman probes was adsorbed to the Au nanoparticle surface prior to adsorbing antibody molecules; this approach is illustrated in Figure 26.5. A capture antibody is attached to a planar Au substrate. After analyte binding, Raman-tagged secondary antibody–Au bioconjugates are added. SERS spectra show the characteristic fingerprint spectra from adsorbed Raman probes, enabling identification of the analyte. By using different Raman probe molecules for different antibodies, two immunoassays could be performed simultaneously. Raman spectra provide narrower bandwidths and a much greater complexity as compared to fluorescence spectra; thus, many more Raman-based dyes can be envisioned. Hence, this bioconjugate SERS-based methodology has potential for multiplexed immunoassays, and detection limits were estimated at ~2 nM [42]. Further improvements may be possible by using Raman probes for which resonant enhancement is possible at the SERS excitation wavelength.

(1) Creation of capture antibody surface

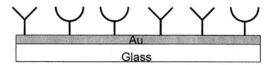

(2) Exposure to analyte

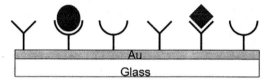

(3) Development with reporter labeled immunogold

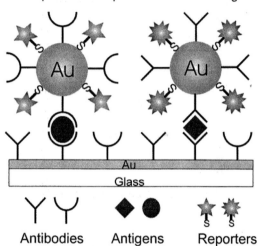

Figure 26.5 SERS immunoassay format used by Porter et al. [42]. Amplification tags comprised of Au nanoparticles coated with both small molecule Raman tags and antibodies for selective biorecognition of analyte. Sandwich immunoassays were probed by SERS. (Reprinted with permission from ACS; © 1999.)

Antibodies   Antigens   Reporters

26.1.3.2 Nucleic Acids

Several groups have studied DNA binding to colloidal Au and Ag particles via SERS [43–45]. Murphy and coworkers explored the binding of intrinsically bent double-stranded DNA sequences to 14 nm-diameter colloidal Au spheres as a means of understand the binding of these sequences to proteins of similar size [45]. Other investigations have been aimed at ultrasensitive DNA detection. For example, Kneipp et al. have demonstrated single molecule detection and identification of single DNA bases on colloidal Ag aggregates [43].

Smith and coworkers have developed a SERRS-based detection strategy for ultrasensitive detection of dye-labeled, single-stranded and double-stranded DNA via SERRS [44]. These authors have sought to improve sensitivity and reduce the standard deviations in signals between experiments by controlling DNA adsorption on the Ag surface. The negative charge on the DNA backbone is electrostatically repelled from the negatively charged colloidal Ag particles. Smith and coworkers addressed this by both incorporating positively charged propargylamino-modified deoxyuridine bases and mixing the modified oligonucleotides with a positively charged polymer, spermine, prior to addition to colloidal Ag nanoparticles. The resulting aggregated Ag suspension yields strong SERRS signals for oligonucleotide-bound dye molecules, and enabled detection limits as low as 8×10^{-13} M [44]. This approach has been used to detect PCR products using SERRS-active primers which contained both a 5′ dye moiety as well as a propynyl-modified sequence to improve adsorption [46].

In 1996, the Mirkin/Letsinger [47] and Alivisatos [48] groups each published methods for single-point attachment of oligonucleotides to Au nanoparticles via self-assembly of thiol groups covalently attached to the 3′ or 5′ terminus of the DNA. These DNA–Au bioconjugates have found a wide range of applications in biosensing [49–51], but have only recently begun to be used in SERS experiments. Franzen and coworkers have used SERS to monitor DNA–Au bioconjugate monolayer morphology and surface aggregation [52].

The potential of Raman probes as tags for multiplexing was exploited recently by the Mirkin group to prepare an elegant multianalyte DNA sensor based on SERRS (Figure 26.6) [53]. In these studies, dye molecules were incorporated into the oligonucleotide sequence prior to assembly of the DNA–Au bioconjugates. A sandwich hybridization assay was performed on a chip surface, such that bioconjugates attached to the surface only when the analyte sequence was present. Rather than measure the Raman signal directly from the surface-bound (dye-harboring) bioconjugates, Mirkin and colleagues chemically reduced Ag onto the Au particles. This process resulted in a much more SERS-active substrate as compared to the 13-nm Au particles alone. The resulting Raman spectra showed good discrimination between six different bioconjugates, as well as detection limits of 20×10^{-15} M. In addition to outstanding sensitivity and multiplexing capability, this approach benefits from the unusual selectivity of DNA–Au bioconjugate hybridization, which permits discrimination of single base mismatches by temperature or salt stringency assays [51, 52].

Mulvaney et al. recently reported a novel SERS-based nanoprobe, in which Raman-active molecules on the surface of Au or Ag nanospheres were protected by a glass shell

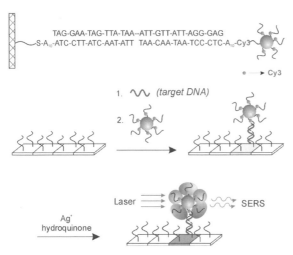

TAG-GAA-TAG-TTA-TAA--ATT-GTT-ATT-AGG-GAG
S-A$_{10}$-ATC-CTT-ATC-AAT-ATT TAA-CAA-TAA-TCC-CTC-A$_{10}$-Cy3

Figure 26.6 SERS-based DNA and RNA detection strategy reported by Mirkin et al. [53]. DNA–Au bioconjugates incorporated Raman-active dye molecules at the 3′, thiol-terminated end of the oligonucleotides. These probes were selectively assembled onto the DNA chip surface via sandwich hybridization assays, followed by chemical reduction of Ag metal selectively onto the Au bioconjugates. SERS was then used to detect and identify the Raman label, and hence the DNA sequence, present in each spot. (Reprinted with permission from AAAS; © 2002.)

[54]. Such particles could potentially be coupled to biomolecules of interest and used as Raman tags in multiplexed bioassays.

26.2
Methods

26.2.1
Planar SPR Substrate Preparation

An excellent resource for SPR is the website of Professor Corn's research group at the University of Wisconsin, as this provides – among other things – software for Fresnel calculations [55]. The preparation of SPR substrates begins with the vacuum evaporation of an Au thin film (approximately 40–50 nm thickness) onto freshly cleaned glass substrates. The surface is then derivatized for selective biomolecule attachment. Different attachment chemistries have been developed, including the use of commercially available chips that include a coupling matrix [15]. These polymer matrices – which are usually a functionalized dextran – are not generally amenable to particle amplification, as the nanoparticles are usually too large to fit into the pores of the matrix. Instead, alkanethiol-based self-assembled monolayers and biotin–streptavidin attachment chemistries are favored for particle-amplified SPR. Streptavidin can be physisorbed onto the Au substrate, and biotinylated DNA strands or proteins can then be attached to the substrate surface. The adsorption of a carboxy- or amino-terminated self-assembled monolayer followed by protein attachment using carbodiimide chemistry (EDC/NHS), has also been successful [56].

26.2.2
Metal Nanoparticles

Recipes for the synthesis of colloidal Au particles of various sizes can be found in Ref. [57]. A slight modification of these protocols for monodisperse 12 nm-diameter Au nanospheres leads to an approximately four-fold increase in particle concentration [58]. The most popular Ag nanoparticle preparation method for SERS is the citrate reduction protocol published by Lee and Meisel in 1982 [37]. Particles prepared following this protocol are typically quite polydisperse but give rise to excellent SERS enhancements. Other Ag particle recipes include EDTA reduction [41] and BH_4 reduction [33]. Synthetic methods also exist for preparation of core-shell nanoparticles including Au core-Ag shell [59] and silica core–Au shell [60]. Au and Ag nanoparticles are commercially available from Ted Pella, Inc. (www.tedpella.com), with or without adsorbed antibodies or streptavidin.

26.2.3
Bioconjugates

Protein–nanoparticle conjugates, prepared by direct adsorption of proteins to colloidal Au, have been used for decades as electron-dense markers in transmission electron microscopy experiments [57]. Detailed methods for preparation of protein–Au conjugates are available [57]. A typical protocol begins with a flocculation assay, which determines the concentration of protein necessary to stabilize nanoparticles from salt-induced aggregation. This procedure has been described in several references, among them the excellent books by Hyat [40, 57, 61].

DNA–nanoparticle conjugates, prepared by self-assembly of 5′ or 3′ terminal thiol- modified oligonucleotides were introduced by the Mirkin–Letsinger and Alivisatos groups in 1996 [47, 48]. Several useful references on preparation, handling, and characterization of these bioconjugates have appeared, including their separation by gel electrophoresis and factors influencing hybridization efficiency [62–64].

26.2.4
General Comments

Other methods are useful for coupling biomolecules to particles that have different surface chemistries. For example, Au nanoclusters with phosphine ligands are commercially available as "nanogold" and undecagold" from Nanoprobes, Inc. (www.nanoprobes.com). These particles are covalently modified with reactive groups for use in coupling reactions with biomolecules of interest.

Researchers new to colloidal metal particles often have difficulty with particle stability. Metal nanoparticles are stabilized against irreversible aggregation by charge repulsion due to adsorbed ions and/or steric hindrance due to large adsorbates (e. g., biomolecules). The addition of even small amounts of salts to charge-stabilized nanoparticles can lead to irreversible aggregation, as evidenced by a dramatic color change and ultimately precipitation of aggregates from solution. All glassware must be scrupulously clean, and all H_2O must be deionized. The addition of biomolecules for particle derivatization must be carried out with care in order to prevent nanoparticle aggregation due, for example, to the buffer solution in which the biomolecules are dissolved, or to crosslinking by the biomolecules themselves [61].

26.3
Future Outlook

Recent years have seen great strides in controlling the size, shape, and monodispersity of colloidal metal particles. It is now possible to produce metal nanoparticles with highly controlled optical properties based on particle size, shape, and composition. Such optical probes will be increasingly employed in ultrasensitive detection strategies such as those described here. Triangular or rod-shaped nanoparticles are particularly attractive for LSPR assays, and may also find application in SERS bioconjugates. For example, these particles could lead to higher SERS intensities for Raman-tagged metal core/SiO_2 shell nanoparticles. Arrays of Ag nanoparticles prepared by nanosphere lithography such as those used for LSPR have recently been shown to give SERS enhancements $\geq 10^8$ [65], and may find application in SERS-based biosensing.

The work described here demonstrates the flexibility of well-defined particle–biomolecule conjugates as amplification tags in a wide variety of sensing formats. Indeed, such conjugates are also finding application in biosensor strategies ranging from light scattering [66] to QCM [67] to electrical detection [51]. Regardless of application details, the preparation of stable, well-defined bioconjugates is critically important. Challenges include nonspecific binding of nanoparticles, which can limit the sensitivity and dynamic range of bioconjugate-based assays. Results from the Mirkin group indicate that excellent rejection of nonspecific binding can be achieved [50, 51]. In addition, the long-term stability of bioconjugates becomes important for commercial applications. It is encouraging to note that bioactive Au–protein conjugates can be purchased, stored for months, and used successfully in tissue staining for transmission electron microscopy; this bodes well for improvements in stability for the many novel bioconjugates now being prepared in laboratories around the world.

We expect vigorous research and increasing commercial applications for nanoparticle–biomolecule conjugates in the coming years. Nanoparticle-amplified SPR has tremendous potential for real-time sensing as well as coupling with spatially-patterned surfaces, such as DNA microarrays, for multiplexing. The most successful SERS strategies in the immediate future will likely be those that, like the work of Porter and of Mirkin, detect analytes indirectly by sensing a SERS-active label. SERS has tremendous potential for ultrasensitive multiplexing, and may find application in, for example, medical diagnostics and the analysis of gene expression.

References

[1] (a) Niemeyer, C. M., *Angew. Chem. Int. Ed.* **2001**, *40*, 4128–4158. (b) Storhoff, J. J., Mirkin, C. A., *Chem. Rev.* **1999**, *99*, 1849–1862. (c) Bangs, L. B., *Pure Appl. Chem.* **1996**, *68*, 1873–1879.

[2] (a) Raether, H., *Surface Plasmons on Smooth and Rough Surfaces and on Gratings*, Vol. *111*, Springer-Verlag, Berlin, **1998**. (b) Sambles, J. R., Bradbery, G. W., Yang, F., *Contemp. Phys.* **1991**, *32*, 173–183. (c) Homola, J., Yee, S. S., Gauglitz, G., *Sens. Actuators B* **1999**, *54*, 3–15.

[3] (a) Kelly K. L., Coronado, E., Zhao L. L., Shatz, G. C., *J. Phys. Chem. B* **2002**, *107*, 668–677. (b) Mulvaney, P., *Langmuir* **1996**, *12*, 788–800. (c) Khlebtsov, N. G., Bogatyrev, V. A., Dykman, L. A., Melnikov, A. G., *J. Coll. Interfac. Sci.* **1996**, *180*, 436–445.

[4] (a) Brandt, E. S., Cotton, T. M., in: Rossiter, B. W., Baetzold, R. C. (eds), *Surface-Enhanced Raman Scattering*, 2nd edn., John Wiley Sons, New York, **1993**, pp. 633–718. (b) Campion, A., Kambhampati, P., *Chem. Soc. Rev.* **1998**, *27*, 241–250. (c) Chang, R. K., Furtak, T. E., *Surface Enhanced Raman Scattering*, Plenum Press, New York, 1982.

[5] Englebienne, P., *J. Mater. Chem.* **1999**, *9*, 1043–1054.

[6] Storhoff, J. J., Lazarides, A. A., Mucic, R. C., Mirkin, C. A., Letsinger, R. L., Shatz, G. C., *J. Am. Chem. Soc.* **2000**, *122*, 4640–4650.

[7] Elghanian, R., Storhoff, J. J., Mucic, R. C., Letsinger, R. L., Mirkin, C. A., *Science* **1997**, *277*, 1078–1081.

[8] Englebienne, P., Van Noonacker, A., Verhas, M., *Analyst* **2001**, *126*, 1645–1651.

[9] Hirsch, L. R., Jackson, J. B., Lee, A., Halas, N. J., West, J. L., *Anal. Chem.* **2003**, *75*, 2377–2381.

[10] Englebienne, P., *Analyst* **1998**, *123*, 1599–1603.

[11] Xu, H., Kll, M., *Sens. Actuators B* **2002**, *87*, 244–249.

[12] Haes, A. J., Van Duyne, R. P., *J. Am. Chem. Soc.* **2002**, *124*, 10596–10604.

[13] Haynes, C. L., Van Duyne, R. P., *J. Phys. Chem. B* **2001**, *105*, 5599–5611.

[14] Riboh, J. C., Haes, A. J., McFarland, A. D., Yonzon, C. R., Van Duyne, R. P., *J. Phys. Chem. B* **2003**, *107*, 1772–1780.

[15] http://www.biacore.com, March 2003.

[16] Wink, T., van Suilen, S. J., Bult, A., van Bennekom, W. P., *Anal. Chem.* 1998, 70, 827–832.

[17] Jordan, C. E., Frutos, A. G., Thiel, A. J., Corn, R. M., *Anal. Chem.* 1997, 69, 4939–4947.

[18] Kubitschko, A., Spinke, J., Brükner, T., Pohl, S., Oranth, N., *Anal. Biochem.* 1997, 253, 112–122.

[19] Stella, B., Arpicco, S., Peracchia, M. T., Desmaéle, D., Hoebeke, J., Renoir, M., D'Angelo, J., Cattel, L., Couvreur, P., *J. Pharm. Sci.* 2000, 89, 1452–1464.

[20] Natan, M. J., Lyon, A. L. Surface Plasmon Resonance Biosensing with Colloidal Au Amplification, in: Feldheim, D. L., Foss, Jr., C. A. (eds), *Metal Nanoparticles: Synthesis, Characterization, and Applications*, Marcel Dekker: New York, 2002, pp. 183–205.

[21] Buckel, P. E., Davies, R. J., Kinning, T., Yeung, D., Edwards, P. R., Pollard-Knight, D., Lowe, C. R., *Biosensors Bioelectron.* 1993, 8, 355–363.

[22] Gu, J. H., Lu, H., Chem, Y. W., Liu, L. Y., Wang, P., Ma, J. M., Lu, Z. H., *Supramolec. Sci.* 1998, 5, 695–698.

[23] Lyon, L. A., Musick, M. D., Natan, M. J., *Anal. Chem.* 1998, 70, 5177–5183.

[24] Sandrock, M. L., Foss, C. A., Jr., *J. Phys. Chem B* 1999, 103, 11398–11406.

[25] Leung, P.-T., Pollard-Knight, D., Malan, G. P., Finlan, M. F., *Sens. Actuat. B* 1994, 22, 175–180.

[26] (a) Hutter, E., Cha, S., Liu, J. F., Park, J., Yi, J., Fendler, J. H., Roy, D., *J. Phys. Chem. B* 2001, 105, 8–12. (b) Chah, S., Hutter, E., Roy, D., Fendler, J. H., Yi, J., *Chem. Phys.* 2001, 272, 127–136.

[27] Lyon, L. A., Pea, D. J., Natan, M. J., *J. Phys. Chem. B* 1999, 103, 5826–5831.

[28] He, L., Musick, M. D., Nicewarner, S. R., Salinas, F. G., Benkovic, S. J., Natan, M. J., Keating, C. D., *J. Am. Chem. Soc.* 2000, 122, 9071–9077.

[29] Campion, A., Ivanecky III, J. E., Child, C. M., Foster, M., *J. Am. Chem. Soc.* 1995, 117, 11807–11808.

[30] (a) Kneipp, K., Kneipp, H., Itzkan, I., Dasari, R. R., Feld, M. J., *Phys. Condens. Matter* 2002, 14, R597–R624. (b) Kneipp, K., Kneipp, H., Itzkan, I., Dasari, R. R., Feld, M., *Chem. Rev.* 1999, 99, 2957–2975.

[31] (a) Holt, R. E., Cotton, T. M., *J. Am. Chem. Soc.* 1989, 111, 2815–2821. (b) Holt, R. E., Cotton, T. M., *J. Am. Chem. Soc.* 1987, 109, 1841–1845.

[32] Cotton, T. M., Schultz, S. G., Van Duyne, R. P., *J. Am. Chem. Soc.* 1980, 102, 7962–7965.

[33] Smulevich, G., Spiro, T. G., *J. Phys. Chem.* 1985, 89, 5168–5173.

[34] MacDonald, I. D. G., Smith, W. E., *Langmuir* 1996, 12, 706–713.

[35] Rospendowshi, B. N., Kelly, K., Wolf, C. R., Smith, W. E., *J. Am. Chem. Soc.* 1991, 113, 1217–1225.

[36] Kelly, K., Rospendowski, B. N., Smith, W. E., Wolf, C. R., *FEBS Lett.* 1987, 222, 120–124.

[37] Lee, P. V., Meisel, D., *J. Phys. Chem.* 1982, 86, 3391–3395.

[38] Broderick, J. B., Natan, M. J., O'Halloran, T. V., Van Duyne, R. P., *Biochemistry* 1993, 32, 13771–13776.

[39] Ahern, A. M., Garrell, R., *Langmuir* 1991, 7, 254–261.

[40] Keating, C. D., Kovaleski, K. M., Natan, M. J., *J. Phys. Chem. B* 1998, 102, 9404–9413.

[41] Keating, C. D., Kovaleski, K. M., Natan, M. J., *J. Phys. Chem. B* 1998, 102, 9414–9425.

[42] Ni, J., Lipert, J., Dawson, C. B., Porter, M. D., *Anal. Chem.* 1999, 71, 4903–4908.

[43] Kneipp, K., Kneipp, H., Kartha, V. B., Manoharan, R., Deinum, G., Itzkan, I., Dasari, R. R., Feld, M. S., *Phys. Rev. E* 1998, 57, R6281–R6284.

[44] Graham, D., Smith, W. E., Linacre, A. M. T., Munro, C. H., Watson, N. D., White, P. C., *Anal. Chem.* 1997, 69, 4703–4707.

[45] Gearheart, L. A., Ploehn, H. J., Murphy, C. J., *J. Phys. Chem. B* 2001, 105, 12609–12615.

[46] Graham, D., Mallinder, B. J., Whitcomb, D., Watson, N. D., Smith, W. E., *Anal. Chem.* 2002, 74, 1069–1074.

[47] Mirkin, C. A., Letsinger, R. L., Mucic, R. C., Storhoff, J. J., *Nature* 1996, 382, 607–609.

[48] Alivisatos, P. A., Johnsson, K. P., Peng, X., Wilson, T. E., Loweth, C. J., Bruchez, M. P. J., Schultz, P. G., *Nature* 1996, 382, 609–611.

[49] (a) Storhoff, J. J., Elghanian, R., Mucic, R. C., Mirkin, C. A., Letsinger, R. L., *J. Am. Chem. Soc.* 1998, 120, 1959–1964. (b) Rey-

nolds, III, R. A., Mirkin, C. A., Letsinger, R. L., *J. Am. Chem. Soc.* **2000**, *122*, 3795–3796.

[50] Taton, T. A., Mirkin, C. A., Letsinger, R. L., *Science* **2000**, *289*, 1757–1760.

[51] Park, S.-J., Taton, T. A., Mirkin, C. A., *Science*, **2002**, *295*, 1503–1506.

[52] Sauthier, M. L., Carroll, R. L., Gorman, C. B., Franzen, S., *Langmuir* **2002**, *18*, 1825–1830.

[53] Cao, Y. C., Jin, R., Mirkin, C. A., *Science* **2002**, *297*, 1536–1540.

[54] Mulvaney, S. P., Musick, M. D., Keating, C. D., Natan, M. J., *Langmuir* **2003**, *19*, 4784–4790.

[55] www.corninfo.chem.wisc.edu, accessed April **2003**.

[56] www.corninfo.chem.wisc.edu/writings/surfacechem.html, accessed April **2003**.

[57] *Colloidal Gold: Principles, Methods, and Applications*, Hayat, M. A. (ed.), Academic Press, San Diego, **1989**, Vols. 1–3.

[58] Grabar, K. C., Freeman, R. G., Hommer, M. B., Natan, M. J., *Anal. Chem.* **1995**, *67*, 735–743.

[59] Cao, Y., Jin, R., Mirkin, C. A., *J. Am. Chem. Soc.* **2001**, *123*, 7961–7962.

[60] Oldenburg, S. J., Averitt, R. D., Westcott, S. L., Halas, N. J., *Chem. Phys. Lett.* **1998**, *288*, 243–247.

[61] Keating, C. D., Musick, M. D., Keefe, M. H., Natan, M. J., *J. Chem. Ed.* **1999**, *76*, 949–956.

[62] Demers, L. M., Mirkin, C. A., Mucic, R. C., Reynolds, III, R. A., Letsinger, R. L., Elghanian, R., Viswanadham, G., *Anal. Chem.* **2000**, *72*, 5535–5541.

[63] Nicewarner-Pena, S. R., Raina, S., Goodrich, G. P., Fedoroff, N. V., Keating, C. D., *J. Am. Chem. Soc.* **2002**, *124*, 7314–7323.

[64] (a) Zanchet, D., Micheel, C. M., Parak, W. J., Gerion, D., Williams, S. C., Alivisatos, A. P., *J. Chem. Phys. B* **2002**, *106*, 11758–11763. (b) Zanchet, D., Michael, C. M., Parak, W. J., Gerion, D., Alivisatos, A. P., *Nano Lett.* **2001**, *1*, 32–35. (c) Parak, W. J., Pellegrino, T., Micheel, C. M., Gerion, D., Williams, S. C., Alivisatos, A. P., *Nano Lett.* **2003**, *3*, 33–36.

[65] Haynes, C. L., Van Duyne, R. P., *J. Phys. Chem. B* **2003**, *107*, 7426–7433.

[66] Taton, T. A., Lu, G., Mirkin, C. A., *J. Am. Chem. Soc.* **2001**, *123*, 5164–5165.

[67] (a) Liu, T., Tang, J., Zhao, H., Deng, Y., Jiang, L., *Langmuir* **2002**, *18*, 5624–5626. (b) Zhao, H. Q., Lin, L., Li, J. R., Tang, J. A., Duan, M. X., Jiang, L., *J. Nanopart. Res.* **2001**, *3*, 321–323.

27
Bioconjugated Silica Nanoparticles for Bioanalytical Applications

Timothy J. Drake, Xiaojun Julia Zhao, and Weihong Tan

27.1
Overview

Fluorescence-based techniques have generated widespread acceptance in many biochemical assays. For detecting small amounts of analyte, fluorescent molecules are often exploited as efficient signal transducers. However, sample analysis in small volumes can be particularly difficult when limited amounts of signaling probes are present. As a consequence, organic fluorophores, dye-doped nanoparticles, metal nanoparticles, and semiconductor nanoparticles are employed to aid in the analysis [1–17]. Improvements in these areas are still needed to make bioassays faster, more accurate and precise, and to remove the need for time-consuming amplification protocols. Nanoparticles show great promise in this regard, largely due to their unique optical properties, high surface-to-volume ratio, and other size-dependent qualities. By using nanoparticles, researchers have been able to increase sensitivity, enhance signal detection, and generate better reproducibility [1, 3, 7, 14–16, 18–20]. In this chapter, we will discuss silica-based nanoparticles and the progress made in bioanalytical and bionanotechnology applications.

Silica-based nanoparticles have been prepared and doped with organic and inorganic dye molecules, as well as being manufactured to contain magnetic cores with a silica coating [2, 3, 20, 21]. These particles have demonstrated great promise in the fields of DNA detection, bacterial detection, and chemical sensing [22–27]. DNA detection has been carried out using various nanoparticle formats [11, 14, 15, 28–34]. Dye-doped silica nanoparticles, as an example, when conjugated to DNA sequences for probing target DNA, contain thousands of dye molecules within one particle. Consequently, a significantly greater signal can result from single binding events, which leads to an increased sensitivity in comparison to single fluorophore-labeled DNA assays [25]. Magnetic silica nanoparticles have also been used to isolate and collect DNA using unique signal-transducing DNA probes [21, 26]. The bioconjugation of proteins has also allowed these particles to be used in various labeling and sensing protocols, such as *Escherichia coli* O157:H7 detection and chemical sensing [22, 27].

The advantages that silica nanoparticles offer in these fluorescence detection methods are largely due to their sensitivity, though the versatility of the silica substrate also

Nanobiotechnology. Edited by Christof Niemeyer, Chad Mirkin
Copyright © 2004 WILEY-VCH Verlag GmbH & Co. KgaA, Weinheim
ISBN 3-527-30658-7

plays a major role. In the development of highly effective nano-luminescent probes, silica has been found to be a more appropriate shell material than polymers. Various dye-doped polymer nanoparticles have been developed [35], and organic dye molecules – due to their hydrophobic properties – are easily incorporated into the polymer matrix to form luminescent nanoparticles [36, 37]. Whilst polymer nanoparticles, due to their hydrophobicity and bioincompatibility, are not very well suited for bioanalysis, silica on the other hand has several characteristics which make it a very attractive alternative substrate. First, it is not subject to microbial attack, and there are no swelling or porosity changes with variations in pH [38]. Silica is also chemically inert and optically transparent. The shell of the silica particles can act as an isolator, limiting the effect of the outside environment on the core of the particles. This is particularly important for dyes which are sensitive to certain solvents and soluble species in the buffer solutions. Photobleaching [39] and thermally induced degradation [39, 40] are primary processes that reduce the operational lifetime of a dye. By encapsulating the dye within the chemically and thermally inert silica shell, photobleaching and photodegradation of the dye can be minimized [3]. For bioconjugation protocols, a large amount of chemical reactions already exist for easy functionalization of silica particles to biomolecules [41–45]. Consequently, the synthesis protocol for the fabrication of silica-based particles and probes is relatively simple and straightforward and requires no special conditions.

27.2
Methods

27.2.1
Fabrication

Typically, two methods may be followed to prepare silica-based nanoparticles: (1) the Stöber method [22, 46, 47]; and (2) a reverse microemulsion method [2, 3,21, 24]. The Stöber method is used primarily to prepare pure silica nanoparticles and organic dye-doped nanoparticles [2, 3,21, 24], while the reverse microemulsion technique can be used for pure, inorganic and organic dye-doped, and magnetic nanoparticles [22, 46, 47].

The reverse or water-in-oil microemulsion (W/O) is a robust and efficient method for nanoparticle fabrication [48] which yields monodispersed particles in the nanometer size range. W/O microemulsions are isotropic and thermodynamically stable single-phase systems consisting of three primary components: water, oil, and a surfactant (Figure 27.1). Water nanodroplets formed in the bulk oil phase act as a confined media, or nanoreactors, for the formation of discrete nanoparticles. The main highlight of this method is size tunability of nanoparticles by varying the water to surfactant molar ratio and dynamic properties of microemulsion [49]. $W_0 = 10$ systems using Triton X-100 can produce silica nanoparticles with 60 nm diameters with minimal size distribution (Figure 27.1).

In the Stöber method [47], suspensions of submicron, silica-based spheres can be obtained, with the particle diameter being controlled through the parameters involved in the process, namely temperature and concentration. Typically, 5.0 mL of ethanol (95 %, v/v) is placed into a conical flask and mixed with 100 µL tetraethylortho-silicate (TEOS).

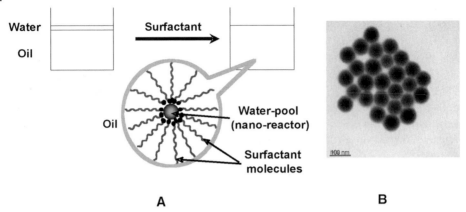

A   **B**

Figure 27.1 (A) Representation of the microemulsion system used for preparing silica nanoparticles. (B) Transmission electron microscopy image of silica nanoparticles prepared by a microemulsion method (~60 nm diameters).

The flask is then cooled to 0 °C and placed floating in an ice-cooled ultrasonication bath. Ammonium hydroxide (NH$_4$OH, 28–30 wt%; 5 mL) is added to the solution while ultrasonication is continued. The particles obtained may range from <100 nm to ~1 μm in diameter (Figure 27.2). Although this method is relatively simple and can be carried out in only a few hours, it is limited by the nonuniformity of the products obtained, and filtration and further separation is needed.

By using these fabrication methods, silica nanoparticles can be easily designed as nanosensors. The versatile silica matrix and modification protocols allow them to be conjugated with various functional groups which allow for specific applications and sensing assays. Pure, magnetic, and dye-doped silica nanoparticles have been prepared and surface-modified for multiple sensing and probing protocols, as described in the following sections.

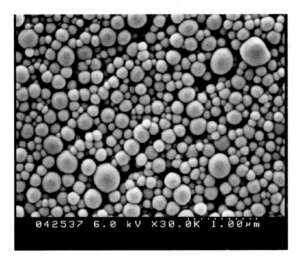

Figure 27.2 Scanning electron microscopy image of silica nanoparticles prepared using the Stöber method.

27.2.2
Particle Probes

Silica nanoparticles for bioanalytical applications have been synthesized and doped with either Tris(2, 2'-bipyridyl)dicholororuthenium(II) (RuBpy), tetramethylrhodamine (TMR), TMR-dextran, rhodamine 6G (R6G), or fluorescein-dextran [2, 3, 24, 46]. The procedures developed allow for the doping of dye molecules in each nanoparticle's silica matrix. Magnetic nanoparticles of Fe_3O_4 and Fe_2O_3 have also been produced using the reverse microemulsion method and further coated with silica for biomolecule immobilization [21, 26].

27.2.2.1 **Dye-doped Silica Nanoparticles**
Dye-doped silica nanoparticles provide highly luminescent signals due to the high quantum yield of the dye molecules and the numerous dye molecules inside the particles. Organic and inorganic dyes can be incorporated into nanoparticles using various techniques [1–3, 35, 46, 50–53]. For inorganic dye-doped silica nanoparticles, RuBpy dye molecules are simply placed into the reverse microemulsion system where they are entrapped in the nanoreactor water pools [3]. The silica matrix then encapsulates the dye molecules as it polymerizes [22]. This method is very effective in the production of very bright and photostable luminescent particles.

Developing improved particle probes could be accomplished by doping the silica matrix with higher quantum yield organic dyes. Higher fluorescent intensities could result and allow for increase sensitivity in various detection techniques where limited detectable events typically exist. However, trapping the organic dye molecules into the W/O microemulsion water pools or dispersing them in the Stöber method proved difficult due to the hydrophobic properties of most organic dyes and the lack of attraction forces between organic molecules and the silica matrix. The solubility of organic dyes is much more prevalent in organic reagents than in aqueous solutions; therefore, dye molecules prefer to stay in the organic phase of the W/O microemulsion system and precipitate in the Stöber method, which results in the formation of pure silica nanoparticles. Unlike RuBpy, most organic dye molecules are neutrally charged, thereby eliminating the electrostatic attraction which occurs between the silica particles and the dye molecules and is believed to aid in the retention of the dye molecules inside the silica matrix. The methods described in the following sections briefly describe the synthesis of R6G- and TMR-doped silica nanoparticles which address these two issues for organic dyes.

A Stöber-based approach has been designed to reduce the hydrophilicity of the silica matrix to allow for hydrophobic R6G molecules to be trapped inside the nanoparticle matrix [20]. TEOS is still used as the hydrophilic precursor, while phenyltriethoxysilane (PTES) is added to aid in trapping the dye molecules. Consequently, the R6G molecules are retained to a greater degree in this mixed-matrix structure, and this results in the formation of R6G-doped nanoparticles. Nanoparticles obtained using this method yield highly fluorescent, photostable probes with minimal dye leakage after prolonged storage in aqueous conditions (Figure 27.3). With the implementation of the Stöber method in the preparation of these particles, relatively large and nonuniform particles are obtained. For more uniform and smaller-sized particles, a method was developed for TMR-doped silica nanoparticles using the W/O microemulsion technique.

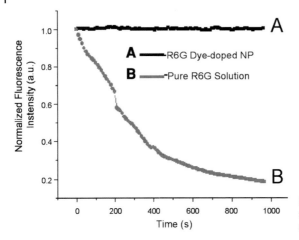

Figure 27.3 Photostability comparison of dye-doped nanoparticle (NP) and pure dye R6G.

Rather than altering the silica matrix, this approach is designed to change the property of the water pool nanoreactors in the microemulsion system to make TMR molecules more hydrophilic and more attractive to the silica matrix during the reaction. The basic approach uses a binary compound, acetic acid (HAc), to form the bulk of the water pool. HAc has both organic and inorganic properties and acts as an aqueous solution to dissolve and retain the TMR molecules inside the water pool. HAc also supplies protons for TMR and makes it positively charged [54]. The addition of HAc to the W/O microemulsion system is a simple and easy method to produce organic-dye doped silica nanoparticles. When compared to their inorganic dye-doped nanoparticle counterparts, the TMR-doped particles provide much higher (~40-fold) fluorescent signals (Figure 27.4) [54]. This is believed to be attributed to the much better quantum yield of organic dyes. Similarly, TMR-dextran-doped silica nanoparticles can be made with W/O microemulsions [55]. By linking a highly hydrophilic dextran molecule to a TMR molecule, the TMR is converted to a hydrophilic molecule and thus trapped inside the water pool in the microemul-

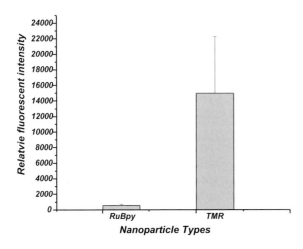

Figure 27.4 Relative fluorescence intensity graph of Organic (TMR) versus Inorganic (RuBpy) -doped silica nanoparticles.

sion system. However, to efficiently trap TMR molecules inside the silica, the TMR–dextran complex needs to become positively charged. By adding a low pH hydrochloric acid solution to the water phase in the microemulsion system, the TMR molecules are firmly entrapped in the silica matrix, and this results in highly fluorescent TMR-doped silica nanoparticles (Figure 27.4).

27.2.2.2 Magnetic Silica Nanoparticles

The synthesis of ultrasmall magnetic nanoparticles with uniform size distribution is very important because of their wide applicability in biology and medicine [56–58]. Examples include magnetic resonance imaging (MRI) contrast agents, the magnetic separation of oligonucleotides, cells, and other biocomponents, and magnetically guided site-specific drug delivery systems. Pure magnetic particles themselves may not be very useful in practical applications because of certain limitations:

- they tend to form large aggregates;
- they undergo structural changes due to instability, and this results in magnetic property changes and
- they undergo rapid biodegradation when directly exposed to biological systems.

Consequently, a suitable coating is essential to prevent such limitations, and the W/O microemulsion has been widely used to synthesize various types of magnetic materials [59–63]. Silica-coated magnetic nanoparticles have also been prepared using such a method [21, 26]. Iron oxide-nanoparticles are first formed by the coprecipitation reaction of ferrous and ferric salts with inorganic bases. The magnetic particles are then coated with silica to allow for biomodification and further used for bioanalysis. The resultant particles can range from 2 to 30 nm in diameter, are very uniform in size, and show magnetic behavior which is close to that of superparamagnetic materials.

27.2.3
Biofunctionalization of Silica Nanoparticles

Although inorganic nanoparticles can be prepared from various materials by several methods, their coupling and functionalization with biological components have been carried out with only a limited number of chemical methods. To prepare such conjugates from nanoparticles and biomolecules, the surface chemistry of the nanoparticle must be such that the ligands are fixed to the nanoparticle and possess terminal functional groups that are available for biochemical coupling reactions. Binding to particle surfaces is frequently carried out via thiol group interaction [64–68], but in most cases a simple thiol bond to the particle surface is not sufficient to accomplish a permanent linkage. Instead, equilibrium is established with dynamic ligand exchange. To avoid this, a shell of silica is often grown on the particle itself by means of sol-gel techniques [47, 69, 70], and the linkage groups pointing outwards are added as functionalized alkoxysilanes during the polycondensation process [71, 72]. The result is a relatively compact silica shell and a tight coating of the surface with coupling groups. Silica-based solid supports have been effectively used for the immobilization of various biomolecules such as enzymes, proteins, and

DNA, for applications ranging from biosensors to interfacial interaction studies [41–45, 73–75].

The above examples demonstrate that some chemical approaches have already been explored for the physical linkage of inorganic and biological materials. However, there remains a great demand for alternative methods which compensate for typical problems that arise in the biofunctionalization of nanoparticles – notably the harsh conditions which often lead to degradation and inactivation of sensitive biological compounds. Ligand exchange reactions that occur at the colloid surface often hinder the formation of stable bioconjugates. Moreover, the synthesis of stoichiometrically defined nanoparticle-biomolecules is a major challenge and is particularly important for generating well-defined nanostructures and nanoscaffolds for bioanalytical applications. A brief description of the methods employed by our laboratories for the bioconjugation of silica nanoparticles is presented in the following sections.

27.2.3.1 Amino-Group Cross-Linkage

An effective bioconjugation approach for the immobilization of proteins (e. g., enzymes and antibodies) onto silica nanoparticles is to chemically modify the particle's surface with amine-reactive groups [22]. Subsequently, free amine groups on the surface of biomolecules can covalently attach to the nanoparticle, leading to chemically bound silica nanoparticles conjugates (Figure 27.5A). The silica nanoparticles are first silanized using trimethoxysilyl-propyldiethylenetriamine (DETA) and then treated with succinic anhydride. The resulting carboxylate-modified particles are washed with deionized water. Two different approaches are then used, based on Bangs Labs' protocols [76], to further modify the nanoparticles. The first approach uses carbodimide hydrochloride in an activation buffer to produce o-acylisourea intermediate on the nanoparticle's surface. The available amine

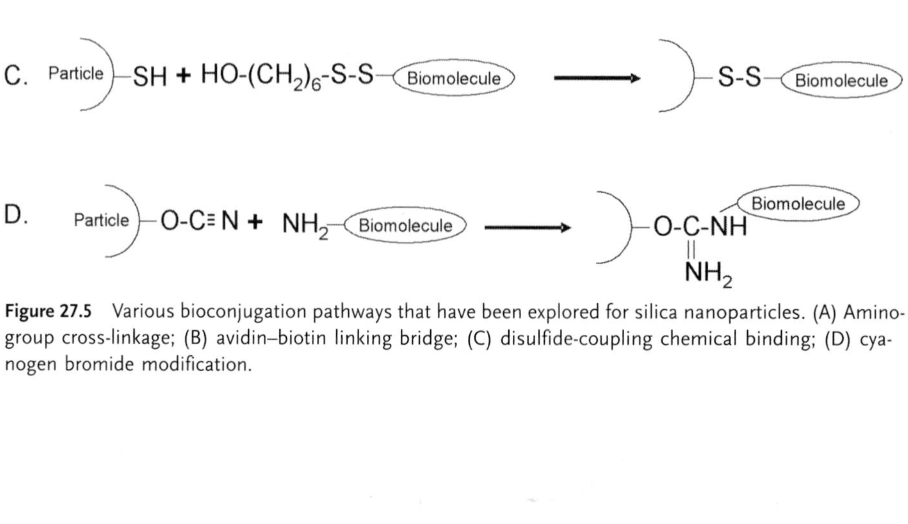

Figure 27.5 Various bioconjugation pathways that have been explored for silica nanoparticles. (A) Amino-group cross-linkage; (B) avidin–biotin linking bridge; (C) disulfide-coupling chemical binding; (D) cyanogen bromide modification.

groups on the biomolecules are then bound to the modified silica nanoparticles through amine bonds by immersion of the nanoparticles in a suitable concentration of the biomolecule solution [22]. The second approach adds a water-soluble *N*-hydroxysuccinimide compound along with carbodimide hydrochloride to the carboxyl-modified silica nanoparticles. The active ester intermediate formed by the N-hydroxy compound replaces the *o*-acylisourea intermediate, which is more stable for hydrolysis, and thus, the coupling efficiency of biomolecules to the nanoparticles is increased [27]. These approaches offer a more stable binding between the silica nanoparticles and biomolecules, which can last for several weeks or until the biomolecules become inactive.

27.2.3.2 Avidin–Biotin Linking Bridge

The affinity constant between avidin and biotin is 10^{15} L mol^{-1}, which is 10^5 to 10^6 times higher than that of regular antibody–antigen interactions. Therefore, the avidin–biotin linkage represents a practical conjugation approach to bind biomolecules onto silica nanoparticles [24–26]. Usually, biotin links easily to biomolecules to form recognition agents, and many commercially available biotinylated biomolecules are produced in this way. As long as avidin molecules are immobilized to the nanoparticles, upon avidin–biotin binding the recognition biomolecules remain on the nanoparticle surface for subsequent binding and identification of target biomolecules.

This avidin immobilization process is very simple and is based on the electrostatic attraction of the avidin for the silica surface (Figure 27.5B). The negatively charged silica surface allows the positively charged avidin to adsorb passively onto the nanoparticle's surface. By incubating appropriate concentrations of avidin with the silica nanoparticles, the avidin is adsorbed onto the nanoparticle surface. To stabilize the adsorption, a subsequent crosslinking step is carried out using a glutaraldehyde solution [24, 25]. The resulting avidin-coated silica nanoparticles conjugate to biotinylated molecules via a strong avidin–biotin affinity in readiness for further applications.

27.2.3.3 Disulfide-coupling Chemical Binding

Disulfide-coupling chemistry has proven to be one of the most efficient methods for the immobilization of oligonucleotides to a substrate [77]. Unlike most other covalent attachment processes, this method does not require preactivation or reduction of the disulfide groups to generate more reactive however unstable thiol species. The disulfide-modified oligonucleotides are directly coupled to the silane-activated silica surface, without any pretreatment (Figure 27.5C). The silica nanoparticles, however, are silanized with 3-mercaptopropyltrimethoxy-silane (MPTS) and then cured for approximately 2 hours in a vacuum oven, thereby forming functional thiol groups on the silica nanoparticle surfaces [23]. Disulfide-modified oligonucleotides are then directly attached onto the nanoparticle surface through disulfide-coupling.

27.2.3.4 Cyanogen Bromide Modification

Silica nanoparticles can also be surface-modified by activating them with sodium carbonate. A solution of cyanogen bromide in acetonitrile is then added to suspension to yield –OCN groups on the particle surface (Figure 27.5D) [22]. The particles are then available for bioconjugation to the biomolecules containing free amino groups.

27.2.4
Bioanlytical Applications for Silica Nanoparticles

The development of nanoparticles has shown great promise in bioanalysis due to their unique optical properties, high surface-to-volume ratio, and other size-dependent qualities. When combined with surface modifications and composition control, these properties provide probes for highly selective and ultra-sensitive bioassays. Of the various fluorescent materials, organic dye molecules are frequently used in bioassays due to their high quantum yield [78, 79]. However, the sensitivity of methods using organic dyes is limited by a low number of recognizable components in trace amounts of biosamples, which results in an inadequate amount of detection signals. Dye molecules also typically suffer from rapid photobleaching [80], resulting in unstable and inaccurate fluorescence signals. The dye-doped nanoparticles described above contain $\sim 10^4$ effective dye molecules for signaling targets and have excellent photostability. For every binding event, the nanoparticle will bring thousands of dye molecules rather than only a few, and this results in an increased sensitivity for most bioanalytical applications.

27.2.4.1 **Cellular Labeling/Detection**
For effective cellular labeling techniques, biomarkers need to have excellent specificity toward biomolecules of interest, and also have optically stable signal transducers. The dye-doped silica nanoparticles described in previous sections of this chapter offer ideal photostability and are easily conjugated through the use of the silica matrix to target-specific molecules, such as antibodies [2, 3,27].

The use of nanoparticles for biomarking has been effectively demonstrated for the labeling of leukemia cells [22]. A mouse anti-human CD-10 antibody is used as the recognition element on CNBr-pretreated nanoparticles. The mononuclear lymphoid cells are incubated with the CD-10 nanoparticles and then washed with phosphate-buffered saline to remove unbound nanoparticles. Fluorescence microscopy is then used to image the leukemia cells, whereupon the brightly fluorescent cells are easily detected and correlate well to optical images.

Another approach in which antibody-labeled particles have shown great promise is in the detection of single bacterium. A rapid bioassay for the precise determination of a single bacterium has been developed and has clear implications in the food and clinical industries, as well as for the identification of bioterrorism agents. Traditional methods used to detect trace amounts of bacteria tend to be laborious and time-consuming due to the complicated assay procedures [81–84]. Rapid single bacterium detection in a large-volume sample is not yet possible, but by using dye-doped silica nanoparticles a nanoprobe molecule has been developed to allow for the rapid, sensitive and accurate detection of single *E. coli* O157:H7 bacterium [27]. Antibodies against *E. coli* O157:H7 are conjugated to RuBpy-doped silica nanoparticles to form the nanoprobe complex, which is used to bind and label the antigen on the *E. coli* O157:H7 cell surface. The resultant fluorescence signals can then be monitored using various techniques. This bioassay takes $\sim$20 minutes to complete and is a convenient, highly selective method. Because the nanoparticle brings a large amount of dye molecules to the surface of the bacterium, single bacterium detection is also possible.

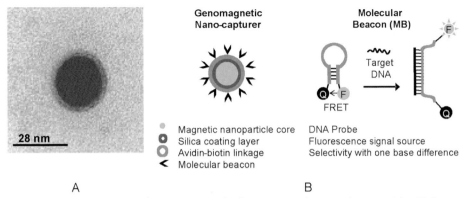

Figure 27.6 (A) Transmission electron micrograph of a magnetic silica-coated nanoparticle. (B) Genomagnetic Nano-capturer structure.

27.2.4.2 DNA Analysis

Magnetic separation is a powerful separation method for biomolecules. Based on the ultrasmall and uniform magnetic nanoparticles synthesized [21], genomagnetic nanocapturers (GMNCs) were developed for the collection of trace amounts of DNA/RNA molecules from complex mixtures [27]. The GMNC can selectively separate a specific DNA sample by hybridization events followed by magnetic separation. The GMNC is constructed with a magnetic nanoparticle, a silica layer, a biotin–avidin linkage, and a molecular beacon [85–89] DNA probe (Figure 27.6), where the magnetic nanoparticles serve as magnetic carriers and molecular beacon (MB) probes act as recognition elements and indicators for specific gene sequences.

There are two major factors that allow the GMNC to be highly effective in DNA analysis. First, the MB's special stem–loop structure is critical for single-base mismatch discrimination; and second, the use of magnetic nanoparticles allow for separation, isolation, and enrichment. The melting profiles of the MB on the GMNC surface allow for efficient isolation of the target DNA from single-base mismatched DNA. By varying the temperature and separating the solution by magnetization, the GMNC is able to separate trace amounts of target DNA/RNA from an artificial complex matrix containing large amounts of random DNAs (100 times more concentrated), as well as proteins (1000 times more concentrated). The target DNA/RNA sequences can be captured down to an initial concentration of 0.3 pM in a complex mixture with high specificity and excellent collection efficiency. The separation and collection of trace amounts of single-base mismatched DNA might potentially represent an effective means of detecting mutant cancer genes and eliminating cancer before it occurs.

27.2.4.3 Ultrasensitive DNA Detection

The amount of dye molecules per nanoparticle can be effectively utilized in DNA hybridization assays, and offers a distinct advantage over traditional fluorescence-based techniques [20, 46, 54, 55]. The assay is based on a sandwich set-up which eliminates the need to label the target (Figure 27.7). Upon hybridization, every binding event will be ide-

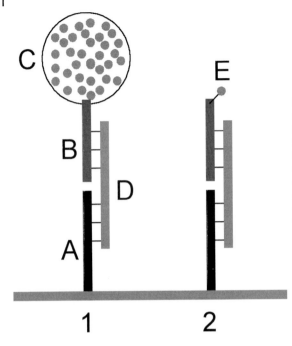

Figure 27.7 DNA detection scheme using a sandwich format assay. (1) Dye-doped nanoparticle assay; (2) single fluorophore-labeled DNA assay. (A) Capture sequence immobilized to a solid support. (B) Probing sequence immobilized to a nanoparticle through avidin–biotin binding. (C) Dye-doped silica nanoparticle which contains thousands of dye molecules per nanoparticle. (D) Unlabelled target sequence. (E) Single fluorophore molecule attached to a probe DNA sequence. Upon hybridization of the capture probe and target sequence, the probe molecule hybridizes with the remaining un-hybridized sequence to attach the signal transducer for detection.

ally reported by a single nanoparticle containing thousands of dye molecules. This results in a significant increase in sensitivity and detection capabilities. For these investigations, RuBpy- and TMR-doped nanoparticles are used with an avidin–biotin conjugation protocol [24, 25]; the resultant assay was able to achieve a sub-picomolar detection limit.

27.3
Outlook

Although significant advances have been made in the area of silica-based nanoparticles, many theoretical and technical problems still need to be solved. These vary from understanding the nanofabrication of molecular-sized probes to the control of photobleaching (a standard problem in fluorescence microscopy). Nanoparticles have already been produced which are highly photostable and approach the molecular size. Currently, investigations are under way to develop a better fundamental understanding of silica nanoparticles and techniques for the bio-imaging of cells and cellular components, of bioassays targeted at biologically relevant diseases, of detection probes for neurochemical monitoring, and of DNA separation and detection techniques. All of these approaches have been designed with nanometer dimensions and may be fully implemented in the ultra-sensitive detection and monitoring of biological events.

Acknowledgments

The authors thank their colleagues at University of Florida for their assistance in the conduct of these studies. The investigations are partially supported by grants NIH NS39891 and NIH NCI CA92581.

References

[1] H. Harma, T. Soukka, T. Lovgren, *Clin. Chem.* **2001**, *47*, 561–568.

[2] S. Santra, P. Zhang, K. M. Wang, R. Tapec, W. H. Tan, *Anal. Chem.* **2001**, *73*, 4988–4993.

[3] S. Santra, K. M. Wang, R. Tapec, W. H. Tan, *J. Biomed. Optics* **2001**, *6*, 160–166.

[4] J. Ji, N. Rosenzweig, I. Jones, Z. Rosenzweig, *Anal. Chem.* **2001**, *73*, 3521–3527.

[5] W. C. W. Chan, D. J. Maxwell, X. H. Gao, R. E. Bailey, M. Y. Han, S. M. Nie, *Curr. Opin. Biotechnol.* **2002**, *13*, 40–46.

[6] W. C. W. Chan, S. M. Nie, *Science* **1998**, *281*, 2016–2018.

[7] X. H. Gao, W. C. W. Chan, S. M. Nie, *J. Biomed. Optics* **2002**, *7*, 532–537.

[8] W. E. Doering, S. M. Nie, *Anal. Chem.* **2003**, *75*, 6171–6176.

[9] A. P. Alinicatos, *Science* **2000**, *289*, 736–737.

[10] D. Gerion, F. Pinaud, S. C. Williams, W. J. Parak, D. Zanchet, S. Weiss, A. P. Alivisatos, *J. Phys. Chem. B* **2001**, *105*, 8861–8871.

[11] H. Cai, C. Xu, P. G. He, Y. Z. Fang, *J. Electroanal. Chem.* **2001**, *510*, 78–85.

[12] Y. W. C. Cao, R. C. Jin, C. A. Mirkin, *Science* **2002**, *297*, 1536–1540.

[13] L. M. Demers, M. Ostblom, H. Zhang, N. H. Jang, B. Liedberg, C. A. Mirkin, *J. Am. Chem. Soc.* **2002**, *124*, 11248–11249.

[14] B. Dubertret, M. Calame, A. J. Libchaber, *Nature Biotechnol.* **2001**, *19*, 365–370.

[15] R. A. Reynolds, C. A. Mirkin, R. L. Letsinger, *Pure Appl. Chem.* **2000**, *72*, 229–235.

[16] R. Elghanian, J. J. Storhoff, R. C. Mucic, R. L. Letsinger, C. A. Mirkin, *Science* **1997**, *277*, 1078–1081.

[17] M. Han, X. Gao, J. Z. Su, S. Nie, *Nature Biotechnol.* **2001**, *19*, 631–635.

[18] X. X. He, K. M. Wang, W. H. Tan, J. Li, X. H. Yang, S. S. Huang, D. Li, D. Xiao, *J. Nanosci. Nanotechnol.* **2002**, *2*, 317–320.

[19] C. M. Niemeyer, *Angew. Chem. Int. Ed.* **2001**, *40*, 4128–4158.

[20] R. Tapec, X. J. J. Zhao, W. H. Tan, *J. Nanosci. Nanotechnol.* **2002**, *2*, 405–409.

[21] S. Santra, R. Tapec, N. Theodoropoulou, J. Dobson, A. Hebard, W. H. Tan, *Langmuir* **2001**, *17*, 2900–2906.

[22] M. Qhobosheane, S. Santra, P. Zhang, W. H. Tan, *Analyst* **2001**, *126*, 1274–1278.

[23] L. R. Hilliard, X. J. Zhao, W. H. Tan, *Anal. Chim. Acta* **2002**, *470*, 51–56.

[24] R. Tapec, X. J. Zhao, G. Yao, S. Santra, W. H. Tan, *Submitted* **2003**.

[25] X. J. Zhao, R. D. Bagwe, W. H. Tan, *Adv. Mat.*, **2003**, in press.

[26] X. J. Zhao, L. R. Hilliard, K. M. Wang, W. H. Tan, *Enc. Nanosci. Nanotech.* **2003**, 1–15.

[27] X. J. Zhao, R. Tapec, K. M. Wang, W. H. Tan, *Anal. Chem.* **2003**, *75*, 3476–3483.

[28] M. H. Charles, M. T. Charreyre, T. Delair, A. Elaissari, C. Pichot, *Stp. Pharma Sci.* **2001**, *11*, 251–263.

[29] D. Gerion, W. J. Parak, S. C. Williams, D. Zanchet, C. M. Micheel, A. P. Alivisatos, *J. Am. Chem. Soc.* **2002**, *124*, 7070–7074.

[30] Y. Weizmann, F. Patolsky, I. Willner, *Analyst* **2001**, *126*, 1502–1504.

[31] A. Gole, C. Dash, C. Soman, S. R. Sainkar, M. Rao, M. Sastry, *Bioconjug. Chem.* **2001**, *12*, 684–690.

[32] J. Reichert, A. Csaki, J. M. Kohler, W. Fritzsche, *Anal. Chem.* **2000**, *72*, 6025–6029.

[33] O. Siiman, K. Gordon, A. Burshteyn, J. A. Maples, J. K. Whitesell, *Cytometry* **2000**, *41*, 298–307.

[34] G. R. Souza, T. J. Miller, *J. Am. Chem. Soc.* **2002**, *123*, 6734–6735.

[35] H. Gao, Y. Zhao, S. Fu, B. Li, M. Li, *Colloid Polymer Sci.* **2002**, *280*, 653.

[36] R. Meallet-Renault, P. Denjean, R. B. Pansu, *Sensors Actuators B-Chemical* **1999**, *59*, 108–112.

[37] R. Meallet-Renault, H. Yoshikawa, Y. Tamaki, T. Asahi, R. B. Pansu, H. Masuhara, *Polymers Adv. Technol.* **2000**, *11*, 772–777.

[38] T. K. Jain, I. Roy, T. K. De, A. Maitra, *J. Am. Chem. Soc.* **1998**, *120*, 11092–11095.

[39] H. K. Kim, S. J. Kang, S. K. Choi, Y. H. Min, C. S. Yoon, *Chem. Mater.* **1999**, *11*, 779.

[40] J. Garcia, E. Ramirez, M. A. Mondragon, R. Ortega, P. Loza, A. Campero, *J. Sol-Gel Sci. Technol.* **1998**, *13*, 657.

[41] C. M. Ingersoll, F. V. Bright, *Chemtech* **1997**, *27*, 26–31.

[42] C. M. Ingersoll, F. V. Bright, *J. Sol-Gel Sci. Technol.* **1998**, *11*, 169–176.

[43] J. Cordek, X. W. Wang, W. H. Tan, *Anal. Chem.* **1999**, *71*, 1529–1533.

[44] X. H. Fang, X. J. Liu, S. Schuster, W. H. Tan, *J. Am. Chem. Soc.* **1999**, *121*, 2921–2922.

[45] X. J. Liu, W. H. Tan, *Mikrochim. Acta* **1999**, *131*, 129–135.

[46] R. Tapec, X. J. Zhao, W. H. Tan, Submitted **2003**.

[47] W. Stober, A. Fink, E. Bohn, *J. Colloid Interface Sci.* **1968**, *26*, 62–69.

[48] K. Osseo-Asare, F. J. Arriagada, *Ceramic Transact.* **1990**, *12*, 3–16.

[49] R. P. Bagwe, K. C. Khilar, *Langmuir* **1997**, *13*, 6432–6438.

[50] J. Schlupen, F. H. Haegel, J. Kuhlmann, H. Geisler, M. J. Schwuger, *Colloids Surfaces A : Physicochem. Engineering Aspects* **1999**, *156*, 335–347.

[51] A. V. Makarova, A. E. Ostafin, H. Miyoshi, J. R. Norris, D. Meisel, *J. Phys. Chem.* **1999**, *103*, 9080.

[52] M. T. Charreyre, P. Zhang, M. A. Winnik, C. Pichot, C. Graillat, *J. Colloid Interface Sci.* **1995**, *170*, 374–382.

[53] M. T. Charreyre, A. Yekta, M. A. Winnik, T. Delair, C. Pichot, *Langmuir* **1995**, *11*, 2423–2428.

[54] X. J. Zhao, W. H. Tan, Submitted **2003**.

[55] X. J. Zhao, W. H. Tan, Submitted **2003**.

[56] S. WinotoMorbach, W. Mullerruchholtz, *Eur. J. Pharmaceut. Biopharmaceut.* **1995**, *41*, 55–61.

[57] V. Tchikov, S. Schutze, M. K. Kronke, *J. Magnetism Magnetic Mater.* **1999**, *194*, 242–247.

[58] L. Babes, B. Denizot, G. Tanguy, J. J. Le Jeune, P. Jallet, *J. Colloid Interface Sci.* **1999**, *212*, 474–482.

[59] P. A. Dresco, V. S. Zaitsev, R. J. Gambino, B. Chu, *Langmuir* **1999**, *15*, 1945–1951.

[60] K. M. Lee, C. M. Sorensen, K. J. Klabunde, G. C. Hadjipanayis, *IEEE Trans. Magnetics* **1992**, *28*, 3180–3182.

[61] C. Liu, B. Zou, A. J. Rondinone, Z. J. Zhang, *J. Am. Chem. Soc.* **2001**, *123*, 4344–4345.

[62] J. A. Lopez-Perez, M. A. Lopez-Quintela, J. Mira, J. Rivas, *IEEE Trans. Magnetics* **1997**, *33*, 4359–4362.

[63] J. A. Lopez Perez, M. A. Lopez Quintela, J. Mira, J. Rivas, S. W. Charles, *J. Phys. Chem. B* **1997**, *101*, 8045–8047.

[64] D. Zanchet, C. M. Micheel, W. J. Parak, D. Gerion, A. P. Alivisatos, *Nano Lett.* **2001**, *1*, 32–35.

[65] R. C. Mucic, J. J. Storhoff, C. A. Mirkin, R. L. Letsinger, *J. Am. Chem. Soc.* **1998**, *120*, 12674–12675.

[66] M. Giersig, P. Mulvaney, *Langmuir* **1993**, *9*, 3408–3413.

[67] T. H. Galow, A. K. Boal, V. Rotello, *Abstracts, Papers Am. Chem. Soc.* **2000**, *220*, U72.

[68] A. P. Alivisatos, K. P. Johnsson, X. G. Peng, T. E. Wilson, C. J. Loweth, M. P. Bruchez, P. G. Schultz, *Nature* **1996**, *382*, 609–611.

[69] Y. Kobayashi, M. A. Correa-Duarte, L. M. Liz-Marzan, *Langmuir* **2001**, *17*, 6375–6379.

[70] T. Ung, L. M. Liz-Marzan, P. Mulvaney, *Langmuir* **1998**, *14*, 3740–3748.

[71] D. Gerion, F. Pinaud, S. C. Williams, W. J. Parak, D. Zanchet, S. Weiss, A. P. Alivisatos, *J. Phys. Chem. B* **2001**, *105*, 8861–8871.

[72] P. A. Buining, B. M. Humbel, A. P. Philipse, A. J. Verkleij, *Langmuir* **1997**, *13*, 3921–3926.

[73] X. J. Liu, W. H. Tan, *Anal. Chem.* **1999**, *71*, 5054–5059.

[74] X. J. Liu, W. Farmerie, S. Schuster, W. H. Tan, *Anal. Biochem.* **2000**, *283*, 56–63.

[75] R. P. Andres, J. D. Bielefeld, J. I. Henderson, D. B. Janes, V. R. Kolagunta, C. P. Kubiak, W. J. Mahoney, R. G. Osifchin, *Science* **1996**, *273*, 1690–1693.

[76] Bangs Laboratories Inc., 9025 Technology Drive Fishers, IN 46038-2866, USA, 1-800-387-0672. **2003**.

[77] Y. H. Rogers, P. Jiang-Baucom, Z. J. Huang, V. Bogdanov, S. Anderson, M. T. Boyce-Jacino, *Anal. Biochem.* **1999**, *266*, 23–30.

[78] J. D. Bui, T. Zelles, H. J. Lou, V. L. Gallion, M. I. Phillips, W. H. Tan, *J. Neurosci. Methods* **1999**, *89*, 9–15.

[79] J. R. Lakowicz, *Principles of Fluorescence Spectroscopy*, 2nd edn., 1999.

[80] D. Gerion, F. Pinaud, S. C. Williams, W. J. Parak, D. Zanchet, S. Weiss, A. P. Alivisatos, *J. Phys. Chem. B* **2001**, *105*, 8861–8871.

[81] R. J. Delves, *Antibody Applications: Essential Techniques*, John Wiley Sons, New York, **1995**.

[82] R. Edwards, *Immunoassays: Essential Data*, John Wiley Sons, New York, **1996**.

[83] D. Gerion, F. Pinaud, S. C. Williams, W. J. Parak, D. Zanchet, S. Weiss, A. P. Alivisatos, *J. Phys. Chem. B* **2001**, *105*, 8861–8871.

[84] S. S. Iqbal, M. W. Mayo, J. G. Bruno, B. V. Bronk, C. A. Batt, J. P. Chambers, *Biosensors Bioelectronics* **2000**, *15*, 549–578.

[85] X. H. Fang, J. W. J. Li, J. Perlette, W. H. Tan, K. M. Wang, *Anal. Chem.* **2000**, *72*, 747A–753A.

[86] B. A. J. Giesendorf, J. A. M. Vet, S. Tyagi, E. J. M. G. Mensink, F. J. M. Trijbels, H. J. Blom, *Clin. Chem.* **1998**, *44*, 482–486.

[87] J. W. J. Li, X. H. Fang, S. M. Schuster, W. H. Tan, *Angew. Chem. Int. Ed.* **2000**, *39*, 1049.

[88] W. H. Tan, X. H. Fang, J. Li, X. J. Liu, *Chemistry-A Eur. J.* **2000**, *6*, 1107–1111.

[89] S. Tyagi, F. R. Kramer, *Nature Biotechnol.* **1996**, *14*, 303–308.

Index

Nanobiotechnology. Edited by Christof Niemeyer, Chad Mirkin
Copyright © 2004 WILEY-VCH Verlag GmbH & Co. KgaA, Weinheim
ISBN 3-527-30658-7